# GAUGE THEORIES IN PARTICLE PHYSICS

## Graduate Student Series in Physics

*Other books in the series*

**Introduction to Gauge Field Theory**
D BAILIN and A LOVE

**Supersymmetric Gauge Field Theory and String Theory**
D BAILIN and A LOVE

**Mechanics of Deformable Media**
A B BHATIA and R N SINGH

**Symmetries in Quantum Mechanics**
M CHAICHIAN and R HAGEDORN

**Hadron Interactions**
P D B COLLINS and A D MARTIN

**The Physics of Structurally Disordered Matter: An Introduction**
N E CUSACK

**Basic Ideas and Concepts in Nuclear Physics**
K HEYDE

**Collective Effects in Solids and Liquids**
N H MARCH and M PARRINELLO

**Geometry, Topology and Physics**
M NAKAHARA

**Supersymmetry, Superfields and Supergravity: An Introduction**
P P SRIVASTAVA

**Superfluidity and Superconductivity**
D R TILLEY and J TILLEY

GRADUATE STUDENT SERIES IN PHYSICS

Series Editor:
Professor Douglas F Brewer, MA, DPhil
*Emeritus Professor of Experimental Physics, University of Sussex*

# GAUGE THEORIES IN PARTICLE PHYSICS

## A PRACTICAL INTRODUCTION
## THIRD EDITION

Volume 2

Non-Abelian Gauge Theories:
QCD and the Electroweak Theory

IAN J R AITCHISON
*Department of Physics*
*University of Oxford*

ANTHONY J G HEY
*Department of Electronics and Computer Science*
*University of Southampton*

IoP

INSTITUTE OF PHYSICS PUBLISHING
Bristol and Philadelphia

*British Library Cataloguing-in-Publication Data*

A catalogue record for this book is available from the British Library.

ISBN 0 7503 0950 4

*Library of Congress Cataloging-in-Publication Data are available*

**Front cover image:** Simulation by the ATLAS experiment of the decay of a Higgs boson into four muons (yellow tracks). © CERN Geneva.

Commissioning Editor: John Navas
Production Editor: Simon Laurenson
Production Control: Sarah Plenty and Leah Fielding
Cover Design: Victoria Le Billon
Marketing: Nicola Newey and Verity Cooke

Published by Institute of Physics Publishing, wholly owned by The Institute of Physics, London

Institute of Physics Publishing, Dirac House, Temple Back, Bristol BS1 6BE, UK

US Office: Institute of Physics Publishing, The Public Ledger Building, Suite 929, 150 South Independence Mall West, Philadelphia, PA 19106, USA

Typeset in LaTeX $2_\varepsilon$ by Text 2 Text Limited, Torquay, Devon
Printed in the UK by MPG Books Ltd, Bodmin, Cornwall

To Jessie
and to
Jean, Katherine and Elizabeth

# CONTENTS

# PREFACE TO VOLUME 2 OF THE THIRD EDITION

Volume 1 of our new two-volume third edition covers relativistic quantum mechanics, electromagnetism as a gauge theory, and introductory quantum field theory, and leads up to the formulation and application of quantum electrodynamics (QED), including renormalization. This second volume is devoted to the remaining two parts of the 'Standard Model' of particle physics, namely quantum chromodynamics (QCD) and the electroweak theory of Glashow, Salam and Weinberg.

It is remarkable that all three parts of the Standard Model are quantum gauge field theories: in fact, QCD and the electroweak theory are certain generalizations of QED. We shall therefore be able to build on the foundations of gauge theory, Feynman graphs and renormalization which were laid in Volume 1. However, QCD and the electroweak theory both require substantial extensions of the theoretical framework developed for QED. Most fundamentally, the discussion of global and local symmetries must be enlarged to include *non-Abelian symmetries*, and *spontaneous symmetry breaking*. At a somewhat more technical level, the *lattice (or path-integral) approach to quantum field theory*, and the *renormalization group* are both needed for access to modern work on QCD. For each of these theoretical elements, a self-contained introduction is provided in this volume. Together with their applications, this leads to a simple four-part structure (the numbering of parts, chapters and appendices continues on from Volume 1):

Part 5 Non-Abelian symmetries
Part 6 QCD and the renormalization group (including lattice field theory)
Part 7 Spontaneous symmetry breaking (including the spontaneous breaking of the approximate global chiral symmetry of QCD)
Part 8 The electroweak theory.

We have already mentioned several topics (path integrals, the renormalization group, and chiral symmetry breaking) which are normally found only in texts pitched at a more advanced level than this one—and which were indeed largely omitted from the preceding (second) edition. Nor, as we shall see, are these topics the only newcomers. With their inclusion in this volume, our book now becomes a comprehensive, practical and *accessible* introduction to the major theoretical and experimental aspects of the Standard Model. The emphasis is crucial: in once again substantially extending the scope of the book, we have tried hard not to compromise the title's fundamental aim—which is, as before, to make the chosen material accessible to the wide readership which the previous editions evidently attracted.

A glance at the contents will suggest that we have set ourselves a considerable challenge. On the other hand, not all of the topics are likely to be of equal interest to every reader. It may therefore be helpful to offer some more detailed guidance, while at the same time *highlighting* those items which are new to this edition.

First, then, non-Abelian symmetry. This refers to the fact that the symmetry transformations are matrices (acting on a set of fields), any two of which will generally not commute with each other, so that the order in which they are applied makes a difference. Much of the necessary mathematics already appears in the simpler case in which the symmetry is a global, rather than a local one. In chapter 12 we introduce global non-Abelian symmetries via the physical examples of the (approximate) SU(2) and SU(3) flavour symmetries of the strong interactions.

The underlying mathematics involved here is group theory. However, we take care to develop everything we need on a 'do-it-yourself' basis as we go along, so that no prior knowledge of group theory is necessary. Nevertheless, we have provided a new and fairly serious *appendix (M) on group theory*, which collects together the main relevant ideas, and shows how they apply to the groups we are dealing with (including the Lorentz group). We hope that this compact summary will be of use to those readers who want a sense of the mathematical unity behind the succession of specific calculations provided in the main text.

A further important global non-Abelian symmetry is also introduced in chapter 12—that of chiral symmetry, which is expected to be relevant if the quark masses are substantially less than typical hadronic scales, as is indeed the case. The apparent non-observation of this expected symmetry creates a puzzle, the resolution of which has to be deferred until part 7.

In chapter 13, the second in part 5, we move on to the local versions of SU(2) and SU(3) symmetry, arriving in section 13.5 at the corresponding non-Abelian gauge field theories which are the main focus of the book, being directly relevant to the electroweak theory and to QCD respectively. Crucial new physical phenomena appear, not present in QED—for example, the self-interactions among the gauge field quanta.

On the mathematical side, the algebraic (or group-theoretic) aspects developed in chapter 12 carry through unchanged into chapter 13, but the 'gauging' of the symmetry brings in some new geometrical concepts, such as *'covariant derivative', 'parallel transport', 'connection', and 'curvature'*. We decided against banishing these matters to an appendix, since they are such a significant part of the conceptual structure of all gauge theories, and moreover their inclusion allows instructive reference to be made to a theory otherwise excluded from mention, namely general relativity. All the same, practically-minded readers may want to pass quickly over sections 13.2 and 13.3, and also section 13.5.3, which explains why obtaining the correct Feynman rules for loops in a non-Abelian gauge theory is such a difficult problem, within the 'canonical' approach to quantum field theory as developed in volume 1.

Immediate application of the formalism can now be made to QCD, and this occupies most of the next three chapters, which form part 6. Chapter 14 introduces 'colour' as a dynamical degree of freedom, and leads on to the QCD Lagrangian. Some simple tree-graph applications are then described, using the techniques learned for QED. These provide a good first orientation to data, following 'parton model' and 'scaling' ideas.

But of course a fundamental question immediately arises: how can such an approach, based on perturbation theory, possibly apply to QCD which, after all, describes the strong interactions between quarks? The answer lies in the profound property (possessed only by non-Abelian gauge theories) called 'asymptotic freedom'—that is, the decrease of the effective interaction strength at high energies or short distances. Crucially, this property cannot be understood in terms of tree graphs: loops must be studied, and this immediately involves renormalization. In fact, perturbation theory becomes useful at high energies only after an infinite series of loop contributions has been effectively re-summed. The technique required to do this goes by the name of *the renormalization group* (RG), and it is described in chapter 15, along with applications to asymptotic freedom, and to the calculation of scaling violations in deep inelastic scattering.

We do not expect the majority of our readers to find chapter 15 easy going. But there is no denying the central importance of the RG in modern field theory, nor its direct relevance to experiment. In section 15.2 we have tried to provide an elementary introduction to the RG, by considering in detail the much simpler case of QED, using no more theory than is contained in chapter 11 of volume 1. Sections 15.4 and 15.5 are less central to the main argument, as is an *appendix (N) on dimensional regularization*.

In chapter 16, the third of part 6, we turn to the problem of how to extract predictions from a quantum field theory (in particular, QCD) in the non-perturbative regime. The available technique is computational, based on the discretized (lattice) version of *Feynman's path-integral formulation of quantum field theory*, to which we provide a simple introduction in section 16.4. A substantial bonus of this formulation is that it allows fruitful analogies to be drawn with the statistical mechanics of spin systems on a lattice. In particular, we hope that readers who may have struggled with the formal manipulations of chapter 15 will be refreshed by seeing RG ideas in action from a different and more physical point of view—that of 'integrating out' short distance degrees of freedom, leading to an *effective theory* valid at longer distances. The chapter ends with some illustrative results from lattice QCD calculations, in section 16.7. An *appendix (O) on Grassmann variables* is provided for those interested in seeing how the path-integral formalism can be made to work for fermions.

At this half-way stage, QCD has been established as the theory of strong interactions, by the success of both RG-improved perturbation theory and non-perturbative numerical computations.

Further progress requires one more fundamental idea—the subtle concept of spontaneous symmetry breaking, which forms the subject of part 7. Chapter 17 sets out *the basic theory of spontaneously broken global symmetries*, and also considers two physical examples in considerable detail, namely *the Bogoliubov superfluid* in section 17.3, and *the BCS superconductor* in section 17.7. It is of course true that these systems are not part of the standard model of particle physics. However, the characteristic methods and concepts developed for such systems provide valuable background for the particle physics applications of the idea, which follow in the next two chapters. In particular, our presentation of *chiral symmetry breaking* in chapter 18 follows Nambu's remarkable original analogy between fermion mass generation and the appearance of an energy gap in a superconductor. Section 18.3, on *linear and nonlinear sigma models*, is rather more optional, as is our brief introduction to *chiral anomalies* in section 18.4. In chapter 19, the third in part 7, we consider the spontaneous breaking of local (gauge) symmetries. Here the fundamental point is that it is possible for gauge quanta to acquire mass, while still preserving the local gauge symmetry of the Lagrangian. We consider applications both to the Abelian case of a superconductor (sections 19.2 and 19.4—once again, a valuable working model of the physics), and to the non-Abelian case required for the electroweak theory.

The way is now clear to develop the electroweak theory, in part 8. Chapter 20 is a self-contained review of weak interaction phenomenology, based on Fermi's 'current–current' model. New material here includes discussion of *the discrete symmetries* **C** *and* **P**, and of lepton number conservation taking into account the possibility that *neutrinos may be Majorana particles*, in support of which we provide *an appendix (P) on Majorana fermions*. Chapter 21 describes what goes wrong with the current–current model, and with theories in which the W and Z bosons are given a 'naive' mass, and suggests why a gauge theory is needed to avoid these difficulties. Finally, in chapter 22, all the pieces are put together in the presentation of the electroweak theory. New additions here include *three-family mixing via the CKM matrix*, together with more detail on *higher order (one-loop) corrections*, the *top quark*, and aspects of *Higgs phenomenology*. The remarkably precise agreement—thus far—between theory and experiment, which depends upon the inclusion of one-loop effects, makes it hard to deny that, when interacting weakly, Nature has indeed made use of the subtle intricacies of a renormalizable, spontaneously broken, non-Abelian chiral gauge theory.

But the story of the Standard Model is not yet quite complete. One vital part—the Higgs sector—remains virtual, and phenomenological. Further progress in understanding the mechanism of electroweak symmetry breaking, and of mass generation, requires input from the next generation of experiments, primarily at the LHC. We hope that we leave our readers with a sound grasp of what is at stake in these experiments, and a lively interest in their outcome.

**Acknowledgments**

Our expression of thanks to friends and colleagues, made in the first volume, applies equally to this one. More particularly: Keith Hamilton was a willing 'guinea pig' for chapters 14–17, and we thank him for his careful reading and encouraging comments; we are grateful to Chris Allton for advice on lattice gauge theory results, for section 16.7; and Nikki Fathers again provided essential help in making the electronic version. Above all, the constant and unstinting help of our good friend George Emmons throughout the genesis and production of the book has, once again, been invaluable.

In addition, we are delighted to thank two new correspondents, whom we look forward to greeting physically, as well as electronically. Paolo Strolin and Peter Williams both worked very carefully through Volume 1, and between them found a good many misprints and infelicities. An up-to-date list will be posted on the book's website:
http://bookmarkphysics.iop.org/bookpge.htm?&book=1130p.
Paolo also read drafts of chapters 12 and 14, and made many useful comments. We wish it had been possible to send him more chapters: however, he made numerous excellent suggestions for improving our treatment of weak interactions in the second edition (parts 4 and 5), and—where feasible—many of them have been incorporated into the present part 8. Errors, old and new, there still will be, of course: we hope readers will draw them to our attention.

**Ian J R Aitchison and Anthony J G Hey**
October 2003

# PART 5

# NON-ABELIAN SYMMETRIES

# 12

# GLOBAL NON-ABELIAN SYMMETRIES

In the preceding volume, a very successful dynamical theory—QED—has been introduced, based on the remarkably simple *gauge principle*: namely that the theory should be invariant under local phase transformations on the wavefunctions (chapter 3) or field operators (chapter 7) of charged particles. Such transformations were characterized as *Abelian* in section 3.6, since the phase factors commuted. The second volume of this book will be largely concerned with the formulation and elementary application of the remaining two dynamical theories within the Standard Model—that is, QCD and the electroweak theory. They are built on a generalization of the gauge principle, in which the transformations involve more than one state, or field, at a time. In that case, the 'phase factors' become matrices, which generally do not commute with each other, and the associated symmetry is called a '*non-Abelian*' one. When the phase factors are independent of the spacetime coordinate $x$, the symmetry is a 'global non-Abelian' one; when they are allowed to depend on $x$, one is led to a non-Abelian gauge theory. Both QCD and the electroweak theory are of the latter type, providing generalizations of the Abelian U(1) gauge theory which is QED. It is a striking fact that all three dynamical theories in the Standard Model are based on a gauge principle of local phase invariance.

In this chapter we shall be mainly concerned with two global non-Abelian symmetries, which lead to useful conservation laws but not to any specific dynamical theory. We begin in section 12.1 with the first non-Abelian symmetry to be used in particle physics, the *hadronic isospin* 'SU(2) symmetry' proposed by Heisenberg (1932) in the context of nuclear physics, and now seen as following from the near equality of the u and d quark masses (on typical hadronic scales), and the flavour independence of the QCD interquark forces. In section 12.2 we extend this to $SU(3)_f$ flavour symmetry, as was first done by Gell-Mann (1961) and Ne'eman (1961)—an extension seen, in its turn, as reflecting the rough equality of the u, d and s quark masses, together with flavour independence of QCD. The 'wavefunction' approach of sections 12.1 and 12.2 is then reformulated in field-theoretic language in section 12.3.

In the last section of this chapter, we shall introduce the idea of a global *chiral* symmetry, which is a symmetry of theories with massless fermions. This may be expected to be a good approximate symmetry for the u and d quarks. But the anticipated observable consequences of this symmetry (for example, nucleon parity doublets) appear to be absent. This puzzle will be resolved in part 7, via the profoundly important concept of 'spontaneous symmetry breaking'.

The formalism introduced in this chapter for SU(2) and SU(3) will be required again in the following one, when we consider the local versions of these non-Abelian symmetries and the associated dynamical gauge theories. The whole modern development of non-Abelian gauge theories began with the attempt by Yang and Mills (1954) (see also Shaw 1955) to make hadronic isospin into a local symmetry. However, the beautiful formalism developed by these authors turned out *not* to describe interactions between hadrons. Instead, it describes the interactions between the *constituents* of the hadrons, namely quarks—and this in two respects. First, a local SU(3) symmetry (called $SU(3)_c$) governs the strong interactions of quarks, binding them into hadrons (see part 6). Second, a local SU(2) symmetry (called *weak isospin*) governs the weak interactions of quarks (and leptons); together with QED, this constitutes the electroweak theory (see part 8). It is important to realize that, despite the fact that each of these two local symmetries is based on the same group as one of the earlier global (flavour) symmetries, the physics involved is completely different. In the case of the strong quark interactions, the $SU(3)_c$ group refers to a new degree of freedom ('colour') which is quite distinct from flavour u, d, s (see chapter 14). In the weak interaction case, since the group is an SU(2), it is natural to use 'isospin language' in talking about it, particularly since flavour degrees of freedom are involved. But we must always remember that it is *weak* isospin, which (as we shall see in chapter 20) is an attribute of leptons as well as of quarks and, hence, physically quite distinct from hadronic spin. Furthermore, it is a parity-violating chiral gauge theory.

Despite the attractive conceptual unity associated with the gauge principle, the way in which each of QCD and the electroweak theory 'works' is actually quite different from QED and from each other. Indeed it is worth emphasizing very strongly that it is, *a priori*, far from obvious why either the strong interactions between quarks or the weak interactions should have anything to do with gauge theories at all. Just as in the U(1) (electromagnetic) case, gauge invariance forbids a mass term in the Lagrangian for non-Abelian gauge fields, as we shall see in chapter 13. Thus it would seem that gauge field quanta are necessarily massless. But this, in turn, would imply that the associated forces must have a long-range (Coulombic) part, due to exchange of these massless quanta—and of course in neither the strong nor the weak interaction case is that what is observed.[1] As regards the former, the gluon quanta are indeed massless but the contradiction is resolved by *non-perturbative* effects which lead to *confinement*, as we indicated in chapter 2. We shall discuss this further in chapter 16. In weak interactions, a third realization appears: the gauge quanta acquire mass via (it is believed) a second instance of *spontaneous symmetry breaking*, as will be explained in part 7. In fact, a further application of this idea is required in the electroweak theory because of

[1] Pauli had independently developed the theory of non-Abelian gauge fields during 1953 but did not publish any of this work because of the seeming physical irrelevancy associated with the masslessness problem (Enz 2002, pp 474-82; Pais 2002, pp 242-5).

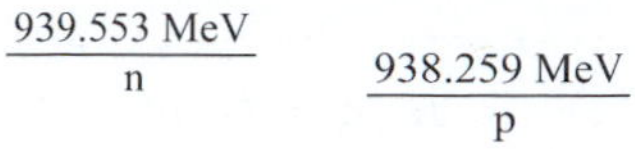


**Figure 12.1.** Early evidence for isospin symmetry.

the chiral nature of the gauge symmetry in this case: the quark and lepton masses also must be 'spontaneously generated'.

## 12.1 The flavour symmetry SU(2)$_\mathrm{f}$

### 12.1.1 The nucleon isospin doublet and the group SU(2)

The transformations initially considered in connection with the gauge principle in section 3.5 were just global phase transformations on a single wavefunction

$$\psi' = \mathrm{e}^{\mathrm{i}\alpha}\psi. \tag{12.1}$$

The generalization to non-Abelian invariances comes when we take the simple step—but one with many ramifications—of considering more than one wavefunction, or state, at a time. Quite generally in quantum mechanics, we know that whenever we have a set of states which are *degenerate* in energy (or mass) there is no unique way of specifying the states: any linear combination of some initially chosen set of states will do just as well, provided the normalization conditions on the states are still satisfied. Consider, for example, the simplest case of just two such states—to be specific, the neutron and proton (figure 12.1). This single near coincidence of the masses was enough to suggest to Heisenberg (1932) that, as far as the strong nuclear forces were concerned (electromagnetism being negligible by comparison), the two states could be regarded as truly degenerate, so that any arbitrary linear combination of neutron and proton wavefunctions would be entirely equivalent, as far as this force was concerned, for a single 'neutron' or single 'proton' wavefunction. This hypothesis became known as the 'charge independence of nuclear forces'. Thus redefinitions of neutron and proton wavefunctions could be allowed, of the form

$$\psi_\mathrm{p} \to \psi_\mathrm{p}' = \alpha\psi_\mathrm{p} + \beta\psi_\mathrm{n} \tag{12.2}$$

$$\psi_\mathrm{n} \to \psi_\mathrm{n}' = \gamma\psi_\mathrm{p} + \delta\psi_\mathrm{n} \tag{12.3}$$

for complex coefficients $\alpha$, $\beta$, $\gamma$ and $\delta$. In particular, since $\psi_\mathrm{p}$ and $\psi_\mathrm{n}$ are degenerate, we have

$$H\psi_\mathrm{p} = E\psi_\mathrm{p} \qquad H\psi_\mathrm{n} = E\psi_\mathrm{n} \tag{12.4}$$

from which it follows that

$$H\psi_\mathrm{p}' = H(\alpha\psi_\mathrm{p} + \beta\psi_\mathrm{n}) = \alpha H\psi_\mathrm{p} + \beta H\psi_\mathrm{n} \tag{12.5}$$

$$= E(\alpha\psi_\mathrm{p} + \beta\psi_\mathrm{n}) = E\psi_\mathrm{p}' \tag{12.6}$$

and, similarly,

$$H\psi_{\mathrm{n}}' = E\psi_{\mathrm{n}}' \tag{12.7}$$

showing that the redefined wavefunctions still describe two states with the same energy degeneracy.

The two-fold degeneracy seen in figure 12.1 is suggestive of that found in spin-$\frac{1}{2}$ systems in the absence of any magnetic field: the $s_z = \pm\frac{1}{2}$ components are degenerate. The analogy can be brought out by introducing the *two-component nucleon isospinor*

$$\psi^{(1/2)} \equiv \begin{pmatrix} \psi_{\mathrm{p}} \\ \psi_{\mathrm{n}} \end{pmatrix} \equiv \psi_{\mathrm{p}}\chi_{\mathrm{p}} + \psi_{\mathrm{n}}\chi_{\mathrm{n}} \tag{12.8}$$

where

$$\chi_{\mathrm{p}} = \begin{pmatrix} 1 \\ 0 \end{pmatrix} \qquad \chi_{\mathrm{n}} = \begin{pmatrix} 0 \\ 1 \end{pmatrix}. \tag{12.9}$$

In $\psi^{(1/2)}$, $\psi_{\mathrm{p}}$ is the amplitude for the nucleon to have 'isospin up' and $\psi_{\mathrm{n}}$ is that for it to have 'isospin down'.

As far as the states are concerned, this terminology arises, of course, from the formal identity between the 'isospinors' of (12.9) and the two-component eigenvectors (4.59) corresponding to eigenvalues $\pm\frac{1}{2}\hbar$ of (true) spin: compare also (4.60) and (12.8). It is important to be clear, however, that the degrees of freedom involved in the two cases are quite distinct; in particular, even though both the proton and the neutron have (true) spin-$\frac{1}{2}$, the transformations (12.2) and (12.3) leave the (true) spin part of their wavefunctions completely untouched. Indeed, we are suppressing the spinor part of both wavefunctions altogether (they are of course 4-component Dirac spinors). As we proceed, the precise mathematical nature of this 'spin-$\frac{1}{2}$' analogy will become clear.

Equations (12.2) and (12.3) can be compactly written in terms of $\psi^{(1/2)}$ as

$$\psi^{(1/2)} \to \psi^{(1/2)\prime} = \mathbf{V}\psi^{(1/2)} \qquad \mathbf{V} = \begin{pmatrix} \alpha & \beta \\ \gamma & \delta \end{pmatrix} \tag{12.10}$$

where $\mathbf{V}$ is the indicated complex $2 \times 2$ matrix. Heisenberg's proposal, then, was that the physics of strong interactions between nucleons remained the same under the transformation (12.10): in other words, a symmetry was involved. We must emphasize that such a symmetry can *only* be exact in the *absence* of electromagnetic interactions: it is, therefore, an intrinsically approximate symmetry, though presumably quite a useful one in view of the relative weakness of electromagnetic interactions as compared to hadronic ones.

We now consider the general form of the matrix $\mathbf{V}$, as constrained by various relevant restrictions: quite remarkably, we shall discover that (after extracting an overall phase) $\mathbf{V}$ has essentially the same mathematical form as the matrix $\mathbf{U}$ of (4.81), which we encountered in the discussion of the transformation of (real) spin wavefunctions under rotations of the (real) space axes. It will be instructive to see how the present discussion leads to the same form (4.81).

We first note that **V** of (12.10) depends on four arbitrary complex numbers or, alternatively, on eight real parameters. By contrast, the matrix **U** of (4.81) depends on only three real parameters: two to describe the axis of rotation represented by the unit vector $\hat{\boldsymbol{n}}$, together with a third for the angle of rotation $\theta$. However, **V** is subject to certain restrictions and these reduce the number of free parameters in **V** to three, as we now discuss. First, in order to preserve the normalization of $\psi^{(1/2)}$, we require

$$\psi^{(1/2)'\dagger}\psi^{(1/2)'} = \psi^{(1/2)\dagger}\mathbf{V}^\dagger\mathbf{V}\psi^{(1/2)} = \psi^{(1/2)\dagger}\psi^{(1/2)} \tag{12.11}$$

which implies that **V** has to be *unitary*:

$$\mathbf{V}^\dagger\mathbf{V} = \mathbf{1}_2 \tag{12.12}$$

where $\mathbf{1}_2$ is the unit $2 \times 2$ matrix. Clearly this unitarity property is in no way restricted to the case of two states—the transformation coefficients for $n$ degenerate states will form the entries of an $n \times n$ unitary matrix. A trivialization is the case $n = 1$, for which, as we noted in section 3.6, **V** reduces to a single phase factor as in (12.1), indicating how all the previous work is going to be contained as a special case of these more general transformations. Indeed, from the elementary properties of determinants, we have

$$\det \mathbf{V}^\dagger\mathbf{V} = \det \mathbf{V}^\dagger \cdot \det \mathbf{V} = \det \mathbf{V}^* \cdot \det \mathbf{V} = |\det \mathbf{V}|^2 = 1 \tag{12.13}$$

so that

$$\det \mathbf{V} = \exp(\mathrm{i}\theta) \tag{12.14}$$

where $\theta$ is a real number. We can separate off such an overall phase factor from the transformations mixing 'p' and 'n', because it corresponds to a rotation of the phase of both p and n wavefunctions by the *same* amount:

$$\psi'_\mathrm{p} = \mathrm{e}^{\mathrm{i}\alpha}\psi_\mathrm{p} \qquad \psi'_\mathrm{n} = \mathrm{e}^{\mathrm{i}\alpha}\psi_\mathrm{n}. \tag{12.15}$$

The **V** corresponding to (12.15) is $\mathbf{V} = \mathrm{e}^{\mathrm{i}\alpha}\mathbf{1}_2$, which has determinant $\exp(2\mathrm{i}\alpha)$ and is, therefore, of the form (12.1) with $\theta = 2\alpha$. In the field-theoretic formalism of section 7.2, such a symmetry can be shown to lead to the conservation of baryon number $N_\mathrm{u} + N_\mathrm{d} - N_{\bar{\mathrm{u}}} - N_{\bar{\mathrm{d}}}$, where bar denotes the anti-particle.

The new physics will lie in the remaining transformations which satisfy

$$\det \mathbf{V} = +1. \tag{12.16}$$

Such a matrix is said to be a *special* unitary matrix—which simply means it has unit determinant. Thus, finally, the **V**'s we are dealing with are *special, unitary,* $2 \times 2$ *matrices*. The set of all such matrices form a *group*. The general defining properties of a group are given in appendix M. In the present case, the elements of the group are all such $2 \times 2$ matrices and the 'law of combination' is just ordinary matrix multiplication. It is straightforward to verify (problem 12.1) that all the defining properties are satisfied here; the group is called 'SU(2)', the 'S' standing for 'special', the 'U' for 'unitary' and the '2' for '2 × 2'.

SU(2) is actually an example of a *Lie group* (see appendix M). Such groups have the important property that their physical consequences may be found by considering 'infinitesimal' transformations, that is—in this case—matrices $\mathbf{V}$ which differ only slightly from the 'no-change' situation corresponding to $\mathbf{V} = \mathbf{1}_2$. For such an infinitesimal SU(2) matrix $\mathbf{V}_{\text{infl}}$, we may therefore write

$$\mathbf{V}_{\text{infl}} = \mathbf{1}_2 + \mathrm{i}\xi \tag{12.17}$$

where $\xi$ is a $2 \times 2$ matrix whose entries are all first-order small quantities. The condition $\det \mathbf{V}_{\text{infl}} = 1$ now reduces, on neglect of second-order terms $O(\xi^2)$, to the condition (see problem 12.2)

$$\operatorname{Tr} \xi = 0. \tag{12.18}$$

The condition that $\mathbf{V}_{\text{infl}}$ be unitary, i.e.

$$(\mathbf{1}_2 + \mathrm{i}\xi)(\mathbf{1}_2 - \mathrm{i}\xi^\dagger) = \mathbf{1}_2 \tag{12.19}$$

similarly reduces (in first order) to the condition

$$\xi = \xi^\dagger. \tag{12.20}$$

Thus $\xi$ is a $2 \times 2$ traceless Hermitian matrix, which means it must have the form

$$\xi = \begin{pmatrix} a & b - \mathrm{i}c \\ b + \mathrm{i}c & -a \end{pmatrix} \tag{12.21}$$

where $a, b, c$ are infinitesimal parameters. Writing

$$a = \epsilon_3/2 \qquad b = \epsilon_1/2 \qquad c = \epsilon_2/2 \tag{12.22}$$

(12.21) can be put in the more suggestive form

$$\xi = \boldsymbol{\epsilon} \cdot \boldsymbol{\tau}/2 \tag{12.23}$$

where $\boldsymbol{\epsilon}$ stands for the three quantities

$$\boldsymbol{\epsilon} = (\epsilon_1, \epsilon_2, \epsilon_3) \tag{12.24}$$

which are all first-order small. The three matrices $\boldsymbol{\tau}$ are just the familiar Hermitian Pauli matrices

$$\tau_1 = \begin{pmatrix} 0 & 1 \\ 1 & 0 \end{pmatrix} \qquad \tau_2 = \begin{pmatrix} 0 & -\mathrm{i} \\ \mathrm{i} & 0 \end{pmatrix} \qquad \tau_3 = \begin{pmatrix} 1 & 0 \\ 0 & -1 \end{pmatrix} \tag{12.25}$$

here called 'tau' precisely in order to distinguish them from the mathematically identical 'sigma' matrices which are associated with the real spin degree of freedom. Hence, a general infinitesimal SU(2) matrix takes the form

$$\mathbf{V}_{\text{infl}} = (\mathbf{1}_2 + \mathrm{i}\boldsymbol{\epsilon} \cdot \boldsymbol{\tau}/2) \tag{12.26}$$

and an infinitesimal SU(2) transformation of the p–n doublet is specified by

$$\begin{pmatrix} \psi'_{\mathrm{p}} \\ \psi'_{\mathrm{n}} \end{pmatrix} = (\mathbf{1}_2 + \mathrm{i}\boldsymbol{\epsilon} \cdot \boldsymbol{\tau}/2) \begin{pmatrix} \psi_{\mathrm{p}} \\ \psi_{\mathrm{n}} \end{pmatrix}. \tag{12.27}$$

The $\boldsymbol{\tau}$-matrices clearly play an important role, since they determine the forms of the three independent infinitesimal SU(2) transformations. They are called the *generators* of infinitesimal SU(2) transformations; more precisely, the matrices $\boldsymbol{\tau}/2$ provide a particular *matrix representation* of the generators, namely the two-dimensional or 'fundamental' one (see appendix M). We note that they do not commute amongst themselves: rather, introducing $\mathbf{T}^{(1/2)} \equiv \boldsymbol{\tau}/2$, we find (see problem 12.3)

$$[T_i^{(1/2)}, T_j^{(1/2)}] = \mathrm{i}\epsilon_{ijk} T_k^{(1/2)}, \tag{12.28}$$

where $i$, $j$ and $k$ run from 1 to 3 and a sum on the repeated index $k$ is understood as usual. The reader will recognize the commutation relations (12.28) as being precisely the same as those of angular momentum operators in quantum mechanics:

$$[J_i, J_j] = \mathrm{i}\epsilon_{ijk} J_k. \tag{12.29}$$

In that case, the choice $J_i = \sigma_i/2 \equiv J_i^{(1/2)}$ would correspond to a (real) spin-$\frac{1}{2}$ system. Here the identity between the tau's and the sigma's gives us a good reason to regard our 'p–n' system as formally analogous to a 'spin-$\frac{1}{2}$' one. Of course, the 'analogy' was made into a mathematical identity by the judicious way in which $\xi$ was parametrized in (12.23).

The form for a *finite* SU(2) transformation $\mathbf{V}$ may then be obtained from the infinitesimal form using the result

$$\mathrm{e}^A = \lim_{n\to\infty} (1 + A/n)^n \tag{12.30}$$

generalized to matrices. Let $\boldsymbol{\epsilon} = \boldsymbol{\alpha}/n$, where $\boldsymbol{\alpha} = (\alpha_1, \alpha_2, \alpha_3)$ are three real finite (not infinitesimal) parameters, apply the infinitesimal transformation $n$ times and let $n$ tend to infinity. We obtain

$$\mathbf{V} = \exp(\mathrm{i}\boldsymbol{\alpha} \cdot \boldsymbol{\tau}/2) \tag{12.31}$$

so that

$$\psi^{(1/2)'} \equiv \begin{pmatrix} \psi'_{\mathrm{p}} \\ \psi'_{\mathrm{n}} \end{pmatrix} = \exp(\mathrm{i}\boldsymbol{\alpha} \cdot \boldsymbol{\tau}/2) \begin{pmatrix} \psi_{\mathrm{p}} \\ \psi_{\mathrm{n}} \end{pmatrix} = \exp(\mathrm{i}\boldsymbol{\alpha} \cdot \boldsymbol{\tau}/2)\psi^{(1/2)}. \tag{12.32}$$

Note that in the finite transformation, the generators appear in the exponent. Indeed, (12.31) has the form

$$\mathbf{V} = \exp(\mathrm{i}G) \tag{12.33}$$

where $G = \boldsymbol{\alpha} \cdot \boldsymbol{\tau}/2$, from which the unitary property of $\mathbf{V}$ easily follows:

$$\mathbf{V}^{\dagger} = \exp(-\mathrm{i}G^{\dagger}) = \exp(-\mathrm{i}G) = \mathbf{V}^{-1}, \tag{12.34}$$

where we used the Hermiticity of the tau's. Equation (12.33) has the general form

$$\text{unitary matrix} = \exp(\text{i Hermitian matrix}), \tag{12.35}$$

where the 'Hermitian matrix' is composed of the generators and the transformation parameters. We shall meet generalizations of this structure in the following section for SU(2), again in section 12.2 for SU(3), and a field-theoretic version of it in section 12.3.

As promised, (12.32) is of essentially the same mathematical form as (4.81). In each case, three real parameters appear: in (4.81) there are three parameters to describe the axis $\hat{\boldsymbol{n}}$ and angle $\theta$ of rotation; in (12.32) there are just the three components of $\boldsymbol{\alpha}$. We can always[2] write $\boldsymbol{\alpha} = |\boldsymbol{\alpha}|\hat{\boldsymbol{\alpha}}$ and identify $|\boldsymbol{\alpha}|$ with $\theta$ and $\hat{\boldsymbol{\alpha}}$ with $\hat{\boldsymbol{n}}$.

In the form (12.32), it is clear that our $2 \times 2$ isospin transformation is a generalization of the global phase transformation of (12.1), except that

(a) there are now *three* 'phase angles' $\boldsymbol{\alpha}$; and
(b) there are non-commuting matrix operators (the $\boldsymbol{\tau}$'s) appearing in the exponent.

The last fact is the reason for the description 'non-Abelian' phase invariance. As the commutation relations for the $\boldsymbol{\tau}$ matrices show, SU(2) is a non-Abelian group in that two SU(2) transformations do not, in general, commute. By contrast, in the case of electric charge or particle number, successive transformations clearly commute: this corresponds to an Abelian phase invariance and, as noted in section 3.6, to an Abelian U(1) group.

We may now put our initial 'spin-$\frac{1}{2}$' analogy on a more precise mathematical footing. In quantum mechanics, states within a degenerate multiplet may conveniently be characterized by the eigenvalues of a complete set of Hermitian operators which commute with the Hamiltonian and with each other. In the case of the p–n doublet, it is easy to see what these operators are. We may write (12.4), (12.6) and (12.7) as

$$H_2\psi^{(1/2)} = E\psi^{(1/2)} \tag{12.36}$$

and

$$H_2\psi^{(1/2)'} = E\psi^{(1/2)'} \tag{12.37}$$

[2] It is not completely obvious that the general SU(2) matrix *can* be parametrized by an angle $\theta$ with $0 \leq \theta \leq 2\pi$, and $\hat{\boldsymbol{n}}$: for further discussion, see appendix M, section M.7.

where $H_2$ is the $2 \times 2$ matrix

$$H_2 = \begin{pmatrix} H & 0 \\ 0 & H \end{pmatrix}. \tag{12.38}$$

Hence $H_2$ is proportional to the unit matrix in this two-dimensional space and it therefore commutes with the tau's:

$$[H_2, \boldsymbol{\tau}] = 0. \tag{12.39}$$

It then also follows that $H_2$ commutes with $\mathbf{V}$ or, equivalently,

$$\mathbf{V} H_2 \mathbf{V}^{-1} = H_2 \tag{12.40}$$

which is the statement that $H_2$ is invariant under the transformation (12.32). Now the tau's are Hermitian and, hence, correspond to possible observables. Equation (12.39) implies that their eigenvalues are constants of the motion (i.e. conserved quantities), associated with the invariance (12.40). But the tau's do not commute amongst themselves and so, according to the general principles of quantum mechanics, we cannot give definite values to more than one of them at a time. The problem of finding a classification of the states which makes the maximum use of (12.39), given the commutation relations (12.28), is easily solved by making use of the formal identity between the operators $\tau_i/2$ and angular momentum operators $J_i$ (cf (12.29)). The answer is[3] that the total squared 'spin'

$$(\mathbf{T}^{(1/2)})^2 = (\tfrac{1}{2}\boldsymbol{\tau})^2 = \tfrac{1}{4}(\tau_1^2 + \tau_2^2 + \tau_3^2) = \tfrac{3}{4}\mathbf{1}_2 \tag{12.41}$$

and one component of spin, say $T_3^{(1/2)} = \frac{1}{2}\tau_3$, can be given definite values simultaneously. The corresponding eigenfunctions are just the $\chi_p$'s and $\chi_n$'s of (12.9), which satisfy

$$\tfrac{1}{4}\boldsymbol{\tau}^2 \chi_p = \tfrac{3}{4}\chi_p \qquad \tfrac{1}{2}\boldsymbol{\tau}_3 \chi_p = \tfrac{1}{2}\chi_p \tag{12.42}$$

$$\tfrac{1}{4}\boldsymbol{\tau}^2 \chi_n = \tfrac{3}{4}\chi_n \qquad \tfrac{1}{2}\boldsymbol{\tau}_3 \chi_n = -\tfrac{1}{2}\chi_n. \tag{12.43}$$

The reason for the 'spin' part of the name 'isospin' should by now be clear: the term is actually a shortened version of the historical one 'isotopic spin'.

In concluding this section we remark that, in this two-dimensional p–n space, the electromagnetic charge operator is represented by the matrix

$$\mathbf{Q}_{em} = \begin{pmatrix} 1 & 0 \\ 0 & 0 \end{pmatrix} = \frac{1}{2}(\mathbf{1}_2 + \tau_3). \tag{12.44}$$

It is clear that although $\mathbf{Q}_{em}$ commutes with $\tau_3$, it does not commute with either $\tau_1$ or $\tau_2$. Thus, as we would expect, electromagnetic corrections to the strong interaction Hamiltonian will violate SU(2) symmetry.

[3] See, for example, Mandl (1992).

### 12.1.2 Larger (higher-dimensional) multiplets of SU(2) in nuclear physics

For the single nucleon states considered so far, the foregoing is really nothing more than the general quantum mechanics of a two-state system, phrased in 'spin-$\frac{1}{2}$' language. The real power of the isospin (SU(2)) symmetry concept becomes more apparent when we consider states of *several* nucleons. For $A$ nucleons in the nucleus, we introduce three 'total isospin operators' $\mathbf{T} = (T_1, T_2, T_3)$ via

$$\mathbf{T} = \tfrac{1}{2}\boldsymbol{\tau}_{(1)} + \tfrac{1}{2}\boldsymbol{\tau}_{(2)} + \cdots + \tfrac{1}{2}\boldsymbol{\tau}_{(A)} \tag{12.45}$$

which are Hermitian. The Hamiltonian $H$ describing the strong interactions of this system is presumed to be invariant under the transformation (12.40) for all the nucleons independently. It then follows that

$$[H, \mathbf{T}] = 0. \tag{12.46}$$

Thus, the eigenvalues of the $\mathbf{T}$ operators are constants of the motion. Further, since the isospin operators for different nucleons commute with each other (they are quite independent), the commutation relations (12.28) for each of the individual $\boldsymbol{\tau}$'s imply (see problem 12.4) that the components of $\mathbf{T}$ defined by (12.45) satisfy the commutation relations

$$[T_i, T_j] = \mathrm{i}\epsilon_{ijk} T_k \tag{12.47}$$

for $i, j, k = 1, 2, 3$, which are simply the standard angular momentum commutation relations, once more. Thus the energy levels of nuclei ought to be characterized—after allowance for electromagnetic effects, and correcting for the slight neutron–proton mass difference—by the eigenvalues of $\mathbf{T}^2$ and $T_3$, say, which can be simultaneously diagonalized along with $H$. These eigenvalues should then be, to a good approximation, 'good quantum numbers' for nuclei, if the assumed isospin invariance is true.

What are the possible eigenvalues? We know that the $\mathbf{T}$'s are Hermitian and satisfy exactly the same commutation relations (12.47) as the angular momentum operators. These conditions are all that are needed to show that the eigenvalues of $\mathbf{T}^2$ are of the form $T(T+1)$, where $T = 0, \frac{1}{2}, 1, \ldots$, and that for a given $T$ the eigenvalues of $T_3$ are $-T, -T+1, \ldots, T-1, T$; that is, there are $2T+1$ *degenerate states* for a given $T$. These states all have the same $A$ value, and since $T_3$ counts $+\frac{1}{2}$ for every proton and $-\frac{1}{2}$ for every neutron, it is clear that successive values of $T_3$ correspond physically to changing one neutron into a proton or *vice versa*. Thus we expect to see 'charge multiplets' of levels in neighbouring nuclear isobars. These are precisely the multiplets of which we have already introduced examples in chapter 1 (see figure 1.8) which we reproduce here as figure 12.2 for convenience. These level schemes (which have been adjusted for Coulomb energy differences, and for the neutron–proton mass difference) provide clear evidence of $T = \frac{1}{2}$ (doublet), $T = 1$ (triplet) and $T = \frac{3}{2}$ (quartet) multiplets. It is important to note that states in the same $T$-multiplet must have the same $J^P$ quantum numbers

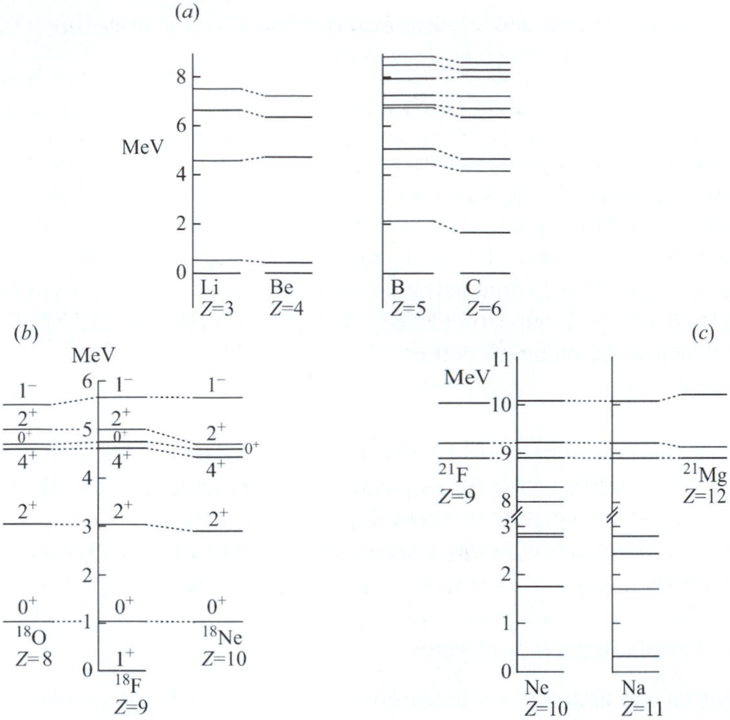


**Figure 12.2.** Energy levels (adjusted for Coulomb energy and neutron–proton mass differences) of nuclei of the same mass number but different charge, showing (*a*) 'mirror' doublets, (*b*) triplets and (*c*) doublets and quartets.

(these are indicated on the levels for $^{18}$F); obviously the nuclear forces will depend on the space and spin degrees of freedom of the nucleons and will only be the same between different nucleons if the space-spin part of the wavefunction is the same. Thus, the assumed invariance of the nucleon–nucleon force produces a richer nuclear multiplet structure, going beyond the original p–n doublet. These higher-dimensional multiplets ($T = 1, \frac{3}{2}, \ldots$) are called 'irreducible representations' of SU(2). The commutation relations (12.47) are called the *Lie algebra* of SU(2)[4] (see appendix M) and the general group-theoretical problem of understanding all *possible* multiplets for SU(2) is equivalent to the problem of finding matrices which satisfy these commutation relations. These are, in fact, precisely the angular momentum matrices of dimension $(2T + 1) \times (2T + 1)$ which are generalizations of the $\boldsymbol{\tau}/2$'s, which themselves correspond to $T = \frac{1}{2}$,

[4] Likewise, the angular momentum commutation relations (12.29) are the Lie algebra of the rotation group SO(3). The Lie algebras of the two groups are, therefore, the same. For an indication of how, nevertheless, the groups do differ, see appendix M, section M.7.

as indicated in the notation $\mathbf{T}^{(\frac{1}{2})}$. For example, the $T = 1$ matrices are $3 \times 3$ and can be compactly summarized by (problem 12.5)

$$(T_i^{(1)})_{jk} = -\mathrm{i}\epsilon_{ijk}, \tag{12.48}$$

where the numbers $-\mathrm{i}\epsilon_{ijk}$ are deliberately chosen to be the *same* numbers (with a minus sign) that specify the algebra in (12.47): the latter are called the *structure constants* of the SU(2) group (see appendix M, sections M.3–M.5). In general, there will be $(2T + 1) \times (2T + 1)$ matrices $\mathbf{T}^{(T)}$ which satisfy (12.47) and correspondingly $(2T + 1)$ dimensional wavefunctions $\psi^{(T)}$ analogous to the two-dimensional ($T = \frac{1}{2}$) case of (12.8). The generalization of (12.32) to these higher-dimensional multiplets is then

$$\psi^{(T)\prime} = \exp(\mathrm{i}\boldsymbol{\alpha} \cdot \mathbf{T}^{(T)})\psi^{(T)} \tag{12.49}$$

which has the general form of (12.35). In this case, the matrices $\mathbf{T}^{(T)}$ provide a $(2T + 1)$-dimensional matrix representation of the generators of SU(2). We shall meet field-theoretic representations of the generators in section 12.3.

We now proceed to consider isospin in our primary area of interest, which is particle physics.

### 12.1.3 Isospin in particle physics

The neutron and proton states themselves are actually only the ground states of a whole series of corresponding $B = 1$ levels with isospin $\frac{1}{2}$ (i.e. doublets), as noted in chapter 1 (see figure 1.10(*a*)). Another series of baryonic levels comes in *four* charge states, as shown in figure 1.10(*b*), corresponding to $T = \frac{3}{2}$; and in the meson sector, the $\pi$'s appear as the lowest states of a sequence of mesonic triplets ($T = 1$), shown in figure 1.11. Many other examples also exist but with one remarkable difference as compared to the nuclear physics case: no baryon states are known with $T > \frac{3}{2}$ nor any meson states with $T > 1$.

The most natural interpretation of these facts is that the observed states are composites of more basic entities which carry different charges but are nearly degenerate in mass, while the forces between these entities are charge-independent, just as in the nuclear (p,n) case. These entities are, of course, the quarks: the n contains (udd), the p is (uud) and the $\Delta$-quartet is (uuu, uud, udd, ddd). The u–d isospin doublet plays the role of the p–n doublet in the nuclear case and this degree of freedom is what we now call SU(2) isospin flavour symmetry at the quark level, denoted by $\mathrm{SU(2)_f}$. We shall denote the u–d quark doublet wavefunction by

$$q = \begin{pmatrix} u \\ d \end{pmatrix} \tag{12.50}$$

omitting now the explicit representation label '$(\frac{1}{2})$' and shortening '$\psi_\mathrm{u}$' to just '$u$', and similarly for '$d$'. Then, under an $\mathrm{SU(2)_f}$ transformation,

$$q \to q' = \mathbf{V}q = \exp(\mathrm{i}\boldsymbol{\alpha} \cdot \boldsymbol{\tau}/2)\, q. \tag{12.51}$$

The limitation $T \le \frac{3}{2}$ for baryonic states can be understood in terms of their being composed of three $T = \frac{1}{2}$ constituents (two of them pair to $T = 1$ or $T = 0$ and the third adds to $T = 1$ to make $T = \frac{3}{2}$ or $T = \frac{1}{2}$ and to $T = 0$ to make $T = \frac{1}{2}$, by the usual angular momentum addition rules). It is, however, a challenge for QCD to explain why—for example—states with four or five quarks should not exist (nor states of one or two quarks!) and why a state of six quarks, for example, appears as the deuteron, which is a loosely bound state of n and p, rather than as a compact $B = 2$ analogue of the n and p themselves.

Meson states such as the pion are formed from a quark and an anti-quark and it is, therefore, appropriate at this point to explain how *anti-particles* are described in isospin terms. An anti-particle is characterized by having the signs of all its additively conserved quantum numbers reversed, relative to those of the corresponding particle. Thus if a u-quark has $B = \frac{1}{3}, T = \frac{1}{2}, T_3 = \frac{1}{2}$, a ū-quark has $B = -\frac{1}{3}, T = \frac{1}{2}, T_3 = -\frac{1}{2}$. Similarly, the d̄ has $B = -\frac{1}{3}, T = \frac{1}{2}$ and $T_3 = \frac{1}{2}$. Note that, while $T_3$ is an additively conserved quantum number, the magnitude of the isospin is not additively conserved: rather, it is 'vectorially' conserved according to the rules of combining angular-momentum-like quantum numbers, as we have seen. Thus, the anti-quarks d̄ and ū form the $T_3 = +\frac{1}{2}$ and $T_3 = -\frac{1}{2}$ members of an $SU(2)_f$ doublet, just as u and d themselves do, and the question arises: given that the $(u, d)$ doublet transforms as in (12.51), how does the $(\bar{u}, \bar{d})$ doublet transform?

The answer is that anti-particles are assigned to the *complex conjugate* of the representation to which the corresponding particles belong. Thus, identifying $\bar{u} \equiv u^*$ and $\bar{d} \equiv d^*$, we have[5]

$$q^{*\prime} = \mathbf{V}^* q^* \qquad \text{or} \begin{pmatrix} \bar{u} \\ \bar{d} \end{pmatrix}' = \exp(-\mathrm{i}\boldsymbol{\alpha} \cdot \boldsymbol{\tau}^*/2) \begin{pmatrix} \bar{u} \\ \bar{d} \end{pmatrix} \tag{12.52}$$

for the $SU(2)_f$ transformation law of the anti-quark doublet. In mathematical terms, this means (compare (12.32)) that the three matrices $-\frac{1}{2}\boldsymbol{\tau}^*$ must represent the generators of $SU(2)_f$ in the $\mathbf{2}^*$ representation (i.e. the complex conjugate of the original two-dimensional representation, which we will now call $\mathbf{2}$). Referring to (12.25), we see that $\tau_1^* = \tau_1$, $\tau_2^* = -\tau_2$ and $\tau_3^* = \tau_3$. It is then straightforward to check that the three matrices $-\tau_1/2, +\tau_2/2$ and $-\tau_3/2$ do indeed satisfy the required commutation relations (12.28) and, thus, provide a valid matrix representation of the SU(2) generators. Also, since the third component of isospin is here represented by $-\tau_3^*/2 = -\tau_3/2$, the desired reversal in sign of the additively conserved eigenvalue does occur.

Although the quark doublet $(u, d)$ and anti-quark doublet $(\bar{u}, \bar{d})$ do transform differently under $SU(2)_f$ transformations, there is nevertheless a sense in which the $\mathbf{2}^*$ and $\mathbf{2}$ representations are somehow the 'same': after all, the quantum

[5] The overbar ($\bar{u}$ etc) here stands only for 'anti-particle', and has nothing to do with the Dirac conjugate $\bar{\psi}$ introduced in section 4.4.

numbers $T = \frac{1}{2}$, $T_3 = \pm\frac{1}{2}$ describe them both. In fact, the two representations are 'unitarily equivalent', in that we can find a unitary matrix $U_C$ such that

$$U_C \exp(-\mathrm{i}\boldsymbol{\alpha}\cdot\boldsymbol{\tau}^*/2)U_C^{-1} = \exp(\mathrm{i}\boldsymbol{\alpha}\cdot\boldsymbol{\tau}/2). \tag{12.53}$$

This requirement is easier to disentangle if we consider infinitesimal transformations, for which (12.53) becomes

$$U_C(-\boldsymbol{\tau}^*)U_C^{-1} = \boldsymbol{\tau} \tag{12.54}$$

or

$$U_C\tau_1U_C^{-1} = -\tau_1 \qquad U_C\tau_2U_C^{-1} = \tau_2 \qquad U_C\tau_3U_C^{-1} = -\tau_3. \tag{12.55}$$

Bearing the commutation relations (12.28) in mind, and the fact that $\tau_i^{-1} = \tau_i$, it is clear that we can choose $U_C$ proportional to $\tau_2$, and set

$$U_C = \mathrm{i}\tau_2 = \begin{pmatrix} 0 & 1 \\ -1 & 0 \end{pmatrix} \tag{12.56}$$

to obtain a convenient unitary form. This implies that the doublet

$$U_C\begin{pmatrix} \bar{u} \\ \bar{d} \end{pmatrix} = \begin{pmatrix} \bar{d} \\ -\bar{u} \end{pmatrix} \tag{12.57}$$

transforms in exactly the same way as $(u, d)$. This result is useful, because it means that we can use the familiar tables of (Clebsch–Gordan) angular momentum coupling coefficients for combining quark and anti-quark states together, *provided* we include the relative minus sign between the $\bar{d}$ and $\bar{u}$ components which has appeared in (12.57). Note that, as expected, the $\bar{d}$ is in the $T_3 = +\frac{1}{2}$ position and the $\bar{u}$ is in the $T_3 = -\frac{1}{2}$ position.

As an application of these results, let us compare the $T = 0$ combination of the p and n states to form the (isoscalar) deuteron, and the combination of $(u, d)$ and $(\bar{u}, \bar{d})$ states to form the isoscalar $\omega$-meson. In the first, the isospin part of the wavefunction is $\frac{1}{\sqrt{2}}(\psi_\mathrm{p}\psi_\mathrm{n} - \psi_\mathrm{n}\psi_\mathrm{p})$, corresponding to the $S = 0$ combination of two spin-$\frac{1}{2}$ particles in quantum mechanics given by $\frac{1}{\sqrt{2}}(|\uparrow\rangle|\downarrow\rangle - |\downarrow\rangle|\uparrow\rangle)$. But, in the second case, the corresponding wavefunction is $\frac{1}{\sqrt{2}}(\bar{d}d - (-\bar{u})u) = \frac{1}{\sqrt{2}}(\bar{d}d + \bar{u}u)$. Similarly, the $T = 1$, $T_3 = 0$ state describing the $\pi^0$ is $\frac{1}{\sqrt{2}}(\bar{d}d + (-\bar{u})u) = \frac{1}{\sqrt{2}}(\bar{d}d - \bar{u}u)$.

There is a very convenient alternative way of obtaining these wavefunctions, which we include here because it generalizes straightforwardly to SU(3): its advantage is that it avoids the use of the explicit C–G coupling coefficients and of their (more complicated) analogues in SU(3).

Bearing in mind the identifications $\bar{u} \equiv u^*$, $\bar{d} \equiv d^*$, we see that the $T = 0$ $\bar{q}q$ combination $\bar{u}u + \bar{d}d$ can be written as $u^*u + d^*d$ which is just $q^\dagger q$, (recall that $^\dagger$ means transpose and complex conjugate). Under an SU(2)$_\mathrm{f}$ transformation, $q \to q' = \mathbf{V}q$, so $q^\dagger \to q'^\dagger = q^\dagger \mathbf{V}^\dagger$ and

$$q^\dagger q \to q'^\dagger q' = q^\dagger \mathbf{V}^\dagger \mathbf{V} q = q^\dagger q \tag{12.58}$$

using $\mathbf{V}^\dagger \mathbf{V} = \mathbf{1}_2$; thus, $q^\dagger q$ is indeed an SU(2)$_\mathrm{f}$ invariant, which means it has $T = 0$ (no multiplet partners).

We may also construct the $T = 1$ $q$–$\bar{q}$ states in a similar way. Consider the three quantities $v_i$ defined by

$$v_i = q^\dagger \tau_i q \qquad i = 1, 2, 3. \tag{12.59}$$

Under an infinitesimal SU(2)$_\mathrm{f}$ transformation

$$q' = (\mathbf{1}_2 + \mathrm{i}\boldsymbol{\epsilon} \cdot \boldsymbol{\tau}/2)q \tag{12.60}$$

the three quantities $v_i$ transform to

$$v_i' = q^\dagger (\mathbf{1}_2 - \mathrm{i}\boldsymbol{\epsilon} \cdot \boldsymbol{\tau}/2)\tau_i(\mathbf{1}_2 + \mathrm{i}\boldsymbol{\epsilon} \cdot \boldsymbol{\tau}/2)q \tag{12.61}$$

where we have used $q'^\dagger = q^\dagger(\mathbf{1}_2 + \mathrm{i}\boldsymbol{\epsilon} \cdot \boldsymbol{\tau}/2)^\dagger$ and then $\boldsymbol{\tau}^\dagger = \boldsymbol{\tau}$. Retaining only the first-order terms in $\boldsymbol{\epsilon}$ gives (problem 12.6)

$$v_i' = v_i + \mathrm{i}\frac{\epsilon_j}{2} q^\dagger(\tau_i \tau_j - \tau_j \tau_i)q \tag{12.62}$$

where the sum on $j = 1, 2, 3$ is understood. But from (12.28) we know the commutator of two $\tau$'s, so that (12.62) becomes

$$\begin{aligned} v_i' &= v_i + \mathrm{i}\frac{\epsilon_j}{2} q^\dagger 2\mathrm{i}\epsilon_{ijk}\tau_k q \qquad (\text{sum on } k = 1, 2, 3) \\ &= v_i - \epsilon_{ijk}\epsilon_j q^\dagger \tau_k q \\ &= v_i - \epsilon_{ijk}\epsilon_j v_k \end{aligned} \tag{12.63}$$

which may also be written in 'vector' notation as

$$\boldsymbol{v}' = \boldsymbol{v} - \boldsymbol{\epsilon} \times \boldsymbol{v}. \tag{12.64}$$

Equation (12.63) states that, under an (infinitesimal) SU(2)$_\mathrm{f}$ transformation, the three quantities $v_i (i = 1, 2, 3)$ transform into *specific linear combinations* of themselves, as determined by the coefficients $\epsilon_{ijk}$ (the $\epsilon$'s are just the parameters of the infinitesimal transformation). This is precisely what is needed for a set of quantities to *form the basis for a representation*. In this case, it is the $T = 1$ representation as we can guess from the multiplicity of three, but we can also directly verify it as follows. Equation (12.49) with $T = 1$, together with (12.48), tell us how a $T = 1$ triplet should transform: namely, under an infinitesimal transformation (with $\mathbf{1}_3$ the unit $3 \times 3$ matrix),

$$
\begin{aligned}
\psi_i^{(1)\prime} &= (\mathbf{1}_3 + \mathrm{i}\boldsymbol{\epsilon} \cdot \mathbf{T}^{(1)})_{ik}\psi_k^{(1)} \qquad \text{(sum on } k = 1, 2, 3) \\
&= (\mathbf{1}_3 + \mathrm{i}\epsilon_j T_j^{(1)})_{ik}\psi_k^{(1)} \qquad \text{(sum on } j = 1, 2, 3) \\
&= (\delta_{ik} + \mathrm{i}\epsilon_j (T_j^{(1)})_{ik}\psi_k^{(1)} \\
&= (\delta_{ik} + \mathrm{i}\epsilon_j . - \mathrm{i}\epsilon_{jik})\psi_k^{(1)} \qquad \text{using (12.48)} \\
&= \psi_i^{(1)} - \epsilon_{ijk}\epsilon_j\psi_k^{(1)} \qquad \text{using the anti-symmetry of } \epsilon_{ijk}.
\end{aligned} \tag{12.65}
$$

which is exactly the same as (12.63).

As an aside, the reader may have been struck by the similarity between the Lorentz 4-vector combination of Dirac spinors given by '$\bar{\psi}\gamma^\mu\psi$' and the present triplet combination '$q^\dagger\boldsymbol{\tau}q$'. Indeed, recalling the close connection between SU(2) and SO(3), we can at once infer from (12.59)–(12.63) that if $\phi$ is a two-component (Pauli) spinor, $\phi^\dagger\boldsymbol{\sigma}\phi$ behaves as a vector (SO(3)-triplet) under real-space rotations. The spatial part $\bar{\psi}\boldsymbol{\gamma}\psi$ generalizes this to four-component spinors (the $\mu = 0$ part $\psi^\dagger\psi$ is rotationally invariant, analogous to $q^\dagger q$); when the transformations are extended to include Lorentz (velocity) transformations, the four combinations $\bar{\psi}\gamma^\mu\psi$ behave as a 4-vector, as we have seen in volume 1.

Returning to the physics of $v_i$, inserting (12.50) into (12.59), we find explicitly that

$$
v_1 = \bar{u}d + \bar{d}u \qquad v_2 = -\mathrm{i}\,\bar{u}d + \mathrm{i}\,\bar{d}u \qquad v_3 = \bar{u}u - \bar{d}d. \tag{12.66}
$$

Apart from the normalization factor of $1/\sqrt{2}$, $v_3$ may, therefore, be identified with the $T_3 = 0$ member of the $T = 1$ triplet, having the quantum numbers of the $\pi^0$. Neither $v_1$ nor $v_2$ has a definite value of $T_3$, however: rather, we need to consider the linear combinations

$$
\tfrac{1}{2}(v_1 + \mathrm{i}v_2) = \bar{u}d \qquad T_3 = -1 \tag{12.67}
$$

and

$$
\tfrac{1}{2}(v_1 - \mathrm{i}v_2) = \bar{d}u \qquad T_3 = +1 \tag{12.68}
$$

which have the quantum numbers of the $\pi^-$ and $\pi^+$. The use of $v_1 \pm \mathrm{i}v_2$ here is precisely analogous to the use of the 'spherical basis' wavefunctions $x \pm \mathrm{i}y = r\sin\theta\mathrm{e}^{\pm\mathrm{i}\phi}$ for $\ell = 1$ states in quantum mechanics, rather than the 'Cartesian' ones $x$ and $y$.

We are now ready to proceed to SU(3).

## 12.2 Flavour SU(3)$_\mathrm{f}$

Larger hadronic multiplets also exist, in which strange particles are grouped with non-strange ones. Gell-Mann (1961) and Ne'eman (1961) (see also Gell-Mann and Ne'eman (1964)) were the first to propose SU(3)$_f$ as the correct generalization of isospin SU(2)$_f$ to include strangeness. Like SU(2), SU(3) is a group whose elements are matrices—in this case, unitary $3 \times 3$ ones, of unit determinant. The general group-theoretic analysis of SU(3) is quite complicated but is fortunately not necessary for the physical applications we require. We can, in fact, develop all the results needed by mimicking the steps followed for SU(2).

We start by finding the general form of an SU(3) matrix. Such matrices obviously act on three-component column vectors, the generalization of the two-component isospinors of SU(2). In more physical terms, we regard the three quark wavefunctions $u, d$ and $s$ as being approximately degenerate and we consider unitary $3 \times 3$ transformations among them via

$$q' = \mathbf{W}q \tag{12.69}$$

where $q$ now stands for the three-component column vector

$$q = \begin{pmatrix} u \\ d \\ s \end{pmatrix} \tag{12.70}$$

and $\mathbf{W}$ is a $3 \times 3$ unitary matrix of determinant 1 (again, an overall phase has been extracted). The representation provided by this triplet of states is called the 'fundamental' representation of SU(3)$_f$ (just as the isospinor representation is the fundamental one of SU(2)$_f$).

To determine the general form of an SU(3) matrix $\mathbf{W}$, we follow exactly the same steps as in the SU(2) case. An infinitesimal SU(3) matrix has the form

$$\mathbf{W}_{\text{infl}} = \mathbf{1}_3 + \mathrm{i}\boldsymbol{\chi} \tag{12.71}$$

where $\boldsymbol{\chi}$ is a $3 \times 3$ traceless Hermitian matrix. Such a matrix involves *eight* independent parameters (problem (12.7)) and can be written as

$$\boldsymbol{\chi} = \boldsymbol{\eta} \cdot \boldsymbol{\lambda}/2 \tag{12.72}$$

where $\boldsymbol{\eta} = (\eta_1, \ldots, \eta_8)$ and the $\boldsymbol{\lambda}$'s are eight matrices generalizing the $\boldsymbol{\tau}$ matrices of (12.25). They are the generators of SU(3) in the three-dimensional fundamental representation and their commutation relations define the *algebra of SU(3)* (compare (12.28) for SU(2)):

$$[\lambda_a/2, \lambda_b/2] = \mathrm{i} f_{abc}\lambda_c/2 \tag{12.73}$$

where $a, b$ and $c$ run from 1 to 8.

The $\lambda$-matrices (often called the *Gell-Mann matrices*), are given in appendix M, along with the *SU(3) structure constants* $f_{abc}$. A finite SU(3) transformation on the quark triplet is then (cf (12.32))

$$q' = \exp(\mathrm{i}\boldsymbol{\alpha} \cdot \boldsymbol{\lambda}/2)q \tag{12.74}$$

which also has the 'generalized phase transformation' character of (12.35), now with *eight* 'phase angles'. Thus, **W** is parametrized as $\mathbf{W} = \exp(\mathrm{i}\boldsymbol{\alpha} \cdot \boldsymbol{\lambda}/2)$.

As in the case of $\mathrm{SU}(2)_\mathrm{f}$, exact symmetry under $\mathrm{SU}(3)_\mathrm{f}$ would imply that the three states u, d and s were degenerate in mass. Actually, of course, this is not the case: in particular, while the u and d quark masses are of order 5–10 MeV, the s-quark mass is much greater, of order 150 MeV. Nevertheless, it is still possible to regard this as relatively small on a typical hadronic mass scale of $\sim$ 1 GeV, so we may proceed to explore the physical consequences of this (approximate) $\mathrm{SU}(3)_\mathrm{f}$ flavour symmetry.

Such a symmetry implies that the eigenvalues of the $\boldsymbol{\lambda}$'s are constants of the motion, but because of the commutation relations (12.73) only a subset of these operators has simultaneous eigenstates. This happened for SU(2) too, but there the very close analogy with SO(3) told us how the states were to be correctly classified, by the eigenvalues of the relevant complete set of mutually commuting operators. Here it is more involved—for a start, there are eight matrices $\lambda_a$. A glance at appendix M, section M.4(v), shows that *two* of the $\lambda$'s are diagonal (in the chosen representation), namely $\lambda_3$ and $\lambda_8$. This means physically that for SU(3) there are *two* additively conserved quantum numbers, which in this case are of course the third component of hadronic isospin (since $\lambda_3$ is simply $\tau_3$ bordered by zeros), and a quantity related to strangeness. Defining the hadronic hyperchange $Y$ by $Y = B + S$, where $B$ is the baryon number ($\frac{1}{3}$ for each quark) and the strangeness values are $S(\mathrm{u}) = S(\mathrm{d}) = 0$, $S(\mathrm{s}) = -1$, we find that the physically required eigenvalues imply that the matrix representing the hypercharge operator is $Y^{(\mathbf{3})} = \frac{1}{\sqrt{3}}\lambda_8$, in this fundamental (three-dimensional) representation, denoted by the symbol **3**. Identifying $T_3^{(\mathbf{3})} = \frac{1}{2}\lambda_3$ then gives the Gell-Mann–Nishijima relation $Q = T_3 + Y/2$ for the quark charges in units of $|e|$.

So $\lambda_3$ and $\lambda_8$ are analogous to $\tau_3$: what about the analogue of $\boldsymbol{\tau}^2$, which is diagonalizable simultaneously with $\tau_3$ in the case of SU(2)? Indeed, (cf (12.41)), $\boldsymbol{\tau}^2$ is a multiple of the $2 \times 2$ unit matrix. In just the same way, one finds that $\boldsymbol{\lambda}^2$ is also proportional to the unit $3 \times 3$ matrix:

$$(\boldsymbol{\lambda}/2)^2 = \sum_{i=1}^{8} (\lambda_a/2)^2 = \frac{4}{3}\mathbf{1}_3 \tag{12.75}$$

as can be verified from the explicit forms of the $\lambda$-matrices given in appendix M, section M.4(v). Thus, we may characterize the 'fundamental triplet' (12.70) by the eigenvalues of $(\boldsymbol{\lambda}/2)^2$, $\lambda_3$ and $\lambda_8$. The conventional way of representing this pictorially is to plot the states in a $Y$–$T_3$ diagram, as shown in figure 12.3.

We may now consider other representations of $\mathrm{SU}(3)_\mathrm{f}$. The first important one is that to which the *anti-quarks* belong. If we denote the fundamental three-dimensional representation accommodating the quarks by **3**, then the anti-quarks have quantum numbers appropriate to the 'complex conjugate' of this

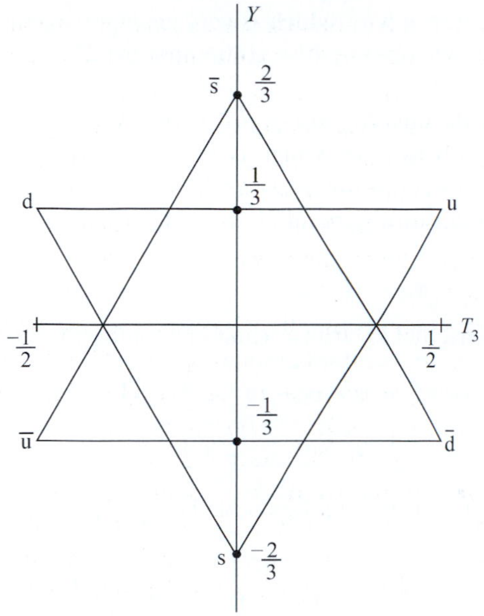


**Figure 12.3.** The $Y$–$T_3$ quantum numbers of the fundamental triplet **3** of quarks and of the anti-triplet **3**$^*$ of anti-quarks.

representation, denoted by **3**$^*$ just as in the SU(2) case. The $\bar{q}$ wavefunctions identified as $\bar{u} \equiv u^*$, $\bar{d} \equiv d^*$ and $\bar{s} \equiv s^*$ then transform by

$$\bar{q}' = \begin{pmatrix} \bar{u} \\ \bar{d} \\ \bar{s} \end{pmatrix}' = \mathbf{W}^* \bar{q} = \exp(-\mathrm{i}\boldsymbol{\alpha} \cdot \boldsymbol{\lambda}^*/2)\bar{q} \tag{12.76}$$

instead of by (12.74). As for the **2**$^*$ representation of SU(2), (12.76) means that the eight quantities $-\boldsymbol{\lambda}^*/2$ represent the SU(3) generators in this **3**$^*$ representation. Referring to appendix M, section M.4(v), one quickly sees that $\lambda_3$ and $\lambda_8$ are real, so that the eigenvalues of the physical observables $T_3^{(\mathbf{3}^*)} = -\lambda_3/2$ and $Y^{(\mathbf{3}^*)} = -\frac{1}{\sqrt{3}}\lambda_8/2$ (in this representation) are reversed relative to those in the **3**, as expected for anti-particles. The ū, d̄ and s̄ states may also be plotted on the $Y$–$T_3$ diagram, figure 12.3, as shown.

Here is already one important difference between SU(3) and SU(2): the fundamental SU(3) representation **3** and its complex conjugate **3**$^*$ are *not* equivalent. This follows immediately from figure 12.3, where it is clear that the extra quantum number $Y$ distinguishes the two representations.

Larger SU(3)$_f$ representations can be created by combining quarks and anti-quarks, as in SU(2)$_f$. For our present purposes, an important one is the eight-dimensional ('octet') representation which appears when one combines the **3**$^*$

and **3** representations, in a way which is very analogous to the three-dimensional ('triplet') representation obtained by combining the $\mathbf{2}^*$ and **2** representations of SU(2).

Consider first the quantity $\bar{u}u + \bar{d}d + \bar{s}s$. As in the SU(2) case, this can be written equivalently as $q^\dagger q$, which is invariant under $q \to q' = \mathbf{W}q$ since $\mathbf{W}^\dagger\mathbf{W} = \mathbf{1}_3$. So this combination is an SU(3) *singlet*. The *octet* coupling is formed by a straightforward generalization of the SU(2) triplet coupling $q^\dagger \boldsymbol{\tau} q$ of (12.59),

$$w_a = q^\dagger \lambda_a q \qquad a = 1, 2, \ldots, 8. \tag{12.77}$$

Under an infinitesimal SU(3)$_\mathrm{f}$ transformation (compare (12.61) and (12.62)),

$$\begin{aligned} w_a \to w'_a &= q^\dagger(\mathbf{1}_3 - \mathrm{i}\boldsymbol{\eta}\cdot\boldsymbol{\lambda}/2)\lambda_a(\mathbf{1}_3 + \mathrm{i}\boldsymbol{\eta}\cdot\boldsymbol{\lambda}/2)q \\ &\approx q^\dagger\lambda_a q + \mathrm{i}\frac{\eta_b}{2} q^\dagger(\lambda_a\lambda_b - \lambda_b\lambda_a)q \end{aligned} \tag{12.78}$$

where the sum on $b = 1$–8 is understood. Using (12.73) for the commutator of two $\lambda$'s we find that

$$w'_a = w_a + \mathrm{i}\frac{\eta_b}{2} q^\dagger \cdot 2\mathrm{i} f_{abc}\lambda_c q \tag{12.79}$$

or

$$w'_a = w_a - f_{abc}\eta_b w_c \tag{12.80}$$

which may usefully be compared with (12.63). Just as in the SU(2)$_\mathrm{f}$ triplet case, equation (12.80) shows that, under an SU(3)$_\mathrm{f}$ transformation, the eight quantities $w_a (a = 1, 2, \ldots, 8)$ transform with specific linear combinations of themselves, as determined by the coefficients $f_{abc}$ (the $\eta$'s are just the parameters of the infinitesimal transformation).

This is, again, precisely what is needed for a set of quantities to form the basis for a representation—in this case, an eight-dimensional representation of SU(3)$_\mathrm{f}$. For a finite SU(3)$_\mathrm{f}$ transformation, we can 'exponentiate' (12.80) to obtain

$$\boldsymbol{w}' = \exp(\mathrm{i}\boldsymbol{\alpha}\cdot\mathbf{G}^{(8)})\boldsymbol{w} \tag{12.81}$$

where $\boldsymbol{w}$ is an eight-component column vector

$$\boldsymbol{w} = \begin{pmatrix} w_1 \\ w_2 \\ \vdots \\ w_8 \end{pmatrix} \tag{12.82}$$

such that $w_a = q^\dagger\lambda_a q$, and where (cf (12.49) for SU(2))$_\mathrm{f}$) the quantities $\mathbf{G}^{(8)} = (G_1^{(8)}, G_2^{(8)}, \ldots, G_8^{(8)})$ are $8 \times 8$ matrices, acting on the eight-component vector $\boldsymbol{w}$ and forming an eight-dimensional representation of the algebra of SU(3): that is to say, the $\mathbf{G}^{(8)}$'s satisfy (cf (12.73))

$$[G_a^{(8)}, G_b^{(8)}] = \mathrm{i} f_{abc} G_c^{(8)}. \tag{12.83}$$

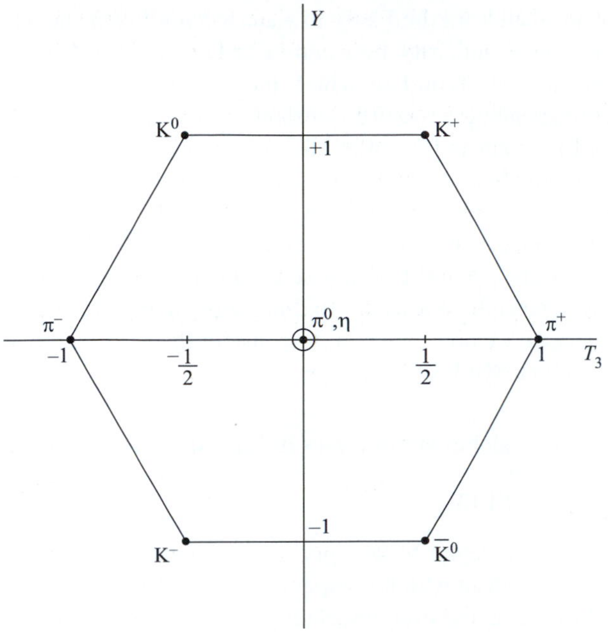


**Figure 12.4.** The $Y$–$T_3$ quantum numbers of the pseudoscalar meson octet.

The actual form of the $G_a^{(8)}$ matrices is given by comparing the infinitesimal version of (12.81) with (12.80):

$$\left(G_a^{(8)}\right)_{bc} = -\mathrm{i} f_{abc} \tag{12.84}$$

as may be checked in problem 12.8, where it is also verified that the matrices specified by (12.84) do obey the commutation relations (12.83).

As in the SU(2)$_f$ case, the eight states generated by the combinations $q^\dagger \lambda_a q$ are not necessarily the ones with the physically desired quantum numbers. To get the $\pi^+$, for example, we again need to form $(w_1 \pm \mathrm{i} w_2)/2$. Similarly, $w_4$ produces $\bar{u}s + \bar{s}u$ and $w_5$ the combination $-\mathrm{i}\,\bar{u}s + \mathrm{i}\,\bar{s}u$, so the $K^\pm$ states are $w_4 \mp \mathrm{i}w_5$. Similarly the $K^0$, $\bar{K}^0$ states are $w_6 - \mathrm{i}w_7$ and $w_6 + \mathrm{i}w_7$, while the $\eta$ (in this simple model) would be $w_8 \sim (\bar{u}u + \bar{d}d - 2\bar{s}s)$, which is orthogonal to both the $\pi^0$ state and the SU(3)$_f$ singlet. In this way all the pseudoscalar octet of $\pi$-partners has been identified, as shown on the $Y$–$T$ diagram of figure 12.4. We say 'octet of $\pi$-partners' but a reader knowing the masses of these particles might well query why we should feel justified in regarding them as (even approximately) degenerate. By contrast, a similar octet of vector ($J^P 1^-$) mesons (the $\omega$, $\rho$, $K^*$ and $\bar{K}^*$) are all much closer in mass, averaging around 800 MeV: in these states the $\bar{q}q$ spins add to $S = 1$, while the orbital angular momentum is still zero. The pion, and to a much lesser extent the kaons, seem to be 'anomalously light' for some reason: we shall learn the likely explanation for this in chapter 18.

There is a deep similarity between (12.84) and (12.48). In both cases, a representation has been found in which the matrix element of a generator is minus the corresponding structure constant. Such a representation is always possible for a Lie group and is called the *adjoint*, or *regular*, representation (see appendix M, section M.5). These representations are of particular importance in gauge theories, as we will see, since gauge quanta always belong to the adjoint representation of the gauged group (for example, the eight gluons in $SU(3)_c$).

Further flavours c, b and t of course exist but the mass differences are now so large that it is generally not useful to think about higher flavour groups such as $SU(4)_f$ etc. Instead, we now move on to consider the field-theoretic formulation of global $SU(2)_f$ and $SU(3)_f$.

## 12.3 Non-Abelian global symmetries in Lagrangian quantum field theory

### 12.3.1 $SU(2)_f$ and $SU(3)_f$

As may already have begun to be apparent in chapter 7, Lagrangian quantum field theory is a formalism which is especially well adapted for the description of symmetries. Without going into any elaborate general theory, we shall now give a few examples showing how global flavour symmetry is very easily built into a Lagrangian, generalizing in a simple way the global U(1) symmetries considered in sections 7.1 and 7.2. This will also prepare the way for the (local) gauge case, to be considered in the following chapter.

Consider, for example, the Lagrangian

$$\hat{\mathcal{L}} = \bar{\hat{u}}(\mathrm{i}\partial\!\!\!/ - m)\hat{u} + \bar{\hat{d}}(\mathrm{i}\partial\!\!\!/ - m)\hat{d} \tag{12.85}$$

describing two free fermions 'u' and 'd' of equal mass $m$, with the overbar now meaning the Dirac conjugate for the four-component spinor fields. As in (12.50), we are using the convenient shorthand $\hat{\psi}_u = \hat{u}$ and $\hat{\psi}_d = \hat{d}$. Let us introduce

$$\hat{q} = \begin{pmatrix} \hat{u} \\ \hat{d} \end{pmatrix} \tag{12.86}$$

so that $\hat{\mathcal{L}}$ can be compactly written as

$$\hat{\mathcal{L}} = \bar{\hat{q}}(\mathrm{i}\partial\!\!\!/ - m)\hat{q}. \tag{12.87}$$

In this form it is obvious that $\hat{\mathcal{L}}$—and, hence, the associated Hamiltonian $\hat{\mathcal{H}}$—are invariant under the global U(1) transformation

$$\hat{q}' = \mathrm{e}^{\mathrm{i}\alpha}\hat{q} \tag{12.88}$$

(cf (12.1)) which is associated with baryon number conservation. It is also invariant under global $SU(2)_f$ transformations acting in the flavour u–d space (cf (12.32)):

$$\hat{q}' = \exp(-\mathrm{i}\boldsymbol{\alpha}\cdot\boldsymbol{\tau}/2)\hat{q} \tag{12.89}$$

(for the change in sign with respect to (12.31), compare sections 7.1 and 7.2 in the U(1) case). In (12.89), the three parameters $\boldsymbol{\alpha}$ are independent of $x$.

What are the conserved quantities associated with the invariance of $\hat{\mathcal{L}}$ under (12.89)? Let us recall the discussion of the simpler U(1) cases studied in sections 7.1 and 7.2. Considering the complex scalar field of section 7.1, the analogue of (12.89) was just $\hat{\phi} \to \hat{\phi}' = \mathrm{e}^{-\mathrm{i}\alpha}\hat{\phi}$ and the conserved quantity was the Hermitian operator $\hat{N}_\phi$ which appeared in the exponent of the unitary operator $\hat{U}$ that effected the transformation $\hat{\phi} \to \hat{\phi}'$ via

$$\hat{\phi}' = \hat{U}\hat{\phi}\hat{U}^\dagger \tag{12.90}$$

with

$$\hat{U} = \exp(\mathrm{i}\alpha\hat{N}_\phi). \tag{12.91}$$

For an infinitesimal $\alpha$, we have

$$\hat{\phi}' \approx (1 - \mathrm{i}\epsilon)\hat{\phi} \qquad \hat{U} \approx 1 + \mathrm{i}\epsilon\hat{N}_\phi \tag{12.92}$$

so that (12.90) becomes

$$(1 - \mathrm{i}\epsilon)\hat{\phi} = (1 + \mathrm{i}\epsilon\hat{N}_\phi)\hat{\phi}(1 - \mathrm{i}\epsilon\hat{N}_\phi) \approx \hat{\phi} + \mathrm{i}\epsilon[\hat{N}_\phi, \hat{\phi}]; \tag{12.93}$$

hence, we require

$$[\hat{N}_\phi, \hat{\phi}] = -\hat{\phi} \tag{12.94}$$

for consistency. Insofar as $\hat{N}_\phi$ determines the form of an infinitesimal version of the unitary transformation operator $\hat{U}$, it seems reasonable to call it the *generator* of these global U(1) transformations (compare the discussion after (12.27) and (12.35) but note that here $\hat{N}_\phi$ is a quantum field operator, not a matrix).

Consider now the $\mathrm{SU(2)_f}$ transformation (12.89), in the infinitesimal case:

$$\hat{q}' = (1 - \mathrm{i}\boldsymbol{\epsilon} \cdot \boldsymbol{\tau}/2)\hat{q}. \tag{12.95}$$

Since the single U(1) parameter $\epsilon$ is now replaced by the three parameters $\boldsymbol{\epsilon} = (\epsilon_1, \epsilon_2, \epsilon_3)$, we shall need three analogues of $\hat{N}_\phi$, which we call

$$\hat{\boldsymbol{T}}^{(\frac{1}{2})} = (\hat{T}_1^{(\frac{1}{2})}, \hat{T}_2^{(\frac{1}{2})}, \hat{T}_3^{(\frac{1}{2})}) \tag{12.96}$$

corresponding to the three independent infinitesimal SU(2) transformations. The generalizations of (12.90) and (12.91) are then

$$\hat{q}' = \hat{U}^{(\frac{1}{2})}\hat{q}\hat{U}^{(\frac{1}{2})\dagger} \tag{12.97}$$

and

$$\hat{U}^{(\frac{1}{2})} = \exp(\mathrm{i}\boldsymbol{\alpha} \cdot \hat{\boldsymbol{T}}^{(\frac{1}{2})}) \tag{12.98}$$

where the $\hat{\boldsymbol{T}}^{(\frac{1}{2})}$ are Hermitian, so that $\hat{U}^{(\frac{1}{2})}$ is unitary (cf (12.35)). It would seem reasonable in this case too to regard the $\hat{\boldsymbol{T}}^{(\frac{1}{2})}$ as providing a *field-theoretic representation* of the generators of $\mathrm{SU}(2)_{\mathrm{f}}$, an interpretation we shall shortly confirm. In the infinitesimal case, (12.97) and (12.98) become

$$(1 - \mathrm{i}\boldsymbol{\epsilon} \cdot \boldsymbol{\tau}/2)\hat{q} = (1 + \mathrm{i}\boldsymbol{\epsilon} \cdot \hat{\boldsymbol{T}}^{(\frac{1}{2})})\hat{q}(1 - \mathrm{i}\boldsymbol{\epsilon} \cdot \hat{\boldsymbol{T}}^{(\frac{1}{2})}) \tag{12.99}$$

using the Hermiticity of the $\hat{\boldsymbol{T}}^{(\frac{1}{2})}$'s. Expanding the right-hand side of (12.99) to first order in $\boldsymbol{\epsilon}$, and equating coefficients of $\boldsymbol{\epsilon}$ on both sides, (12.99) reduces to (problem 12.9)

$$[\hat{\boldsymbol{T}}^{(\frac{1}{2})}, \hat{q}] = -(\boldsymbol{\tau}/2)\hat{q} \tag{12.100}$$

which is the analogue of (12.94). Equation (12.100) expresses a very specific *commutation* property of the operators $\hat{\boldsymbol{T}}^{(\frac{1}{2})}$, which turns out to be satisfied by the expression

$$\hat{\boldsymbol{T}}^{(\frac{1}{2})} = \int \hat{q}^{\dagger}(\boldsymbol{\tau}/2)\hat{q}\,\mathrm{d}^3\boldsymbol{x} \tag{12.101}$$

as can be checked (problem 12.10) from the anti-commutation relations of the fermionic fields in $\hat{q}$. We shall derive (12.101) from Noether's theorem in a little while. Note that if '$\boldsymbol{\tau}/2$' is replaced by 1, (12.101) reduces to the sum of the u and d number operators, as required for the one-parameter U(1) case. The '$\hat{q}^{\dagger}\boldsymbol{\tau}\hat{q}$' combination is precisely the field-theoretic version of the $q^{\dagger}\boldsymbol{\tau}q$ coupling we discussed in section 12.1.3. This means that the three operators $\hat{\boldsymbol{T}}^{(\frac{1}{2})}$ themselves belong to a $T = 1$ triplet of $\mathrm{SU}(2)_{\mathrm{f}}$.

It is possible to verify that these $\hat{\boldsymbol{T}}^{(\frac{1}{2})}$'s do indeed commute with the Hamiltonian $\hat{H}$:

$$\mathrm{d}\hat{\boldsymbol{T}}^{(\frac{1}{2})}/\mathrm{d}t = -\mathrm{i}[\hat{\boldsymbol{T}}^{(\frac{1}{2})}, \hat{H}] = 0 \tag{12.102}$$

so that their eigenvalues are conserved. That the $\hat{\boldsymbol{T}}^{(\frac{1}{2})}$'s are, as already suggested, a field-theoretic representation of the generators of SU(2), appropriate to the case $T = \frac{1}{2}$, follows from the fact that they obey the SU(2) algebra (problem 12.11):

$$[\hat{T}_i^{(\frac{1}{2})}, \hat{T}_j^{(\frac{1}{2})}] = \mathrm{i}\epsilon_{ijk}\hat{T}_k^{(\frac{1}{2})}. \tag{12.103}$$

For many purposes it is more useful to consider the raising and lowering operators

$$\hat{T}_{\pm}^{(\frac{1}{2})} = (\hat{T}_1^{(\frac{1}{2})} \pm \mathrm{i}\hat{T}_2^{(\frac{1}{2})}). \tag{12.104}$$

For example, we easily find

$$\hat{T}_{+}^{(\frac{1}{2})} = \int \hat{u}^{\dagger}\hat{d}\,\mathrm{d}^3\boldsymbol{x} \tag{12.105}$$

which destroys a d quark and creates a u, or destroys a ū and creates a $\bar{\mathrm{d}}$, in either case raising the $\hat{T}_3^{(\frac{1}{2})}$ eigenvalue by +1, since

$$\hat{T}_3^{(\frac{1}{2})} = \frac{1}{2}\int (\hat{u}^\dagger \hat{u} - \hat{d}^\dagger \hat{d})\, \mathrm{d}^3\boldsymbol{x} \tag{12.106}$$

which counts $+\frac{1}{2}$ for each u (or $\bar{\mathrm{d}}$) and $-\frac{1}{2}$ for each d (or ū). Thus, these operators certainly 'do the job' expected of field-theoretic isospin operators, in this isospin-$\frac{1}{2}$ case.

In the U(1) case, considering now the fermionic example of section 7.2 for variety, we could go further and associate the conserved operator $\hat{N}_\psi$ with a *conserved current* $\hat{N}_\psi^\mu$:

$$\hat{N}_\psi = \int \hat{N}_\psi^0\, \mathrm{d}^3\boldsymbol{x} \qquad \hat{N}_\psi^\mu = \bar{\hat{\psi}}\gamma^\mu\hat{\psi} \tag{12.107}$$

where

$$\partial_\mu \hat{N}_\psi^\mu = 0. \tag{12.108}$$

The obvious generalization appropriate to (12.101) is

$$\hat{\boldsymbol{T}}^{(\frac{1}{2})} = \int \hat{\boldsymbol{T}}^{(\frac{1}{2})0} \mathrm{d}^3\boldsymbol{x} \qquad \hat{\boldsymbol{T}}^{(\frac{1}{2})\mu} = \bar{\hat{q}}\gamma^\mu\frac{\boldsymbol{\tau}}{2}\hat{q}. \tag{12.109}$$

Note that both $\hat{N}_\psi^\mu$ and $\hat{\boldsymbol{T}}^{(\frac{1}{2})\mu}$ are of course functions of the spacetime coordinate $x$, via the (suppressed) dependence of the $\hat{q}$-fields on $x$. Indeed one can verify from the equations of motion that

$$\partial_\mu \hat{\boldsymbol{T}}^{(\frac{1}{2})\mu} = 0. \tag{12.110}$$

Thus $\hat{\boldsymbol{T}}^{(\frac{1}{2})\mu}$ is a *conserved isospin current operator* appropriate to the $T = \frac{1}{2}$ (u, d) system: it transforms as a 4-vector under Lorentz transformations and as a $T = 1$ triplet under $\mathrm{SU(2)_f}$ transformations.

Clearly there should be some general formalism for dealing with all this more efficiently and it is provided by a generalization of the steps followed, in the U(1) case, in equations (7.6)–(7.8). Suppose the Lagrangian involves a set of fields $\hat{\psi}_r$ (they could be bosons or fermions) and suppose that it is *invariant* under the infinitesimal transformation

$$\delta\hat{\psi}_r = -\mathrm{i}\epsilon T_{rs}\hat{\psi}_s \tag{12.111}$$

for some set of numerical coefficients $T_{rs}$. Equation (12.111) generalizes (7.5). Then since $\hat{\mathcal{L}}$ is invariant under this change,

$$0 = \delta\hat{\mathcal{L}} = \frac{\partial\hat{\mathcal{L}}}{\partial\hat{\psi}_r}\delta\hat{\psi}_r + \frac{\partial\hat{\mathcal{L}}}{\partial(\partial^\mu\hat{\psi}_r)}\partial^\mu(\delta\hat{\psi}_r). \tag{12.112}$$

But

$$\frac{\partial\hat{\mathcal{L}}}{\partial\hat{\psi}_r} = \partial^\mu\left(\frac{\partial\hat{\mathcal{L}}}{\partial(\partial^\mu\hat{\psi}_r)}\right) \qquad (12.113)$$

from the equations of motion. Hence,

$$\partial^\mu\left(\frac{\partial\hat{\mathcal{L}}}{\partial(\partial^\mu\hat{\psi}_r)}\delta\hat{\psi}_r\right) = 0 \qquad (12.114)$$

which is precisely a current conservation law of the form

$$\partial^\mu\hat{j}_\mu = 0. \qquad (12.115)$$

Indeed, disregarding the irrelevant constant small parameter $\epsilon$, the conserved current is

$$\hat{j}_\mu = -\mathrm{i}\frac{\partial\hat{\mathcal{L}}}{\partial(\partial^\mu\hat{\psi}_r)}T_{rs}\hat{\psi}_s. \qquad (12.116)$$

Let us try this out on (12.87) with

$$\delta\hat{q} = (-\mathrm{i}\boldsymbol{\epsilon}\cdot\boldsymbol{\tau}/2)\hat{q}. \qquad (12.117)$$

As we know already, there are now three $\epsilon$'s and so three $T_{rs}$'s, namely $\frac{1}{2}(\tau_1)_{rs}$, $\frac{1}{2}(\tau_2)_{rs}$, $\frac{1}{2}(\tau_3)_{rs}$. For each one we have a current, for example

$$\hat{T}^{(\frac{1}{2})}_{1\mu} = -\mathrm{i}\frac{\partial\hat{\mathcal{L}}}{\partial(\partial^\mu\hat{q})}\frac{\tau_1}{2}\hat{q} = \bar{\hat{q}}\gamma_\mu\frac{\tau_1}{2}\hat{q} \qquad (12.118)$$

and similarly for the other $\tau$'s and so we recover (12.109). From the invariance of the Lagrangian under the transformation (12.117), there follows the conservation of an associated symmetry current. This is the quantum field theory version of Noether's theorem (Noether 1918).

This theorem is of fundamental significance as it tells us how to relate symmetries (under transformations of the general form (12.111)) to 'current' conservation laws (of the form (12.115), and it constructs the actual currents for us. In gauge theories, the *dynamics* is generated from a symmetry, in the sense that (as we have seen in the local U(1) of electromagnetism) the symmetry currents are the dynamical currents that drive the equations for the force field. Thus, the symmetries of the Lagrangian are basic to gauge field theories.

Let us look at another example, this time involving spin-0 fields. Suppose we have three spin-0 fields all with the same mass, and take

$$\hat{\mathcal{L}} = \tfrac{1}{2}\partial_\mu\hat{\phi}_1\partial^\mu\hat{\phi}_1 + \tfrac{1}{2}\partial_\mu\hat{\phi}_2\partial^\mu\hat{\phi}_2 + \tfrac{1}{2}\partial_\mu\hat{\phi}_3\partial^\mu\hat{\phi}_3 - \tfrac{1}{2}m^2(\hat{\phi}_1^2 + \hat{\phi}_2^2 + \hat{\phi}_3^2). \qquad (12.119)$$

It is obvious that $\hat{\mathcal{L}}$ is invariant under an arbitrary rotation of the three $\hat{\phi}$'s among themselves, generalizing the 'rotation about the three-axis' considered for the $\hat{\phi}_1$–$\hat{\phi}_2$ system of section 7.1. An infinitesimal such rotation is (cf (12.64), and noting the sign change in the field theory case)

$$\hat{\boldsymbol{\phi}}' = \hat{\boldsymbol{\phi}} + \boldsymbol{\epsilon} \times \hat{\boldsymbol{\phi}} \tag{12.120}$$

which implies

$$\delta\hat{\phi}_r = -\mathrm{i}\epsilon_a T^{(1)}_{ars}\hat{\phi}_s \tag{12.121}$$

with

$$T^{(1)}_{ars} = -\mathrm{i}\epsilon_{ars} \tag{12.122}$$

as in (12.48). There are, of course, three conserved $\hat{\boldsymbol{T}}$ operators again and three $\hat{\boldsymbol{T}}^{\mu}$'s, which we call $\hat{\boldsymbol{T}}^{(1)}$ and $\hat{\boldsymbol{T}}^{(1)\mu}$ respectively, since we are now dealing with a $T = 1$ isospin case. The $a = 1$ component of the conserved current in this case is, from (12.116),

$$\hat{T}^{(1)\mu}_1 = \hat{\phi}_2\partial^{\mu}\hat{\phi}_3 - \hat{\phi}_3\partial^{\mu}\hat{\phi}_2. \tag{12.123}$$

Cyclic permutations give us the other components which can be summarized as

$$\hat{\boldsymbol{T}}^{(1)\mu} = \mathrm{i}(\hat{\phi}^{(1)\mathrm{tr}}\mathbf{T}^{(1)}\partial^{\mu}\hat{\phi}^{(1)} - (\partial^{\mu}\hat{\phi}^{(1)})^{\mathrm{tr}}\mathbf{T}^{(1)}\hat{\phi}^{(1)}) \tag{12.124}$$

where we have written

$$\hat{\phi}^{(1)} = \begin{pmatrix} \hat{\phi}_1 \\ \hat{\phi}_2 \\ \hat{\phi}_3 \end{pmatrix} \tag{12.125}$$

and $^{\mathrm{tr}}$ denotes transpose. Equation (8.76) has the form expected of a bosonic spin-0 current but with the matrices $\mathbf{T}^{(1)}$ appearing, appropriate to the $T = 1$ (triplet) representation of $\mathrm{SU}(2)_{\mathrm{f}}$.

The general form of such SU(2) currents should now be clear. For an isospin $T$-multiplet of bosons, we shall have the form

$$\mathrm{i}(\hat{\phi}^{(T)\dagger}\mathbf{T}^{(T)}\partial^{\mu}\hat{\phi}^{(T)}) - (\partial^{\mu}\hat{\phi}^{(T)})^{\dagger}\mathbf{T}^{(T)}\hat{\phi}^{(T)}) \tag{12.126}$$

where we have put the $\dagger$ to allow for possibly complex fields; and for an isospin $T$-multiplet of fermions we shall have

$$\bar{\hat{\psi}}^{(T)}\gamma^{\mu}\mathbf{T}^{(T)}\hat{\psi}^{(T)}, \tag{12.127}$$

where, in each case, the $(2T + 1)$ components of $\hat{\phi}$ or $\hat{\psi}$ transforms as a $T$-multiplet under SU(2), i.e.

$$\hat{\psi}^{(T)\prime} = \exp(-\mathrm{i}\boldsymbol{\alpha}\cdot\mathbf{T}^{(T)})\hat{\psi}^{(T)} \tag{12.128}$$

and similarly for $\hat{\phi}^{(T)}$, where $\mathbf{T}^{(T)}$ are the $2T + 1 \times 2T + 1$ matrices representing the generators of $\mathrm{SU}(2)_{\mathrm{f}}$ in this representation. In all cases, the integral over all space of the $\mu = 0$ component of these currents results in a triplet of isospin operators obeying the SU(2) algebra (12.47), as in (12.103).

The cases considered so far have all been *free* field theories but SU(2)-invariant interactions can be easily formed. For example, the interaction $g_1\bar{\hat{\psi}}\boldsymbol{\tau}\hat{\psi}\cdot\hat{\boldsymbol{\phi}}$ describes SU(2)-invariant interactions between a $T=\frac{1}{2}$ isospinor (spin-$\frac{1}{2}$) field $\hat{\psi}$ and a $T=1$ isotriplet (Lorentz scalar) $\hat{\boldsymbol{\phi}}$. An effective interaction between pions and nucleons could take the form $g_\pi\bar{\hat{\psi}}\boldsymbol{\tau}\gamma_5\hat{\psi}\cdot\hat{\boldsymbol{\phi}}$, allowing for the pseudoscalar nature of the pions (we shall see in the following section that $\bar{\hat{\psi}}\gamma_5\hat{\psi}$ is a pseudoscalar, so the product is a true scalar as is required for a parity-conserving strong interaction). In these examples the 'vector' analogy for the $T=1$ states allows us to see that the 'dot product' will be invariant. A similar dot product occurs in the interaction between the isospinor $\hat{\psi}^{(\frac{1}{2})}$ and the weak SU(2) gauge field $\hat{\boldsymbol{W}}_\mu$, which has the form

$$g\bar{\hat{q}}\gamma^\mu\frac{\boldsymbol{\tau}}{2}\hat{q}\cdot\hat{\boldsymbol{W}}_\mu \tag{12.129}$$

as will be discussed in the following chapter. This is just the SU(2) dot product of the symmetry current (12.109) and the gauge field triplet, both of which are in the adjoint ($T=1$) representation of SU(2).

All of the foregoing can be generalized straightforwardly to $\mathrm{SU(3)_f}$. For example, the Lagrangian

$$\hat{\mathcal{L}}=\bar{\hat{q}}(\mathrm{i}\not{\partial}-m)\hat{q} \tag{12.130}$$

with $\hat{q}$ now extended to

$$\hat{q}=\begin{pmatrix}\hat{u}\\ \hat{d}\\ \hat{s}\end{pmatrix} \tag{12.131}$$

describes free u, d and s quarks of equal mass $m$. $\hat{\mathcal{L}}$ is clearly invariant under global $\mathrm{SU(3)_f}$ transformations

$$\hat{q}'=\exp(-\mathrm{i}\boldsymbol{\alpha}\cdot\boldsymbol{\lambda}/2)\hat{q} \tag{12.132}$$

as well as the usual global U(1) transformation associated with quark number conservation. The associated Noether currents are (in somewhat informal notation)

$$\hat{G}_a^{(\mathrm{q})\mu}=\bar{\hat{q}}\gamma^\mu\frac{\lambda_a}{2}\hat{q}\qquad a=1,2,\ldots,8 \tag{12.133}$$

(note that there are eight of them) and the associated conserved 'charge operators' are

$$\hat{G}_a^{(\mathrm{q})}=\int\hat{G}_a^{(\mathrm{q})0}\mathrm{d}^3\boldsymbol{x}=\int\hat{q}^\dagger\frac{\lambda_a}{2}\hat{q}\qquad a=1,2,\ldots,8 \tag{12.134}$$

which obey the SU(3) commutation relations

$$[\hat{G}_a^{(\mathrm{q})},\hat{G}_b^{(\mathrm{q})}]=\mathrm{i}f_{abc}\hat{G}_c^{(\mathrm{q})}. \tag{12.135}$$

SU(3)-invariant interactions can also be formed. A particularly important one is the 'SU(3) dot-product' of two octets (the analogues of the SU(2) triplets), which arises in the quark–gluon vertex of QCD (see chapters 13 and 14):

$$-\mathrm{i}g_{\mathrm{s}}\sum_{\mathrm{f}}\bar{\hat{q}}_{\mathrm{f}}\gamma^{\mu}\frac{\lambda_a}{2}\hat{q}_{\mathrm{f}}\hat{A}^{a}_{\mu}. \tag{12.136}$$

In (12.136), $\hat{q}_{\mathrm{f}}$ stands for the $\mathrm{SU}(3)_{\mathrm{c}}$ *colour* triplet

$$\hat{q}_{\mathrm{f}} = \begin{pmatrix} \hat{f}_{\mathrm{r}} \\ \hat{f}_{\mathrm{b}} \\ \hat{f}_{\mathrm{g}} \end{pmatrix} \tag{12.137}$$

where '$\hat{f}$' is any of the six quark flavour fields $\hat{u}, \hat{d}, \hat{c}, \hat{s}, \hat{t}, \hat{b}$ and $\hat{A}^{a}_{\mu}$ are the eight ($a = 1, 2, \ldots, 8$) gluon fields. Once again, (12.136) has the form 'symmetry current×gauge field' characteristic of all gauge interactions.

### 12.3.2 Chiral symmetry

As our final example of a global non-Abelian symmetry, we shall introduce the idea of *chiral symmetry*, which is an exact symmetry for fermions in the limit in which their masses may be neglected. We have seen that the u and d quarks have indeed very small masses ($\lesssim$ 10 MeV) on hadronic scales and even the s quark ($\sim$150 MeV) is relatively small. Thus, we may certainly expect some physical signs of the symmetry associated with $m_{\mathrm{u}} \approx m_{\mathrm{d}} \approx 0$, and possibly also of the larger symmetry holding when $m_{\mathrm{u}} \approx m_{\mathrm{d}} \approx m_{\mathrm{s}} \approx 0$. As we shall see, however, this expectation leads to a puzzle, the resolution of which will have to be postponed until the concept of 'spontaneous symmetry breaking' has been developed in part 7.

We begin with the simplest case of just one fermion. Since we are interested in the 'small mass' regime, it is sensible to use the representations (4.97) of the Dirac matrices, in which the momentum part of the Dirac Hamiltonian is 'diagonal' and the mass appears as an 'off-diagonal' coupling (compare problem 4.15):

$$\boldsymbol{\alpha} = \begin{pmatrix} \boldsymbol{\sigma} & 0 \\ 0 & -\boldsymbol{\sigma} \end{pmatrix} \qquad \beta = \begin{pmatrix} 0 & 1 \\ 1 & 0 \end{pmatrix}. \tag{12.138}$$

Writing the general Dirac spinor $\omega$ as

$$\omega = \begin{pmatrix} \phi \\ \chi \end{pmatrix} \tag{12.139}$$

we have (as in (4.98), (4.99))

$$E\phi = \boldsymbol{\sigma}\cdot\boldsymbol{p}\phi + m\chi \tag{12.140}$$

$$E\chi = -\boldsymbol{\sigma}\cdot\boldsymbol{p}\chi + m\phi. \tag{12.141}$$

We now introduce the matrix $\gamma_5$ defined, in this representation, as

$$\gamma_5 = \begin{pmatrix} 1 & 0 \\ 0 & -1 \end{pmatrix}. \tag{12.142}$$

The matrix $\gamma_5$ plays a prominent role in chiral symmetry, as we shall see. Its defining property is that it anti-commutes with the $\gamma^\mu$ matrices:

$$\{\gamma_5, \gamma^\mu\} = 0. \tag{12.143}$$

With the choice (12.138) for $\boldsymbol{\alpha}$ and $\beta$, we have

$$\boldsymbol{\gamma} = \begin{pmatrix} 0 & -\boldsymbol{\sigma} \\ \boldsymbol{\sigma} & 0 \end{pmatrix} \qquad \gamma^0 = \beta = \begin{pmatrix} 0 & 1 \\ 1 & 0 \end{pmatrix} \tag{12.144}$$

and (12.143) can easily be verified. In a general representation, $\gamma_5$ is defined by

$$\gamma_5 = \mathrm{i}\gamma^0\gamma^1\gamma^2\gamma^3 \tag{12.145}$$

which reduces to (12.142) in the present case.

'Chirality' means 'handedness' from the Greek word for hand, $\chi\epsilon i\rho$. Its use here stems from the fact that, in the limit $m \to 0$, the two-component spinors $\phi, \chi$ become helicity eigenstates (cf problem 4.15), having definite 'handedness'. As $m \to 0$, we have $E \to |\boldsymbol{p}|$ and (12.140) and (12.141) reduce to

$$(\boldsymbol{\sigma} \cdot \boldsymbol{p}/|\boldsymbol{p}|)\tilde{\phi} = \tilde{\phi} \tag{12.146}$$

$$(\boldsymbol{\sigma} \cdot \boldsymbol{p}/|\boldsymbol{p}|)\tilde{\chi} = -\tilde{\chi} \tag{12.147}$$

so that the limiting spinor $\tilde{\phi}$ has positive helicity, and $\tilde{\chi}$ negative helicity (cf (4.67) and (4.68)). In this $m \to 0$ limit, the two helicity spinors are *decoupled*, reflecting the fact that no Lorentz transformation can reverse the helicity of a massless particle. Also in this limit, the Dirac energy operator is

$$\boldsymbol{\alpha} \cdot \boldsymbol{p} = \begin{pmatrix} \boldsymbol{\sigma} \cdot \boldsymbol{p} & 0 \\ 0 & -\boldsymbol{\sigma} \cdot \boldsymbol{p} \end{pmatrix} \tag{12.148}$$

which is easily seen to commute with $\gamma_5$. Thus, the massless states may equivalently be classified by the eigenvalues of $\gamma_5$, which are clearly $\pm 1$ since $\gamma_5^2 = I$.

Consider then a massless fermion with positive helicity. It is described by the '$u$'-spinor $\begin{pmatrix} \tilde{\phi} \\ 0 \end{pmatrix}$ which is an eigenstate of $\gamma_5$ with eigenvalue $+1$. Similarly, a fermion with negative helicity is described by $\begin{pmatrix} 0 \\ \tilde{\chi} \end{pmatrix}$ which has $\gamma_5 = -1$. Thus, for these states chirality equals helicity. We have to be more careful for anti-fermions, however. A physical anti-fermion of energy $E$ and momentum $\boldsymbol{p}$

is described by a '$v$'-spinor corresponding to $-E$ and $-\boldsymbol{p}$; but with $m = 0$ in (12.140) and (12.141) the equations for $\phi$ and $\chi$ remain the same for $-E, -\boldsymbol{p}$ as for $E, \boldsymbol{p}$. Consider the spin, however. If the physical anti-particle has positive helicity, with $\boldsymbol{p}$ along the $z$-axis say, then $s_z = +\frac{1}{2}$. The corresponding $v$-spinor must then have $s_z = -\frac{1}{2}$ (see section 4.5.3) and must, therefore, be of $\tilde{\chi}$ type (12.147). So the $v$-spinor for this anti-fermion of positive helicity is $\begin{pmatrix} 0 \\ \tilde{\chi} \end{pmatrix}$ which has $\gamma_5 = -1$. In summary, for fermions the $\gamma_5$ eigenvalue is equal to the helicity and for anti-fermions it is equal to minus the helicity. It is the $\gamma_5$ eigenvalue that is called the 'chirality'.

In the massless limit, the chirality of $\tilde{\phi}$ and $\tilde{\chi}$ is a good quantum number ($\gamma_5$ commuting with the energy operator) and we may say that 'chirality is conserved' in this massless limit. However, the massive spinor $\omega$ is clearly *not* an eigenstate of chirality:

$$\gamma_5\omega = \begin{pmatrix} \phi \\ -\chi \end{pmatrix} \neq \lambda \begin{pmatrix} \phi \\ \chi \end{pmatrix}. \tag{12.149}$$

Referring to (12.140) and (12.141), we may therefore regard the mass terms as 'coupling the states of different chirality'.

It is usual to introduce operators $P_{\mathrm{R,L}} = ((1 \pm \gamma_5)/2)$ which 'project' out states of definite chirality from $\omega$:

$$\omega = \left(\frac{1+\gamma_5}{2}\right)\omega + \left(\frac{1-\gamma_5}{2}\right)\omega \equiv P_{\mathrm{R}}\omega + P_{\mathrm{L}}\omega \equiv \omega_{\mathrm{R}} + \omega_{\mathrm{L}} \tag{12.150}$$

so that

$$\omega_{\mathrm{R}} = \begin{pmatrix} 1 & 0 \\ 0 & 0 \end{pmatrix}\begin{pmatrix} \phi \\ \chi \end{pmatrix} = \begin{pmatrix} \phi \\ 0 \end{pmatrix} \qquad \omega_{\mathrm{L}} = \begin{pmatrix} 0 \\ \chi \end{pmatrix}. \tag{12.151}$$

Then clearly $\gamma_5\omega_{\mathrm{R}} = \omega_{\mathrm{R}}$ and $\gamma_5\omega_{\mathrm{L}} = -\omega_{\mathrm{L}}$; slightly confusingly, the notation 'R', 'L' is used for the *chirality* eigenvalue.

We now reformulate this in field-theoretic terms. The Dirac Lagrangian for a single massless fermion is

$$\hat{\mathcal{L}}_0 = \bar{\hat{\psi}}\mathrm{i}\partial\!\!\!/\hat{\psi}. \tag{12.152}$$

This is invariant not only under the now familiar global U(1) transformation $\hat{\psi} \to \hat{\psi}' = \mathrm{e}^{-\mathrm{i}\alpha}\hat{\psi}$ but also under the 'global *chiral* U(1)' transformation

$$\hat{\psi} \to \hat{\psi}' = \mathrm{e}^{-\mathrm{i}\beta\gamma_5}\hat{\psi} \tag{12.153}$$

where $\beta$ is an arbitrary ($x$-independent) real parameter. The invariance is easily verified: using $\{\gamma^0, \gamma_5\} = 0$, we have

$$\bar{\hat{\psi}}' = \hat{\psi}'^{\dagger}\gamma^0 = \hat{\psi}^{\dagger}\mathrm{e}^{\mathrm{i}\beta\gamma_5}\gamma^0 = \hat{\psi}^{\dagger}\gamma^0\mathrm{e}^{-\mathrm{i}\beta\gamma_5} = \bar{\hat{\psi}}\mathrm{e}^{-\mathrm{i}\beta\gamma_5}, \tag{12.154}$$

and then using $\{\gamma^\mu, \gamma_5\} = 0$,

$$\begin{aligned}\bar{\hat{\psi}}'\gamma^\mu\partial_\mu\hat{\psi}' &= \bar{\hat{\psi}}\mathrm{e}^{-\mathrm{i}\beta\gamma_5}\gamma^\mu\partial_\mu\mathrm{e}^{-\mathrm{i}\beta\gamma_5}\hat{\psi} \\ &= \bar{\hat{\psi}}\gamma^\mu\mathrm{e}^{\mathrm{i}\beta\gamma_5}\partial_\mu\mathrm{e}^{-\mathrm{i}\beta\gamma_5}\hat{\psi} \\ &= \bar{\hat{\psi}}\gamma^\mu\partial_\mu\hat{\psi}\end{aligned} \tag{12.155}$$

as required. The corresponding Noether current is

$$\hat{j}_5^\mu = \bar{\hat{\psi}}\gamma^\mu\gamma_5\hat{\psi} \tag{12.156}$$

and the spatial integral of its $\mu = 0$ component is the (conserved) chirality operator

$$\hat{Q}_5 = \int \hat{\psi}^\dagger\gamma_5\hat{\psi}\,\mathrm{d}^3\boldsymbol{x} = \int (\hat{\phi}^\dagger\hat{\phi} - \hat{\chi}^\dagger\hat{\chi})\,\mathrm{d}^3\boldsymbol{x}. \tag{12.157}$$

We denote this chiral U(1) by $\mathrm{U}(1)_5$.

It is interesting to compare the form of $\hat{Q}_5$ with that of the corresponding operator $\int \hat{\psi}^\dagger\hat{\psi}\mathrm{d}^3\boldsymbol{x}$ in the non-chiral case (cf (7.48)). The difference has to do with their behaviour under a transformation briefly considered in section 4.4, namely *parity*. Under the parity transformation $\boldsymbol{p} \to -\boldsymbol{p}$ and thus, for (12.140) and (12.141) to be covariant under parity, we require $\phi \to \chi, \chi \to \phi$; this will ensure (as we saw at the end of section 4.4) that the Dirac equation in the parity-transformed frame will be consistent with the one in the original frame. In the representation (12.138), this is equivalent to saying that the spinor $\omega_{\mathrm{P}}$ in the parity-transformed frame is given by

$$\omega_{\mathrm{P}} = \gamma^0\omega. \tag{12.158}$$

which implies $\phi_{\mathrm{P}} = \chi, \chi_{\mathrm{P}} = \phi$. All this carries over to the field-theoretic case, with $\hat{\psi}_{\mathrm{P}} = \gamma^0\hat{\psi}$. Consider, then, the operator $\hat{Q}_5$ in the parity-transformed frame:

$$(\hat{Q}_5)_{\mathrm{P}} = \int \hat{\psi}_{\mathrm{P}}^\dagger\gamma_5\psi_{\mathrm{P}}\,\mathrm{d}^3\boldsymbol{x} = \int \hat{\psi}^\dagger\gamma^0\gamma_5\gamma^0\hat{\psi}\,\mathrm{d}^3\boldsymbol{x} = -\int \hat{\psi}^\dagger\gamma_5\hat{\psi}\,\mathrm{d}^3\boldsymbol{x} = -\hat{Q}_5 \tag{12.159}$$

where we used $\{\gamma^0, \gamma_5\} = 0$ and $(\gamma^0)^2 = 1$. Hence, $\hat{Q}_5$ is a 'pseudoscalar' operator, meaning that it changes sign in the parity-transformed frame. We can also see this directly from (12.157), making the interchange $\hat{\phi} \leftrightarrow \hat{\chi}$. In contrast, the non-chiral operator $\int \hat{\psi}^\dagger\hat{\psi}\mathrm{d}^3\boldsymbol{x}$ is a (true) scalar, remaining the same in the parity-transformed frame.

In a similar way, the appearance of the $\gamma_5$ in the current operator $\hat{j}_5^\mu = \bar{\hat{\psi}}\gamma^\mu\gamma_5\hat{\psi}$ affects its parity properties: for example, the $\mu = 0$ component $\hat{\psi}^\dagger\gamma_5\hat{\psi}$ is a pseudoscalar, as we have seen. The spatial parts $\bar{\hat{\psi}}\boldsymbol{\gamma}\gamma_5\hat{\psi}$ behave as an *axial vector* rather than a normal (*polar*) vector under parity; that is, they behave like $\boldsymbol{r} \times \boldsymbol{p}$ for example, rather than like $\boldsymbol{r}$, in that they do *not* reverse sign under parity.

(Polar and axial vectors will be discussed again in section 20.2.) Such a current is referred to generally as an 'axial vector current', as opposed to the ordinary vector currents with no $\gamma_5$.

As a consequence of (12.159), the operator $\hat{Q}_5$ changes the parity of any state on which it acts. We can see this formally by introducing the (unitary) parity operator $\hat{P}$ in field theory, such that states of definite parity $|+\rangle$, $|-\rangle$ satisfy

$$\hat{P}|+\rangle = |+\rangle \qquad \hat{P}|-\rangle = -|-\rangle. \tag{12.160}$$

Equation (12.159) then implies that $\hat{P}\hat{Q}_5\hat{P}^{-1} = -\hat{Q}_5$, following the normal rule for operator transformations in quantum mechanics. Consider now the state $\hat{Q}_5|+\rangle$. We have

$$\begin{aligned}\hat{P}\hat{Q}_5|+\rangle &= (\hat{P}\hat{Q}_5\hat{P}^{-1})\hat{P}|+\rangle \\ &= -\,\hat{Q}_5|+\rangle\end{aligned} \tag{12.161}$$

showing that $\hat{Q}_5|+\rangle$ is an eigenstate of $\hat{P}$ with the opposite eigenvalue, $-1$.

A very important physical consequence now follows from the fact that (in this simple $m = 0$ model) $\hat{Q}_5$ is a symmetry operator commuting with the Hamiltonian $\hat{H}$. We have

$$\hat{H}\hat{Q}_5|\psi\rangle = \hat{Q}_5\hat{H}|\psi\rangle = E\hat{Q}_5|\psi\rangle. \tag{12.162}$$

Hence, for every state $|\psi\rangle$ with energy eigenvalue $E$, there should exist a state $\hat{Q}_5|\psi\rangle$ with the same eigenvalue $E$ and the opposite parity; that is, chiral symmetry apparently implies the existence of 'parity doublets'.

Of course, it may reasonably be objected that all of this refers not only to the massless but also the *non-interacting* case. However, this is just where the analysis begins to get interesting. Suppose we allow the fermion field $\hat{\psi}$ to interact with a U(1)-gauge field $\hat{A}^\mu$ via the standard electromagnetic coupling

$$\hat{\mathcal{L}}_{\text{int}} = q\bar{\hat{\psi}}\gamma^\mu\hat{\psi}\hat{A}_\mu. \tag{12.163}$$

Remarkably enough, $\hat{\mathcal{L}}_{\text{int}}$ is *also* invariant under the chiral transformation (12.153), for the simple reason that the 'Dirac' structure of (12.163) is exactly the same as that of the free kinetic term $\bar{\hat{\psi}}\not{\partial}\hat{\psi}$: the 'covariant derivative' prescription $\partial^\mu \to D^\mu = \partial^\mu + \mathrm{i}q\hat{A}^\mu$ automatically means that any 'Dirac' (e.g. $\gamma_5$) symmetry of the kinetic part will be preserved when the gauge interaction is included. Thus chirality remains a 'good symmetry' in the presence of a U(1)-gauge interaction.

The generalization of this to the more physical $m_\mathrm{u} \approx m_\mathrm{d} \approx 0$ case is quite straightforward. The Lagrangian (12.87) becomes

$$\hat{\mathcal{L}} = \bar{\hat{q}}\,\mathrm{i}\not{\partial}\hat{q} \tag{12.164}$$

as $m \to 0$, which is invariant under the $\gamma_5$-version of (12.89),[6] namely

$$\hat{q}' = \exp(-\mathrm{i}\boldsymbol{\beta}\cdot\boldsymbol{\tau}/2\gamma_5)\hat{q}. \tag{12.165}$$

[6] $\hat{\mathcal{L}}_0$ is also invariant under $\hat{q}' = \mathrm{e}^{-\mathrm{i}\beta\gamma_5}\hat{q}$ which is an 'axial' version of the global U(1) associated with quark number conservation. We shall discuss this additional U(1)-symmetry in section 18.1.1.

There are three associated Noether currents (compare (12.109))

$$\hat{\boldsymbol{T}}_5^{(\frac{1}{2})\mu} = \bar{\hat{q}}\gamma^\mu\gamma_5\frac{\boldsymbol{\tau}}{2}\hat{q} \tag{12.166}$$

which are axial vectors and three associated 'charge' operators

$$\hat{\boldsymbol{T}}_5^{(\frac{1}{2})} = \int \hat{q}^\dagger\gamma_5\frac{\boldsymbol{\tau}}{2}\hat{q}\,\mathrm{d}^3\boldsymbol{x} \tag{12.167}$$

which are pseudoscalars, belonging to the $T = 1$ representation of SU(2). We have a new non-Abelian global symmetry, called chiral $\mathrm{SU(2)_f}$, which we shall denote by $\mathrm{SU(2)_{f5}}$. As far as their action in the isospinor u–d space is concerned, these chiral charges have exactly the same effect as the ordinary flavour isospin operators of (12.109). But they are pseudoscalars rather than scalars and, hence, they flip the parity of a state on which they act. Thus, whereas the isospin-raising operator $\hat{T}_+^{(\frac{1}{2})}$ is such that

$$\hat{T}_+^{(\frac{1}{2})}|d\rangle = |u\rangle \tag{12.168}$$

$\hat{T}_{+5}^{(\frac{1}{2})}$ will also produce a u-type state from a d-type one via

$$\hat{T}_{+5}^{(\frac{1}{2})}|d\rangle = |\tilde{u}\rangle \tag{12.169}$$

but the $|\tilde{u}\rangle$ state will have opposite parity from $|u\rangle$. Further, since $[\hat{T}_{+5}^{(\frac{1}{2})}, \hat{H}] = 0$, this state $|\tilde{u}\rangle$ will be degenerate with $|d\rangle$. Similarly, the state $|\tilde{d}\rangle$ produced via $\hat{T}_{-5}^{(\frac{1}{2})}|u\rangle$ will have opposite parity from $|d\rangle$, and will be degenerate with $|u\rangle$. The upshot is that we have two massless states $|u\rangle$, $|d\rangle$ of (say) positive parity and a further two massless states $|\tilde{u}\rangle$, $|\tilde{d}\rangle$ of negative parity, in this simple model.

Suppose we now let the quarks interact, for example by an interaction of the QCD type already indicated in (12.136). In that case, the interaction terms have the form

$$\bar{\hat{u}}\gamma^\mu\frac{\lambda_a}{2}\hat{u}\hat{A}_\mu^a + \bar{\hat{d}}\gamma^\mu\frac{\lambda_a}{2}\hat{d}\hat{A}_\mu^a \tag{12.170}$$

where

$$\hat{u} = \begin{pmatrix} \hat{u}_\mathrm{r} \\ \hat{u}_\mathrm{b} \\ \hat{u}_\mathrm{g} \end{pmatrix} \qquad \hat{d} = \begin{pmatrix} \hat{d}_\mathrm{r} \\ \hat{d}_\mathrm{b} \\ \hat{d}_\mathrm{g} \end{pmatrix} \tag{12.171}$$

and the $3 \times 3$ $\lambda$'s act in the r–b–g space. Just as in the previous U(1) case, the interaction (12.170) is invariant under the global $\mathrm{SU(2)_{f5}}$ chiral symmetry (12.165), acting in the u–d space. Note that, somewhat confusingly, (12.170) is *not* a simple 'gauging' of (12.164): a covariant derivative is being introduced but in the space of a new (colour) degree of freedom, not in flavour space. In fact, the flavour degrees of freedom are 'inert' in (12.170), so that it is invariant under

$SU(2)_f$ transformations, while the Dirac structure implies that it is also invariant under chiral $SU(2)_{f5}$ transformations (12.165). All the foregoing can be extended unchanged to chiral $SU(3)_{f5}$, given that QCD is 'flavour blind' and supposing that $m_s \approx 0$.

The effect of the QCD interactions must be to bind the quark into nucleons such as the proton $(uud)$ and neutron $(udd)$. But what about the equally possible states $(\tilde{u}\tilde{u}\tilde{d})$ and $(\tilde{u}\tilde{d}\tilde{d})$, for example? These would have to be degenerate in mass with $(uud)$ and $(udd)$, and of opposite parity. Yet such 'parity doublet' partners of the physical p and n are not observed and so we have a puzzle.

One might feel that this whole discussion is unrealistic, based as it is on massless quarks. Are the baryons then supposed to be massless too? If so, perhaps the discussion is idle, as they are evidently by no means massless. But it is not necessary to suppose that the mass of a relativistic bound state has any very simple relation to the masses of its constituents: its mass may derive, in part at least, from the interaction energy in the fields. Alternatively, one might suppose that somehow the finite mass of the u and d quarks, which of course breaks the chiral symmetry, splits the degeneracy of the nucleon parity doublets, promoting the negative-parity 'nucleon' state to an acceptably high mass. But this seems very implausible in view of the actual magnitudes of $m_u$ and $m_d$ compared to the nucleon masses.

In short, we have here a situation in which a *symmetry of the Lagrangian* (to an apparently good approximation) does *not* seem to result in the expected *multiplet structure of the states*. The resolution of this puzzle will have to await our discussion of 'spontaneous symmetry breaking', in part 7.

In conclusion, we note an important feature of the flavour symmetry currents $\hat{\boldsymbol{T}}^{(\frac{1}{2})\mu}$ and $\hat{\boldsymbol{T}}_5^{(\frac{1}{2})\mu}$ discussed in this and the preceding section. Although these currents have been introduced entirely within the context of *strong* interaction symmetries, it is a remarkable fact that exactly these currents also appear in strangeness-conserving semileptonic *weak* interactions such as $\beta$-decay, as we shall see in chapter 20. (The fact that *both* appear is precisely a manifestation of *parity violation* in weak interactions.) Thus some of the physical consequences of 'spontaneously broken chiral symmetry' will involve weak interaction quantities.

## Problems

**12.1** Verify that the set of all unitary $2 \times 2$ matrices with determinant equal to +1 form a group, the law of combination being matrix multiplication.

**12.2** Derive (12.18).

**12.3** Check the commutation relations (12.28).

**12.4** Show that the $T_i$'s defined by (12.45) satisfy (12.47).

**12.5** Write out each of the $3 \times 3$ matrices $T_i^{(1)} (i = 1, 2, 3)$ whose matrix elements are given by (12.48) and verify that they satisfy the SU(2) commutation relations (12.47).

**12.6** Verify (12.62).

**12.7** Show that a general Hermitian traceless $3 \times 3$ matrix is parametrized by eight real numbers.

**12.8** Check that (12.84) is consistent with (12.80) and the infinitesimal form of (12.81) and verify that the matrices $G_a^{(8)}$ defined by (12.84) satisfy the commutation relations (12.83).

**12.9** Verify, by comparing the coefficients of $\epsilon_1$, $\epsilon_2$ and $\epsilon_3$ on both sides of (12.99), that (12.100) follows from (12.99).

**12.10** Verify that the operators $\hat{\boldsymbol{T}}^{(\frac{1}{2})}$ defined by (12.101) satisfy (12.100). (Note: use the anti-commutation relations of the fermionic operators.)

**12.11** Verify that the operators $\hat{\boldsymbol{T}}^{(\frac{1}{2})}$ given by (12.101) satisfy the commutation relations (12.103).

# 13

# LOCAL NON-ABELIAN (GAUGE) SYMMETRIES

> The difference between a neutron and a proton is then a purely arbitrary process. As usually conceived, however, this arbitrariness is subject to the following limitations: once one chooses what to call a proton, what a neutron, at one spacetime point, one is then not free to make any choices at other spacetime points.
>
> It seems that this is not consistent with the localized field concept that underlies the usual physical theories. In the present paper we wish to explore the possibility of requiring all interactions to be invariant under *independent* rotations of the isotopic spin at all spacetime points...
>
> Yang and Mills (1954)

Consider the global SU(2) isospinor transformation (12.32), written here again,

$$\psi^{(\frac{1}{2})\prime}(x) = \exp(\mathrm{i}\boldsymbol{\alpha} \cdot \boldsymbol{\tau}/2)\psi^{(\frac{1}{2})}(x) \tag{13.1}$$

for an isospin doublet wavefunction $\psi^{(\frac{1}{2})}(x)$. The dependence of $\psi^{(\frac{1}{2})}(x)$ on the spacetime coordinate $x$ has now been included explicitly but the parameters $\boldsymbol{\alpha}$ are independent of $x$, which is why the transformation is called a 'global' one. As we have seen in the previous chapter, invariance under this transformation amounts to the assertion that the choice of *which* two base states—$(p, n)$, $(u, d)$, ...—to use is a matter of convention: any such non-Abelian phase transformation on a chosen pair produces another equally good pair. However, the choice cannot be made independently at all spacetime points, only *globally*. To Yang and Mills (1954) (cf the quotation above) this seemed somehow an unaesthetic limitation of symmetry: 'Once one chooses what to call a proton, what a neutron, at one spacetime point, one is then not free to make any choices at other spacetime points.' They even suggested that this could be viewed as 'inconsistent with the localized field concept' and they, therefore, 'explored the possibility' of replacing this global (spacetime independent) phase transformation by the local (spacetime dependent) one

$$\psi^{(\frac{1}{2})\prime}(x) = \exp[\mathrm{i}g\boldsymbol{\tau} \cdot \boldsymbol{\alpha}(x)/2]\psi^{(\frac{1}{2})}(x) \tag{13.2}$$

in which the phase parameters $\boldsymbol{\alpha}(x)$ are also now functions of $x = (t, \boldsymbol{x})$ as indicated. Note that we have inserted a parameter $g$ in the exponent to make the analogy with the electromagnetic U(1) case

$$\psi'(x) = \exp[\mathrm{i}q\chi(x)]\psi(x) \tag{13.3}$$

even stronger: $g$ will be a coupling strength, analogous to the electromagnetic charge $q$. The consideration of theories based on (13.2) was the fundamental step taken by Yang and Mills (1954); see also Shaw (1955).

Global symmetries and their associated (possibly approximate) conservation laws are certainly important but they do not have the *dynamical* significance of local symmetries. We saw in section 7.4 how the 'requirement' of local U(1) phase invariance led almost automatically to the local gauge theory of QED, in which the conserved current $\bar{\hat{\psi}}\gamma^{\mu}\hat{\psi}$ of the global U(1) symmetry is 'promoted' to the role of dynamical current which, when dotted into the gauge field $\hat{A}^{\mu}$, gave the interaction term in $\hat{\mathcal{L}}_{\text{QED}}$. A similar link between symmetry and dynamics appears if—following Yang and Mills—we generalize the non-Abelian global symmetries of the preceding chapter to local non-Abelian symmetries, which are the subject of the present one.

However, as mentioned in the introduction to chapter 12, the original Yang–Mills attempt to get a theory of hadronic interactions by 'localizing' the flavour symmetry group SU(2) turned out not to be phenomenologically viable (although a remarkable attempt was made to push the idea further by Sakurai (1960)). In the event, the successful application of a local SU(2) symmetry was to the *weak* interactions. But this is complicated by the fact that the symmetry is 'spontaneously broken' and, consequently, we shall delay the discussion of this application until after QCD—which *is* the theory of strong interactions but at the quark rather than the composite (hadronic) level. QCD is based on the local form of an SU(3) symmetry—once again, however, it is *not* the flavour SU(3) of section 12.2 but a symmetry with respect to a totally new degree of freedom, colour. This will be introduced in the following chapter.

Although the application of local SU(2) symmetry to the weak interactions will follow that of local SU(3) to the strong, we shall begin our discussion of local non-Abelian symmetries with the local SU(2) case, since the group theory is more familiar. We shall also start with the 'wavefunction' formalism, deferring the field-theory treatment until section 13.5.

## 13.1 Local SU(2) symmetry: the covariant derivative and interactions with matter

In this section we shall introduce the main ideas of the non-Abelian SU(2) gauge theory which results from the demand of invariance, or covariance, under transformations such as (13.2). We shall generally use the language of isospin when referring to the physical states and operators, bearing in mind that this will eventually mean *weak* isospin.

We shall mimic as literally as possible the discussion of electromagnetic gauge covariance in sections 3.4 and 3.5 of volume 1. As in that case, no free-particle wave equation can be covariant under the transformation (13.2) (taking the isospinor example for definiteness), since the gradient terms in the equation will act on the phase factor $\boldsymbol{\alpha}(x)$. However, wave equations with a suitably defined *covariant derivative* can be covariant under (13.2); physically this means that, just as for electromagnetism, covariance under local non-Abelian phase transformations requires the introduction of a definite force field.

In the electromagnetic case, the covariant derivative is

$$D^{\mu} = \partial^{\mu} + \mathrm{i}qA^{\mu}(x). \tag{13.4}$$

For convenience, we recall here the crucial property of $D^{\mu}$. Under a local U(1) phase transformation, a wavefunction transforms as (cf (13.3))

$$\psi(x) \to \psi'(x) = \exp(\mathrm{i}q\chi(x))\psi(x) \tag{13.5}$$

from which it easily follows that the derivative (gradient) of $\psi$ transforms as

$$\partial^{\mu}\psi(x) \to \partial^{\mu}\psi'(x) = \exp(\mathrm{i}q\chi(x))\partial^{\mu}\psi(x) + \mathrm{i}q\partial^{\mu}\chi(x)\exp(\mathrm{i}q\chi(x))\psi(x). \tag{13.6}$$

Comparing (13.6) with (13.5), we see that, in addition to the expected first term on the right-hand side of (13.6), which has the same form as the right-hand side of (13.5), there is an *extra* term in (13.6). By contrast, the covariant derivative of $\psi$ transforms as (see section 3.4 of volume 1)

$$D^{\mu}\psi(x) \to D'^{\mu}\psi'(x) = \exp(\mathrm{i}q\chi(x))D^{\mu}\psi(x) \tag{13.7}$$

exactly as in (13.5), with no additional term on the right-hand side. Note that $D^{\mu}$ has to carry a prime also, since it contains $A^{\mu}$ which transforms to $A'^{\mu} = A^{\mu} - \partial^{\mu}\chi(x)$ when $\psi$ transforms by (13.5). The property (13.7) ensures the gauge covariance of wave equations in the U(1) case; the similar property in the quantum field case meant that a globally U(1)-invariant Lagrangian could be converted immediately to a locally U(1)-invariant one by replacing $\partial^{\mu}$ by $\hat{D}^{\mu}$ (section 7.4).

In appendix D of volume 1 we introduced the idea of 'covariance' in the context of coordinate transformations of 3- and 4-vectors. The essential notion was of something 'maintaining the same form' or 'transforming the same way'. The transformations being considered here are gauge transformations rather than coordinate ones; nevertheless, it is true that, under them, $D^{\mu}\psi$ transforms in the same way as $\psi$, while $\partial^{\mu}\psi$ does not. Thus, the term covariant derivative seems appropriate. In fact, there is a much closer analogy between the 'coordinate' and the 'gauge' cases, which we did not present in volume 1 but shall discuss in the following section.

We need the local SU(2) generalization of (13.4), appropriate to the local SU(2) transformation (13.2). Just as in the U(1) case (13.6), the ordinary gradient acting on $\psi^{(\frac{1}{2})}(x)$ does not transform in the same way as $\psi^{(\frac{1}{2})}(x)$: taking $\partial^\mu$ of (13.2) leads to

$$\partial^\mu \psi^{(\frac{1}{2})\prime}(x) = \exp[\mathrm{i}g\boldsymbol{\tau}\cdot\boldsymbol{\alpha}(x)/2]\partial^\mu\psi^{(\frac{1}{2})}(x) + \mathrm{i}g\boldsymbol{\tau}\cdot\partial^\mu\boldsymbol{\alpha}(x)/2\,\exp[\mathrm{i}g\boldsymbol{\tau}\cdot\boldsymbol{\alpha}(x)/2]\psi^{(\frac{1}{2})}(x) \tag{13.8}$$

as can be checked by writing the matrix exponential exp[A] as the series

$$\exp[\mathrm{A}] = \sum_{n=0}^{\infty} \mathrm{A}^n/n!$$

and differentiating term by term. By analogy with (13.7), the key property we demand for our *SU(2) covariant derivative* $D^\mu\psi^{(\frac{1}{2})}$ is that this quantity should transform like $\psi^{(\frac{1}{2})}$—i.e. without the second term in (13.8). So we require

$$(D'^\mu\psi^{(\frac{1}{2})\prime}(x)) = \exp[\mathrm{i}g\boldsymbol{\tau}\cdot\boldsymbol{\alpha}(x)/2](D^\mu\psi^{(\frac{1}{2})}(x)). \tag{13.9}$$

The definition of $D^\mu$ which generalizes (13.4) so as to fulfil this requirement is

$$D^\mu(\text{acting on an isospinor}) = \partial^\mu + \mathrm{i}g\boldsymbol{\tau}\cdot\boldsymbol{W}^\mu(x)/2. \tag{13.10}$$

The definition (13.10), as indicated on the left-hand side, is only appropriate for isospinors $\psi^{(\frac{1}{2})}$: it has to be suitably generalized for other $\psi^{(t)}$'s (see (13.44)).

We now discuss (13.9) and (13.10) in detail. The $\partial^\mu$ is multiplied implicitly by the unit 2 matrix and the $\boldsymbol{\tau}$'s act on the two-component space of $\psi^{(\frac{1}{2})}$. The $\boldsymbol{W}^\mu(x)$ are *three* independent gauge fields

$$\boldsymbol{W}^\mu = (W_1^\mu, W_2^\mu, W_3^\mu) \tag{13.11}$$

generalizing the single electromagnetic gauge field $A^\mu$. They are called SU(2) gauge fields or, more generally, *Yang–Mills fields*. The term $\boldsymbol{\tau}\cdot\boldsymbol{W}^\mu$ is then the $2\times 2$ matrix

$$\boldsymbol{\tau}\cdot\boldsymbol{W}^\mu = \begin{pmatrix} W_3^\mu & W_1^\mu - \mathrm{i}W_2^\mu \\ W_1^\mu + \mathrm{i}W_2^\mu & -W_3^\mu \end{pmatrix} \tag{13.12}$$

using the $\tau$'s of (12.25): the $x$-dependence of the $W^\mu$'s is understood. Let us 'decode' the desired property (13.9), for the algebraically simpler case of an infinitesimal local SU(2) transformation with parameters $\boldsymbol{\epsilon}(x)$, which are of course functions of $x$ since the transformation is local. In this case, $\psi^{(\frac{1}{2})}$ transforms by

$$\psi^{(\frac{1}{2})\prime} = (1 + \mathrm{i}g\boldsymbol{\tau}\cdot\boldsymbol{\epsilon}(x)/2)\psi^{(\frac{1}{2})} \tag{13.13}$$

and the 'uncovariant' derivative $\partial^\mu\psi^{(\frac{1}{2})}$ transforms by

$$\partial^\mu\psi^{(\frac{1}{2})\prime} = (1 + \mathrm{i}g\boldsymbol{\tau}\cdot\boldsymbol{\epsilon}(x)/2)\partial^\mu\psi^{(\frac{1}{2})} + \mathrm{i}g\boldsymbol{\tau}\cdot\partial^\mu\boldsymbol{\epsilon}(x)/2\,\psi^{(\frac{1}{2})} \tag{13.14}$$

where we have retained only the terms linear in $\boldsymbol{\epsilon}$ from an expansion of (13.8) with $\boldsymbol{\alpha} \to \boldsymbol{\epsilon}$. We have now dropped the $x$-dependence of the $\psi^{(\frac{1}{2})}$'s but kept that of $\boldsymbol{\epsilon}(x)$ and we have used the simple '1' for the unit matrix in the two-dimensional isospace. Equation (13.14) exhibits again an 'extra piece' on the right-hand side, as compared to (13.13). However, inserting (13.10) and (13.13) into our covariant derivative requirement (13.9) yields, for the left-hand side in the infinitesimal case,

$$D'^{\mu}\psi^{(\frac{1}{2})\prime} = (\partial^{\mu} + \mathrm{i}g\boldsymbol{\tau}\cdot\boldsymbol{W}'^{\mu}/2)[1 + \mathrm{i}g\boldsymbol{\tau}\cdot\boldsymbol{\epsilon}(x)/2]\psi^{(\frac{1}{2})} \tag{13.15}$$

while the right-hand side is

$$[1 + \mathrm{i}g\boldsymbol{\tau}\cdot\boldsymbol{\epsilon}(x)/2](\partial^{\mu} + \mathrm{i}g\boldsymbol{\tau}\cdot\boldsymbol{W}^{\mu}/2)\psi^{(\frac{1}{2})}. \tag{13.16}$$

In order to verify that these are the same, however, we would need to know $\boldsymbol{W}'^{\mu}$—that is, the transformation law for the three $\boldsymbol{W}^{\mu}$ fields. Instead, we shall proceed 'in reverse', and use the *imposed* equality between (13.15) and (13.16) to determine the transformation law of $\boldsymbol{W}^{\mu}$.

Suppose that, under this infinitesimal transformation,

$$\boldsymbol{W}^{\mu} \to \boldsymbol{W}'^{\mu} = \boldsymbol{W}^{\mu} + \delta\boldsymbol{W}^{\mu}. \tag{13.17}$$

Then the condition of equality is

$$\begin{aligned}[\partial^{\mu} + \mathrm{i}g\boldsymbol{\tau}/2\cdot(\boldsymbol{W}^{\mu} + \delta\boldsymbol{W}^{\mu})][1 + \mathrm{i}g\boldsymbol{\tau}\cdot\boldsymbol{\epsilon}(x)/2]\psi^{(\frac{1}{2})}& \\ = [1 + \mathrm{i}g\boldsymbol{\tau}\cdot\boldsymbol{\epsilon}(x)/2](\partial^{\mu} + \mathrm{i}g\boldsymbol{\tau}\cdot\boldsymbol{W}^{\mu}/2)\psi^{(\frac{1}{2})}&.\end{aligned} \tag{13.18}$$

Multiplying out the terms, neglecting the term of second order involving the product of $\delta\boldsymbol{W}^{\mu}$ and $\boldsymbol{\epsilon}$ and noting that

$$\partial^{\mu}(\boldsymbol{\epsilon}\psi) = (\partial^{\mu}\boldsymbol{\epsilon})\psi + \boldsymbol{\epsilon}(\partial^{\mu}\psi) \tag{13.19}$$

we find that many terms cancel and we are left with

$$\begin{aligned}\mathrm{i}g\frac{\boldsymbol{\tau}\cdot\delta\boldsymbol{W}^{\mu}}{2} = &-\mathrm{i}g\frac{\boldsymbol{\tau}\cdot\partial^{\mu}\boldsymbol{\epsilon}(x)}{2} \\ &+ (\mathrm{i}g)^2\left[\left(\frac{\boldsymbol{\tau}\cdot\boldsymbol{\epsilon}(x)}{2}\right)\left(\frac{\boldsymbol{\tau}\cdot\boldsymbol{W}^{\mu}}{2}\right) - \left(\frac{\boldsymbol{\tau}\cdot\boldsymbol{W}^{\mu}}{2}\right)\left(\frac{\boldsymbol{\tau}\cdot\boldsymbol{\epsilon}(x)}{2}\right)\right].\end{aligned} \tag{13.20}$$

Using the identity for Pauli matrices (see problem 4.4(*b*))

$$\boldsymbol{\sigma}\cdot\boldsymbol{a}\,\boldsymbol{\sigma}\cdot\boldsymbol{b} = \boldsymbol{a}\cdot\boldsymbol{b} + \mathrm{i}\boldsymbol{\sigma}\cdot\boldsymbol{a}\times\boldsymbol{b} \tag{13.21}$$

this yields

$$\boldsymbol{\tau}\cdot\delta\boldsymbol{W}^{\mu} = -\boldsymbol{\tau}\cdot\partial^{\mu}\boldsymbol{\epsilon}(x) - g\boldsymbol{\tau}\cdot(\boldsymbol{\epsilon}(x)\times\boldsymbol{W}^{\mu}). \tag{13.22}$$

Equating components of $\boldsymbol{\tau}$ on both sides, we deduce

$$\boxed{\delta \boldsymbol{W}^\mu = -\partial^\mu \boldsymbol{\epsilon}(x) - g[\boldsymbol{\epsilon}(x) \times \boldsymbol{W}^\mu].} \tag{13.23}$$

The reader may note the close similarity between these manipulations and those encountered in section 12.1.3.

Equation (13.23) defines the way in which the SU(2) gauge fields $\boldsymbol{W}^\mu$ transform under an infinitesimal SU(2) gauge transformation. If it were not for the presence of the first term $\partial^\mu \boldsymbol{\epsilon}(x)$ on the right-hand side, (13.23) would be simply the (infinitesimal) transformation law for the $T = 1$ triplet representation of SU(2)—see (12.64) and (12.65) in section 12.1.3. As mentioned at the end of section 12.2, the $T = 1$ representation is the 'adjoint', or 'regular', representation of SU(2) and this is the one to which gauge fields belong, in general. But there is the extra term $-\partial^\mu \boldsymbol{\epsilon}(x)$. Clearly this is directly analogous to the $-\partial^\mu \chi(x)$ term in the transformation of the U(1) gauge field $A^\mu$; here, an independent infinitesimal function $\epsilon_i(x)$ is required for each component $W_i^\mu(x)$. If the $\epsilon$'s were independent of $x$, then $\partial^\mu \boldsymbol{\epsilon}(x)$ would of course vanish and the transformation law (13.23) would indeed be just that of an SU(2) triplet. Thus, we can say that under global SU(2) transformations, the $\boldsymbol{W}^\mu$ behave as a normal triplet. But under *local* SU(2) transformations they acquire the additional $-\partial^\mu \boldsymbol{\epsilon}(x)$ piece and, thus, no longer transform 'properly' as an SU(2) triplet. In exactly the same way, $\partial^\mu \psi^{(\frac{1}{2})}$ did not transform 'properly' as an SU(2) doublet, under a local SU(2) transformation, because of the second term in (13.14), which also involves $\partial^\mu \boldsymbol{\epsilon}(x)$. The remarkable result behind the fact that $D^\mu \psi^{(\frac{1}{2})}$ *does* transform 'properly' under local SU(2) transformations is that the extra term in (13.23) precisely cancels that in (13.14)!

To summarize progress so far: we have shown that, for infinitesimal transformations, the relation

$$(D'^\mu \psi^{(\frac{1}{2})\prime}) = [1 + \mathrm{i}g\boldsymbol{\tau} \cdot \boldsymbol{\epsilon}(x)/2](D^\mu \psi^{(\frac{1}{2})}) \tag{13.24}$$

(where $D^\mu$ is given by (13.10)) holds true if in addition to the infinitesimal local SU(2) phase transformation on $\psi^{(\frac{1}{2})}$

$$\psi^{(\frac{1}{2})\prime} = [1 + \mathrm{i}g\boldsymbol{\tau} \cdot \boldsymbol{\epsilon}(x)/2]\psi^{(\frac{1}{2})} \tag{13.25}$$

the gauge fields transform according to

$$\boldsymbol{W}'^\mu = \boldsymbol{W}^\mu - \partial^\mu \boldsymbol{\epsilon}(x) - g[\boldsymbol{\epsilon}(x) \times \boldsymbol{W}^\mu]. \tag{13.26}$$

In obtaining these results, the form (13.10) for the covariant derivative has been assumed and only the infinitesimal version of (13.2) has been treated explicitly. It turns out that (13.10) is still appropriate for the finite (non-infinitesimal) transformation (13.2) but the associated transformation law for the gauge fields is then slightly more complicated than (13.26). Let us write

$$\mathrm{U}(\boldsymbol{\alpha}(x)) \equiv \exp[\mathrm{i}g\boldsymbol{\tau} \cdot \boldsymbol{\alpha}(x)/2] \tag{13.27}$$

so that $\psi^{(\frac{1}{2})}$ transforms by

$$\psi^{(\frac{1}{2})\prime} = \mathbf{U}(\boldsymbol{\alpha}(x))\psi^{(\frac{1}{2})}. \tag{13.28}$$

Then we require

$$D'^{\mu}\psi^{(\frac{1}{2})\prime} = \mathbf{U}(\boldsymbol{\alpha}(x))D^{\mu}\psi^{(\frac{1}{2})}. \tag{13.29}$$

The left-hand side is

$$\begin{aligned}(\partial^{\mu} + \mathrm{i}g\boldsymbol{\tau}\cdot\boldsymbol{W}'^{\mu}/2)\mathbf{U}(\boldsymbol{\alpha}(x))\psi^{(\frac{1}{2})} \\ = (\partial^{\mu}\mathbf{U})\psi^{(\frac{1}{2})} + \mathbf{U}\partial^{\mu}\psi^{(\frac{1}{2})} + \mathrm{i}g\boldsymbol{\tau}\cdot\boldsymbol{W}'^{\mu}/2\,\mathbf{U}\psi^{(\frac{1}{2})}\end{aligned} \tag{13.30}$$

while the right-hand side is

$$\mathbf{U}(\partial^{\mu} + \mathrm{i}g\boldsymbol{\tau}\cdot\boldsymbol{W}^{\mu}/2)\psi^{(\frac{1}{2})}. \tag{13.31}$$

The $\mathbf{U}\partial^{\mu}\psi^{(\frac{1}{2})}$ terms cancel leaving

$$(\partial^{\mu}\mathbf{U})\psi^{(\frac{1}{2})} + \mathrm{i}g\boldsymbol{\tau}\cdot\boldsymbol{W}'^{\mu}/2\,\mathbf{U}\psi^{(\frac{1}{2})} = \mathbf{U}\mathrm{i}g\boldsymbol{\tau}\cdot\boldsymbol{W}^{\mu}/2\,\psi^{(\frac{1}{2})}. \tag{13.32}$$

Since this has to be true for all (two-component) $\psi^{(\frac{1}{2})}$'s, we can treat it as an operator equation acting in the space of $\psi^{(\frac{1}{2})}$'s to give

$$\partial^{\mu}\mathbf{U} + \mathrm{i}g\boldsymbol{\tau}\cdot\boldsymbol{W}'^{\mu}/2\,\mathbf{U} = \mathbf{U}\mathrm{i}g\boldsymbol{\tau}\cdot\boldsymbol{W}^{\mu}/2 \tag{13.33}$$

or, equivalently,

$$\frac{1}{2}\boldsymbol{\tau}\cdot\boldsymbol{W}'^{\mu} = \frac{\mathrm{i}}{g}(\partial^{\mu}\mathbf{U})\mathbf{U}^{-1} + \mathbf{U}\frac{1}{2}\boldsymbol{\tau}\cdot\boldsymbol{W}^{\mu}\mathbf{U}^{-1} \tag{13.34}$$

which defines the (finite) transformation law for SU(2) gauge fields. Problem 13.1 verifies that (13.34) reduces to (13.26) in the infinitesimal case $\boldsymbol{\alpha}(x) \to \boldsymbol{\epsilon}(x)$.

Suppose now that we consider a Dirac equation for $\psi^{(\frac{1}{2})}$:

$$(\mathrm{i}\gamma_{\mu}\partial^{\mu} - m)\psi^{(\frac{1}{2})} = 0 \tag{13.35}$$

where both the 'isospinor' components of $\psi^{(\frac{1}{2})}$ are four-component Dirac spinors. We assert that we can ensure *local SU(2) gauge covariance by replacing* $\partial^{\mu}$ *in this equation by the covariant derivative of* (13.10). Indeed, we have

$$\begin{aligned}\mathbf{U}(\boldsymbol{\alpha}(x))[\mathrm{i}\gamma_{\mu}D^{\mu} - m]\psi^{(\frac{1}{2})} &= \mathrm{i}\gamma_{\mu}\mathbf{U}(\boldsymbol{\alpha}(x))D^{\mu}\psi^{(\frac{1}{2})} - m\boldsymbol{U}(\boldsymbol{\alpha}(x))\psi^{(\frac{1}{2})} \\ &= \mathrm{i}\gamma_{\mu}D'^{\mu}\psi^{(\frac{1}{2})\prime} - m\psi^{(\frac{1}{2})\prime}\end{aligned} \tag{13.36}$$

using equations (13.28) and (13.29). Thus, if

$$(\mathrm{i}\gamma_{\mu}D^{\mu} - m)\psi^{(\frac{1}{2})} = 0 \tag{13.37}$$

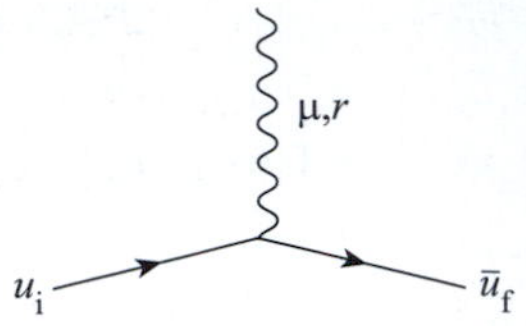


**Figure 13.1.** Vertex for isospinor-W interaction.

then

$$(\mathrm{i}\gamma_\mu D'^\mu - m)\psi^{(\frac{1}{2})\prime} = 0 \tag{13.38}$$

proving the asserted covariance. In the same way, any free-particle wave equation satisfied by an 'isospinor' $\psi^{(\frac{1}{2})}$—the relevant equation is determined by the Lorentz spin of the particles involved—can be made locally covariant by the use of the covariant derivative $D^\mu$, just as in the U(1) case.

The essential point here, of course, is that the locally covariant form includes *interactions* between the $\psi^{(\frac{1}{2})}$'s and the gauge fields $\boldsymbol{W}^\mu$, which are determined by the local phase invariance requirement (the 'gauge principle'). Indeed, we can already begin to find some of the Feynman rules appropriate to tree graphs for SU(2) gauge theories. Consider again the case of an SU(2) isospinor fermion, $\psi^{(\frac{1}{2})}$, obeying equation (13.38). This can be written as

$$(\mathrm{i}\partial\!\!\!/ - m)\psi^{(\frac{1}{2})} = g(\boldsymbol{\tau}/2)\cdot \boldsymbol{W}\!\!\!\!/\,\psi^{(\frac{1}{2})}. \tag{13.39}$$

In lowest-order perturbation theory the one-W emission/absorption process is given by the amplitude (cf (8.39)) for the electromagnetic case)

$$-\mathrm{i}g\int \bar{\psi}_{\mathrm{f}}^{(\frac{1}{2})}(\boldsymbol{\tau}/2)\gamma_\mu\psi_{\mathrm{i}}^{(\frac{1}{2})}\cdot \boldsymbol{W}^\mu\,\mathrm{d}^4x \tag{13.40}$$

exactly as advertised (for the field-theoretic vertex) in (12.129). The matrix degree of freedom in the $\tau$'s is sandwiched between the two-component isospinors $\psi^{(\frac{1}{2})}$: the $\gamma$ matrix acts on the four-component (Dirac) parts of $\psi^{(\frac{1}{2})}$. The external $\boldsymbol{W}^\mu$ field is now specified by a spin-1 polarization vector $\epsilon^\mu$, like a photon, and by an 'SU(2) polarization vector' $a^r$ $(r = 1, 2, 3)$ which tells us which of the three SU(2) W-states is participating. The Feynman rule for figure 13.1 is, therefore,

$$-\mathrm{i}g(\tau^r/2)\gamma_\mu \tag{13.41}$$

which is to be sandwiched between spinors/isospinors $u_{\mathrm{i}}$, $\bar{u}_{\mathrm{f}}$ and dotted into $\epsilon^\mu$ and $a^r$. (13.41) is a very economical generalization of rule (ii) in comment (3) of section 8.3.

The foregoing is easily generalized to SU(2) multiplets other than doublets. We shall change the notation slightly to use $t$ instead of $T$ for the 'isospin' quantum number, so as to emphasize that it is *not* the hadronic isospin, for which we retain $T$: $t$ will be the symbol used for the *weak isospin* to be introduced in chapter 20. The general local SU(2) transformation for a $t$-multiplet is then

$$\psi^{(t)} \rightarrow \psi^{(t)\prime} = \exp[\mathrm{i}g\boldsymbol{\alpha}(x).\mathbf{T}^{(t)}]\psi^{(t)}, \tag{13.42}$$

where the $(2t+1)\times(2t+1)$ matrices $T_i^{(t)} (i = 1, 2, 3)$ satisfy (cf (12.47))

$$[T_i^{(t)}, T_j^{(t)}] = \mathrm{i}\epsilon_{ijk}T_k^{(t)}. \tag{13.43}$$

The appropriate covariant derivative is

$$D^\mu = \partial^\mu + \mathrm{i}g\mathbf{T}^{(t)}\cdot\boldsymbol{W}^\mu \tag{13.44}$$

which is a $(2t+1)\times(2t+1)$ matrix acting on the $(2t+1)$ components of $\psi^{(t)}$. The gauge fields interact with such 'isomultiplets' in a *universal* way—only one $g$, the same for all the particles—which is prescribed by the local covariance requirement to be simply that interaction which is generated by the covariant derivatives. The fermion vertex corresponding to (13.44) is obtained by replacing $\boldsymbol{\tau}/2$ in (13.40) by $\boldsymbol{T}^{(t)}$.

We end this section with some comments:

(a) It is a remarkable fact that only one constant $g$ is needed. This is *not* the same as in electromagnetism. There, each charged field interacts with the gauge field $A^\mu$ via a coupling whose strength is its charge $(e, -e, 2e, -5e, )$. The crucial point is the appearance of the quadratic $g^2$ multiplying the *commutator* of the $\boldsymbol{\tau}$'s, $[\boldsymbol{\tau}\cdot\boldsymbol{\epsilon}, \boldsymbol{\tau}\cdot\boldsymbol{W}]$, in the $\boldsymbol{W}^\mu$ transformation (equation (13.20)). In the electromagnetic case, there is no such commutator—the associated U(1) phase group is Abelian. As signalled by the presence of $g^2$, a commutator is a nonlinear quantity, and the scale of quantities appearing in such commutation relations is not arbitrary. It is an instructive exercise to check that, once $\delta\boldsymbol{W}^\mu$ is given by equation (13.23)—in the SU(2) case—then the $g$'s appearing in $\psi^{(\frac{1}{2})\prime}$ (equation (13.13)) and $\psi^{(t)\prime}$ (via the infinitesimal version of equation (13.42)) must be the *same* as the one appearing in $\delta\boldsymbol{W}^\mu$.

(b) According to the foregoing argument, it is actually a mystery why electric charge should be quantized. Since it is the coupling constant of an Abelian group, each charged field could have an arbitrary charge from this point of view: there are no commutators to fix the scale. This is one of the motivations of attempts to embed the electromagnetic gauge transformations inside a larger non-Abelian group structure. Such is the case, for example, in 'grand unified theories' of strong, weak and electromagnetic interactions.

(c) Finally we draw attention to the extremely important physical significance of the second term $\delta\boldsymbol{W}^\mu$ (equation (13.23)). The gauge fields themselves are not 'inert' as far as the gauge group is concerned: in the SU(2) case they have isospin 1, while for a general group they belong to the regular representation of the group. This is profoundly different from the electromagnetic case, where the gauge field $A^{\mu}$ for the photon is of course uncharged: quite simply, $e = 0$ for a photon and the second term in (13.23) is absent for $A^{\mu}$. The fact that non-Abelian (Yang–Mills) gauge fields carry non-Abelian 'charge' degrees of freedom means that, since they are also the quanta of the force field, *they will necessarily interact with themselves.* Thus, a non-Abelian gauge theory of gauge fields alone, with no 'matter' fields, has non-trivial interactions and is not a free theory.

We shall examine the form of these 'self-interactions' in section 13.5.2. First, we explore further the geometrical analogy, already hinted at, for the (gauge) covariant derivative. This will ultimately lead us to an important new quantity in non-Abelian gauge theories, the analogue of the Maxwell field strength tensor $F^{\mu\nu}$.

## 13.2 Covariant derivatives and coordinate transformations

Let us go back to the U(1) case, equations (13.4)–(13.7). There, the introduction of the (gauge) covariant derivative $D^{\mu}$ produced an object, $D^{\mu}\psi(x)$, which transformed like $\psi(x)$ under local U(1) phase transformations, unlike the ordinary derivative $\partial^{\mu}\psi(x)$ which acquired an 'extra' piece when transformed. This followed from simple calculus, of course—but there is a slightly different way of thinking about it. The derivative involves not only $\psi(x)$ at the point $x$ but also $\psi$ at the infinitesimally close, but different, point $x + \mathrm{d}x$; and the transformation law of $\psi(x)$ involves $\alpha(x)$, while that of $\psi(x + \mathrm{d}x)$ would involve the different function $\alpha(x + \mathrm{d}x)$. Thus, we may perhaps expect something to 'go wrong' with the transformation law for the gradient.

To bring out the geometrical analogy we are seeking, let us split $\psi$ into its real and imaginary parts $\psi = \psi_{\mathrm{R}} + \mathrm{i}\psi_{\mathrm{I}}$, and write $\alpha(x) = q\chi(x)$ so that (13.3) becomes (cf (3.63))

$$\begin{aligned} \psi'_{\mathrm{R}}(x) &= \cos\alpha(x)\psi_{\mathrm{R}}(x) - \sin\alpha(x)\psi_{\mathrm{I}}(x) \\ \psi'_{\mathrm{I}}(x) &= \sin\alpha(x)\psi_{\mathrm{R}}(x) + \cos\alpha(x)\psi_{\mathrm{I}}(x). \end{aligned} \tag{13.45}$$

If we think of $\psi_{\mathrm{R}}(x)$ and $\psi_{\mathrm{I}}(x)$ as being the components of a 'vector' $\vec{\psi}(x)$ along the $\vec{e}_{\mathrm{R}}$ and $\vec{e}_{\mathrm{I}}$ axes, respectively, then (13.45) would represent the components of $\vec{\psi}(x)$ as referred to new axes $\vec{e}'_{\mathrm{R}}$ and $\vec{e}'_{\mathrm{I}}$, which have been rotated by $-\alpha(x)$ about an axis in the direction $\vec{e}_{\mathrm{R}} \times \vec{e}_{\mathrm{I}}$ (i.e. normal to the $\vec{e}_{\mathrm{R}}$–$\vec{e}_{\mathrm{I}}$ plane), as shown in figure 13.2. Other such 'vectors' $\vec{\phi}_1(x), \vec{\phi}_2(x), \ldots$ (i.e. other wavefunctions for particles of the same charge $q$) *when evaluated at the same point* $x$ will have 'components' transforming the same as (13.45) under the axis rotation $\vec{e}_{\mathrm{R}}, \vec{e}_{\mathrm{I}} \rightarrow \vec{e}'_{\mathrm{R}}, \vec{e}'_{\mathrm{I}}$. But the components of the vector $\vec{\psi}(x + \mathrm{d}x)$ will behave differently. The transformation law (13.45) when written at $x + \mathrm{d}x$ will involve $\alpha(x + \mathrm{d}x)$, which

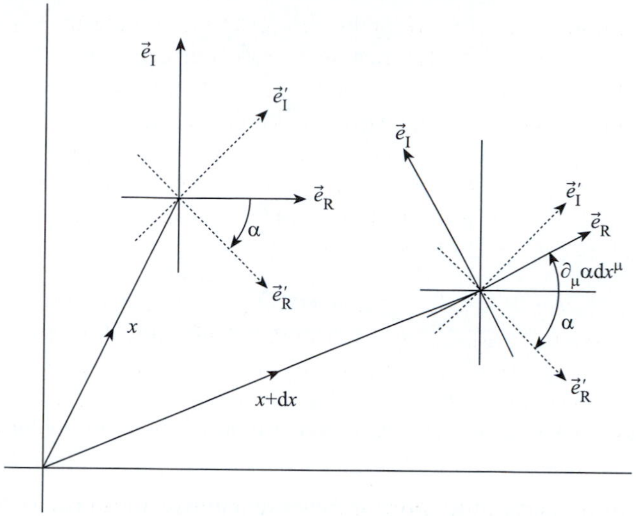


**Figure 13.2.** Geometrical analogy for a U(1) gauge transformation.

(to first order in d$x$) is $\alpha(x) + \partial_\mu \alpha(x)\,dx^\mu$. Thus, for $\psi'_R(x + dx)$ and $\psi'_I(x + dx)$, the rotation angle is $\alpha(x) + \partial_\mu \alpha(x)\,dx^\mu$ rather than $\alpha(x)$. Now comes the key step in the analogy: we may think of the additional angle $\partial_\mu \alpha(x)\,dx^\mu$ as coming about because, in going from $x$ to $x + dx$, the coordinate basis vectors $\vec{e}_R$ and $\vec{e}_I$ have been rotated through $+\partial_\mu \alpha(x)\,dx^\mu$ (see figure 13.3)! But that would mean that our 'naive' approach to rotations of the derivative of $\vec{\psi}(x)$ amounts to using one set of axes at $x$, and another at $x + dx$, which is likely to lead to 'trouble'.

Consider now an elementary example (from Schutz (1988, chapter 5)) where just this kind of problem arises, namely the use of polar coordinate basis vectors $\vec{e}_r$ and $\vec{e}_\theta$, which point in the $r$ and $\theta$ directions respectively. We have, as usual,

$$x = r\cos\theta \qquad y = r\sin\theta \tag{13.46}$$

and in a (real!) Cartesian basis $d\vec{r}$ is given by

$$d\vec{r} = dx\vec{i} + dy\vec{j}. \tag{13.47}$$

Using (13.46) in (13.47) we find

$$\begin{aligned} d\vec{r} &= (dr\cos\theta - r\sin\theta\, d\theta)\vec{i} + (dr\sin\theta + r\cos\theta\, d\theta)\vec{j} \\ &= dr\vec{e}_r + d\theta\vec{e}_\theta \end{aligned} \tag{13.48}$$

where

$$\vec{e}_r = \cos\theta\,\vec{i} + \sin\theta\,\vec{j} \qquad \vec{e}_\theta = -r\sin\theta\,\vec{i} + r\cos\theta\,\vec{j}. \tag{13.49}$$

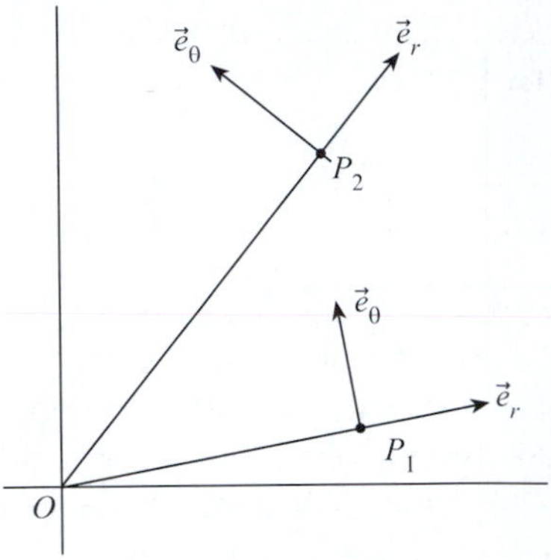


**Figure 13.3.** Changes in the basis vectors $\vec{e}_r$ and $\vec{e}_\theta$ of polar coordinates.

Plainly, $\vec{e}_r$ and $\vec{e}_\theta$ change direction (and even magnitude, for $\vec{e}_\theta$) as we move about in the $x$–$y$ plane, as shown in figure 13.3. So at each point $(r, \theta)$ we have *different* axes $\vec{e}_r$, $\vec{e}_\theta$.

Now suppose that we wish to describe a vector field $\vec{V}$ in terms of $\vec{e}_r$ and $\vec{e}_\theta$ via

$$\vec{V} = V^r\vec{e}_r + V^\theta\vec{e}_\theta \equiv V^\alpha\vec{e}_\alpha \qquad \text{(sum on } \alpha = r, \theta) \tag{13.50}$$

and that we are also interested in the derivatives of $\vec{V}$, in this basis. Let us calculate $\partial\vec{V}/\partial r$, for example, by brute force:

$$\frac{\partial\vec{V}}{\partial r} = \frac{\partial V^r}{\partial r}\vec{e}_r + \frac{\partial V^\theta}{\partial r}\vec{e}_\theta + V^r\frac{\partial\vec{e}_r}{\partial r} + V^\theta\frac{\partial\vec{e}_\theta}{\partial r} \tag{13.51}$$

where we have included the derivatives of $\vec{e}_r$ and $\vec{e}_\theta$ to allow for the fact that *these vectors are not constant*. From (13.49) we find

$$\frac{\partial\vec{e}_r}{\partial r} = 0 \qquad \frac{\partial\vec{e}_\theta}{\partial r} = -\sin\theta\,\vec{i} + \cos\theta\,\vec{j} = \frac{1}{r}\vec{e}_\theta \tag{13.52}$$

which allows the last two terms in (13.51) to be evaluated. Similarly, we can calculate $\partial\vec{V}/\partial\theta$. In general, we may write these results as

$$\frac{\partial\vec{V}}{\partial q^\beta} = \frac{\partial V^\alpha}{\partial q^\beta}\vec{e}_\alpha + V^\alpha\frac{\partial\vec{e}_\alpha}{\partial q^\beta} \tag{13.53}$$

where $\beta = 1, 2$ with $q^1 = r$, $q^2 = \theta$ and $\alpha = r, \theta$.

In the present case, we were able to calculate $\partial\vec{e}_\alpha/\partial q^\beta$ explicitly from (13.49), as in (13.52). But whatever the nature of the coordinate system, $\partial\vec{e}_\alpha/\partial q^\beta$ is some vector and must be expressible as a linear combination of the basis vectors via an expression of the form

$$\frac{\partial\vec{e}_\alpha}{\partial q^\beta} = \Gamma^\gamma{}_{\alpha\beta}\vec{e}_\gamma \tag{13.54}$$

where the repeated index $\gamma$ is summed over as usual ($\gamma = r, \theta$). Inserting (13.54) into (13.53) and interchanging the 'dummy' (i.e. summed over) indices $\alpha$ and $\gamma$ gives finally

$$\frac{\partial \vec{V}}{\partial q^\beta} = \left(\frac{\partial V^\alpha}{\partial q^\beta} + \Gamma^\alpha{}_{\gamma\beta} V^\gamma\right) \vec{e}_\alpha. \tag{13.55}$$

This is a very important result: it shows that, whereas the components of $\vec{V}$ in the basis $\vec{e}_\alpha$ are just $V^\alpha$, the components of the derivative of $\vec{V}$ are not simply $\partial V^\alpha / \partial q^\beta$ but *contain an additional term*: the 'components of the derivative of a vector' are not just the 'derivatives of the components of the vector'.

Let us abbreviate $\partial/\partial q^\beta$ to $\partial_\beta$: then (13.55) tells us that in the $\vec{e}_\alpha$ basis, as used in (13.55), the components of the $\partial_\beta$ derivative of $\vec{V}$ are

$$\partial_\beta V^\alpha + \Gamma^\alpha{}_{\gamma\beta} V^\gamma \equiv D_\beta V^\alpha. \tag{13.56}$$

The expression (13.56) is called the 'covariant derivative' of $V^\alpha$ within the context of the mathematics of general coordinate systems: it is denoted (as in (13.56)) by $D_\beta V^\alpha$ or, often, by $V^\alpha{}_{;\beta}$ (in the latter notation, $\partial_\beta V^\alpha$ is $V^\alpha{}_{,\beta}$). The most important property of $D_\beta V^\alpha$ is its transformation character under general coordinate transformations. Crucially, it transforms as a *tensor* $T^\alpha_\beta$ (see appendix D of volume 1) with the indicated 'one up, one down' indices: we shall not prove this here, referring instead to Schutz (1988), for example. This property is the reason for the name 'covariant derivative', meaning in this case essentially that it transforms the way its indices would have you believe it should. By contrast, and despite appearances, $\partial_\beta V^\alpha$ by itself does *not* transform as a '$T^\alpha_\beta$' tensor and, in a similar way, $\Gamma^\alpha{}_{\gamma\beta}$ is *not* a '$T^\alpha{}_{\gamma\beta}$'-type tensor: only the combined object $D_\beta V^\alpha$ is a '$T^\alpha_\beta$'.

This circumstance is highly reminiscent of the situation we found in the case of gauge transformations. Consider the simplest case, that of U(1), for which $D_\mu \psi = \partial_\mu \psi + \mathrm{i} q A_\mu \psi$. The quantity $D_\mu \psi$ transforms under a gauge transformation in the same way as $\psi$ itself but $\partial_\mu \psi$ does not. There is, thus, a close analogy between the 'good' transformation properties of $D_\beta V^\alpha$ and of $D_\mu \psi$. Further, the structure of $D_\mu \psi$ is very similar to that of $D_\beta V^\alpha$. There are two pieces, the first of which is the straightforward derivative, while the second involves a new field ($\Gamma$ or $A$) and is also proportional to the original field. The 'i' of course is a big difference, showing that in the gauge symmetry case the transformations mix the real and imaginary parts of the wavefunction, rather than actual spatial components of a vector.

Indeed, the analogy is even closer in the non-Abelian—e.g. local SU(2)—case. As we have seen, $\partial^\mu \psi^{(\frac{1}{2})}$ does not transform as an SU(2) isospinor because of the extra piece involving $\partial^\mu \boldsymbol{\epsilon}$; nor do the gauge fields $\boldsymbol{W}^\mu$ transform as pure $T = 1$ states, also because of a $\partial^\mu \boldsymbol{\epsilon}$ term. But the gauge covariant combination $(\partial^\mu + \mathrm{i} g \boldsymbol{\tau} \cdot \boldsymbol{W}^\mu/2)\psi^{(\frac{1}{2})}$ does transform as an isospinor under local SU(2) transformations, the two 'extra' $\partial^\mu \boldsymbol{\epsilon}$ pieces cancelling each other out.

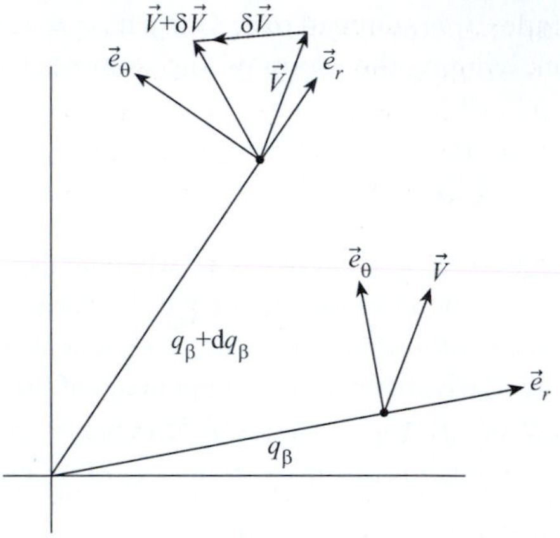


**Figure 13.4.** Parallel transport of a vector $\vec{V}$ in a polar coordinate basis.

There is a useful way of thinking about the two contributions to $D_\beta V^\alpha$ (or $D_\mu \psi$). Let us multiply (13.56) by $\mathrm{d}q^\beta$ and sum over $\beta$ so as to obtain

$$DV^\alpha \equiv \partial_\beta V^\alpha \,\mathrm{d}q^\beta + \Gamma^\alpha{}_{\gamma\beta} V^\alpha \,\mathrm{d}q^\beta. \tag{13.57}$$

The first term on the right-hand side of (13.57) is $\frac{\partial V^\alpha}{\partial q^\beta}\,\mathrm{d}q^\beta$ which is just the conventional differential $\mathrm{d}V^\alpha$, representing the change in $V^\alpha$ in moving from $q^\beta$ to $q^\beta + \mathrm{d}q^\beta$: $\mathrm{d}V^\alpha = [V^\alpha(q^\beta + \mathrm{d}q^\beta) - V^\alpha(q^\beta)]$. Again, despite appearances, the quantities $\mathrm{d}V^\alpha$ do not form the components of a vector and the reason is that $V^\alpha(q^\beta + \mathrm{d}q^\beta)$ are components with respect to axes at $q^\beta + \mathrm{d}q^\beta$, while $V^\alpha(q^\beta)$ are components with respect to *different* axes at $q^\beta$. To form a 'good' differential $DV^\alpha$, transforming as a vector, we must subtract quantities defined in the *same* coordinate system. This means that we need some way of 'carrying' $V^\alpha(q^\beta)$ to $q^\beta + \mathrm{d}q^\beta$, while keeping it somehow 'the same' as it was at $q^\beta$.

A reasonable definition of such a 'preserved' vector field is one that is unchanged in length and has the same orientation relative to the axes at $q^\beta + \mathrm{d}q^\beta$ as it had relative to the axes at $q^\beta$ (see figure 13.4). In other words, $\vec{V}$ is 'dragged around' with the changing coordinate frame, a process called *parallel transport*. Such a definition of 'no change' of course implies that change *has* occurred, in general, with respect to the *original* axes at $q^\beta$. Let us denote by $\delta V^\alpha$ the difference between the components of $\vec{V}$ after parallel transport to $q^\beta + \mathrm{d}q^\beta$ and the components of $\vec{V}$ at $q^\beta$ (see figure 13.4). Then a reasonable definition of the 'good' differential of $V^\alpha$ would be $V^\alpha(q^\beta + \mathrm{d}q^\beta) - (V^\alpha(q^\beta) + \delta V^\alpha) = \mathrm{d}V^\alpha - \delta V^\alpha$. We interpret this as the covariant differential $DV^\alpha$ of (13.57) and, accordingly, make the identification

$$\delta V^\alpha = -\Gamma^\alpha{}_{\gamma\beta} V^\gamma \,\mathrm{d}q^\beta. \tag{13.58}$$

On this interpretation, then, the coefficients $\Gamma^{\alpha}{}_{\gamma\beta}$ connect the components of a vector at one point with its components at a nearby point, after the vector has been carried by 'parallel transport' from one point to the other: they are often called 'connection coefficients' or just 'the connection'.

In an analogous way we can write, in the U(1) gauge case,

$$\begin{aligned} D\psi \equiv D^{\mu}\psi \mathrm{d}x_{\mu} &= \partial^{\mu}\psi \mathrm{d}x_{\mu} + \mathrm{i}eA^{\mu}\psi \mathrm{d}x_{\mu} \\ &\equiv \mathrm{d}\psi - \delta\psi \end{aligned} \tag{13.59}$$

with

$$\delta\psi = -\mathrm{i}eA^{\mu}\psi \, \mathrm{d}x_{\mu}. \tag{13.60}$$

Equation (13.60) has a very similar structure to (13.58), suggesting that the electromagnetic potential $A^{\mu}$ might well be referred to as a 'gauge connection', as indeed it is in some quarters. Equations (13.59) and (13.60) generalize straightforwardly for $D\psi^{(\frac{1}{2})}$ and $\delta\psi^{(\frac{1}{2})}$.

We can relate (13.60) in a very satisfactory way to our original discussion of electromagnetism as a gauge theory in chapter 3 and, in particular, to (3.8.2). For transport restricted to the three spatial directions, (13.60) reduces to

$$\delta\psi(\boldsymbol{x}) = \mathrm{i}e\boldsymbol{A} \cdot \mathrm{d}\boldsymbol{x}\psi(x). \tag{13.61}$$

However, the solution (3.82) gives

$$\psi(\boldsymbol{x}) = \exp\left(\mathrm{i}e\int_{-\infty}^{\boldsymbol{x}} \boldsymbol{A} \cdot \mathrm{d}\boldsymbol{\ell}\right)\psi(\boldsymbol{A} = \boldsymbol{0}, \boldsymbol{x}) \tag{13.62}$$

replacing $q$ by $e$. So

$$\begin{aligned} \psi(\boldsymbol{x} + \mathrm{d}\boldsymbol{x}) &= \exp\left(\mathrm{i}e\int_{-\infty}^{\boldsymbol{x}+\mathrm{d}\boldsymbol{x}} \boldsymbol{A} \cdot \mathrm{d}\boldsymbol{\ell}\right)\psi(\boldsymbol{A} = \boldsymbol{0}, \boldsymbol{x} + \mathrm{d}\boldsymbol{x}) \\ &= \exp\left(\mathrm{i}e\int_{\boldsymbol{x}}^{\boldsymbol{x}+\mathrm{d}\boldsymbol{x}} \boldsymbol{A} \cdot \mathrm{d}\boldsymbol{\ell}\right)\exp\left(\mathrm{i}e\int_{-\infty}^{\boldsymbol{x}} \boldsymbol{A} \cdot \mathrm{d}\boldsymbol{\ell}\right)\psi(\boldsymbol{A} = \boldsymbol{0}, \boldsymbol{x} + \mathrm{d}\boldsymbol{x}) \\ &\approx (1 + \mathrm{i}e\boldsymbol{A} \cdot \mathrm{d}\boldsymbol{x})\exp\left(\mathrm{i}e\int_{-\infty}^{\boldsymbol{x}} \boldsymbol{A} \cdot \mathrm{d}\boldsymbol{\ell}\right)[\psi(\boldsymbol{A} = \boldsymbol{0}, \boldsymbol{x}) \\ &\quad + \nabla\psi(\boldsymbol{A} = \boldsymbol{0}, \boldsymbol{x}) \cdot \mathrm{d}\boldsymbol{x}] \\ &\approx \psi(\boldsymbol{x}) + \mathrm{i}e\boldsymbol{A} \cdot \mathrm{d}\boldsymbol{x}\psi(\boldsymbol{x}) \\ &\quad + \exp\left(\mathrm{i}e\int_{-\infty}^{\boldsymbol{x}} \boldsymbol{A} \cdot \mathrm{d}\boldsymbol{\ell}\right)\nabla\psi(\boldsymbol{A} = \boldsymbol{0}, \boldsymbol{x}) \cdot \mathrm{d}\boldsymbol{x} \end{aligned} \tag{13.63}$$

to first order in $\mathrm{d}\boldsymbol{x}$. On the right-hand side of (13.63), we see (i) the change $\delta\psi$ of (13.61), due to 'parallel transport' as prescribed by the gauge connection $\boldsymbol{A}$, and (ii) the change in $\psi$ viewed as a function of $\boldsymbol{x}$, in the absence of $\boldsymbol{A}$. The solution (13.62) gives, in fact, the 'integrated' form of the small displacement law (13.63).

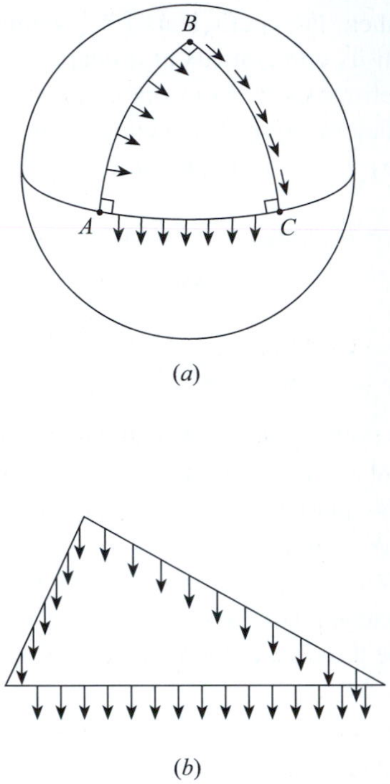


**Figure 13.5.** Parallel transport (*a*) round a curved triangle on the surface of a sphere and (*b*) round a triangle in a flat plane.

At this point the reader might object, going back to the $\vec{e}_r$, $\vec{e}_\theta$ example, that we had made a lot of fuss about nothing: after all, no one forced us to use the $\vec{e}_r$, $\vec{e}_\theta$ basis, and if we had simply used the $\vec{i}$, $\vec{j}$ basis (which is constant throughout the plane) we would have had no such 'trouble'. This is a fair point, provided that we somehow knew that we are really doing physics in a 'flat' space—such as the Euclidean plane. But suppose instead that our two-dimensional space was the surface of a sphere. Then, an intuitively plausible definition of parallel transport is shown in figure 13.5(*a*), in which transport is carried out around a closed path consisting of three great circle arcs A $\to$ B, B $\to$ C, C $\to$ A, with the rule that, at each stage, the vector is drawn 'as parallel as possible' to the previous one. It is clear from the figure that the vector we end up with at A, after this circuit, is no longer parallel to the vector with which we started; in fact, it has rotated by $\pi/2$ in this example, in which one-eighth of the surface area of the unit sphere is enclosed by the triangle ABC. By contrast, the parallel transport of a vector round a flat triangle in the Euclidean plane leads to no such net change in the vector (figure 13.5(*b*)).

It seems reasonable to suppose that the information about whether the space

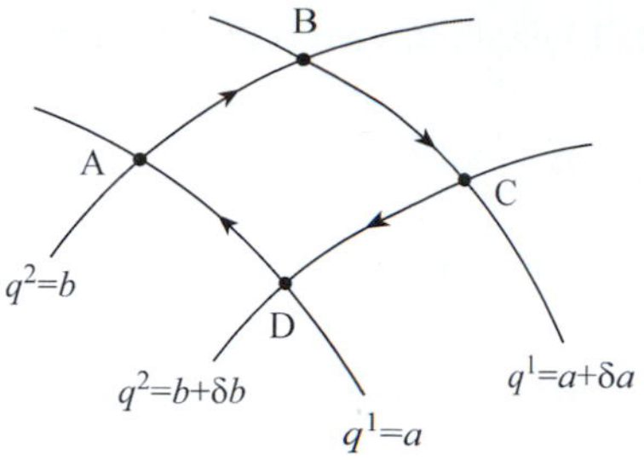


**Figure 13.6.** Closed loop ABCD in $q^1$–$q^2$ space.

we are dealing with is 'flat' or 'curved' is contained in the connection $\Gamma^{\alpha}{}_{\gamma\beta}$. In a similar way, in the gauge case the analogy we have built up so far would lead us to expect that there are potentials $A^{\mu}$ which are somehow 'flat' ($\boldsymbol{E} = \boldsymbol{B} = \boldsymbol{0}$) and others which represent 'curvature' (non-zero $\boldsymbol{E}, \boldsymbol{B}$). This is what we discuss next.

## 13.3 Geometrical curvature and the gauge field strength tensor

Consider a small closed loop in our (possibly curved) two-dimensional space—see figure 13.6—whose four sides are the coordinate lines $q^1 = a$, $q^1 = a + \delta a$, $q^2 = b$, $q^2 = b + \delta b$. We want to calculate the net change (if any) in $\delta V^{\alpha}$ as we parallel transport $\vec{V}$ around the loop. The change along A → B is

$$\begin{aligned}(\delta V^{\alpha})_{\mathrm{AB}} &= -\int_{q^2=b,q^1=a}^{q^2=b,q^1=a+\delta a} \Gamma^{\alpha}{}_{\gamma 1} V^{\gamma}\,\mathrm{d}q^1 \\ &\approx -\delta a \Gamma^{\alpha}{}_{\gamma 1}(a,b) V^{\gamma}(a,b)\end{aligned} \tag{13.64}$$

to first order in $\delta a$, while that along C → D is

$$\begin{aligned}(\delta V^{\alpha})_{\mathrm{CD}} &= -\int_{q^2=b+\delta b,q^1=a+\delta a}^{q^2=b+\delta b,q^1=a} \Gamma^{\alpha}{}_{\gamma 1} V^{\gamma}\,\mathrm{d}q^1 \\ &= +\int_{q^2=b+\delta b,q^1=a}^{q^2=b+\delta b,q^1=a+\delta a} \Gamma^{\alpha}{}_{\gamma 1} V^{\gamma}\,\mathrm{d}q^1 \\ &\approx \delta a \Gamma^{\alpha}{}_{\gamma 1}(a, b+\delta b) V^{\gamma}(a, b+\delta b).\end{aligned} \tag{13.65}$$

Now

$$\Gamma^{\alpha}{}_{\gamma 1}(a, b+\delta b) \approx \Gamma^{\alpha}{}_{\gamma 1}(a,b) + \delta b \frac{\partial \Gamma^{\alpha}{}_{\gamma 1}}{\partial q^2} \tag{13.66}$$

and, remembering that we are parallel-transporting $\vec{V}$,

$$V^{\gamma}(a, b+\delta b) \approx V^{\gamma}(a,b) - \Gamma^{\gamma}{}_{\delta 2} V^{\delta} \delta b. \tag{13.67}$$

Combining (13.64) and (13.65) to lowest order, we find that

$$(\delta V^\alpha)_{\mathrm{AB}} + (\delta V^\alpha)_{\mathrm{CD}} \approx \delta a \delta b \left[ \frac{\partial \Gamma^\alpha{}_{\gamma 1}}{\partial q^2} V^\gamma - \Gamma^\alpha{}_{\gamma 1} \Gamma^\gamma{}_{\delta 2} V^\delta \right] \tag{13.68}$$

or, interchanging dummy (summed) indices $\gamma$ and $\delta$ in the last term,

$$(\delta V^\alpha)_{\mathrm{AB}} + (\delta V^\alpha)_{\mathrm{CD}} \approx \delta a \delta b \left[ \frac{\partial \Gamma^\alpha{}_{\gamma 1}}{\partial q^2} - \Gamma^\alpha{}_{\delta 1} \Gamma^\delta{}_{\gamma 2} \right] V^\gamma . \tag{13.69}$$

Similarly,

$$(\delta V^\alpha)_{\mathrm{BC}} + (\delta V^\alpha)_{\mathrm{DA}} \approx \delta a \delta b \left[ -\frac{\partial \Gamma^\alpha{}_{\gamma 2}}{\partial q^1} + \Gamma^\alpha{}_{\delta 2} \Gamma^\delta{}_{\gamma 1} \right] V^\gamma \tag{13.70}$$

and so the net change around the whole small loop is

$$(\delta V^\alpha)_{\mathrm{ABCD}} \approx \delta a \delta b \left[ \frac{\partial \Gamma^\alpha{}_{\gamma 1}}{\partial q^2} - \frac{\partial \Gamma^\alpha{}_{\gamma 2}}{\partial q^1} + \Gamma^\alpha{}_{\delta 2} \Gamma^\delta{}_{\gamma 1} - \Gamma^\alpha{}_{\delta 1} \Gamma^\delta{}_{\gamma 2} \right] V^\gamma . \tag{13.71}$$

The indices '1' and '2' appear explicitly because the loop was chosen to go along these directions. In general, (13.71) would take the form

$$(\delta V^\alpha)_{\mathrm{loop}} \approx \left[ \frac{\partial \Gamma^\alpha{}_{\gamma\beta}}{\partial q^\sigma} - \frac{\partial \Gamma^\alpha{}_{\gamma\sigma}}{\partial q^\beta} + \Gamma^\alpha{}_{\delta\sigma} \Gamma^\delta{}_{\gamma\beta} - \Gamma^\alpha{}_{\delta\beta} \Gamma^\delta{}_{\gamma\sigma} \right] V^\gamma \mathrm{d}A^{\beta\sigma} \tag{13.72}$$

where $\mathrm{d}A^{\beta\sigma}$ is the area element. The quantity in brackets in (13.72) is the *Reimann curvature tensor* $R^\alpha{}_{\gamma\beta\sigma}$, which can clearly be calculated once the connection coefficients are known. A flat space is one for which all components $R^\alpha{}_{\gamma\beta\sigma} = 0$; the reader may verify that this is the case for our polar basis $\vec{e}_r, \vec{e}_\theta$ in the Euclidean plane. A non-zero value for any component of $R^\alpha{}_{\gamma\beta\sigma}$ means the space is curved.

We now follow exactly similar steps to calculate the net change in $\delta\psi$ as given by (13.60), around the small two-dimensional rectangle defined by the coordinate lines $x_1 = a$, $x_1 = a + \delta a$, $x_2 = b$, $x_2 = b + \delta b$, labelled as in figure 13.6 but with $q^1$ replaced by $x_1$ and $q^2$ by $x_2$. Then

$$(\delta\psi)_{\mathrm{AB}} = -\mathrm{i}eA^1(a,b)\psi(a,b)\delta a \tag{13.73}$$

and

$$\begin{aligned}(\delta\psi)_{\mathrm{CD}} &= +\mathrm{i}eA^1(a, b+\delta b)\psi(a, b+\delta b)\delta a \\ &\approx \mathrm{i}e \left( A^1(a,b) + \frac{\partial A^1}{\partial x_2}\delta b \right) [\psi(a,b) - \mathrm{i}eA^2(a,b)\psi(a,b)\delta b]\delta a \\ &\approx \mathrm{i}eA^1(a,b)\psi(a,b)\delta a \\ &\quad + \mathrm{i}e \left[ \frac{\partial A^1}{\partial x_2}\psi(a,b) - \mathrm{i}eA^1(a,b)A^2(a,b)\psi(a,b) \right] \delta a \delta b. \end{aligned} \tag{13.74}$$

Combining (13.73) and (13.74) we find

$$(\delta\psi)_{\mathrm{AB}} + (\delta\psi)_{\mathrm{CD}} \approx \left[ \mathrm{i}e \frac{\partial A^1}{\partial x_2}\psi + e^2 A^1 A^2 \psi \right] \delta a \delta b. \tag{13.75}$$

Similarly,

$$(\delta\psi)_{\mathrm{BC}} + (\delta\psi)_{\mathrm{DA}} \approx \left[ -\mathrm{i}e \frac{\partial A^2}{\partial x_1}\psi - e^2 A^1 A^2 \psi \right] \delta a \delta b \tag{13.76}$$

with the result that the net change around the loop is

$$(\delta\psi)_{\mathrm{ABCD}} \approx \mathrm{i}e \left( \frac{\partial A^1}{\partial x_2} - \frac{\partial A^2}{\partial x_1} \right) \psi \delta a \delta b. \tag{13.77}$$

For a general loop, (13.77) is replaced by

$$\begin{aligned} (\delta\psi)_{\mathrm{loop}} &= \mathrm{i}e \left( \frac{\partial A^\mu}{\partial x_\nu} - \frac{\partial A^\nu}{\partial x_\mu} \right) \psi \, \mathrm{d}x_\mu \, \mathrm{d}x_\nu \\ &= \mathrm{i}e F^{\mu\nu} \psi \, \mathrm{d}x_\mu \, \mathrm{d}x_\nu, \end{aligned} \tag{13.78}$$

where $F^{\mu\nu} = \partial^\nu A^\mu - \partial^\mu A^\nu$ is the familiar field strength tensor of QED.

The analogy we have been pursuing would, therefore, suggest that $F^{\mu\nu} = 0$ indicates 'no physical effect', while $F^{\mu\nu} \neq 0$ implies the presence of a physical effect. Indeed, when $A^\mu$ has the 'pure gauge' form $A^\mu = \partial^\mu \chi$ the associated $F^{\mu\nu}$ is zero: this is because such an $A^\mu$ can clearly be reduced to zero by a gauge transformation (and also, consistently, because $(\partial^\mu\partial^\nu - \partial^\nu\partial^\mu)\chi = 0$). If $A^\mu$ is not expressible as the 4-gradient of a scalar, then $F^{\mu\nu} \neq 0$ and an electromagnetic field is present, analogous to the spatial curvature revealed by $R^\alpha_{\gamma\beta\sigma} \neq 0$.

Once again, there is a satisfying consistency between this 'geometrical' viewpoint and the discussion of the Aharonov–Bohm effect in section 3.6. As in our remarks at the end of the previous section and equations (13.61)–(13.63), equation (3.83) can be regarded as the integrated form of (13.78), for spatial loops. Transport round such a loop results in a non-trivial net phase change if non-zero $\boldsymbol{B}$ flux is enclosed, and this can be observed.

From this point of view there is undoubtedly a strong conceptual link between Einstein's theory of gravity and quantum gauge theories. In the former, matter (or energy) is regarded as the source of curvature of spacetime, causing the spacetime axes themselves to vary from point to point, and determining the trajectories of massive particles; in the latter, charge is the source of curvature in an 'internal' space (the complex $\psi$-plane, in the U(1) case), a curvature which we call an electromagnetic field and which has observable physical effects.

The reader may consider repeating, for the local SU(2) case, the closed-loop transport calculation of (13.73)–(13.77). It will lead to an expression for the non-Abelian field strength tensor. A closely related, and (for the non-Abelian case) slightly simpler, way of obtaining the result is to consider the commutator of two-covariant derivatives. Consider first the U(1) case. Then

$$[D^\mu, D^\nu]\psi \equiv (D^\mu D^\nu - D^\nu D^\mu)\psi = \mathrm{i}eF^{\mu\nu}\psi \tag{13.79}$$

as is verified in problem 13.2. (A similar result holds in the spacetime coordinate transformation case, where the curvature tensor appears on the right-hand side.) Equation (13.79) suggests that we will find the SU(2) analogue of $F^{\mu\nu}$ by evaluating

$$[D^\mu, D^\nu]\psi^{(\frac{1}{2})} \tag{13.80}$$

where, as usual,

$$D^\mu(\text{on } \psi^{(\frac{1}{2})}) = \partial^\mu + \mathrm{i}g\boldsymbol{\tau}\cdot A^\mu/2. \tag{13.81}$$

Problem 13.3 confirms that the result is

$$[D^\mu, D^\nu]\psi^{(\frac{1}{2})} = \mathrm{i}g\boldsymbol{\tau}/2\cdot(\partial^\mu \boldsymbol{W}^\nu - \partial^\nu \boldsymbol{W}^\mu - g\boldsymbol{W}^\mu \times \boldsymbol{W}^\nu)\psi^{(\frac{1}{2})}; \tag{13.82}$$

the manipulations are very similar to those in (13.20)–(13.23). Noting the analogy between the right-hand side of (13.82) and (13.79), we accordingly expect the SU(2) 'curvature', or field strength tensor, to be given by

$$\boldsymbol{F}^{\mu\nu} = \partial^\mu \boldsymbol{W}^\nu - \partial^\nu \boldsymbol{W}^\mu - g\boldsymbol{W}^\mu \times \boldsymbol{W}^\nu \tag{13.83}$$

or, in component notation,

$$F_i^{\mu\nu} = \partial^\mu W_i^\nu - \partial^\nu W_i^\mu - g\epsilon_{ijk} W_j^\mu W_k^\nu. \tag{13.84}$$

This tensor is of fundamental importance in a (non-Abelian) gauge theory. Since it arises from the commutator of two gauge-covariant derivatives, we are guaranteed that it itself is gauge covariant—that is to say, 'it transforms under local SU(2) transformations in the way its SU(2) structure would indicate'. Now $\boldsymbol{F}^{\mu\nu}$ has clearly three SU(2) components and must be an SU(2) triplet: indeed, it is true that under an infinitesimal local SU(2) transformation

$$\boldsymbol{F}'^{\mu\nu} = \boldsymbol{F}^{\mu\nu} - g\boldsymbol{\epsilon}(x) \times \boldsymbol{F}^{\mu\nu} \tag{13.85}$$

which is the expected law (cf (12.64)) for an SU(2) triplet. Problem 13.4 verifies that (13.85) follows from (13.83) and the transformation law (13.23) for the $\boldsymbol{W}^\mu$ fields. Note particularly that $\boldsymbol{F}^{\mu\nu}$ transforms 'properly', as an SU(2) triplet should, *without* the $\partial^\mu$ part which appears in $\delta\boldsymbol{W}^\mu$.

This non-Abelian $\boldsymbol{F}^{\mu\nu}$ is a much more interesting object than the Abelian $F^{\mu\nu}$ (which is actually U(1)-gauge *invariant*, of course: $F'^{\mu\nu} = F^{\mu\nu}$). $\boldsymbol{F}^{\mu\nu}$ contains the gauge coupling constant $g$, confirming (cf comment (c) in section 13.1.1) that the gauge fields themselves carry SU(2) 'charge' and act as sources for the field strength.

It is now straightforward to move to the quantum field case and construct the SU(2) Yang–Mills analogue of the Maxwell Lagrangian $-\frac{1}{4}\hat{F}_{\mu\nu}\hat{F}^{\mu\nu}$. It is simply $-\frac{1}{4}\hat{\boldsymbol{F}}_{\mu\nu}\cdot\hat{\boldsymbol{F}}^{\mu\nu}$, the SU(2) 'dot product' ensuring SU(2) invariance (see problem 13.5), even under *local* transformation, in view of the transformation law (13.85). But before proceeding in this way, we first need to introduce local SU(3) symmetry.

## 13.4 Local SU(3) symmetry

Using what has been done for global SU(3) symmetry in section 12.2, and the preceding discussion of how to make a global SU(2) into a local one, it is straightforward to develop the corresponding theory of local SU(3). This is the gauge group of QCD, the three degrees of freedom of the fundamental quark triplet now referring to 'colour', as will be further discussed in chapter 14. We denote the basic triplet by $\psi$, which transforms under a local SU(3) transformation according to

$$\psi' = \exp[\mathrm{i}g_{\mathrm{s}}\boldsymbol{\lambda}\cdot\boldsymbol{\alpha}(x)/2]\psi \tag{13.86}$$

which is the same as the global transformation (12.74) but with the eight constant parameters $\boldsymbol{\alpha}$ replaced by $x$-dependent ones, and with a coupling strength $g_{\mathrm{s}}$ inserted. The SU(3)-covariant derivative, when acting on an SU(3) triplet $\psi$, is given by the indicated generalization of (13.10), namely

$$D^{\mu}(\text{acting on SU(3) triplet}) = \partial^{\mu} + \mathrm{i}g_{\mathrm{s}}\boldsymbol{\lambda}/2\cdot\boldsymbol{A}^{\mu} \tag{13.87}$$

where $A_1^{\mu}, A_2^{\mu}, \ldots, A_8^{\mu}$ are eight gauge fields, the quanta of which are called *gluons*. The coupling is denoted by '$g_{\mathrm{s}}$' in anticipation of the application to strong interactions via QCD.

The infinitesimal version of (13.86) is (cf (13.13))

$$\psi' = (1 + \mathrm{i}g_{\mathrm{s}}\boldsymbol{\lambda}\cdot\boldsymbol{\eta}(x)/2)\psi \tag{13.88}$$

where '1' stands for the unit matrix in the three-dimensional space of components of the triplet $\psi$. As in (13.14), it is clear that $\partial^{\mu}\psi'$ will involve an 'unwanted' term $\partial^{\mu}\boldsymbol{\eta}(x)$. By contrast, the desired covariant derivative $D^{\mu}\psi$ should transform according to

$$D'^{\mu}\psi' = (1 + \mathrm{i}g_{\mathrm{s}}\boldsymbol{\lambda}\cdot\boldsymbol{\eta}(x)/2)D^{\mu}\psi \tag{13.89}$$

without the $\partial^{\mu}\boldsymbol{\eta}(x)$ term. Problem 13.6 verifies that this is fulfilled by having the gauge fields transform by

$$A_a'^{\mu} = A_a^{\mu} - \partial^{\mu}\eta_a(x) - g_{\mathrm{s}}f_{abc}\eta_b(x)A_c^{\mu}. \tag{13.90}$$

Comparing (13.90) with (12.80) we can identify the term in $f_{abc}$ as telling us that the eight fields $A_a^{\mu}$ transform as an SU(3) octet, the $\eta$'s now depending on $x$, of course. This is the adjoint or regular representation of SU(3)—as we have now

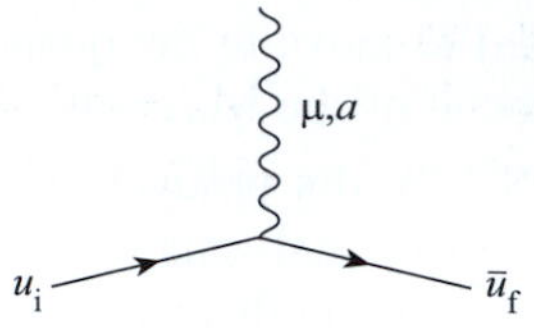


**Figure 13.7.** Quark–gluon vertex.

come to expect for gauge fields. However, the $\partial^\mu \eta_a(x)$ piece spoils this simple transformation property under local transformations. But it *is* just what is needed to cancel the corresponding $\partial^\mu \boldsymbol{\eta}(x)$ term in $\partial^\mu \psi'$, leaving $D^\mu \psi$ transforming as a proper triplet via (13.89). The finite version of (13.90) can be derived as in section 13.1 for SU(2) but we shall not need the result here.

As in the SU(2) case, the free Dirac equation for an SU(3)-triplet $\psi$,

$$(\mathrm{i}\gamma_\mu \partial^\mu - m)\psi = 0 \tag{13.91}$$

can be 'promoted' into one which is covariant under local SU(3) transformations by replacing $\partial^\mu$ by $D^\mu$ of (13.87), leading to

$$(\mathrm{i}\not\partial - m)\psi = g_s \boldsymbol{\lambda}/2 \cdot \not{\boldsymbol{A}}\psi \tag{13.92}$$

(compare (13.39)). This leads immediately to the one gluon emission amplitude (see figure 13.7)

$$-\mathrm{i}g_s \int \bar{\psi}_f \boldsymbol{\lambda}/2\gamma^\mu \psi_i \cdot \boldsymbol{A}_\mu \, \mathrm{d}^4 x \tag{13.93}$$

as already suggested in section 12.3.1: the SU(3) current of (12.133)—but this time in *colour* space—is 'dotted' with the gauge field. The Feynman rule for figure 13.7 is, therefore,

$$-\mathrm{i}g_s \lambda_a/2\gamma^\mu. \tag{13.94}$$

The SU(3) field strength tensor can be calculated by evaluating the commutator of two $D$'s of the form (13.87): the result (problem 13.7) is

$$F_a^{\mu\nu} = \partial^\mu A_a^\nu - \partial^\nu A_a^\mu - g_s f_{abc} A_b^\mu A_c^\nu \tag{13.95}$$

which is closely analogous to the SU(2) case (13.84) ( the structure constants of SU(2) are given by the $\epsilon_{ijk}$ symbol and of SU(3) by $f_{abc}$). Once again, the crucial property of $F_a^{\mu\nu}$ is that, under *local* SU(3) transformations, it develops no '$\partial^\mu \eta_a$' part but transforms as a 'proper' octet:

$$F_a^{\prime\mu\nu} = F_a^{\mu\nu} - g_s f_{abc} \eta_b(x) F_c^{\mu\nu}. \tag{13.96}$$

This allows us to write down a locally SU(3)-invariant analogue of the Maxwell Lagrangian

$$-\tfrac{1}{4} F_a^{\mu\nu} F_{a\mu\nu} \tag{13.97}$$

by dotting the two octets together.

It is now time to consider locally SU(2)- and SU(3)-invariant quantum field Lagrangians and, in particular, the resulting self-interactions among the gauge quanta.

## 13.5 Local non-Abelian symmetries in Lagrangian quantum field theory

### 13.5.1 Local SU(2) and SU(3) Lagrangians

We consider here only the particular examples relevant to the strong and electroweak interactions of quarks: namely, a (weak) SU(2) doublet of fermions interacting with SU(2) gauge fields $W_i^\mu$ and a (strong) SU(3) triplet of fermions interacting with the gauge fields $A_a^\mu$. We follow the same steps as in the U(1) case of chapter 7, noting again that for quantum fields the sign of the exponents in (13.28) and (13.86) is reversed, by convention; thus (12.89) is replaced by its local version

$$\hat{q}' = \exp(-\mathrm{i}g\hat{\boldsymbol{\alpha}}(x)\cdot\boldsymbol{\tau}/2)\hat{q} \tag{13.98}$$

and (12.132) by

$$\hat{q}' = \exp(-\mathrm{i}g_{\mathrm{s}}\hat{\boldsymbol{\alpha}}(x)\cdot\boldsymbol{\lambda}/2)\hat{q}. \tag{13.99}$$

The globally SU(2)-invariant Lagrangian (12.87) becomes locally SU(2)-invariant if we replaced $\partial^\mu$ by $D^\mu$ of (13.10), with $\hat{W}^\mu$ now a quantum field:

$$\begin{aligned}\hat{\mathcal{L}}_{\mathrm{D,local\ SU(2)}} &= \bar{\hat{q}}(\mathrm{i}\hat{\not{D}} - m)\hat{q} \\ &= \bar{\hat{q}}(\mathrm{i}\not{\partial} - m)\hat{q} - g\bar{\hat{q}}\gamma^\mu\boldsymbol{\tau}/2\hat{q}\cdot\hat{\boldsymbol{W}}_\mu\end{aligned} \tag{13.100}$$

with an interaction of the form 'symmetry current (12.109) dotted into the gauge field'. To this we must add the SU(2) Yang–Mills term

$$\mathcal{L}_{\mathrm{Y-M,SU(2)}} = -\tfrac{1}{4}\hat{\boldsymbol{F}}_{\mu\nu}\cdot\hat{\boldsymbol{F}}^{\mu\nu} \tag{13.101}$$

to get the local SU(2) analogue of $\mathcal{L}_{\mathrm{QED}}$. It is *not* possible to add a mass term for the gauge fields of the form $\frac{1}{2}\hat{\boldsymbol{W}}^\mu\cdot\hat{\boldsymbol{W}}_\mu$, since such a term would not be invariant under the gauge transformations (13.26) or (13.34) of the W-fields. Thus, just as in the U(1) (electromagnetic) case, the W-quanta of this theory are *massless*. We presumably also need a gauge-fixing term for the gauge fields, as in section 7.3, which we can take to be[1]

$$\mathcal{L}_{\mathrm{gf}} = -\frac{1}{2\xi}(\partial_\mu\hat{\boldsymbol{W}}^\mu\cdot\partial_\nu\hat{\boldsymbol{W}}^\nu). \tag{13.102}$$

[1] We shall see in section 13.5.3 that in the non-Abelian case this gauge-fixing term does *not* completely solve the problem of quantizing such gauge fields; however, it is adequate for tree graphs.

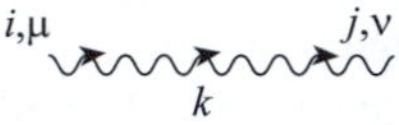


**Figure 13.8.** SU(2) gauge-boson propagator.

The Feynman rule for the fermion-W vertex is then the same as already given in (13.41), while the W-propagator is (figure 13.8)

$$\frac{\mathrm{i}[-g^{\mu\nu} + (1-\xi)k^\mu k^\nu/k^2]}{k^2 + \mathrm{i}\epsilon}\delta^{ij}. \tag{13.103}$$

Before proceeding to the SU(3) case, we must now emphasize three respects in which our local SU(2) Lagrangian is not suitable (yet) for describing weak interactions. First, weak interactions violate parity, in fact 'maximally', by which is meant that only the 'left-handed' part $\hat{\psi}_{\mathrm{L}}$ of the fermion field enters the interactions with the $\boldsymbol{W}^\mu$ fields, where $\hat{\psi}_{\mathrm{L}} \equiv ((1-\gamma_5)/2)\hat{\psi}$; for this reason the weak isospin group is called $\mathrm{SU(2)_L}$. Second, the physical $\mathrm{W}^\pm$ are, of course, not massless and, therefore, cannot be described by propagators of the form (13.103). And third, the *fermion* mass term violates the 'left-handed' SU(2) gauge symmetry, as the discussion in section 12.3.2 shows. In this case, however, the chiral symmetry which is broken by fermion masses in the Lagrangian is a local, or gauge, symmetry (in section 12.3.2 the chiral flavour symmetry was a global symmetry). If we want to preserve the chiral gauge symmetry $\mathrm{SU(2)_L}$–and it is necessary for renormalizability—then we shall have to replace the simple fermion mass term in (13.100) by something else, as will be explained in chapter 22.

The locally $\mathrm{SU(3)_c}$-invariant Lagrangian for one quark triplet (cf (12.137))

$$\hat{q}_{\mathrm{f}} = \begin{pmatrix} \hat{f}_{\mathrm{r}} \\ \hat{f}_{\mathrm{b}} \\ \hat{f}_{\mathrm{g}} \end{pmatrix} \tag{13.104}$$

where 'f' stands for 'flavour', and 'r, b, and g' for 'red, blue and green', is

$$\bar{\hat{q}}_{\mathrm{f}}(\mathrm{i}\hat{\not{D}} - m_{\mathrm{f}})\hat{q}_{\mathrm{f}} - \frac{1}{4}\hat{F}_{a\mu\nu}\hat{F}_a^{\mu\nu} - \frac{1}{2\xi}(\partial_\mu \hat{A}_a^\mu)(\partial_\nu \hat{A}_a^\nu) \tag{13.105}$$

where $\hat{D}^\mu$ is given by (13.87) with $\boldsymbol{A}^\mu$ replaced by $\hat{\boldsymbol{A}}^\mu$ and the footnote before equation (13.102) also applies here. This leads to the interaction term (cf (13.93))

$$-g_{\mathrm{s}}\bar{\hat{q}}_{\mathrm{f}}\gamma^\mu \boldsymbol{\lambda}/2\hat{q}_{\mathrm{f}} \cdot \hat{\boldsymbol{A}}_\mu \tag{13.106}$$

and the Feynman rule (13.94) for figure 13.7. Once again, the gluon quanta must be *massless* and their propagator is the same as (13.103), with $\delta_{ij} \to \delta_{ab}(a, b = 1, 2, \ldots, 8)$. The different quark flavours are included by simply repeating the first term of (13.105) for all flavours:

$$\sum_{\mathrm{f}} \bar{\hat{q}}_{\mathrm{f}}(\mathrm{i}\hat{\not{D}} - m_{\mathrm{f}})\hat{q}_{\mathrm{f}} \tag{13.107}$$

which incorporates the hypothesis that the $SU(3)_c$-gauge interaction is 'flavour-blind', i.e. exactly the same for each flavour. Note that although the flavour masses are different, the masses of different 'coloured' quarks of the same flavour are the same ($m_u \neq m_d$, $m_{u,r} = m_{u,b} = m_{u,g}$).

The Lagrangians (13.100)–(13.102) and (13.105), though easily written down after all this preparation, are unfortunately not adequate for anything but tree graphs. We shall indicate why this is so in section 13.5.3. Before that, we want to discuss in more detail the nature of the gauge-field self-interactions contained in the Yang–Mills pieces.

### 13.5.2 Gauge field self-interactions

We start by pointing out an interesting ambiguity in the prescription for 'covariantizing' wave equations which we have followed, namely 'replace $\partial^\mu$ by $D^\mu$'. Suppose we wished to consider the electromagnetic interactions of charged massless spin-1 particles, call them X's, carrying charge $e$. The standard wave equation for such free massless vector particles would be the same as for $A^\mu$, namely

$$\Box X^\mu - \partial^\mu \partial^\nu X_\nu = 0. \tag{13.108}$$

To 'covariantize' this (i.e. introduce the electromagnetic coupling), we would replace $\partial^\mu$ by $D^\mu = \partial^\mu + \mathrm{i}eA^\mu$ so as to obtain

$$D^2 X^\mu - D^\mu D^\nu X_\nu = 0. \tag{13.109}$$

But this procedure is not unique: if we had started from the perfectly equivalent free particle wave equation

$$\Box X^\mu - \partial^\nu \partial^\mu X_\nu = 0 \tag{13.110}$$

we would have arrived at

$$D^2 X^\mu - D^\nu D^\mu X_\nu = 0 \tag{13.111}$$

which is not the same as (13.109), since (cf (13.79))

$$[D^\mu, D^\nu] = \mathrm{i}eF^{\mu\nu}. \tag{13.112}$$

The simple prescription $\partial^\mu \rightarrow D^\mu$ has, in this case, failed to produce a unique wave equation. We can allow for this ambiguity by introducing an arbitrary parameter $\delta$ in the wave equation, which we write as

$$D^2 X^\mu - D^\nu D^\mu X_\nu + \mathrm{i}e\delta F^{\mu\nu} X_\nu = 0. \tag{13.113}$$

The $\delta$ term in (13.113) contributes to the magnetic moment coupling of the X-particle to the electromagnetic field and is called the 'ambiguous magnetic moment'. Just such an ambiguity would seem to arise in the case of the charged weak interaction quanta $W^{\pm}$ (their masses do not affect this argument). For the photon itself, of course, $e = 0$ and there is no such ambiguity.

It is important to be clear that (13.113) is fully U(1) gauge-covariant, so that $\delta$ cannot be fixed by further appeal to the local U(1) symmetry. Moreover, it turns out that the theory for arbitrary $\delta$ is *not renormalizable* (though we shall not show this here): thus, the quantum electrodynamics of charged massless vector bosons is, in general, non-renormalizable.

However, the theory *is* renormalizable if—to continue with the present terminology—the photon, the X-particle and its anti-particle, the $\bar{\mathrm{X}}$, are the members of an SU(2)-gauge triplet (like the W's), with gauge coupling constant $e$. This is, indeed, very much how the photon and the $W^{\pm}$ are 'unified' but there is a complication (as always!) in that case, having to do with the necessity for finding room in the scheme for the neutral weak boson $Z^0$ as well. We shall see how this works in chapter 19: meanwhile we continue with this X–$\gamma$ model. We shall show that when the X–$\gamma$ interaction contained in (13.113) is regarded as a 3–X vertex in a local SU(2) gauge theory, the value of $\delta$ has to equal one; for this value the theory is renormalizable. In this interpretation, the $X^\mu$ wavefunction is identified with '$\frac{1}{\sqrt{2}}(X_1^\mu + \mathrm{i}X_2^\mu)$' and $\bar{X}^\mu$ with '$\frac{1}{\sqrt{2}}(X_1^\mu - \mathrm{i}X_2^\mu)$' in terms of components of the SU(2) triplet $X_i^\mu$, while $A^\mu$ is identified with $X_3^\mu$.

Consider then equation (13.113) written in the form[2]

$$\Box X^\mu - \partial^\nu \partial^\mu X_\nu = \hat{V} X^\mu \tag{13.114}$$

where

$$\begin{aligned} \hat{V} X^\mu = & -\mathrm{i}e\{[\partial^\nu(A_\nu X^\mu) + A^\nu \partial_\nu X^\mu] \\ & - (1+\delta)[\partial^\nu(A^\mu X_\nu) + A^\nu \partial^\mu X_\nu] \\ & + \delta[\partial^\mu(A^\nu X_\nu) + A^\mu \partial^\nu X_\nu]\} \end{aligned} \tag{13.115}$$

and we have dropped terms of $O(e^2)$ which appear in the '$D^2$' term: we shall come back to them later. The terms inside the { } brackets have been written in such a way that each [ ] bracket has the structure

$$\partial(AX) + A(\partial X) \tag{13.116}$$

which will be convenient for the following evaluation.

The lowest-order ($O(e)$) perturbation theory amplitude for 'X → X' under the potential $\hat{V}$ is then

$$-\mathrm{i}\int X_\mu^*(\mathrm{f})\hat{V} X^\mu(\mathrm{i})\,\mathrm{d}^4x. \tag{13.117}$$

Inserting (13.115) into (13.117) clearly gives something involving two '$X$'-wavefunctions and one '$A$' one, i.e. a triple-X vertex (with $A^\mu \equiv X_3^\mu$), shown

[2] The sign chosen for $\hat{V}$ here apparently *differs* from that in the KG case (4.133) but it does agree when allowance is made, in the amplitude (13.117), for the fact that the dot product of the polarization vectors is negative (cf (7.84)).

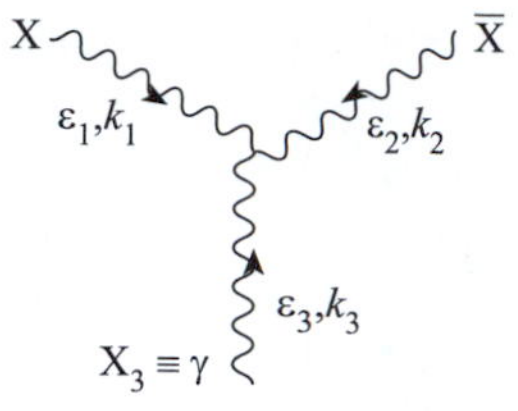


**Figure 13.9.** Triple-X vertex.

in figure 13.9. To obtain the rule for this vertex from (13.117), consider the first [ ] bracket in (13.115). It contributes

$$-\mathrm{i}(-\mathrm{i}e)\int \bar{X}^*_\mu(2)\{\partial^\nu(X_{3\nu}(3)X^\mu(1)) + X^\nu_3(3)\partial_\nu X^\mu(1)\}\,\mathrm{d}^4x \tag{13.118}$$

where the (1), (2), (3) refer to the momenta as shown in figure 13.9, and for reasons of symmetry are all taken to be ingoing; thus,

$$X^\mu_3(1) = \epsilon^\mu_3 \exp(-\mathrm{i}k_3\cdot x) \tag{13.119}$$

for example. The first term in (13.118) can be evaluated by a partial integration to turn the $\partial^\nu$ onto the $\bar{X}^*_\mu(2)$, while in the second term $\partial_\nu$ acts straightforwardly on $X^\mu(1)$. Omitting the usual $(2\pi)^4\delta^4$ energy–momentum conserving factor, we find (problem 13.8) that (13.118) leads to the amplitude

$$\mathrm{i}e\epsilon_1\cdot\epsilon_2\,(k_1-k_2)\cdot\epsilon_3. \tag{13.120}$$

In a similar way, the other terms in (13.117) give

$$-\mathrm{i}e\delta(\epsilon_1\cdot\epsilon_3\,\epsilon_2\cdot k_2 - \epsilon_2\cdot\epsilon_3\,\epsilon_1\cdot k_1) \tag{13.121}$$

and

$$+\mathrm{i}e(1+\delta)(\epsilon_2\cdot\epsilon_3\,\epsilon_1\cdot k_2 - \epsilon_1\cdot\epsilon_3\,\epsilon_2\cdot k_1). \tag{13.122}$$

Adding all the terms up and using the 4-momentum conservation condition

$$k_1+k_2+k_3=0 \tag{13.123}$$

we obtain the vertex

$$+\mathrm{i}e\{\epsilon_1\cdot\epsilon_2(k_1-k_2)\cdot\epsilon_3+\epsilon_2\cdot\epsilon_3(\delta k_2-k_3)\cdot\epsilon_1+\epsilon_3\cdot\epsilon_1(k_3-\delta k_1)\cdot\epsilon_2\}. \tag{13.124}$$

It is quite evident from (13.124) that the value $\delta = 1$ has a privileged role and we strongly suspect that this will be the value selected by the proposed SU(2) gauge symmetry of this model. We shall check this in two ways: in the first, we consider a 'physical' process involving the vertex (13.124) and show how

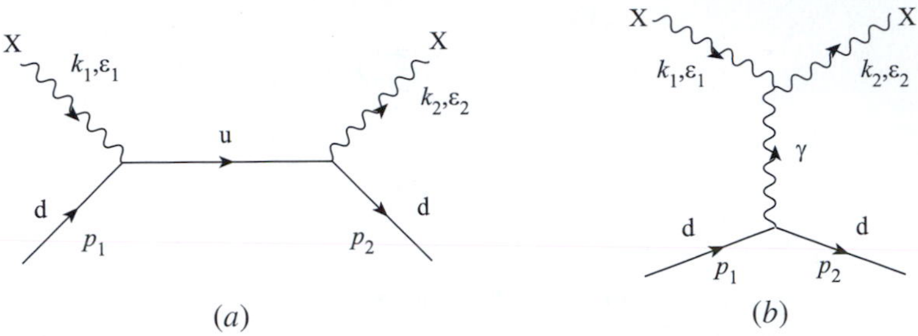


**Figure 13.10.** Tree graphs contributing to $X + d \to X + d$.

requiring it to be SU(2)-gauge invariant fixes $\delta$ to be 1; in the second, we 'unpack' the relevant vertex from the compact Yang–Mills Lagrangian $-\frac{1}{4}\hat{\boldsymbol{X}}_{\mu\nu}\cdot\hat{\boldsymbol{X}}^{\mu\nu}$.

The process we shall choose is $X + d \to X + d$ where d is a fermion (which we call a quark) transforming as the $T_3 = -\frac{1}{2}$ component of a doublet under the SU(2) gauge group, its $T_3 = +\frac{1}{2}$ partner being the u. There are two contributing Feynman graphs, shown in figures 13.10(*a*) and (*b*). Consider, first, the amplitude for figure 13.10(*a*). We use the rule of figure 13.1, with the $\tau$-matrix combination $\tau_+ = (\tau_1 + \mathrm{i}\tau_2)/\sqrt{2}$ corresponding to the absorption of the positively charged X and $\tau_- = (\tau_1 - \mathrm{i}\tau_2)/\sqrt{2}$ for the emission of the X. Then figure 13.10(*a*) is

$$(-\mathrm{i}e)^2\bar{\psi}^{(\frac{1}{2})}(p_2)\frac{\tau_-}{2}\not{\epsilon}_2\frac{\mathrm{i}}{\not{p}_1+\not{k}_1-m}\frac{\tau_+}{2}\not{\epsilon}_1\psi^{(\frac{1}{2})}(p_1), \tag{13.125}$$

where

$$\psi^{(\frac{1}{2})} = \begin{pmatrix} u \\ d \end{pmatrix} \tag{13.126}$$

and we have chosen real polarization vectors. Using the explicit forms (12.25) for the $\tau$-matrices, (13.125) becomes

$$(-\mathrm{i}e)^2\bar{d}(p_2)\frac{1}{\sqrt{2}}\not{\epsilon}_2\frac{\mathrm{i}}{\not{p}_1+\not{k}_1-m}\frac{1}{\sqrt{2}}\not{\epsilon}_1 d(p_1). \tag{13.127}$$

We must now discuss how to implement gauge invariance. In the QED case of electron Compton scattering (section 8.6.2), the test of gauge invariance was that the amplitude should vanish if any photon polarization vector $\epsilon^\mu(k)$ was replaced by $k^\mu$—see (8.168). This requirement was derived from the fact that a gauge transformation on the photon $A^\mu$ took the form $A^\mu \to A'^\mu = A^\mu - \partial^\mu\chi$, so that, consistently with the Lorentz condition, $\epsilon^\mu$ could be replaced by $\epsilon'^\mu = \epsilon^\mu + \beta k^\mu$ (cf 8.165) without changing the physics. But the SU(2) analogue of the U(1)-gauge transformation is given by (13.26), for infinitesimal $\epsilon$'s, and although there is indeed an analogous '$-\partial^\mu\boldsymbol{\epsilon}$' part, there is also an additional part (with $g \to e$ in our case) expressing the fact that the X's carry SU(2) charge. However, this extra part does involve the coupling $e$. Hence, if we were to make the *full* change corresponding to (13.26) in a tree graph of order $e^2$, the extra part

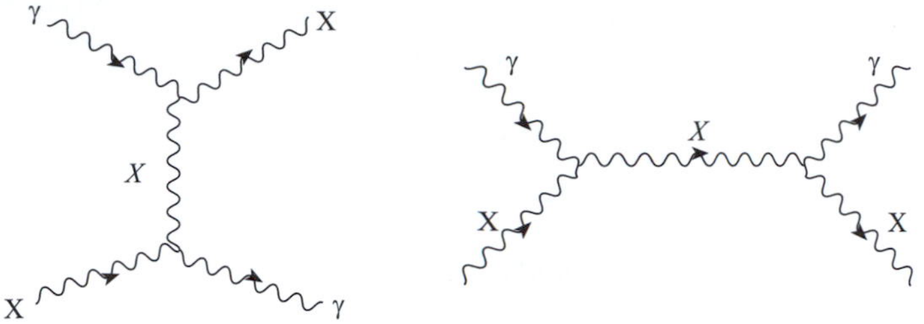


**Figure 13.11.** Tree graphs contributing to $\gamma + \mathrm{X} \to \gamma + \mathrm{X}$.

would produce a term of order $e^3$. We shall take the view that gauge invariance should hold at each order of perturbation theory separately; thus, we shall demand that the tree graphs for X–d scattering, for example, should be invariant under $\epsilon^\mu \to k^\mu$ for any $\epsilon$.

The replacement $\epsilon_1 \to k_1$ in (13.127) produces the result (problem 13.9)

$$(-\mathrm{i}e)^2 \frac{\mathrm{i}}{2} \bar{d}(p_2) \not{\epsilon}_2 d(p_1) \tag{13.128}$$

where we have used the Dirac equation for the quark spinors of mass $m$. The term (13.128) is certainly not zero—but we must of course also include the amplitude for figure 13.10(*b*). Using the vertex of (13.124) with suitable sign changes of momenta, and the photon propagator of (7.119), and remembering that d has $\tau_3 = -1$, the amplitude for figure 13.10(*b*) is

$$\begin{aligned} &\mathrm{i}e[\epsilon_1 \cdot \epsilon_2 (k_1 + k_2)_\mu + \epsilon_{2\mu} \epsilon_1 \cdot (-\delta k_2 - k_2 + k_1) + \epsilon_{1\mu} \epsilon_2 \cdot (k_2 - k_1 - \delta k_1)] \\ &\quad \times \frac{-\mathrm{i}g^{\mu\nu}}{q^2} \left[ -\mathrm{i}e\bar{d}(p_2) \left( -\frac{1}{2} \right) \gamma_\nu d(p_1) \right], \end{aligned} \tag{13.129}$$

where $q^2 = (k_1 - k_2)^2 = -2k_1 \cdot k_2$ using $k_1^2 = k_2^2 = 0$, and where the $\xi$-dependent part of the $\gamma$-propagator vanishes since $\bar{d}(p_2) \not{q} d(p_1) = 0$. We now leave it as an exercise (problem 13.10) to verify that, when $\epsilon_1 \to k_1$ in (13.129), the resulting amplitude does exactly cancel the contribution (13.128), *provided that* $\delta = 1$. Thus, the X–$\bar{\mathrm{X}}$–$\gamma$ vertex is, assuming the SU(2)-gauge symmetry,

$$\mathrm{i}e[\epsilon_1 \cdot \epsilon_2 (k_1 - k_2) \cdot \epsilon_3 + \epsilon_2 \cdot \epsilon_3 (k_2 - k_3) \cdot \epsilon_1 + \epsilon_3 \cdot \epsilon_1 \ (k_3 - k_1) \cdot \epsilon_2]. \tag{13.130}$$

The verification of this non-Abelian gauge invariance to order $e^2$ is, of course, not a proof that the entire theory of massless X quanta, $\gamma$'s and quark isospinors will be gauge invariant if $\delta = 1$. Indeed, having obtained the X–X–$\gamma$ vertex, we immediately have something new to check: we can see if the lowest order $\gamma$–X scattering amplitude is gauge invariant. The X–X–$\gamma$ vertex will generate the $O(e^2)$ graphs shown in figure 13.11 and the dedicated reader may check that the sum of these amplitudes is *not* gauge invariant, again in the

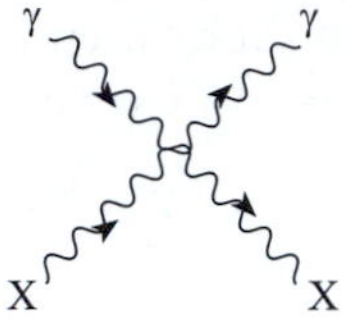


**Figure 13.12.** $\gamma$–$\gamma$–X–X vertex.

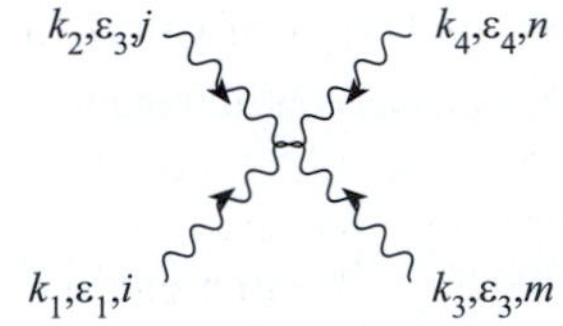


**Figure 13.13.** 4–X vertex.

(tree graph) sense of not vanishing when any $\epsilon$ is replaced by the corresponding $k$. But this is actually correct. In obtaining the X–X–$\gamma$ vertex we dropped an $O(e^2)$ term involving the three fields $A$, $A$ and $X$, in going from (13.115) to (13.124): this will generate an $O(e^2)$ $\gamma$–$\gamma$–X–X interaction, figure 13.12, when used in lowest-order perturbation theory. One can find the amplitude for figure 13.12 by the gauge-invariance requirement applied to figures 13.11 and 13.12, but it has to be admitted that this approach is becoming laborious. It is, of course, far more efficient to deduce the vertices from the compact Yang–Mills Lagrangian $-\frac{1}{4}\hat{\boldsymbol{X}}_{\mu\nu}\cdot\hat{\boldsymbol{X}}^{\mu\nu}$, which we shall now do; nevertheless, some of the physical implications of those couplings, such as we have discussed earlier, are worth exposing.

The SU(2) Yang–Mills Lagrangian for the SU(2) triplet of gauge fields $\hat{\boldsymbol{X}}^{\mu}$ is

$$\hat{\mathcal{L}}_{2,\mathrm{YM}} = -\tfrac{1}{4}\hat{\boldsymbol{X}}_{\mu\nu}\cdot\hat{\boldsymbol{X}}^{\mu\nu} \tag{13.131}$$

where

$$\hat{\boldsymbol{X}}^{\mu\nu} = \partial^{\mu}\hat{\boldsymbol{X}}^{\nu} - \partial^{\nu}\hat{\boldsymbol{X}}^{\mu} - \hat{\boldsymbol{X}}^{\mu}\times\hat{\boldsymbol{X}}^{\nu}. \tag{13.132}$$

$\hat{\mathcal{L}}_{2,\mathrm{YM}}$ can be unpacked a bit into

$$\begin{aligned} &-\tfrac{1}{2}(\partial_{\mu}\hat{\boldsymbol{X}}_{\nu} - \partial_{\nu}\hat{\boldsymbol{X}}_{\mu})\cdot(\partial^{\mu}\hat{\boldsymbol{X}}^{\nu}) + e(\hat{\boldsymbol{X}}_{\mu}\times\hat{\boldsymbol{X}}_{\nu})\cdot\partial^{\mu}\hat{\boldsymbol{X}}^{\nu} \\ &\quad - \tfrac{1}{4}e^{2}[(\hat{\boldsymbol{X}}^{\mu}\cdot\hat{\boldsymbol{X}}_{\mu})^{2} - (\hat{\boldsymbol{X}}^{\mu}\cdot\hat{\boldsymbol{X}}^{\nu})(\hat{\boldsymbol{X}}_{\mu}\cdot\hat{\boldsymbol{X}}_{\nu})]. \end{aligned} \tag{13.133}$$

The X–X–$\gamma$ vertex is in the '$e$' term, the X–X–$\gamma$–$\gamma$ one in the '$e^2$' term. We give the form of the latter using SU(2) '$i$, $j$, $k$' labels, as shown in figure 13.13:

$$\begin{aligned} &-\mathrm{i}e^{2}[\epsilon_{ij\ell}\epsilon_{mn\ell}(\epsilon_{1}\cdot\epsilon_{3}\epsilon_{2}\cdot\epsilon_{4} - \epsilon_{1}\cdot\epsilon_{4}\epsilon_{2}\cdot\epsilon_{3}) \\ &\quad + \epsilon_{in\ell}\epsilon_{jm\ell}(\epsilon_{1}\cdot\epsilon_{2}\epsilon_{3}\cdot\epsilon_{4} - \epsilon_{1}\cdot\epsilon_{3}\epsilon_{2}\cdot\epsilon_{4}) \\ &\quad + \epsilon_{im\ell}\epsilon_{nj\ell}(\epsilon_{1}\cdot\epsilon_{4}\epsilon_{2}\cdot\epsilon_{3} - \epsilon_{1}\cdot\epsilon_{2}\epsilon_{3}\cdot\epsilon_{4})]. \end{aligned} \tag{13.134}$$

The reason for the collection of terms seen in (13.130) and (13.134) can be understood as follows. Consider the 3–X vertex

$$\langle k_2, \epsilon_2, j; k_3, \epsilon_3, k | e(\hat{\boldsymbol{X}}_\mu \times \hat{\boldsymbol{X}}_\nu) \cdot \partial^\mu \hat{\boldsymbol{X}}^\nu | k_1, \epsilon_1, i \rangle \tag{13.135}$$

for example. When each $\hat{\boldsymbol{X}}$ is expressed as a mode expansion and the initial and final states are also written in terms of appropriate $\hat{a}$'s and $\hat{a}^\dagger$'s, the amplitude will be a vacuum expectation value (vev) of six $\hat{a}$'s and $\hat{a}^\dagger$'s; the different terms in (13.130) arise from the different ways of getting a non-zero value for this vev, by manipulations similar to those in section 6.3.

We end this chapter by presenting an introduction to the problem of quantizing non-Abelian gauge field theories. Our aim will be, first, to indicate where the approach followed for the Abelian gauge field $\hat{A}^\mu$ in section 7.3.2 fails and then to show how the assumption (nevertheless) that the Feynman rules we have established for tree graphs work for loops as well, leads to violations of unitarity. This calculation will indicate a very curious way of remedying the situation 'by hand', through the introduction of *ghost particles*, only present in loops.

### 13.5.3 Quantizing non-Abelian gauge fields

We consider for definiteness the SU(2)-gauge theory with massless gauge fields $\hat{\boldsymbol{W}}^\mu(x)$, which we shall call gluons, by a slight abuse of language. We try to carry through for the Yang–Mills Lagrangian

$$\hat{\mathcal{L}}_2 = -\tfrac{1}{4} \hat{\boldsymbol{F}}_{\mu\nu} \cdot \hat{\boldsymbol{F}}^{\mu\nu} \tag{13.136}$$

where

$$\hat{\boldsymbol{F}}_{\mu\nu} = \partial_\mu \hat{\boldsymbol{W}}_\nu - \partial_\nu \hat{\boldsymbol{W}}_\mu - g \hat{\boldsymbol{W}}_\mu \times \hat{\boldsymbol{W}}_\nu, \tag{13.137}$$

the same steps we followed for the Maxwell one in section 7.3.2.

We begin by re-formulating the prescription arrived at in (7.116), which we reproduce again here for convenience:

$$\hat{\mathcal{L}}_\xi = -\tfrac{1}{4} \hat{F}_{\mu\nu} \hat{F}^{\mu\nu} - \frac{1}{2\xi} (\partial_\mu \hat{A}^\mu)^2. \tag{13.138}$$

$\hat{\mathcal{L}}_\xi$ leads to the equation of motion

$$\Box \hat{A}^\mu - \partial^\mu \partial_\nu \hat{A}^\nu + \frac{1}{\xi} \partial^\mu \partial_\nu \hat{A}^\nu = 0. \tag{13.139}$$

This has the drawback that the limit $\xi \to 0$ appears to be singular (though the propagator (7.119) is well behaved as $\xi \to 0$). To avoid this unpleasantness, consider the Lagrangian (Lautrup 1967)

$$\hat{\mathcal{L}}_{\xi B} = -\tfrac{1}{4} \hat{F}_{\mu\nu} \hat{F}^{\mu\nu} + \hat{B} \partial_\mu \hat{A}^\mu + \tfrac{1}{2} \xi \hat{B}^2 \tag{13.140}$$

where $\hat{B}$ is a scalar field. We may think of the '$\hat{B}\partial \cdot \hat{A}$' term as a field-theory analogue of the procedure followed in classical Lagrangian mechanics, whereby a constraint (in this case the gauge-fixing one $\partial \cdot \hat{A} = 0$) is brought into the Lagrangian with a 'Lagrange multiplier' (here the field $\hat{B}$). The momentum conjugate to $\hat{A}^0$ is now

$$\hat{\pi}^0 = \hat{B} \tag{13.141}$$

while the Euler–Lagrange equations for $\hat{A}^\mu$ read as

$$\Box \hat{A}^\mu - \partial^\mu \partial_\nu \hat{A}^\nu = \partial^\mu \hat{B} \tag{13.142}$$

and for $\hat{B}$ yield

$$\partial_\mu \hat{A}^\mu + \xi \hat{B} = 0. \tag{13.143}$$

Eliminating $\hat{B}$ from (13.140) by means of (13.143), we recover (13.138). Taking $\partial_\mu$ of (13.142) we learn that $\Box \hat{B} = 0$, so that $\hat{B}$ is a free massless field. Applying $\Box$ to (13.143) then shows that $\Box \partial_\mu \hat{A}^\mu = 0$, so that $\partial_\mu \hat{A}^\mu$ is also a free massless field.

In this formulation, the appropriate subsidiary condition for getting rid of the unphysical (non-transverse) degrees of freedom is (cf (7.108))

$$\hat{B}^{(+)}(x)|\Psi\rangle = 0. \tag{13.144}$$

Kugo and Ojima (1979) have shown that (13.144) provides a satisfactory definition of the Hilbert space of states. In addition to this, it is also essential to prove that all physical results are independent of the gauge parameter $\xi$.

We now try to generalize the foregoing in a straightforward way to (13.136). The obvious analogue of (13.140) would be to consider

$$\hat{\mathcal{L}}_{2,\xi B} = -\tfrac{1}{4}\hat{\boldsymbol{F}}_{\mu\nu} \cdot \hat{\boldsymbol{F}}^{\mu\nu} + \hat{\boldsymbol{B}} \cdot (\partial_\mu \hat{\boldsymbol{W}}^\mu) + \tfrac{1}{2}\xi \hat{\boldsymbol{B}} \cdot \hat{\boldsymbol{B}} \tag{13.145}$$

where $\hat{\boldsymbol{B}}$ is an SU(2) triplet of scalar fields. Equation (13.145) gives (cf (13.142))

$$(\hat{D}^\nu)_{ij} \hat{F}_{j\mu\nu} + \partial_\mu \hat{B}_i = 0 \tag{13.146}$$

where the covariant derivative is now the one appropriate to the SU(2) triplet $\boldsymbol{F}_{\mu\nu}$ (see (13.44) with $t = 1$, and (12.48)) and $i, j$ are the SU(2) labels. Similarly, (13.143) becomes

$$\partial_\mu \hat{\boldsymbol{W}}^\mu + \xi \hat{\boldsymbol{B}} = \boldsymbol{0}. \tag{13.147}$$

It is possible to verify that

$$(\hat{D}^\mu)_{ki} (\hat{D}^\nu)_{ij} \hat{F}_{j\mu\nu} = 0 \tag{13.148}$$

where $i, j, k$ are the SU(2) matrix indices, which implies that

$$(\hat{D}^\mu)_{ki} \partial_\mu \hat{B}_i = 0. \tag{13.149}$$

This is the crucial result: it implies that the auxiliary field $\hat{\boldsymbol{B}}$ is *not* a free field in this non-Abelian case and so neither (from (13.147)) is $\partial_\mu \hat{\boldsymbol{W}}^\mu$. In consequence, the obvious generalizations of (7.108) or (13.144) cannot be used to define the physical (transverse) states. The reason is that a condition like (13.144) must hold for all times, and only if the field is free is its time variation known (and essentially trivial).

Let us press ahead nevertheless and assume that the rules we have derived so far are the correct Feynman rules for this gauge theory. We will see that this leads to physically unacceptable consequences, namely to the *violation of unitarity*.

In fact, this is a problem which threatens all gauge theories if the gauge field is treated covariantly, i.e. as a 4-vector. As we saw in section 7.3.2, this introduces *unphysical degrees of freedom* which must somehow be eliminated from the theory—or at least prevented from affecting physical processes. In QED we do this by imposing the condition (7.108), or (13.144), but as we have seen the analogous conditions will not work in the non-Abelian case, and so unphysical states may make their presence felt, for example in the 'sum over intermediate states' which arises in the unitarity relation. This relation determines the imaginary part of an amplitude via an equation of the form (cf (11.63))

$$2\,\mathrm{Im}\langle \mathrm{f}|\mathcal{M}|\mathrm{i}\rangle = \int \sum_{\mathrm{n}} \langle \mathrm{f}|\mathcal{M}|\mathrm{n}\rangle \langle \mathrm{n}|\mathcal{M}^\dagger|\mathrm{i}\rangle \,\mathrm{d}\rho_{\mathrm{n}} \tag{13.150}$$

where $\langle \mathrm{f}|\mathcal{M}|\mathrm{i}\rangle$ is the (Feynman) amplitude for the process $\mathrm{i} \to \mathrm{f}$, and the sum is over a complete set of physical intermediate states $|\mathrm{n}\rangle$, which can enter at the given energy; $\mathrm{d}\rho_{\mathrm{n}}$ represents the phase space element for the general intermediate state $|\mathrm{n}\rangle$. Consider now the possibility of gauge quanta appearing in the states $|\mathrm{n}\rangle$. Since unitarity deals only with physical states, such quanta can have only the two degrees of freedom (polarizations) allowed for a physical massless gauge field (cf section 7.3.1). Now part of the power of the 'Feynman rules' approach to perturbation theory is that it is manifestly covariant. But there is no completely covariant way of selecting out just the two physical components of a massless polarization vector $\epsilon_\mu$, from the four originally introduced precisely for reasons of covariance. In fact, when gauge quanta appear as virtual particles in *intermediate* states in Feynman graphs, they will not be restricted to having only two polarization states (as we shall see explicitly in a moment). Hence, there is a real chance that when the imaginary part of such graphs is calculated, a contribution from the unphysical polarization states will be found, which has no counterpart at all in the physical unitarity relation, so that unitarity will not be satisfied. Since unitarity is an expression of conservation of probability, its violation is a serious disease indeed.

Consider, for example, the process $\mathrm{q}\bar{\mathrm{q}} \to \mathrm{q}\bar{\mathrm{q}}$ (where the 'quarks' are an SU(2) doublet) whose imaginary part has a contribution from a state containing two gluons (figure 13.14):

$$2\,\mathrm{Im}\langle \mathrm{q}\bar{\mathrm{q}}|\mathcal{M}|\mathrm{q}\bar{\mathrm{q}} >= \int \sum \langle \mathrm{q}\bar{\mathrm{q}}|\mathcal{M}|\mathrm{gg}\rangle \langle \mathrm{gg}|\mathcal{M}^\dagger|\mathrm{q}\bar{\mathrm{q}}\rangle \,\mathrm{d}\rho_2 \tag{13.151}$$

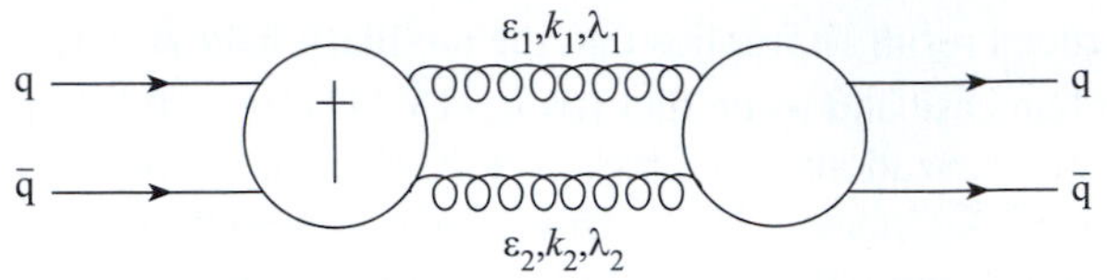


**Figure 13.14.** Two-gluon intermediate state in the unitarity relation for the amplitude for $q\bar{q} \to q\bar{q}$.

where $d\rho_2$ is the two-body phase space for the g-g state. The two-gluon amplitudes in (13.151) must have the form

$$\mathcal{M}_{\mu_1\nu_1}\epsilon_1^{\mu_1}(k_1,\lambda_1)\epsilon_2^{\nu_1}(k_2,\lambda_2) \tag{13.152}$$

where $\epsilon^\mu(k,\lambda)$ is the polarization vector for the gluon with polarization $\lambda$ and 4-momentum $k$. The sum in (13.151) is then to be performed over $\lambda_1 = 1, 2$ and $\lambda_2 = 1, 2$ which are the physical polarization states (cf section 7.3.1). Thus (13.151) becomes

$$2\,\mathrm{Im}\,\mathcal{M}_{q\bar{q}\to q\bar{q}} = \int \sum_{\lambda_1=1,2;\lambda_2=1,2} \mathcal{M}_{\mu_1\nu_1}\epsilon_1^{\mu_1}(k_1,\lambda_1)\epsilon_2^{\nu_1}(k_2,\lambda_2)$$
$$\times\,\mathcal{M}^*_{\mu_2\nu_2}\epsilon_1^{\mu_2}(k_1,\lambda_1)\epsilon_2^{\nu_2}(k_2,\lambda_2)\,d\rho_2. \tag{13.153}$$

For later convenience we are using real polarization vectors as in (7.78) and (7.79): $\epsilon(k_i, \lambda_i = +1) = (0, 1, 0, 0)$, $\epsilon(k_i, \lambda_i = -1) = (0, 0, 1, 0)$ and, of course, $k_1^2 = k_2^2 = 0$.

We now wish to find out whether or not a result of the form (13.153) will hold when the $\mathcal{M}$'s represent some suitable Feynman graphs. We first note that we want the unitarity relation (13.153) to be satisfied order by order in perturbation theory: that is to say, when the $\mathcal{M}$'s on both sides are expanded in powers of the coupling strengths (as in the usual Feynman graph expansion), the coefficients of corresponding powers on each side should be equal. Since each emission or absorption of a gluon produces one power of the SU(2) coupling $g$, the right-hand side of (13.153) involves at least the power $g^4$. Thus the lowest-order process in which (13.153) may be tested is for the fourth-order amplitude $\mathcal{M}^{(4)}_{q\bar{q}\to q\bar{q}}$. There are quite a number of contributions to $\mathcal{M}^{(4)}_{q\bar{q}\to q\bar{q}}$, some of which are shown in figure 13.15: all contain a loop. On the right-hand side of (13.153), each $\mathcal{M}$ involves two polarization vectors, and so each must represent the $O(g^2)$ contribution to $q\bar{q} \to gg$, which we call $\mathcal{M}^{(2)}_{\mu\nu}$; thus, both sides are consistently of order $g^4$. There are three contributions to $\mathcal{M}^{(2)}_{\mu\nu}$ shown in figure 13.16: when these are placed in (13.153), contributions to the imaginary part of $\mathcal{M}^{(4)}_{q\bar{q}\to q\bar{q}}$ are generated, which should agree with the imaginary part of the total $O(g^4)$ loop-graph contribution. Let us see if this works out.

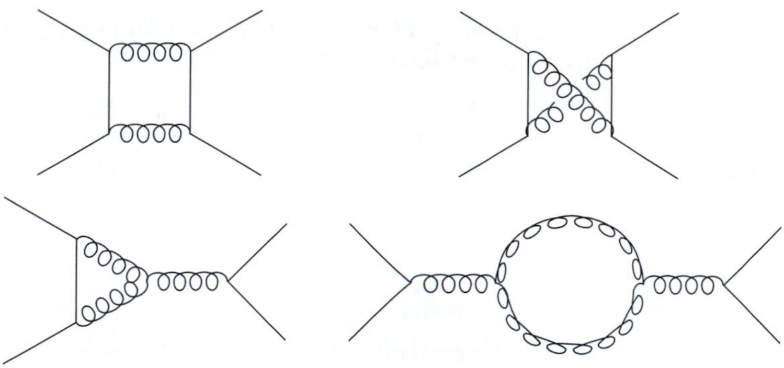

**Figure 13.15.** Some $O(g^4)$ contributions to $q\bar{q} \to q\bar{q}$.

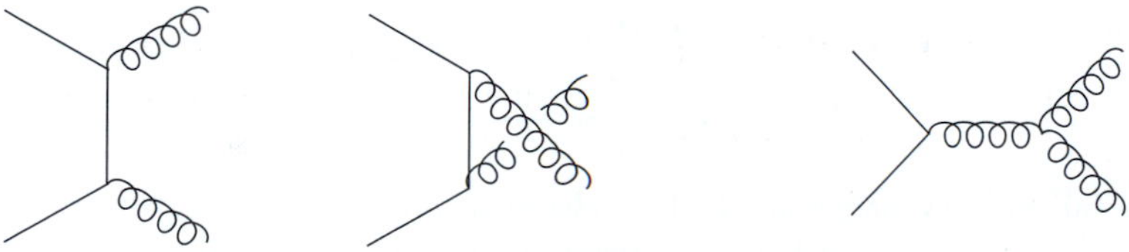

**Figure 13.16.** $O(g^2)$ contributions to $q\bar{q} \to gg$.

We choose to work in the gauge $\xi = 1$, so that the gluon propagator takes the familiar form $-ig^{\mu\nu}\delta_{ij}/k^2$. According to the rules for propagators and vertices already given, each of the loop amplitudes $\mathcal{M}^{(4)}_{q\bar{q}\to q\bar{q}}$ (e.g. those of figure 13.15) will be proportional to the product of the propagators for the quarks and the gluons, together with appropriate '$\gamma$' and '$\tau$' vertex factors, the whole being integrated over the loop momentum. The extraction of the imaginary part of a Feynman diagram is a technical matter, having to do with careful consideration of the '$i\epsilon$' in the propagators. Rules for doing this exist (Eden *et al* 1966, section 2.9) and in the present case the result is that, to compute the imaginary part of the amplitudes of figure 13.15, one replaces each gluon propagator of momentum $k$ by

$$\pi(-g^{\mu\nu})\delta(k^2)\theta(k_0)\delta_{ij}. \qquad (13.154)$$

That is, the propagator is replaced by a condition stating that, in evaluating the imaginary part of the diagram, the gluon's mass is constrained to have the physical (free-field) value of zero, instead of varying freely as the loop momentum varies, and its energy is positive. These conditions (one for each gluon) have the effect of converting the loop integral with a standard two-body phase space integral for the gg intermediate state, so that eventually

$$\operatorname{Im}\mathcal{M}^{(4)}_{q\bar{q}\to q\bar{q}} = \int \mathcal{M}^{(2)}_{\mu_1\nu_1}(-g^{\mu_1\nu_1})\mathcal{M}^{(2)}_{\mu_2\nu_2}(-g^{\mu_2\nu_2})\,d\rho_2 \qquad (13.155)$$

where $\mathcal{M}^{(2)}_{\mu_1\nu_1}$ is the sum of the three $O(g^2)$ tree graphs shown in figure 13.16, with all external legs satisfying the 'mass-shell' conditions.

So, the imaginary part of the loop contribution to $\mathcal{M}^{(4)}_{q\bar{q}\to q\bar{q}}$ does seem to have the form (13.150) as required by unitarity, with $|n\rangle$ the gg intermediate state as in (13.153). But there is one essential difference between (13.155) and (13.153): the place of the factor $-g^{\mu\nu}$ in (13.155) is taken in (13.153) by the gluon polarization sum

$$P^{\mu\nu}(k) \equiv \sum_{\lambda=1,2} \epsilon^{\mu}(k,\lambda)\epsilon^{\nu}(k,\lambda) \tag{13.156}$$

for $k = k_1, k_2$ and $\lambda = \lambda_1, \lambda_2$ respectively. Thus, we have to investigate whether this difference matters.

To proceed further, it is helpful to have an explicit expression for $P^{\mu\nu}$. We might think of calculating the necessary sum over $\lambda$ by brute force, using two $\epsilon$'s specified by the conditions (cf (7.84))

$$\epsilon^{\mu}(k,\lambda)\epsilon_{\mu}(k,\lambda') = -\delta_{\lambda\lambda'} \qquad \epsilon\cdot k = 0. \tag{13.157}$$

The trouble is that conditions (13.157) *do not fix the* $\epsilon$'s *uniquely if* $k^2 = 0$. (Note the $\delta(k^2)$ in (13.154).) Indeed, it is precisely the fact that any given $\epsilon_\mu$ satisfying (13.157) can be replaced by $\epsilon_\mu + \lambda k_\mu$ that both reduces the degrees of freedom to two (as we saw in section 7.3.1) and evinces the essential arbitrariness in the $\epsilon_\mu$ specified only by (13.157). In order to calculate (13.156), we need to put another condition on $\epsilon_\mu$, so as to fix it uniquely. A standard choice (see, e.g., Taylor 1976, pp 14–15) is to supplement (13.157) with the further condition

$$t\cdot\epsilon = 0 \tag{13.158}$$

where $t$ is some 4-vector. This certainly fixes $\epsilon_\mu$ and enables us to calculate (13.156) but, of course, now two further difficulties have appeared: namely, the physical results seem to depend on $t_\mu$; and have we not lost Lorentz covariance because the theory involves a special 4-vector $t_\mu$?

Setting these questions aside for the moment, we can calculate (13.156) using the conditions (13.157) and (13.158), finding (problem 13.11)

$$P_{\mu\nu} = -g_{\mu\nu} - [t^2 k_\mu k_\nu - k\cdot t(k_\mu t_\nu + k_\nu t_\mu)]/(k\cdot t)^2. \tag{13.159}$$

But only the *first* term on the right-hand side of (13.159) is to be seen in (13.155). A crucial quantity is clearly

$$\begin{aligned} U_{\mu\nu}(k,t) &\equiv -g_{\mu\nu} - P_{\mu\nu} \\ &= [t^2 k_\mu k_\nu - k\cdot t(k_\mu t_\nu + k_\nu t_\mu)]/(k\cdot t)^2. \end{aligned} \tag{13.160}$$

We note that whereas

$$k^\mu P_{\mu\nu} = k^\nu P_{\mu\nu} = 0 \tag{13.161}$$

(from the condition $k \cdot \epsilon = 0$), the same is *not* true of $k^\mu U_{\mu\nu}$—in fact,

$$k^\mu U_{\mu\nu} = -k_\nu \tag{13.162}$$

where we have used $k^2 = 0$. It follows that $U_{\mu\nu}$ may be regarded as including polarization states for which $\epsilon \cdot k \neq 0$. In physical terms, therefore, a gluon appearing internally in a Feynman graph has to be regarded as existing in more than just the two polarization states available to an external gluon (cf section 7.3.1). $U_{\mu\nu}$ characterizes the contribution of these unphysical polarization states.

The discrepancy between (13.155) and (13.153) is then

$$\mathrm{Im}\,\mathcal{M}^{(4)}_{\mathrm{q\bar{q}\to q\bar{q}}} = \int \mathcal{M}^{(2)}_{\mu_1\nu_1}[U^{\mu_1\nu_1}(k_1)]\mathcal{M}^{(2)}_{\mu_2\nu_2}[U^{\mu_2\nu_2}(k_2, t_2)]\,\mathrm{d}\rho_2 \tag{13.163}$$

together with similar terms involving one $P$ and one $U$. It follows that these unwanted contributions will, in fact, vanish if

$$k_1^{\mu_1}\mathcal{M}^{(2)}_{\mu_1\nu_1} = 0 \tag{13.164}$$

and similarly for $k_2$. This will also ensure that amplitudes are independent of $t_\mu$.

Condition (13.164) is apparently the same as the U(1)-gauge-invariance requirement of (8.165), already recalled in the previous section. As discussed there, it can be interpreted here also as expressing gauge invariance in the non-Abelian case, working to this given order in perturbation theory. Indeed, the diagrams in figure 13.16 are essentially 'crossed' versions of those in figure 13.10. However, there is one crucial difference here. In figure 13.10, both the X's were physical, their polarizations satisfying the condition $\epsilon \cdot k = 0$. In figure 13.16, by contrast, neither of the gluons, in the discrepant contribution (13.163), satisfies $\epsilon \cdot k = 0$—see the sentence following (13.162). Thus, the crucial point is that (13.164) must be true for each gluon, *even when the other gluon has* $\epsilon \cdot k \neq 0$. And, in fact, we shall now see that whereas the (crossed) version of (13.164) did hold for our dX $\to$ dX amplitudes of section 13.3.2, (13.164) *fails* for states with $\epsilon \cdot k \neq 0$.

The three graphs of figure 13.16 together yield

$$\begin{aligned}
&\mathcal{M}^{(2)}_{\mu_1\nu_1}\epsilon_1^{\mu}(k_1,\lambda_1)\epsilon_2^{\nu_1}(k_2,\lambda_2) \\
&\quad = g^2\bar{v}(p_2)\frac{\tau_j}{2}\not{\epsilon}_2 a_{2j}\frac{1}{\not{p}_1 - \not{k}_1 - m}\frac{\tau_i}{2}a_{1i}\not{\epsilon}_1 u(p_1) \\
&\quad\quad + g^2\bar{v}(p_2)\frac{\tau_i}{2}a_{1i}\not{\epsilon}_1\frac{1}{\not{p}_1 - \not{k}_2 - m}\frac{\tau_j}{2}a_{2j}\not{\epsilon}_2 u(p_1) \\
&\quad\quad + (-\mathrm{i})g^2\epsilon_{kij}[(p_1+p_2+k_1)^{\nu_1}g^{\mu_1\rho} + (-k_2-p_1-p_2)^{\mu_1}g^{\rho\nu_1} \\
&\quad\quad + (-k_1+k_2)^\rho g^{\mu_1\nu_1}]\epsilon_{1\mu_1}a_{1i}a_{2j}\epsilon_{2\nu_1}\frac{-1}{(p_1+p_2)^2}\bar{v}(p_2)\frac{\tau_k}{2}\gamma_\rho u(p_1)
\end{aligned} \tag{13.165}$$

where we have written the gluon polarization vectors as a product of a Lorentz 4-vector $\epsilon_\mu$ and an 'SU(2) polarization vector' $a_i$ to specify the triplet state label. Now replace $\epsilon_1$, say, by $k_1$. Using the Dirac equation for $u(p_1)$ and $\bar{v}(p_2)$, the first two terms reduce to (cf (13.128))

$$\begin{aligned} &g^2\bar{v}(p_2)\not\epsilon_2[\tau_i/2, \tau_j/2]u(p_1)a_{1i}a_{2j} \\ &\quad = \mathrm{i}g^2\bar{v}(p_2)\not\epsilon_2\epsilon_{ijk}(\tau_k/2)u(p_1)a_{1i}a_{2j} \end{aligned} \tag{13.166}$$

using the SU(2) algebra of the $\tau$'s. The third term in (13.165) gives

$$-\mathrm{i}g^2\epsilon_{ijk}\bar{v}(p_2)\not\epsilon_2(\tau_k/2)u(p_1)a_{1i}a_{2j} \tag{13.167}$$

$$+\mathrm{i}g^2\frac{\epsilon_{ijk}}{2k_1\cdot k_2}\bar{v}(p_2)\not k_1(\tau_k/2)u(p_1)k_2\cdot\epsilon_2 a_{1i}a_{2j}. \tag{13.168}$$

We see that the first part (13.167) certainly does cancel (13.166) but there remains the second piece (13.168), *which only vanishes if* $k_2\cdot\epsilon_2 = 0$. This is not sufficient to guarantee the absence of all unphysical contributions to the imaginary part of the two-gluon graphs, as the preceding discussion shows. *We conclude that loop diagrams involving two (or, in fact, more) gluons, if constructed according to the simple rules for tree diagrams, will violate unitarity.*

The correct rule for loops must be such as to satisfy unitarity. Since there seems no other way in which the offending piece in (13.168) can be removed, we must infer that the rule for loops will have to involve some extra term, or terms, over and above the simple tree-type constructions, which will cancel the contributions of unphysical polarization states. To get an intuitive idea of what such extra terms might be, we return to expression (13.160) for the sum over unphysical polarization states $U_{\mu\nu}$ and make a specific choice for $t$. We take $t_\mu = \bar{k}_\mu$, where the 4-vector $\bar{k}$ is defined by $\bar{k} = (-|\boldsymbol{k}|, \boldsymbol{k})$ and $\boldsymbol{k} = (0, 0, |\boldsymbol{k}|)$. This choice obviously satisfies (13.158). Then

$$U_{\mu\nu}(k, \bar{k}) = (k_\mu\bar{k}_\nu + k_\nu\bar{k}_\mu)/(2|\boldsymbol{k}|^2) \tag{13.169}$$

and unitarity (cf (13.163)) requires

$$\int \mathcal{M}^{(2)}_{\mu_1\nu_1}\mathcal{M}^{(2)}_{\mu_2\nu_2}\frac{(k_1^{\mu_1}\bar{k}_1^{\mu_2} + k_1^{\mu_2}\bar{k}_1^{\mu_1})}{2|\boldsymbol{k}_1|^2}\frac{(k_2^{\nu_1}\bar{k}_2^{\nu_2} + k_2^{\nu_2}\bar{k}_2^{\nu_1})}{2|\boldsymbol{k}_2|^2}\,\mathrm{d}\rho_2 \tag{13.170}$$

to vanish; but it does not. Let us work in the centre of momentum (CM) frame of the two gluons, with $k_1 = (|\boldsymbol{k}|, 0, 0, |\boldsymbol{k}|)$, $k_2 = (|\boldsymbol{k}|, 0, 0, -|\boldsymbol{k}|)$, $\bar{k}_1 = (-|\boldsymbol{k}|, 0, 0, |\boldsymbol{k}|)$, $\bar{k}_2 = (-|\boldsymbol{k}|, 0, 0, -|\boldsymbol{k}|)$, and consider for definiteness the contractions with the $\mathcal{M}^{(2)}_{\mu_1\nu_1}$ term. These are $\mathcal{M}^{(2)}_{\mu_1\nu_1}k_1^{\mu_1}k_2^{\nu_1}$, $\mathcal{M}^{(2)}_{\mu_1\nu_1}k_1^{\mu_1}\bar{k}_2^{\nu_1}$ etc. Such quantities can be calculated from expression (13.165) by setting $\epsilon_1 = k_1, \epsilon_2 = k_2$ for the first, $\epsilon_1 = k_1, \epsilon_2 = \bar{k}_2$ for the second, and so on. We have already obtained the result of putting $\epsilon_1 = k_1$. From (13.168) it is clear that a term in which $\epsilon_2$ is replaced by $k_2$ as well as $\epsilon_1$ by $k_1$ will vanish, since $k_2^2 = 0$.

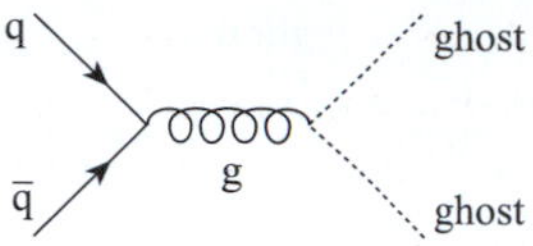


**Figure 13.17.** Tree-graph interpretation of the expression (13.172).

A typical non-vanishing term is of the form $\mathcal{M}^{(2)}_{\mu_1\nu_1} k_1^{\mu_1} \bar{k}_2^{\nu_1}/2|\boldsymbol{k}|^2$. From (13.168) this reduces to

$$-\mathrm{i}g^2 \frac{\epsilon_{ijk}}{2k_1 \cdot k_2} \bar{v}(p_2) \not{k}_1 (\tau_k/2) u(p_1) a_{1i} a_{2j} \tag{13.171}$$

using $k_2 \cdot \bar{k}_2/2|\boldsymbol{k}|^2 = -1$. We may rewrite (13.171) as

$$j_{\mu k} \frac{-g^{\mu\nu}\delta_{k\ell}}{(k_1+k_2)^2} \mathrm{i}g\epsilon_{ij\ell} a_{1i} a_{2j} k_{1\nu} \tag{13.172}$$

where

$$j_{\mu k} = g\bar{v}(p_2)\gamma_\mu(\tau_k/2)u(p_1) \tag{13.173}$$

is the SU(2) current associated with the $\mathrm{q}\bar{\mathrm{q}}$ pair.

The unwanted terms of the form (13.172) can be eliminated if we adopt the following rule (on the grounds of 'forcing the theory to make sense'). In addition to the fourth-order diagrams of the type shown in figure 13.15, constructed according to the simple 'tree' prescriptions, there must exist a previously unknown fourth-order contribution, *only present in loops*, such that it has an imaginary part which is non-zero in the same physical region as the two-gluon intermediate state and, moreover, is of just the right magnitude to cancel all the contributions to (13.170) from terms like (13.172). Now (13.172) has the appearance of a one-gluon intermediate state amplitude. The $\mathrm{q}\bar{\mathrm{q}} \rightarrow \mathrm{g}$ vertex is represented by the current (13.173), the gluon propagator appears in Feynman gauge $\xi = 1$ and the rest of the expression would have the interpretation of a coupling between the intermediate gluon and two scalar particles with SU(2) polarizations $a_{1i}, a_{2j}$. Thus, (13.172) can be interpreted as the amplitude for the tree graph shown in figure 13.17, where the dotted lines represent the scalar particles. It seems plausible, therefore, that the fourth-order graph we are looking for has the form shown in figure 13.18. The new scalar particles must be massless, so that this new amplitude has an imaginary part in the same physical region as the gg state. When the imaginary part of figure 13.18 is calculated in the usual way, it will involve contributions from the tree graph of figure 13.17 and these can be arranged to cancel the unphysical polarization pieces like (13.172).

For this cancellation to work, the scalar particle loop graph of figure 13.18 must enter with the opposite sign from the three-gluon loop graph of figure 13.15, which in retrospect was the cause of all the trouble. Such a relative minus sign between single closed-loop graphs would be expected if the scalar particles in figure 13.18 were, in fact, fermions! (Recall the rule given in section 11.3 and

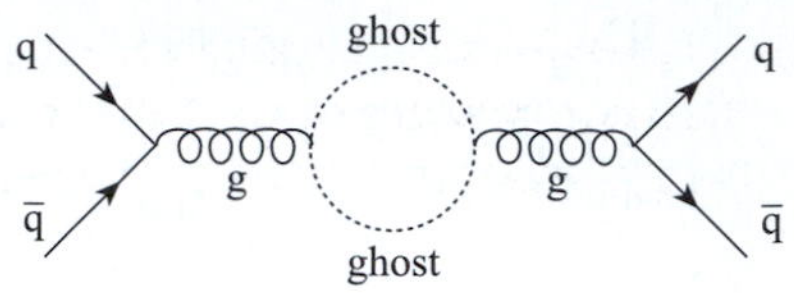


**Figure 13.18.** Ghost loop diagram contributing in fourth order to $q\bar{q} \to q\bar{q}$.

problem 11.2.) Thus, we appear to need *scalar* particles obeying *Fermi* statistics. Such particles are called 'ghosts'. We must emphasize that although we have introduced the tree graph of figure 13.17, which apparently involves ghosts as external lines, in reality the ghosts are always confined to loops, their function being to cancel unphysical contributions from intermediate gluons.

The preceding discussion has, of course, been entirely heuristic. It can be followed through so as to yield the correct prescription for eliminating unphysical contributions from a single closed gluon loop. But, as Feynman recognized (1963, 1977), unitarity alone is not a sufficient constraint to provide the prescription for more than one closed gluon loop. Clearly what is required is some additional term in the Lagrangian which will do the job in general. Such a term indeed exists and was first derived using the path integral form of quantum field theory (see chapter 16) by Faddeev and Popov (1967). The result is that the covariant gauge-fixing term (13.102) must be supplemented by the 'ghost Lagrangian'

$$\hat{\mathcal{L}}_{\mathrm{g}} = \partial_\mu \hat{\eta}_i^\dagger \hat{D}_{ij}^\mu \hat{\eta}_j \qquad (13.174)$$

where the $\eta$ field is an SU(2) triplet and spinless, but obeying *anti*-commutation relations: the covariant derivative is the one appropriate for an SU(2) triplet, namely (from (13.44) and (12.48))

$$\hat{D}_{ij}^\mu = \partial^\mu \delta_{ij} + g\epsilon_{kij} \hat{W}_k^\mu \qquad (13.175)$$

in this case. The result (13.174) is derived in standard books of quantum field theory, for example Cheng and Li (1984), Peskin and Schroeder (1995) or Ryder (1996). We should add the caution that the form of the ghost Lagrangian depends on the choice of the gauge-fixing term: there are gauges in which the ghosts are absent. The complete Feynman rules for non-Abelian gauge field theories are given in Cheng and Li (1984), for example. We give the rules for tree diagrams, for which there are no problems with ghosts, in appendix Q.

## Problems

**13.1** Verify that (13.34) reduces to (13.26) in the infinitesimal case.

**13.2** Verify equation (13.79).

**13.3** Using the expression for $D^\mu$ in (13.81), verify (13.82).

**13.4** Verify the transformation law (13.85) of $\boldsymbol{F}^{\mu\nu}$ under local SU(2) transformations.

**13.5** Verify that $\boldsymbol{F}_{\mu\nu} \cdot \boldsymbol{F}^{\mu\nu}$ is invariant under local SU(2) transformations.

**13.6** Verify that the (infinitesimal) transformation law (13.90) for the SU(3) gauge field $A_a^\mu$ is consistent with (13.89).

**13.7** By considering the commutator of two $D^\mu$'s of the form (13.87), verify (13.95).

**13.8** Verify that (13.118) reduces to (13.120) (omitting the $(2\pi)^4\delta^4$ factors).

**13.9** Verify that the replacement of $\epsilon_1$ by $k_1$ in (13.127) leads to (13.128).

**13.10** Verify that when $\epsilon_1$ is replaced by $k_1$ in (13.129), the resulting amplitude cancels the contribution (13.128), provided that $\delta = 1$.

**13.11** Show that $P^{\mu\nu}$ of (13.156), with the $\epsilon$'s specified by the conditions (13.157) and (13.158), is given by (13.159).

# PART 6

# QCD AND THE RENORMALIZATION GROUP

# 14

# QCD I: INTRODUCTION AND TREE-GRAPH PREDICTIONS

In the previous chapter we have introduced the elementary concepts and formalism associated with non-Abelian quantum gauge field theories. There are now many indications that the strong interactions between quarks are described by a theory of this type, in which the gauge group is an $SU(3)_c$, acting on a degree of freedom called 'colour' (indicated by the subscript c). This theory is called quantum chromodynamics or QCD for short. QCD will be our first application of the theory developed in chapter 13, and we shall devote the next two chapters, and much of chapter 16, to it.

In the present chapter we introduce QCD and discuss some of its simpler experimental consequences. We briefly review the evidence for the 'colour' degree of freedom in section 14.1, and then proceed to the dynamics of colour, and the QCD Lagrangian, in section 14.2. Perhaps the most remarkable thing about the dynamics of QCD is that, despite its being a theory of the *strong* interactions, there are certain kinematic regimes—roughly speaking, short distances or high energies—in which it is effectively a quite *weakly* interacting theory. This is a consequence of a fundamental property—possessed only by non-Abelian gauge theories—whereby the effective interaction strength becomes progressively smaller in such regimes. This property is called 'asymptotic freedom' and has already been mentioned in section 11.5.3 of volume 1. In appropriate cases, therefore, the lowest-order perturbation theory amplitudes (tree graphs) provide a very convincing qualitative, or even 'semi-quantitative', orientation to the data. In sections 14.3 and 14.4 we shall see how the tree-graph techniques acquired for QED in volume 1 produce more useful physics when applied to QCD.

However, most of the quantitative experimental support for QCD now comes from comparison with predictions which include higher-order QCD corrections; indeed, the asymptotic freedom property itself emerges from summing a whole class of higher-order contributions, as we shall indicate at the beginning of chapter 15. This immediately involves all the apparatus of *renormalization*. The necessary calculations quite rapidly become too technical for the intended scope of this book but in chapter 15 we shall try to provide an elementary introduction to the issues involved, and to the necessary techniques, by building on the discussion of renormalization given in chapters 10 and 11 of volume 1. The main new concept will be the *renormalization group* (and related ideas), which is an essential tool in the modern confrontation of perturbative QCD with data. Some of the simpler predictions of the renormalization group technique will be compared with experimental data in the last part of chapter 15.

In chapter 16 we work towards understanding some non-perturbative aspects of QCD. As a natural concomitant of asymptotic freedom, it is to be expected that the effective coupling strength becomes progressively larger at longer distances or lower energies, ultimately being strong enough to lead (presumably) to the confinement of quarks and gluons: this is sometimes referred to as 'infrared slavery'. In this regime, perturbation theory clearly fails. An alternative, purely numerical, approach is available however, namely the method of 'lattice' QCD, which involves replacing the spacetime continuum by a *discrete lattice* of points. At first sight, this may seem to be a topic rather disconnected from everything that has preceded it. But we shall see that, in fact, it provides some powerful new insights into several aspects of quantum field theory in general, and in particular of renormalization, by revisiting it in coordinate (rather than momentum) space. Chapter 16 therefore serves as a useful 'retrospective' on our conceptual progress thus far.

## 14.1 The colour degree of freedom

The first intimation of a new, unrevealed degree of freedom of matter came from baryon spectroscopy (Greenberg 1964; see also Han and Nambu 1965 and Tavkhelidze 1965). For a baryon made of three spin-$\frac{1}{2}$ quarks, the original non-relativistic quark model wavefunction took the form

$$\psi_{3q} = \psi_{\text{space}}\psi_{\text{spin}}\psi_{\text{flavour}}. \tag{14.1}$$

It was soon realized (e.g. Dalitz 1965) that the product of these space, spin and flavour wavefunctions for the ground-state baryons was *symmetric* under interchange of any two quarks. For example, the $\Delta^{++}$ state mentioned in section 1.2.3 is made of three u quarks (flavour symmetric) in the $J^P = \frac{3}{2}^+$ state, which has zero orbital angular momentum and is hence spatially symmetric and a symmetric $S = \frac{3}{2}$ spin wavefunction. But we saw in section 7.2 that quantum field theory requires fermions to obey the exclusion principle—i.e. the wavefunction $\psi_{3q}$ should be *anti*-symmetric with respect to quark interchange. A simple way of implementing this requirement is to suppose that the quarks carry a further degree of freedom, called colour, with respect to which the 3q wavefunction can be anti-symmetrized, as follows. We introduce a *colour wavefunction* with colour index $\alpha$:

$$\psi_\alpha \qquad (\alpha = 1, 2, 3).$$

We are here writing the three labels as '1, 2, 3' but they are often referred to by colour names such as 'red, blue, green'; it should be understood that this is merely a picturesque way of referring to the three basic states of this degree of freedom and has nothing to do with real colour! With the addition of this degree of freedom, we can certainly form a three-quark wavefunction which is anti-symmetric in colour by using the anti-symmetric symbol $\epsilon_{\alpha\beta\gamma}$, namely

$$\psi_{3\text{q,colour}} = \epsilon_{\alpha\beta\gamma}\psi_\alpha\psi_\beta\psi_\gamma \tag{14.2}$$

and this must then be multiplied into (14.1) to give the full 3q wavefunction. To date, *all* known baryon states can be described in this way, i.e. the symmetry of the 'traditional' space–spin–flavour wavefunction (14.1) is symmetric overall, while the required anti-symmetry is restored by the additional factor (14.2). As far as meson ($\bar{\text{q}}$q) states are concerned, what was previously a $\pi^+$ wavefunction $d^*u$ is now

$$\frac{1}{\sqrt{3}}(d_1^*u_1 + d_2^*u_2 + d_3^*u_3) \tag{14.3}$$

which we write in general as $(1/\sqrt{3})d_\alpha^\dagger u_\alpha$. We shall shortly see the group-theoretical significance of this 'neutral superposition', and of (14.2). Meanwhile, we note that (14.2) is actually the *only* way of making an anti-symmetric combination of the three $\psi$'s; it is therefore called a (colour) *singlet*. It is reassuring that there is only one way of doing this—otherwise, we would have obtained more baryon states than are physically observed. As we shall see in section 14.2.1, (14.3) is also a colour singlet combination.

This would seem a somewhat artificial device unless there were some physical consequences of this increase in the number of quark types—and there are. In any process which we can describe in terms of creation or annihilation of quarks, the *multiplicity* of quark types will enter into the relevant observable cross-section or decay rate. For example, at high energies the ratio

$$R = \frac{\sigma(\text{e}^+\text{e}^- \to \text{hadrons})}{\sigma(\text{e}^+\text{e}^- \to \mu^+\mu^-)} \tag{14.4}$$

will, in the quark parton model (see section 9.5), reflect the magnitudes of the individual quark couplings to the photon:

$$R = \sum_a e_a^2 \tag{14.5}$$

where $a$ runs over all quark types. For five quarks u, d, s, c, b with respective charges $\frac{2}{3}, -\frac{1}{3}, -\frac{1}{3}, \frac{2}{3}, -\frac{1}{3}$, this yields

$$R_{\text{no colour}} = \tfrac{11}{9} \tag{14.6}$$

and

$$R_{\text{colour}} = \tfrac{11}{3} \tag{14.7}$$

for the two cases, as we saw in section 9.5. The data (figure 14.1) rule out (14.6) and are in good agreement with (14.7) at energies well above the b threshold and well below the $\text{Z}^0$ resonance peak. There is an indication that the data tend to lie

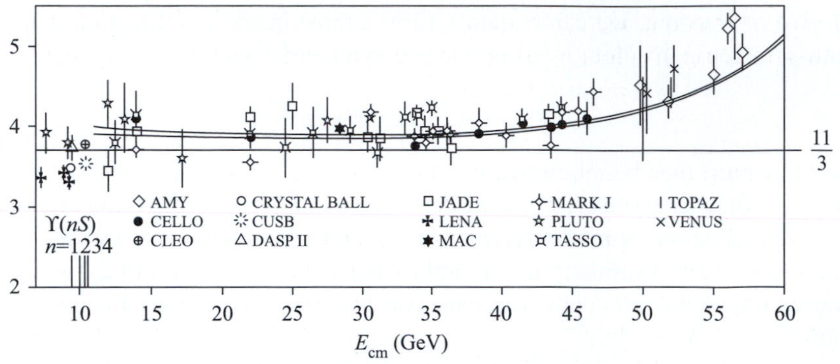


**Figure 14.1.** The ratio $R$ (see (14.4)) (Montanet *et al* 1994).

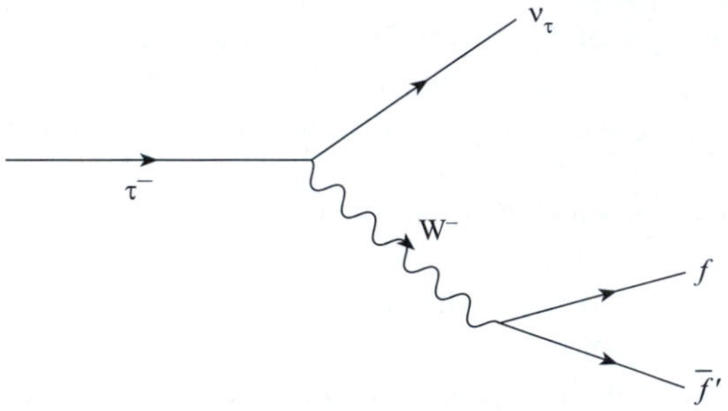


**Figure 14.2.** $\tau$ decay.

*above* the parton model prediction: this is actually predicted by QCD via higher-order corrections, as will be discussed in section 15.1.

A number of branching fractions also provide simple ways of measuring the number of colours $N_c$. For example, consider the branching fraction for $\tau^- \to e^-\bar{\nu}_e\nu_\tau$ (i.e. the ratio of the rate for $\tau^- \to e^-\bar{\nu}_e\nu_\tau$ to that for all other decays). $\tau^-$ decays proceed via the weak process shown in figure 14.2, where the final fermions can be $e^-\bar{\nu}_e$, $\mu^-\bar{\nu}_\mu$, or $\bar{u}d$, the last with multiplicity $N_c$. Thus

$$B(\tau^- \to e^-\bar{\nu}_e\nu_\tau) \approx \frac{1}{2+N_c}. \tag{14.8}$$

Experiments give $B \approx 18\%$ and hence $N_c \approx 3$. Similarly, the branching fraction $B(W^- \to e^-\bar{\nu}_e)$ is $\sim \frac{1}{3+2N_c}$ (from $f \equiv e, \mu, \tau$, u and c). Experiment gives a value of 10.7%, so again $N_c \approx 3$.

In chapter 9 we also discussed the Drell–Yan process in the quark parton model: it involves the subprocess $q\bar{q} \to l\bar{l}$ which is the inverse of the one in (14.4). We mentioned that a factor of $\frac{1}{3}$ appears in this case: this arises because we must average over the nine possible initial $q\bar{q}$ combinations (factor $\frac{1}{9}$) and

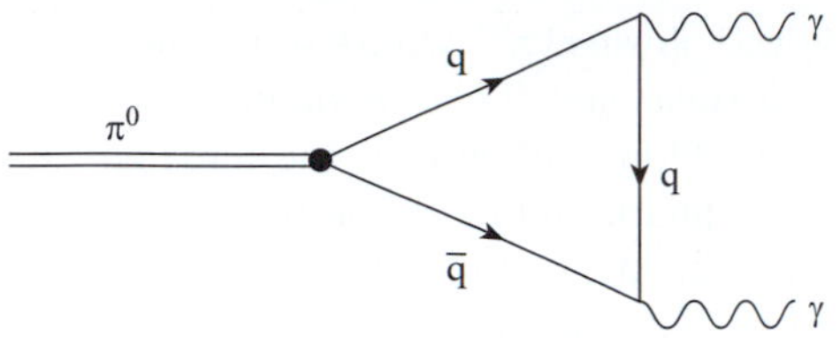


**Figure 14.3.** Triangle graph for $\pi^0$ decay.

then sum over the number of such states that lead to the colour neutral photon, which is three ($\bar{q}_1 q_1$, $\bar{q}_2 q_2$ and $\bar{q}_3 q_3$). With this factor, and using quark distribution functions consistent with deep inelastic scattering, the parton model gives a good first approximation to the data.

Finally, we mention the rate for $\pi^0 \to \gamma\gamma$. As will be discussed in section 18.4, this process is entirely calculable from the graph shown in figure 14.3 (and the one with the $\gamma$'s 'crossed'), where 'q' is u or d. The amplitude is proportional to the square of the quark charges but because the $\pi^0$ is an isovector, the contributions from the $u\bar{u}$ and $d\bar{d}$ states have opposite signs (see section 12.1.3). Thus, the rate contains a factor

$$((\tfrac{2}{3})^2 - (\tfrac{1}{3})^2)^2 = \tfrac{1}{9}. \tag{14.9}$$

However, the original calculation of this rate by Steinberger (1949) used a model in which the proton and neutron replaced the u and d in the loop, in which case the factor corresponding to (14.9) is just one (since the n has zero charge). Experimentally the rate agrees well with Steinberger's calculation, indicating that (14.9) needs to be multiplied by nine, which corresponds to $N_c = 3$ identical amplitudes of the form shown in figure 14.3.

## 14.2 The dynamics of colour

### 14.2.1 Colour as an SU(3) group

We now want to consider the possible dynamical role of colour—in other words, the way in which the forces between quarks depend on their colours. We have seen that we seem to need three different quark types for each given flavour. They must all have the same mass or else we would observe some 'fine structure' in the hadronic levels. Furthermore, and for the same reason, 'colour' must be an exact symmetry of the Hamiltonian governing the quark dynamics. What symmetry group is involved? We shall consider how some empirical facts suggest that the answer is $SU(3)_c$.

To begin with, it is certainly clear that the interquark force must depend on colour, since we do *not* observe 'colour multiplicity' of hadronic states: for example we do not see eight other coloured $\pi^+$'s ($d_1^* u_2, d_3^* u_1, \ldots$) degenerate with the one 'colourless' physical $\pi^+$ whose wavefunction was given previously. The observed hadronic states are all *colour singlets* and the force must somehow be responsible for this. More particularly, the force has to produce only those very restricted *types* of quark configuration which are observed in the hadron spectrum. Consider again the analogy drawn in section 1.2 between isospin multiplets in nuclear physics and in particle physics. There is one very striking difference in the latter case: for mesons *only* $T = 0, \frac{1}{2}$ and 1 occur, and for baryons *only* $T = 0, \frac{1}{2}$, 1 and $\frac{3}{2}$, while in nuclei there is nothing in principle to stop us finding $T = \frac{5}{2}, 3, \ldots,$ states. (In fact, such nuclear states are hard to identify experimentally, because they occur at high excitation energy for some of the isobars—cf figure 1.8($c$)—where the levels are very dense.) The same restriction holds for $SU(3)_f$ also—only **1**'s and **8**'s occur for mesons and only **1**'s, **8**'s and **10**'s for baryons. In quark terms, this of course is what is translated into the recipe: 'mesons are $\bar{q}q$, baryons are qqq'. It is as if we said, in nuclear physics, that only $A = 2$ and $A = 3$ nuclei exist! Thus, the quark forces must have a dramatic saturation property: apparently no $\bar{q}qq$, no qqqq, qqqqq, ... states exist. Furthermore, no qq or $\bar{q}\bar{q}$ states exist either—nor, for that matter, do single q's or $\bar{q}$'s. All this can be summarized by saying that the quark colour degree of freedom must be *confined*, a property we shall now assume and return to in chapter 16.

If we assume that only colour singlet states exist and that the strong interquark force depends only on colour, the fact that $\bar{q}q$ states are seen but qq and $\bar{q}\bar{q}$ are not, gives us an important clue as to what group to associate with colour. One simple possibility might be that the three colours correspond to the components of an $SU(2)_c$ triplet '$\boldsymbol{\psi}$'. The anti-symmetric, colour singlet, three-quark baryon wavefunction of (14.2) is then just the triple scalar product $\boldsymbol{\psi}_1 \cdot \boldsymbol{\psi}_2 \times \boldsymbol{\psi}_3$, which seems satisfactory. But what about the meson wavefunction? Mesons are formed of quarks and anti-quarks, and we recall from sections 12.1.3 and 12.2 that anti-quarks belong to the complex conjugate of the representation (or multiplet) to which quarks belong. Thus, if a quark colour triplet wavefunction $\psi_\alpha$ transforms under a colour transformation as

$$\psi_\alpha \to \psi'_\alpha = V^{(1)}_{\alpha\beta}\psi_\beta \tag{14.10}$$

where $\mathbf{V}^{(1)}$ is a $3 \times 3$ unitary matrix appropriate to the $T = 1$ representation of SU(2) (cf (12.48) and (12.49)), then the wavefunction for the 'anti'-triplet is $\psi^*_\alpha$, which transforms as

$$\psi^*_\alpha \to \psi^{*\prime}_\alpha = V^{(1)*}_{\alpha\beta}\psi^*_\beta. \tag{14.11}$$

Given this information, we can now construct colour singlet wavefunctions for mesons, built from $\bar{q}q$. Consider the quantity (cf (14.3)) $\sum_\alpha \psi^*_\alpha\psi_\alpha$ where $\psi^*$ represents the anti-quark and $\psi$ the quark. This may be written in matrix notation as $\psi^\dagger\psi$ where the $\psi^\dagger$ as usual denotes the transpose of the complex conjugate of the column vector $\psi$. Then, taking the transpose of (14.11), we find that $\psi^\dagger$

transforms by

$$\psi^{\dagger} \rightarrow \psi^{\dagger\prime} = \psi^{\dagger}\mathbf{V}^{(1)\dagger} \tag{14.12}$$

so that the combination $\psi^{\dagger}\psi$ transforms as

$$\psi^{\dagger}\psi \rightarrow \psi^{\dagger\prime}\psi^{\prime} = \psi^{\dagger}\mathbf{V}^{(1)\dagger}\mathbf{V}^{(1)}\psi = \psi^{\dagger}\psi \tag{14.13}$$

where the last step follows since $\mathbf{V}^{(1)}$ is unitary (compare (12.58)). Thus, the product is *invariant* under (14.10) and (14.11)—that is, it is a colour singlet, as required. This is the meaning of the superposition (14.3).

All this may seem fine, but there is a problem. The three-dimensional representation of $SU(2)_c$ which we are using here has a very special nature: the matrix $\mathbf{V}^{(1)}$ can be chosen to be *real*. This can be understood 'physically' if we make use of the great similarity between SU(2) and the group of rotations in three dimensions (which is the reason for the geometrical language of isospin 'rotations', and so on). We know very well how real three-dimensional vectors transform—namely by an orthogonal $3 \times 3$ matrix. It is the same in SU(2). It is always possible to choose the wavefunctions $\boldsymbol{\psi}$ to be real and the transformation matrix $\mathbf{V}^{(1)}$ to be real also. Since $\mathbf{V}^{(1)}$ is, in general, unitary, this means that it must be orthogonal. But now the basic difficulty appears: there is no distinction between $\boldsymbol{\psi}$ and $\boldsymbol{\psi}^*$! They both transform by the real matrix $\mathbf{V}^{(1)}$. This means that we can make SU(2) invariant (colour singlet) combinations for $\bar{q}\bar{q}$ states, and for qq states just as well as for $\bar{q}q$ states—indeed they are formally identical. But such 'diquark' (or 'anti-diquark') states are not found and hence—by assumption—should *not* be colour singlets.

The next simplest possibility seems to be that the three colours correspond to the components of an $SU(3)_c$ triplet. In this case the quark colour wavefunction $\psi_{\alpha}$ transforms as (cf (12.74))

$$\psi \rightarrow \psi^{\prime} = \mathbf{W}\psi \tag{14.14}$$

where $\mathbf{W}$ is a special unitary $3 \times 3$ matrix parametrized as

$$\mathbf{W} = \exp(\mathrm{i}\boldsymbol{\alpha} \cdot \boldsymbol{\lambda}/2), \tag{14.15}$$

and $\psi^{\dagger}$ transforms as

$$\psi^{\dagger} \rightarrow \psi^{\dagger\prime} = \psi^{\dagger}\mathbf{W}^{\dagger}. \tag{14.16}$$

The proof of the invariance of $\psi^{\dagger}\psi$ goes through as in (14.13), and it can be shown (problem 14.1(*a*)) that the anti-symmetric 3q combination (14.2) is also an $SU(3)_c$ invariant. Thus, both the proposed meson and baryon states are colour singlets. It is *not* possible to choose the $\boldsymbol{\lambda}$'s to be pure imaginary in (14.15) and thus the $3 \times 3$ $\mathbf{W}$ matrices of $SU(3)_c$ cannot be real, so that there is a distinction between $\psi$ and $\psi^*$, as we learned in section 12.2. Indeed, it can be shown (Carruthers (1966, chapter 3), Jones (1990, chapter 8), and see also problem 14.1(*b*)) that, unlike the case of $SU(2)_c$ triplets, it is not possible to form an $SU(3)_c$ colour singlet combination out of two colour triplets qq or antitriplets $\bar{q}\bar{q}$. Thus, $SU(3)_c$ seems to be a possible and economical choice for the colour group.

### 14.2.2 Global $SU(3)_c$ invariance and 'scalar gluons'

As previously stated, we are assuming, on empirical grounds, that the only physically observed hadronic states are colour singlets—and this now means singlets under $SU(3)_c$. What sort of interquark force could produce this dramatic result? Consider an SU(2) analogy again: the interaction of two nucleons belonging to the lowest (doublet) representation of SU(2). Labelling the states by an isospin $T$, the possible $T$ values for two nucleons are $T = 1$ (triplet) and $T = 0$ (singlet). There is a simple isospin-dependent force which can produce a splitting between these states—namely $V\boldsymbol{\tau}_1 \cdot \boldsymbol{\tau}_2$, where the '1' and '2' refer to the two nucleons. The total isospin is $\boldsymbol{T} = \frac{1}{2}(\boldsymbol{\tau}_1 + \boldsymbol{\tau}_2)$ and we have

$$\boldsymbol{T}^2 = \tfrac{1}{4}(\boldsymbol{\tau}_1^2 + 2\boldsymbol{\tau}_1 \cdot \boldsymbol{\tau}_2 + \boldsymbol{\tau}_2^2) = \tfrac{1}{4}(3 + 2\boldsymbol{\tau}_1 \cdot \boldsymbol{\tau}_2 + 3) \tag{14.17}$$

whence

$$\boldsymbol{\tau}_1 \cdot \boldsymbol{\tau}_2 = 2\boldsymbol{T}^2 - 3. \tag{14.18}$$

In the triplet state $\boldsymbol{T}^2 = 2$ and in the singlet state $\boldsymbol{T}^2 = 0$. Thus,

$$(\boldsymbol{\tau}_1 \cdot \boldsymbol{\tau}_2)_{T=1} = 1 \tag{14.19}$$

$$(\boldsymbol{\tau}_1 \cdot \boldsymbol{\tau}_2)_{T=0} = -3 \tag{14.20}$$

and if $V$ is positive, the $T = 0$ state is pulled down. A similar thing happens in $SU(3)_c$. Suppose this interquark force depended on the quark colours via a term proportional to

$$\boldsymbol{\lambda}_1 \cdot \boldsymbol{\lambda}_2. \tag{14.21}$$

Then, in just the same way, we can introduce the total colour operator

$$\boldsymbol{F} = \tfrac{1}{2}(\boldsymbol{\lambda}_1 + \boldsymbol{\lambda}_2) \tag{14.22}$$

so that

$$\boldsymbol{F}^2 = \tfrac{1}{4}(\boldsymbol{\lambda}_1^2 + 2\boldsymbol{\lambda}_1 \cdot \boldsymbol{\lambda}_2 + \boldsymbol{\lambda}_2^2) \tag{14.23}$$

and

$$\boldsymbol{\lambda}_1 \cdot \boldsymbol{\lambda}_2 = 2\boldsymbol{F}^2 - \boldsymbol{\lambda}^2 \tag{14.24}$$

where $\boldsymbol{\lambda}_1^2 = \boldsymbol{\lambda}_2^2 = \boldsymbol{\lambda}^2$, say. Here $\boldsymbol{\lambda}^2 \equiv \sum_{a=1}^{8} (\lambda_a)^2$ is found (see (12.75)) to have the value 16/3 (the unit matrix being understood). The operator $\boldsymbol{F}^2$ commutes with all components of $\boldsymbol{\lambda}_1$ and $\boldsymbol{\lambda}_2$ (as $\boldsymbol{T}^2$ does with $\boldsymbol{\tau}_1$ and $\boldsymbol{\tau}_2$) and represents the quadratic Casimir operator $\hat{C}_2$ of $SU(3)_c$ (see section M.5 of appendix M), in the colour space of the two quarks considered here. The eigenvalues of $\hat{C}_2$ play a very important role in $SU(3)_c$, analogous to that of the total spin/angular momentum in SU(2). They depend on the $SU(3)_c$ representation—indeed, they are one of the defining labels of SU(3) representations in general (see section M.5). Two quarks, each in the representation $\mathbf{3}_c$, combine to give a $\mathbf{6}_c$-dimensional representation and a $\mathbf{3}_c^*$ (see problem 14.1(*b*), and Jones (1990, chapter 8)). The value of $\hat{C}_2$ for the sextet $\mathbf{6}_c$ representation is 10/3 and for the $\mathbf{3}_c^*$ representation is 4/3. Thus the '$\boldsymbol{\lambda}_1 \cdot \boldsymbol{\lambda}_2$' interaction will produce a negative (attractive) eigenvalue $-8/3$ in the $\mathbf{3}_c^*$ states, but a repulsive eigenvalue +4/3 in the $\mathbf{6}_c$ states for two quarks.

The maximum attraction will clearly be for states in which $\boldsymbol{F}^2$ is zero. This is the singlet representation $\mathbf{1}_c$. Two quarks cannot combine to give a colour singlet state, but we have seen in section 12.2 that a quark and an anti-quark can: they combine to give $\mathbf{1}_c$ and $\mathbf{8}_c$. In this case (14.24) is replaced by

$$\boldsymbol{\lambda}_1 \cdot \boldsymbol{\lambda}_2 = 2\boldsymbol{F}^2 - \tfrac{1}{2}(\boldsymbol{\lambda}_1^2 + \boldsymbol{\lambda}_2^2) \tag{14.25}$$

where '1' refers to the quark and '2' to the anti-quark. Thus, the '$\boldsymbol{\lambda}_1 \cdot \boldsymbol{\lambda}_2$' interaction will give a repulsive eigenvalue +2/3 in the $\mathbf{8}_c$ channel, for which $\hat{C}_2 = 3$, and a 'maximally attractive' eigenvalue $-16/3$ in the $\mathbf{1}_c$ channel, for a quark and an anti-quark.

In the case of baryons, built from three quarks, we have seen that when two of them are coupled to the $\mathbf{3}_c^*$ state, the eigenvalue of $\boldsymbol{\lambda}_1 \cdot \boldsymbol{\lambda}_2$ is $-8/3$, one-half of the attraction in the $\bar{q}q$ colour singlet state, but still strongly attractive. The $(qq)$ pair in the $\mathbf{3}_c^*$ state can then couple to the remaining third quark to make the overall colour singlet state (14.2), with maximum binding.

Of course, such a simple potential model does not imply that the energy difference between the $\mathbf{1}_c$ states and all coloured states is *infinite*, as our strict 'colour singlets only' hypothesis would demand, and which would be one (rather crude) way of interpreting confinement. Nevertheless, we can ask: what single particle exchange process between quark (or anti-quark) colour triplets produces a $\boldsymbol{\lambda}_1 \cdot \boldsymbol{\lambda}_2$ type of term? The answer is the exchange of an SU(3)$_c$ octet ($\mathbf{8}_c$) of particles, which (anticipating somewhat) we shall call gluons. Since colour is an exact symmetry, the quark wave equation describing the colour interactions must be SU(3)$_c$ covariant. A simple such equation is

$$(\mathrm{i}\not{\partial} - m)\psi = g_s \frac{\lambda_a}{2} A_a \psi \tag{14.26}$$

where $g_s$ is a 'strong charge' and $A_a$ ($a = 1, 2, \ldots, 8$) is an octet of *scalar* 'gluon potentials'. Equation (14.26) may be compared with (13.92): in the latter, $\not{A}_a$ appears on the right-hand side, because the gauge field quanta are vectors rather than scalars. In (14.26), we are dealing at this stage only with a *global* SU(3) symmetry, not a local SU(3) gauge symmetry, and so the potentials may be taken to be scalars, for simplicity. As in (13.94), the vertex corresponding to (14.26) is

$$-\mathrm{i}g_s\lambda_a/2. \tag{14.27}$$

(14.27) differs from (13.94) simply in the absence of the $\gamma^\mu$ factor, due to the assumed scalar, rather than vector, nature of the 'gluon' here. When we put two such vertices together and join them with a gluon propagator (figure 14.4), the SU(3)$_c$ structure of the amplitude will be

$$\frac{\lambda_{1a}}{2}\delta_{ab}\frac{\lambda_{2b}}{2} = \frac{\boldsymbol{\lambda}_1}{2} \cdot \frac{\boldsymbol{\lambda}_2}{2} \tag{14.28}$$

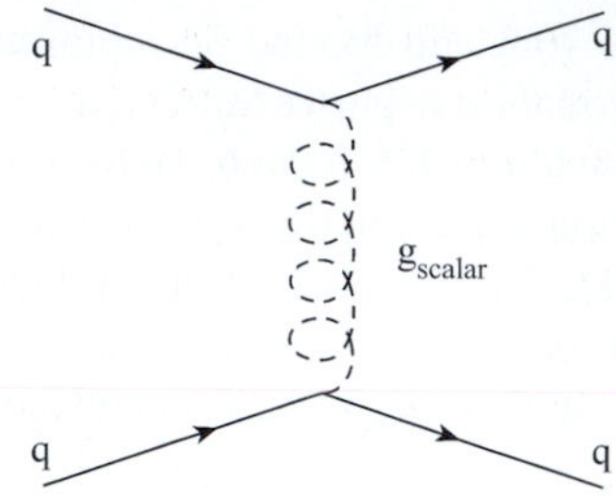


**Figure 14.4.** Scalar gluon exchange between two quarks.

the $\delta_{ab}$ arising from the fact that the freely propagating gluon does not change its colour. This interaction has exactly the required '$\boldsymbol{\lambda}_1 \cdot \boldsymbol{\lambda}_2$' character in the colour space.

### 14.2.3 Local SU(3)$_c$ invariance: the QCD Lagrangian

It is, of course, tempting to suppose that the 'scalar gluons' introduced in (14.26) are, in fact, vector particles, like the photons of QED. Equation (14.26) then becomes

$$(\mathrm{i}\partial\!\!\!/ - m)\psi = g_s \frac{\lambda_a}{2} A\!\!\!/_a \psi \tag{14.29}$$

as in (13.92 ), and the vertex (14.27) becomes

$$-\mathrm{i}g_s \frac{\lambda_a}{2} \gamma^\mu \tag{14.30}$$

as in (13.94). One motivation for this is the desire to make the colour dynamics as much as possible like the highly successful theory of QED, and to derive the dynamics from a gauge principle. As we have seen in the last chapter, this involves the simple but deep step of supposing that the quark wave equation is covariant under *local* SU(3)$_c$ transformations of the form

$$\psi \to \psi' = \exp(\mathrm{i}g_s \boldsymbol{\alpha}(x) \cdot \boldsymbol{\lambda}/2)\psi. \tag{14.31}$$

This is implemented by the replacement

$$\partial_\mu \to \partial_\mu + \mathrm{i}g_s \frac{\lambda_a}{2} A_{a\mu}(x) \tag{14.32}$$

in the Dirac equation for the quarks, which leads immediately to (14.29) and the vertex (14.30).

Of course, the assumption of local SU(3)$_c$ covariance leads to a great deal more: for example, it implies that the gluons are *massless vector* (spin-1) particles and that they interact with themselves via *three-gluon* and *four-gluon* vertices, which are the SU(3)$_c$ analogues of the SU(2) vertices discussed in section 13.5.2.

The most compact way of summarizing all this structure is via the Lagrangian, most of which we have already introduced in chapter 13. Gathering together (13.105) and (13.174) (adapted to SU(3)$_c$), we write it out here for convenience:

$$\mathcal{L}_{\text{QCD}} = \sum_{\text{flavours f}} \bar{\hat{q}}_{\text{f},\alpha}(\mathrm{i}\hat{\not{D}} - m_{\text{f}})_{\alpha\beta}\hat{q}_{\text{f},\beta} - \tfrac{1}{4}\hat{F}_{a\mu\nu}\hat{F}_a^{\mu\nu}$$
$$- \frac{1}{2\xi}(\partial_\nu \hat{A}_a^\mu)(\partial_\nu \hat{A}_a^\nu) + \partial_\mu \hat{\eta}_a^\dagger \hat{D}_{ab}^\mu \hat{\eta}_b. \quad (14.33)$$

In (14.33), repeated indices are, as usual, summed over: $\alpha$ and $\beta$ are SU(3)$_c$-triplet indices running from 1 to 3, and $a, b$ are SU(3)$_c$-octet indices running from 1 to 8. The covariant derivatives are defined by

$$(\hat{D}_\mu)_{\alpha\beta} = \partial_\mu \delta_{\alpha\beta} + \mathrm{i}g_s \tfrac{1}{2}(\lambda_a)_{\alpha\beta}\hat{A}_{a\mu} \quad (14.34)$$

when acting on the quark SU(3)$_c$ triplet, as in (13.87), and by

$$(\hat{D}_\mu)_{ab} = \partial_\mu \delta_{ab} + g_s f_{cab}\hat{A}_{c\mu} \quad (14.35)$$

when acting on the octet of ghost fields. For the second of these, note that the matrices representing the SU(3) generators in the octet representation are as given in (12.84), and these take the place of the '$\lambda/2$' in (14.34) (compare (13.175) in the SU(2) case). We remind the reader that the last two terms in (14.33) are the gauge-fixing and ghost terms, respectively, appropriate to a gauge field propagator of the form (13.103) (with $\delta_{ij}$ replaced by $\delta_{ab}$ here). The Feynman rules for tree graphs following from (14.33) are given in appendix Q.

In arriving at (14.33) we have relied essentially on the 'gauge principle' (invariance under a local symmetry) and the requirement of renormalizability (to forbid the presence of terms with mass dimension higher than four). The renormalizability of such a theory was proved by 't Hooft (1971a, b). However, there is, in fact, one more gauge invariant term of mass dimension four which can be written down, namely

$$\hat{\mathcal{L}}_\theta = \frac{\theta g_s^2}{64\pi^2}\epsilon_{\mu\nu\rho\sigma}\hat{F}_a^{\mu\nu}\hat{F}_a^{\rho\sigma}. \quad (14.36)$$

The factors in front of the '$\epsilon FF$' are chosen conventionally. In the U(1) case of QED, where the summation on $a$ is absent, such a term is proportional to $\boldsymbol{E}\cdot\boldsymbol{B}$ (problem 14.2). This violates both parity and time-reversal symmetry: under $P$, $\boldsymbol{E} \to -\boldsymbol{E}$ and $\boldsymbol{B} \to \boldsymbol{B}$, while under $T$, $\boldsymbol{E} \to \boldsymbol{E}$ and $\boldsymbol{B} \to -\boldsymbol{B}$. The same is true of (14.36). Experimentally, however, we know that strong interactions conserve both $P$ and $T$ to a high degree of accuracy. The coefficient $\theta$ is therefore required to be small.

The reader may wonder whether the '$\theta$ term' (14.36) should give rise to a new Feynman rule. The answer to this is that (14.36) can actually be written as a total divergence:

$$\epsilon_{\mu\nu\rho\sigma}\hat{F}_a^{\mu\nu}\hat{F}_a^{\rho\sigma} = \partial_\mu \hat{K}^\mu \quad (14.37)$$

where

$$\hat{K}^{\mu} = 2\epsilon^{\mu\nu\rho\sigma}\hat{A}_{a\nu}(\partial_{\rho}\hat{A}_{a\sigma} - \tfrac{2}{3}g_{s}f_{abc}\hat{A}_{b\rho}\hat{A}_{c\sigma}). \tag{14.38}$$

Any total divergence in the Lagrangian can be integrated to give only a 'surface' term in the quantum action, which can usually be discarded, making conventional assumptions about the fields going to zero at infinity. There are, however, field configurations ('instantons') such that the contribution of the $\theta$ term does not vanish. These configurations are not reachable in perturbation theory, so no perturbative Feynman rules are associated with (14.36). We refer the reader to Cheng and Li (1984, chapter 16), or to Weinberg (1996, section 23.6), for further discussion of the $\theta$ term.

Elegant and powerful as the gauge principle may be, however, any theory must ultimately stand or fall by its success, or otherwise, in explaining the experimental facts. And this brings us to a central difficulty. We have one well-understood and reliable calculational procedure, namely renormalized perturbation theory. However, we can only use perturbation theory for relatively weak interactions, whereas QCD is supposed to be a strong interaction theory. How can our perturbative QED techniques possibly be used for QCD? Despite the considerable formal similarities between the two theories, which we have emphasized, they differ in at least one crucial respect: the fundamental quanta of QED (leptons and photons) are observed as free particles but those of QCD (quarks and gluons) are not. It seems that in order to compare QCD with data we shall inevitably have to reckon with the complex non-perturbative strong interaction processes ('confinement') which bind quarks and gluons into hadrons—and the underlying simplicity of the QCD structure will be lost.

We must now recall from chapter 7 the very considerable empirical success of the parton model, in which the interactions between the partons (now interpreted as quarks and gluons) were totally ignored! Somehow it does seem to be the case that in deep inelastic scattering—or, more generally, 'hard', high-energy, wide-angle collisions—the hadron constituents are very nearly free and the effective interaction is *relatively weak*. However, we are faced with an almost paradoxical situation, because we also know that the forces are indeed so *strong* that no-one has yet succeeded in separating completely either a quark or a gluon from a hadron, so that they emerge as free particles. The resolution of this unprecedented mystery lies in the fundamental feature of non-Abelian gauge theories called 'asymptotic freedom', whereby the effective coupling strength becomes progressively smaller at short distances or high energies. This property is the most compelling theoretical motivation for choosing a non-Abelian gauge theory for the strong interactions, and it enables a quantitative perturbative approach to be followed (in appropriate circumstances) even in strong interaction physics.

A proper understanding of how this works necessitates a considerable detour, however, into the physics of renormalization. In particular, we need to understand the important group of ideas going under the general heading of the 'renormalization group' and this will be the topic of chapter 15. For the moment we proceed with a discussion of the perturbative applications of QCD at the tree-level, justification being provided by the *assumed* property of asymptotic freedom.

## 14.3 Hard scattering processes and QCD tree graphs

### 14.3.1 Two-jet events in $\bar{\mathrm{p}}\mathrm{p}$ collisions

In chapter 9 of volume 1 we introduced the parton model and discussed how it successfully interpreted deep inelastic and large-$Q^2$ data in terms of the free point-like hadronic 'partons'. This was a model rather than a theory: the theme of most of the rest of this chapter, and the following one, will be the way in which the theory of QCD both justifies the parton model and predicts observable corrections to it. In other words, the partons are now to be identified precisely with the QCD quanta (quarks, anti-quarks and gluons). We shall usually continue to use the language of partons, however, rather than—say—that of hadronic 'constituents', for the following reason. It is only at relatively low energies and/or momentum transfers that the (essentially non-relativistic) concept of a fixed number of constituents in a bound state is meaningful. At relativistic energies and short distances, pair creation and other fluctuation phenomena are so important that it no longer makes sense to think so literally of 'what the bound state is made of'—as we shall see, when we look more closely at it (with larger $Q^2$), more and more 'bits' are revealed. In this situation we prefer 'partons' to 'constituents', since the latter term seems to carry with it more of the traditional connotation of a fixed number.

In section 9.5 we briefly introduced the idea of *jets* in $\mathrm{e}^+\mathrm{e}^-$ physics: well-collimated sprays of hadrons, apparently created as a quark–anti-quark pair, separate from each other at high speed. The dynamics at the parton level, $\mathrm{e}^+\mathrm{e}^- \rightarrow \bar{\mathrm{q}}\mathrm{q}$, was governed by QED. We also saw, in section 9.4, how in hadron–hadron collisions the hadrons acted as beams of partons—quarks, anti-quarks and gluons—which could produce $\bar{\mathrm{l}}\mathrm{l}$ pairs by the inverse process $\bar{\mathrm{q}}\mathrm{q} \rightarrow \bar{\mathrm{l}}\mathrm{l}$: the force acting here is electromagnetic, and well described in the lowest order of perturbative QED. However, it is clear that collisions between the hadronic partons should by no means be limited to QED-induced processes. On the contrary, we expect to see strong interactions between the partons, determined by QCD. In general, therefore, the data will be complicated and hard to interpret, due to all these non-perturbative strong interactions. But the asymptotic freedom property (assumed for the moment) implies that at short distances we can use perturbation theory even for 'strong interactions'. Thus we might hope that the identification and analysis of short-distance parton–parton collisions will lead to direct tests of the tree-graph structure of QCD.

How are short-distance collisions to be identified experimentally? The answer is: in just the same way as Rutherford distinguished the presence of a

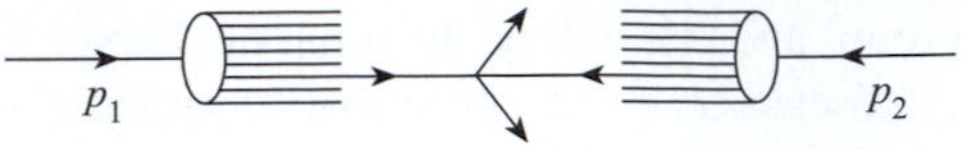


**Figure 14.5.** Parton–parton collision.

small heavy scattering centre (the nucleus) in the atom—by looking at secondary particles emerging at large angles with respect to the beam direction. For each secondary particle we can define a transverse momentum $p_T = p \sin\theta$ where $p$ is the particle momentum and $\theta$ is the emission angle with respect to the beam axis. If hadronic matter were smooth and uniform (cf the Thomson atom), the distribution of events in $p_T$ would be expected to fall off very rapidly at large $p_T$ values—perhaps exponentially. This is just what is observed in the vast majority of events: the average value of $p_T$ measured for charged particles is very low ($\langle p_T \rangle \sim 0.4$ GeV) but in a small fraction of collisions the emission of high-$p_T$ secondaries is observed. They were first seen (Büsser *et al* 1972, 1973, Alper *et al* 1973, Banner *et al* 1983) at the CERN ISR (CMS energies 30–62 GeV) and were interpreted in parton terms as follows. The physical process is viewed as a two-step one. In the first stage (figure 14.5) a parton from one hadron undergoes a short-distance collision with a parton from the other, leading in lowest-order perturbation theory to two wide-angle partons emerging at high speed from the collision volume. This is a 'hard-scattering' process.

As the two partons separate, the effective interaction strength increases and the second stage is entered, that in which the coloured partons turn themselves—under the action of the strong colour-confining force—into colour singlet hadrons. As yet we do not have a quantitative dynamical understanding of this second (non-perturbative) stage, which is called parton *fragmentation*. Nevertheless, we can argue that for the forces to be strong enough to produce the observed hadrons, the dominant processes in the fragmentation stage must involve small momentum transfer. Thus we have a picture in which two fairly well-collimated *jets* of hadrons occur, each having a total 4-momentum approximately equal to that of the parent parton (figure 14.6). Jets will be the observed hadronic manifestation of the underlying QCD processes—just as they are in the analogous QED process of $e^+e^-$ annihilation into hadrons, discussed in section 9.5.

We now face the experimental problem of picking out from the enormous multiplicity of total events just these hard scattering ones, in order to analyze them further. Early experiments used a trigger based on the detection of a single high-$p_T$ particle. But it turns out that such triggering really reduces the probability of observing jets, since the probability that a single hadron in a jet will actually carry most of the jet's total transverse momentum is quite small (Jacob and Landshoff 1978, Collins and Martin 1984, chapter 5). It is much better to surround the collision volume with an array of calorimeters which measure the total energy deposited. *Wide-angle jets* can then be identified by the occurrence of a large amount of total transverse energy deposited in a number of adjacent calorimeter

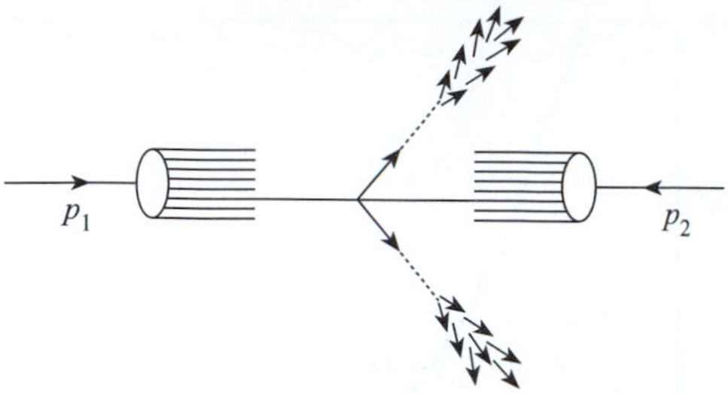


**Figure 14.6.** Parton fragmentation into jets.

cells: this is then a 'jet trigger'. The importance of calorimetric triggers was first emphasized by Bjorken (1973), following earlier work by Berman *et al* (1971). The application of this method to the detection and analysis of wide-angle jets was first reported by the UA2 collaboration at the CERN $\bar{p}p$ collider (Banner *et al* 1982). An impressive body of quite remarkably clean jet data was subsequently accumulated by both the UA1 and UA2 collaborations (at $\sqrt{s} = 546$ and 630 GeV) and by the CDF and D0 collaborations at the FNAL Tevatron collider ($\sqrt{s} = 1.8$ TeV).

For each event the total transverse energy $\sum E_{\mathrm{T}}$ is measured where

$$\sum E_{\mathrm{T}} = \sum_i E_i \sin\theta_i. \tag{14.39}$$

$E_i$ is the energy deposited in the $i$th calorimeter cell and $\theta_i$ is the polar angle of the cell centre: the sum extends over all cells. Figure 14.7 shows the $\sum E_{\mathrm{T}}$ distribution observed by UA2: it follows the 'soft' exponential form for $\sum E_{\mathrm{T}} \leq 60$ GeV but thereafter departs from it, showing clear evidence of the wide-angle collisions characteristic of hard processes.

As we shall see shortly, the majority of 'hard' events are of two-jet type, with the jets sharing the $\sum E_{\mathrm{T}}$ approximately equally. Thus, a 'local' trigger set to select events with localized transverse energy $\geq 30$ GeV and/or a 'global' trigger set at $\geq 60$ GeV can be used. At $\sqrt{s} \geq 500$–600 GeV, there is plenty of energy available to produce such events.

The total $\sqrt{s}$ value is important for another reason. Consider the kinematics of the two-parton collision (figure 14.5) in the $\bar{p}p$ CMS. As in the Drell–Yan process of section 9.4, the right-moving parton has 4-momentum

$$x_1 p_1 = x_1(P, 0, 0, P) \tag{14.40}$$

and the left-moving one

$$x_2 p_2 = x_2(P, 0, 0, -P) \tag{14.41}$$

where $P = \sqrt{s}/2$ and we are neglecting parton transverse momenta, which are approximately limited by the observed $\langle p_{\mathrm{T}} \rangle$ value ($\sim 0.4$ GeV, and thus negligible

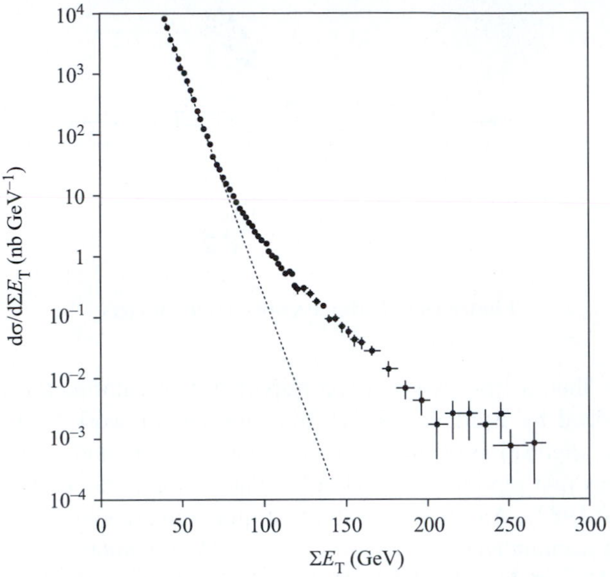


**Figure 14.7.** Distribution of the total transverse energy $\sum E_T$ observed in the UA2 central calorimeter (DiLella 1985).

on this energy scale). Consider the simple case of 90° scattering, which requires (for massless partons) $x_1 = x_2$, equal to $x$ say. The total outgoing transverse energy is then $2xP = x\sqrt{s}$. If this is to be greater than 50 GeV, then partons with $x \geq 0.1$ will contribute to the process. The parton distribution functions are large at these relatively small $x$ values, due to sea quarks (section 9.3) and gluons (figure 14.13), and thus we expect to obtain a reasonable cross-section.

What are the characteristics of jet events? When $\sum E_T$ is large enough ($\geq 150$ GeV), it is found that essentially all of the transverse energy is indeed split roughly equally between two approximately back-to-back jets. A typical such event is shown in figure 14.8. Returning to the kinematics of (14.40) and (14.41), $x_1$ will not, in general, be equal to $x_2$, so that—as is apparent in figure 14.8—the jets will not be collinear. However, to the extent that the transverse parton momenta can be neglected, the jets will be coplanar with the beam direction, i.e. their relative azimuthal angle will be 180°. Figure 14.9 shows a number of examples in which the distribution of the transverse energy over the calorimeter cells is analysed as a function of the jet opening angle $\theta$ and the azimuthal angle $\phi$. It is strikingly evident that we are seeing precisely a kind of 'Rutherford' process or—to vary the analogy—we might say that hadronic jets are acting as the modern counterpart of Faraday's iron filings, in rendering visible the underlying field dynamics!

We may now consider more detailed features of these two-jet events—in

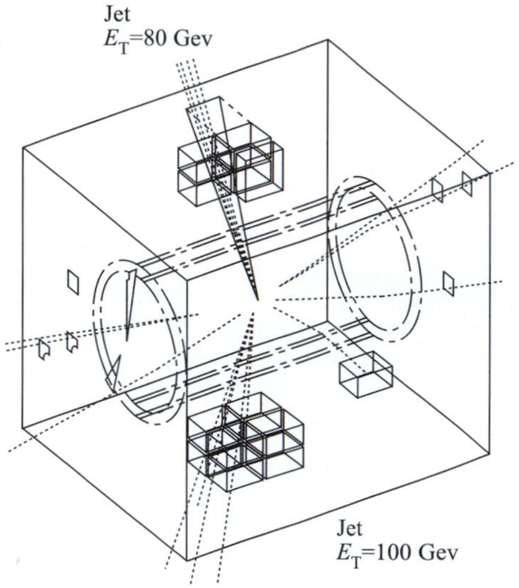


**Figure 14.8.** Two-jet event. Two tightly collimated groups of reconstructed charged tracks can be seen in the cylindrical central detector of UA1, associated with two large clusters of calorimeter energy depositions (Geer 1986).

particular, the expectations based on QCD tree graphs. The initial hadrons provide wide-band beams[1] of quarks, anti-quarks and gluons; thus, we shall have many parton subprocesses, such as $\mathrm{qq} \to \mathrm{qq}$, $\mathrm{q\bar{q}} \to \mathrm{q\bar{q}}$, $\mathrm{q\bar{q}} \to \mathrm{gg}$, $\mathrm{gg} \to \mathrm{gg}$, etc. The most important, numerically, for a $\mathrm{p\bar{p}}$ collider are $\mathrm{q\bar{q}} \to \mathrm{q\bar{q}}$, $\mathrm{gq} \to \mathrm{gq}$ and $\mathrm{gg} \to \mathrm{gg}$. The cross-section will be given, in the parton model, by a formula of the Drell–Yan type, except that the electromagnetic annihilation cross-section

$$\sigma(\mathrm{q\bar{q}} \to \mu^+\mu^-) = 4\pi\alpha^2/3q^2 \tag{14.42}$$

is replaced by the various QCD subprocess cross-sections, each one being weighted by the appropriate distribution functions. At first sight this seems to be a very complicated story, with so many contributing parton processes. But a significant simplification comes from the fact that in the CMS of the parton collision, all processes involving one gluon exchange will lead to essentially the same dominant angular distribution of Rutherford-type, $\sim \sin^{-4}\theta/2$, where $\theta$ is the parton CMS scattering angle (recall section 2.6!). This is illustrated in table 14.1 (adapted from Combridge *et al* (1977)), which lists the different relevant spin-averaged, squared, one-gluon-exchange matrix elements $|\mathcal{M}|^2$,

[1] In the sense that the partons in hadrons have momentum or energy distributions, which are characteristic of their localization to hadronic dimensions.

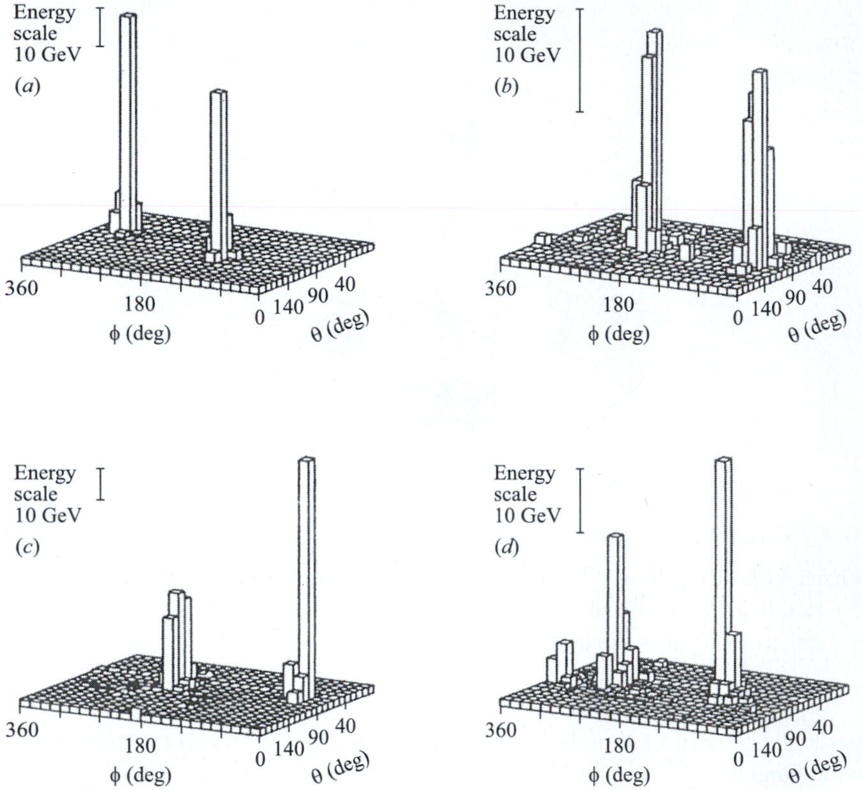


**Figure 14.9.** Four transverse energy distributions for events with $\sum E_{\mathrm{T}} > 100$ GeV, in the $\theta, \phi$ plane (UA2, DiLella 1985). Each bin represents a cell of the UA2 calorimeter. Note that the sum of the $\phi$'s equals 180° (mod 360°).

where the parton differential cross-section is given by (cf (6.129))

$$\frac{\mathrm{d}\sigma}{\mathrm{d}\cos\theta} = \frac{\pi\alpha_{\mathrm{s}}^2}{2\hat{s}}|\mathcal{M}|^2. \tag{14.43}$$

Here $\hat{s}$, $\hat{t}$ and $\hat{u}$ are the subprocess invariants, so that

$$\hat{s} = (x_1 p_1 + x_2 p_2)^2 = x_1 x_2 s \qquad \text{(cf (9.85))}. \tag{14.44}$$

Continuing to neglect the parton transverse momenta, the initial parton configuration shown in figure 14.5 can be brought to the parton CMS by a Lorentz transformation along the beam direction, the outgoing partons then emerging back-to-back at an angle $\theta$ to the beam axis, so $\hat{t} \propto (1 - \cos\theta) \propto \sin^2\theta/2$. Only the terms in $(\hat{t})^{-2} \sim \sin^{-4}\theta/2$ are given in table 14.1. We note that the $\hat{s}, \hat{t}, \hat{u}$ dependence of these terms is the same for the three types of process (and is, in

**Table 14.1.** Spin-averaged squared matrix elements for one-gluon exchange ($\hat{t}$-channel) processes.

| Subprocess | $\lvert\mathcal{M}\rvert^2$ |
|---|---|
| $qq \to qq$ <br> $q\bar{q} \to q\bar{q}$ | $\frac{4}{9}\left(\frac{\hat{s}^2+\hat{u}^2}{\hat{t}^2}\right)$ |
| $qg \to qg$ | $\frac{\hat{s}^2+\hat{u}^2}{\hat{t}^2} + \cdots$ |
| $gg \to gg$ | $\frac{9}{4}\left(\frac{\hat{s}^2+\hat{u}^2}{\hat{t}^2}\right) + \cdots$ |

fact, the same as that found for the $1\gamma$ exchange process $e^+e^- \to \mu^+\mu^-$—see problem 8.18, converting $d\sigma/dt$ into $d\sigma/d\cos\theta$). Figure 14.10 shows the two-jet angular distribution measured by UA1 (Arnison *et al* 1985). The broken curve is the exact angular distribution predicted by all the QCD tree graphs—it actually follows the $\sin^{-4}\theta/2$ shape quite closely.

It is interesting to compare this angular distribution with the one predicted on the assumption that the exchanged gluon is a spinless particle, so that the vertices have the form '$\bar{u}u$' rather than '$\bar{u}\gamma_\mu u$'. Problem 14.3 shows that, in this case, the $1/\hat{t}^2$ factor in the cross-section is completely cancelled, thus ruling out such a model.

This analysis surely constitutes compelling evidence for elementary hard scattering events proceeding via the exchange of a massless vector quantum. It is possible to go much further, in fact. Anticipating our later discussion, the small discrepancy between 'tree-graph' theory (which is labelled 'leading-order QCD scaling curve' in figure 14.10) and experiment can be accounted for by including corrections which are of higher order in $\alpha_s$. To study such deviations from the 'Rutherford' behaviour it is convenient (Combridge and Maxwell 1984) to plot the data in terms of the variable

$$\chi = \frac{1+\cos\theta}{1-\cos\theta} \tag{14.45}$$

which is such that

$$\frac{d\sigma}{d\chi} = \frac{d\sigma}{d\cos\theta}\frac{d\cos\theta}{d\chi} = -\frac{1}{2}(1-\cos\theta)^2\frac{d\sigma}{d\cos\theta}. \tag{14.46}$$

The singular Rutherford term in $d\sigma/d\cos\theta$ is therefore removed, and as $\theta \to 0$, $d\sigma/d\chi \to$ constant. Figure 14.11 shows a jet–jet angular distribution from the D0 collaboration (H Weerts 1994), plotted this way. The broken curve is the 'naive'

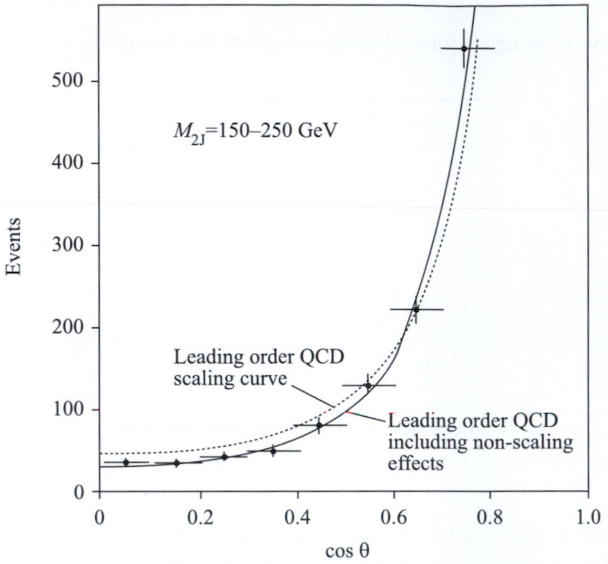


**Figure 14.10.** Two-jet angular distribution plotted against $\cos\theta$ (Arnison *et al* 1985).

parton model prediction and is clearly not in agreement with the data. The full curve includes QCD corrections beyond the tree level (Ellis *et al* 1992), involving the 'running' of the coupling constant $\alpha_s$ and 'scaling violation' in the parton distributions, both of which effects will be discussed later. The corrections lead to good agreement with experiment.

The fact that the angular distributions of all the subprocesses are so similar allows further information to be extracted from these two-jet data. In general, the parton model cross-section will have the form (cf (9.92))

$$\frac{\mathrm{d}^3\sigma}{\mathrm{d}x_1\,\mathrm{d}x_2\,\mathrm{d}\cos\theta} = \sum_{\mathrm{a,b}} \frac{F_{\mathrm{a}}(x_1)}{x_1}\frac{F_{\mathrm{b}}(x_2)}{x_2}\sum_{\mathrm{c,d}}\frac{\mathrm{d}\sigma_{\mathrm{ab}\to\mathrm{cd}}}{\mathrm{d}\cos\theta} \tag{14.47}$$

where $F_{\mathrm{a}}(x_1)/x_1$ is the distribution function for partons of type 'a' (q, $\bar{\mathrm{q}}$ or g), and similarly for $F_{\mathrm{b}}(x_2)/x_2$. Using the near identity of all $\mathrm{d}\sigma/\mathrm{d}\cos\theta$'s, and noting the numerical factors in table 14.1, the sums over parton types reduce to

$$\tfrac{9}{4}\{g(x_1)+\tfrac{4}{9}[q(x_1)+\bar{q}(x_1)]\}\{g(x_2)+\tfrac{4}{9}[q(x_2)+\bar{q}(x_2)]\} \tag{14.48}$$

where $g(x)$, $q(x)$ and $\bar{q}(x)$ are the gluon, quark and anti-quark distribution functions. Thus, effectively, the weighted distribution function[2]

$$\frac{F(x)}{x} = g(x)+\tfrac{4}{9}[q(x)+\bar{q}(x)] \tag{14.49}$$

[2] The $\frac{4}{9}$ reflects the relative strengths of the quark–gluon and gluon–gluon couplings in QCD.

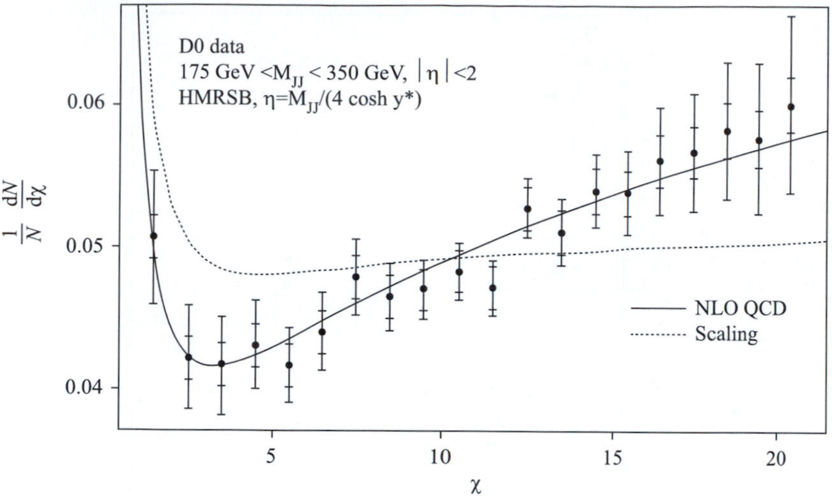


**Figure 14.11.** Distribution of $\chi$ from the D0 collaboration (Weerts 1994) compared with QCD predictions (Ellis *et al* 1992; figure from Ellis *et al* 1996).

is measured (Combridge and Maxwell 1984); in fact, with the weights as in (14.48),

$$\frac{d^3\sigma}{dx_1\,dx_2\,d\cos\theta} = \frac{F(x_1)}{x_1}\frac{F(x_2)}{x_2}\frac{d\sigma_{gg\to gg}}{d\cos\theta}. \tag{14.50}$$

$x_1$ and $x_2$ are kinematically determined from the measured jet variables: from (14.44),

$$x_1 x_2 = \hat{s}/s \tag{14.51}$$

where $\hat{s}$ is the invariant [mass]$^2$ of the two-jet system and

$$x_1 - x_2 = 2P_L/\sqrt{s} \qquad \text{(cf (9.83))} \tag{14.52}$$

with $P_L$ the total two-jet longitudinal momentum. Figure 14.12 shows $F(x)/x$ obtained in the UA1 (Arnison *et al* 1984) and UA2 (Bagnaia *et al* 1984) experiments. Also shown in this figure is the expected $F(x)$ based on contemporary fits to the deep inelastic neutrino scattering data at $Q^2 = 20$ GeV$^2$ and 2000 GeV$^2$ (Abramovicz *et al* 1982a, b, 1983)—the reason for the change with $Q^2$ will be discussed in section 15.7. The agreement is qualitatively very satisfactory. Subtracting the distributions for quarks and anti-quarks as found in deep inelastic lepton scattering, UA1 were able to deduce the gluon distribution function $g(x)$ shown in figure 14.13. It is clear that gluon processes will dominate at small $x$—and even at larger $x$ they will be important because of the colour factors in table 14.1.

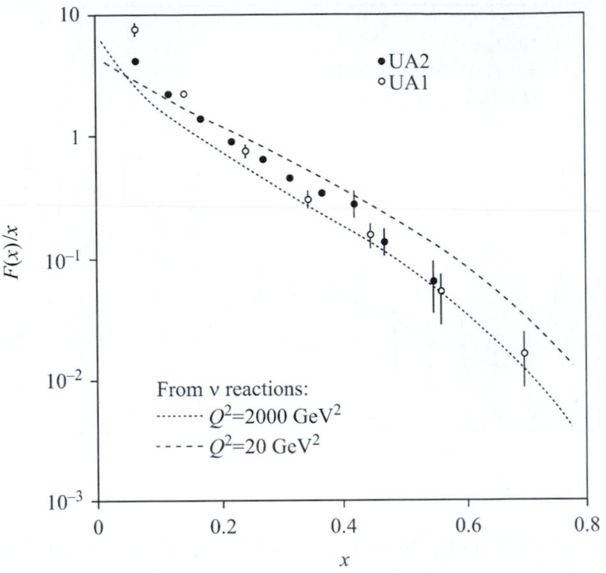


**Figure 14.12.** Effective distribution function measured from two-jet events (Arnison *et al* 1984 and Bagnaia *et al* 1984). The broken and chain curves are obtained from deep inelastic neutrino scattering. Taken from DiLella (1985).

### 14.3.2 Three-jet events

Although most of the high $\sum E_{\mathrm{T}}$ events at hadron colliders are two-jet events, in some 10–30% of the cases the energy is shared between three jets. An example is included as (*d*) in the collection of figure 14.9: a clearer one is shown in figure 14.14. In QCD such events are interpreted as arising from a 2 parton → 2 parton + 1 gluon process of the type gg → ggg, gq → ggq, etc. Once again, one can calculate (Kunszt and Piétarinen 1980, Gottschalk and Sivers 1980, Berends *et al* 1981) all possible contributing tree graphs, of the kind shown in figure 14.15, which should dominate at small $\alpha_{\mathrm{s}}$. They are collectively known as QCD single-bremsstrahlung diagrams. Analysis of triple jets which are well separated both from each other and from the beam directions shows that the data are in good agreement with these lowest-order QCD predictions. For example, figure 14.16 shows the production angular distribution of UA2 (Appel *et al* 1986) as a function of $\cos\theta^*$, where $\theta^*$ is the angle between the leading (most energetic) jet momentum and the beam axis, in the three-jet CMS. It follows just the same $\sin^{-4}\theta^*/2$ curve as in the two-jet case (the data for which are also shown in the figure), as expected for massless quantum exchange—the particular curve is for the representative process gg → ggg.

Another qualitative feature is that the ratio of three-jet to two-jet events is controlled, roughly, by $\alpha_{\mathrm{s}}$ (compare figure 14.15 with the one-gluon exchange

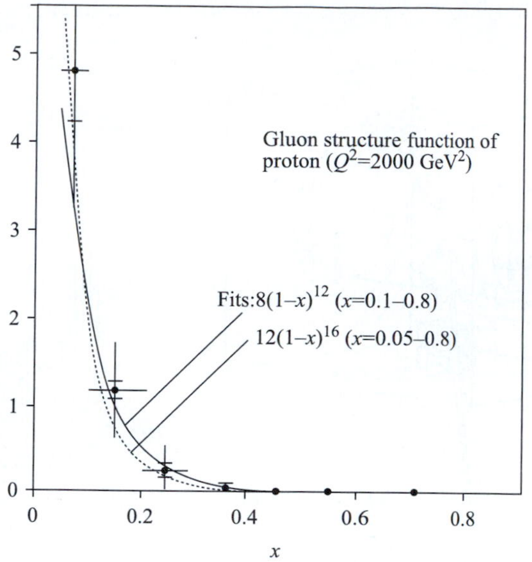


**Figure 14.13.** The gluon dustribution function $g(x)$ extracted from the effective structure function $F(x)$ by subtracting the expected contribution from the quarks and anti-quarks (Geer 1986).

amplitudes of table 14.1). Thus, a crude estimate of $\alpha_s$ could be obtained by comparing the rates of three-jet to two-jet events in $\bar{p}p$ collisions (see figure 14.22 for a similar ratio in $e^+e^-$ annihilation). Other interesting predictions concern the characteristics of the three-jet final state (for example, the distributions in the jet energy variables). At this point, however, it is convenient to leave $\bar{p}p$ collisions and consider instead three-jet events in $e^+e^-$ collisions, for which the complications associated with the initial state hadrons are absent.

## 14.4 Three-jet events in $e^+e^-$ annihilation

Three-jet events in $e^+e^-$ collisions originate, according to QCD, from gluon bremsstrahlung corrections to the two-jet parton model process $e^+e^- \to \gamma^* \to q\bar{q}$, as shown in figure 14.17.[3] This phenomenon was predicted by Ellis *et al* (1976) and subsequently observed by Brandelik *et al* (1979) with the TASSO detector at PETRA and Barber *et al* (1979) with MARK-J at PETRA, thus providing early encouragement for QCD. The situation here is, in many ways, simpler and cleaner than in the $\bar{p}p$ case: the initial state 'partons' are perfectly physical QED quanta and their total 4-momentum is zero, so that the three jets

[3] This is assuming that the total $e^+e^-$ energy is far from the $Z^0$ mass; if not, the contribution from the intermediate $Z^0$ must be added to that from the photon.

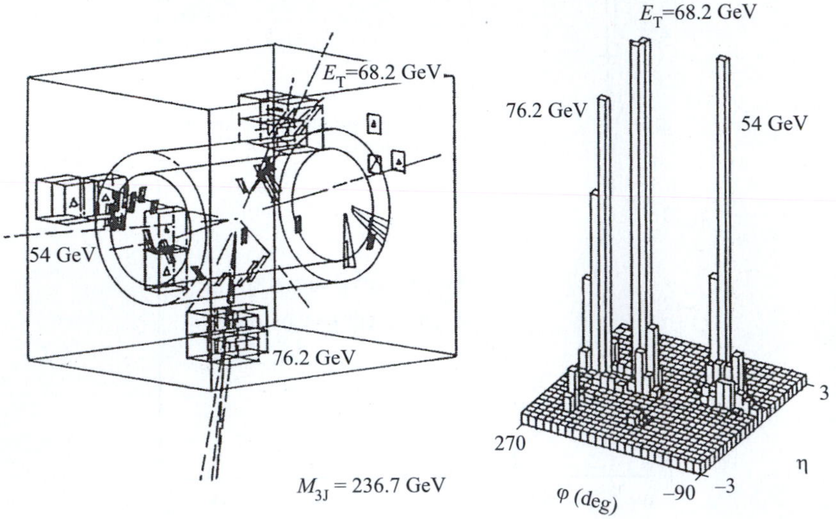


**Figure 14.14.** Three-jet event in the UA1 detector, and the associated transverse energy flow plot (Geer 1986).

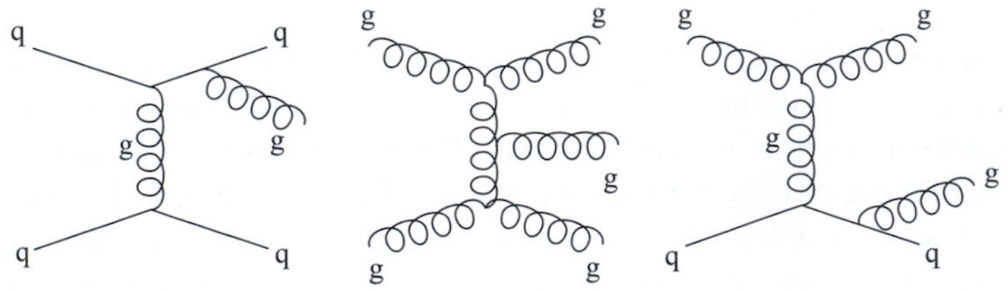


**Figure 14.15.** Some tree graphs associated with three-jet events.

have to be coplanar; further, there is only one type of diagram compared to the large number in the $\bar{p}p$ case and much of that diagram involves the easier vertices of QED. Since the calculation of the cross-section predicted from figure 14.17 is not only relevant to three-jet production in $e^+e^-$ collisions but also to QCD corrections to the *total* $e^+e^-$ annihilation cross-section, and to scaling violations in deep inelastic scattering as well, we shall now consider it in some detail. It is important to emphasize at the outset that *quark masses will be neglected* in this calculation.

The quark, anti-quark and gluon 4-momenta are $p_1$, $p_2$ and $p_3$ respectively, as shown in figure 14.17; the $e^-$ and $e^+$ 4-momenta are $k_1$ and $k_2$. The cross-

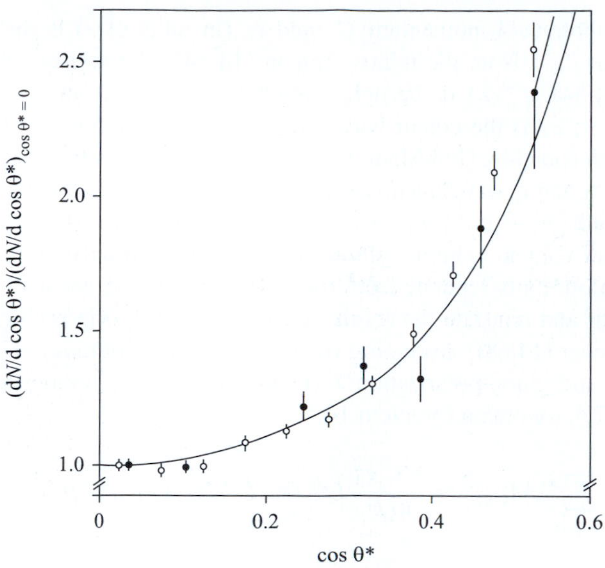


**Figure 14.16.** The distribution of $\cos\theta^*$ (•), the angle of the leading jet with respect to the beam line (normalized to unity at $\cos\theta^* = 0$), for three-jet events in $\bar{p}p$ collisions (Appel *et al* 1986). The distribution for two-jet events is also shown (∘). The full curve is a parton model calculation using the tree-graph amplitudes for gg → ggg, and cut-offs in transverse momentum and angular separation to eliminate divergences (see remarks following equation (14.68)).

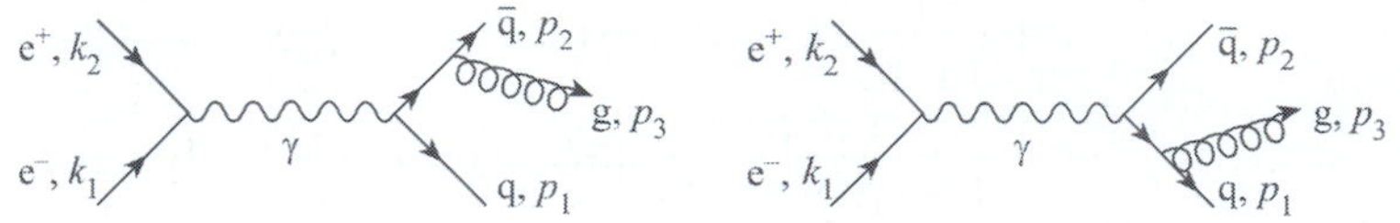


**Figure 14.17.** Gluon brehmsstrahlung corrections to two-jet parton model process.

section is then (cf (6.110) and (6.112))

$$d\sigma = \frac{1}{(2\pi)^5}\delta^4(k_1 + k_2 - p_1 - p_2 - p_3)\frac{|\mathcal{M}_{q\bar{q}g}|^2}{2Q^2}\frac{d^3p_1}{2E_1}\frac{d^3p_2}{2E_2}\frac{d^3p_3}{2E_3} \tag{14.53}$$

where (neglecting all masses)

$$\mathcal{M}_{q\bar{q}g} = \frac{e_a e^2 g_s}{Q^2}\bar{v}(k_2)\gamma^\mu u(k_1)\left(\bar{u}(p_1)\gamma_\nu\frac{\lambda_c}{2}\cdot\frac{(\not{p}_1 + \not{p}_3)}{2p_1\cdot p_3}\cdot\gamma_\mu v(p_2)\right.$$
$$\left. - \bar{u}(p_1)\gamma_\mu\frac{\lambda_c}{2}\cdot\frac{(\not{p}_2 + \not{p}_3)}{2p_2\cdot p_3}\cdot\gamma_\nu v(p_2)\right)\epsilon^{*\nu}(\lambda)a_c \tag{14.54}$$

and $Q^2 = 4E^2$ is the square of the total e$^+$e$^-$ energy, and also the square of the virtual photon's 4-momentum $Q$, and $e_a$ (in units of $e$) is the charge of a quark of type 'a'. Note the minus sign in (14.54): the anti-quark coupling is $-g_s$. In (14.54), $\epsilon^{*\nu}(\lambda)$ is the polarization vector of the outgoing gluon with polarization $\lambda$; $a_c$ is the colour wavefunction of the gluon ($c = 1, \cdots, 8$) and $\lambda_c$ is the corresponding Gell-Mann matrix introduced in section 12.2; the colour parts of the q and $\bar{q}$ wavefunctions are understood to be included in the $u$ and $v$ factors; and $(\not{p}_1 + \not{p}_3)/2p_1 \cdot p_3$ is the virtual quark propagator (cf (L.6) in appendix L of volume 1) before gluon radiation, and similarly for the anti-quark. Since the colour parts separate from the Dirac trace parts, we shall ignore them to begin with and reinstate the result of the colour sum (via problem (14.4)) in the final answer (14.68). Averaging over $e^{\pm}$ spins and summing over final-state quark spins and gluon polarization $\lambda$ (using (8.170), and noting the discussion after (13.127)), we obtain (problem 14.5)

$$\frac{1}{4}\sum_{\text{spins},\lambda} |\mathcal{M}_{q\bar{q}g}|^2 = \frac{e^4 e_a^2 g_s^2}{8Q^4} L^{\mu\nu}(k_1, k_2) H_{\mu\nu}(p_1, p_2, p_3) \tag{14.55}$$

where the lepton tensor is, as usual (equation (8.118)),

$$L^{\mu\nu}(k_1, k_2) = 2(k_1^\mu k_2^\nu + k_1^\nu k_2^\mu - k_1 \cdot k_2 g^{\mu\nu}) \tag{14.56}$$

and the hadron tensor is

$$\begin{aligned} H_{\mu\nu}(p_1, p_2, p_3) = {} & \frac{1}{p_1 \cdot p_2}[L_{\mu\nu}(p_2, p_3) - L_{\mu\nu}(p_1, p_1) + L_{\mu\nu}(p_1, p_2)] \\ & + \frac{1}{p_2 \cdot p_3}[L_{\mu\nu}(p_1, p_3) - L_{\mu\nu}(p_2, p_2) + L_{\mu\nu}(p_1, p_2)] \\ & + \frac{p_1 \cdot p_2}{(p_1 \cdot p_3)(p_2 \cdot p_3)}[2L_{\mu\nu}(p_1, p_2) + L_{\mu\nu}(p_1, p_3) \\ & + L_{\mu\nu}(p_2, p_3)] \end{aligned} \tag{14.57}$$

Combining (14.56) and (14.57) allows complete expressions for the five-fold differential cross-section to be obtained (Ellis *et al* 1976).

Data are generally not extensive enough to permit such differential cross-sections to be studied and so one integrates over three angles describing the orientation (relative to the beam axis) of the production plane containing the three jets. After this integration, the (doubly differential) cross-section is a function of two independent Lorentz invariant variables, which are conveniently taken to be two of the three $s_{ij}$ defined by

$$s_{ij} = (p_i + p_j)^2. \tag{14.58}$$

Since we are considering the massless case $p_i^2 = 0$ throughout, we may also write

$$s_{ij} = 2p_i \cdot p_j. \tag{14.59}$$

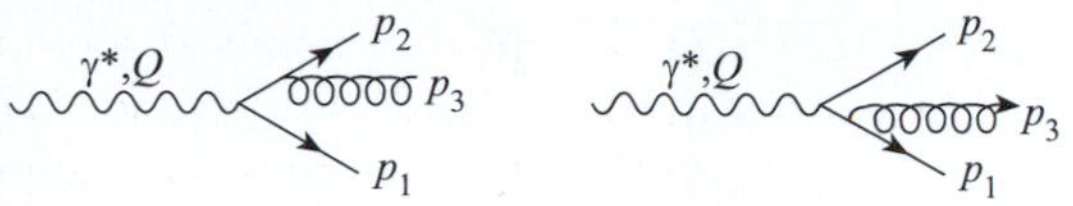


**Figure 14.18.** Virtual photon decaying to $q\bar{q}g$.

These variables are linearly related by

$$2(p_1 \cdot p_2 + p_2 \cdot p_3 + p_3 \cdot p_1) = Q^2 \tag{14.60}$$

as follows from

$$(p_1 + p_2 + p_3)^2 = Q^2 \tag{14.61}$$

and $p_i^2 = 0$. The integration yields (Ellis *et al* 1976)

$$\frac{\mathrm{d}^2\sigma}{\mathrm{d}s_{13}\,\mathrm{d}s_{23}} = \frac{2}{3}\alpha^2 e_a^2 \alpha_s \frac{1}{(Q^2)^3}\left(\frac{s_{13}}{s_{23}} + \frac{s_{23}}{s_{13}} + \frac{2Q^2 s_{12}}{s_{13}s_{23}}\right) \tag{14.62}$$

where $\alpha_s = g_s^2/4\pi$.

We may understand the form of this result in a simple way, as follows. It seems plausible that after integrating over the production angles, the lepton tensor will be proportional to $Q^2 g^{\mu\nu}$, all directional knowledge of the $k_1$ having been lost. Indeed, if we use $-g^{\mu\nu}L_{\mu\nu}(p, p') = 4p \cdot p'$ together with (14.57), we easily find that

$$-\frac{1}{4}g^{\mu\nu}H_{\mu\nu} = \frac{p_1 \cdot p_3}{p_2 \cdot p_3} + \frac{p_2 \cdot p_3}{p_1 \cdot p_3} + \frac{p_1 \cdot p_2 Q^2}{(p_1 \cdot p_3)(p_2 \cdot p_3)} = \frac{s_{13}}{s_{23}} + \frac{s_{23}}{s_{13}} + \frac{2Q^2 s_{12}}{s_{13}s_{23}} \tag{14.63}$$

exactly the factor appearing in (14.62). In turn, the result may be given a simple physical interpretation. From (7.115) we note that we can replace $-g^{\mu\nu}$ by $\sum_{\lambda'} \epsilon^\mu(\lambda')\epsilon^{\nu*}(\lambda')$ for a virtual photon of polarization $\lambda'$, the $\lambda' = 0$ state contributing negatively. Thus, effectively, the result of doing the angular integration is (up to constants and $Q^2$ factors) to replace the lepton factor $\bar{v}(k_2)\gamma^\mu u(k_1)$ by $-\mathrm{i}\epsilon^\mu(\lambda')$, so that $F$ is proportional to the $\gamma^* \to q\bar{q}g$ processes shown in figure 14.18. But these are basically the same amplitudes as the ones we already met in Compton scattering (section 8.6). To compare with section 8.6.3, we convert the initial-state fermion (electron/quark) into a final-state anti-fermion (positron/anti-quark) by $p \to -p$, and then identify the variables of figure 14.18 with those of figure 8.14(*a*) by

$$\begin{gathered} p' \to p_1 \qquad k' \to p_3 \qquad -p \to p_2 \qquad s \to 2p_1 \cdot p_3 = s_{13} \\ t \to 2p_1 \cdot p_2 = s_{12} \qquad u \to 2p_2 \cdot p_3 = s_{23}. \end{gathered} \tag{14.64}$$

Remembering that in (8.180) the virtual $\gamma$ had squared 4-momentum $-Q^2$, we see that the Compton '$\sum |\mathcal{M}|^2$' of (8.180) indeed becomes proportional to the factor (14.63), as expected.

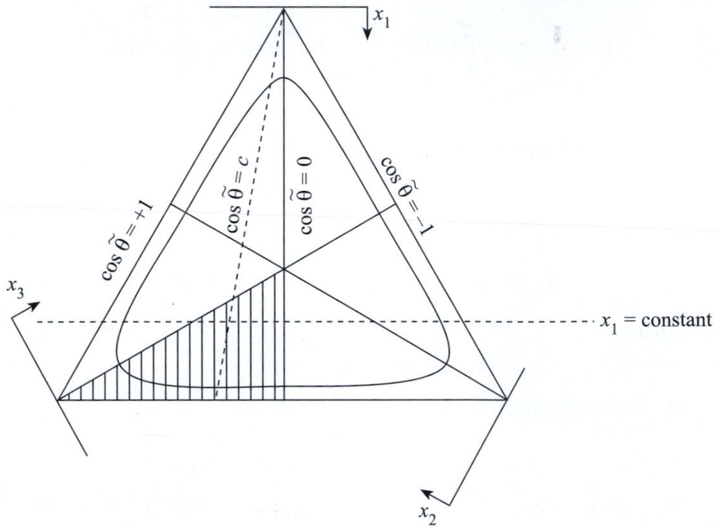


**Figure 14.19.** The kinematically allowed region in $(x_i)$ is the interior of the equilateral triangle.

In three-body final states of the type under discussion here, it is often convenient to preserve the symmetry between the $s_{ij}$'s and use *three* (dimensionless) variables $x_i$ defined by

$$s_{23} = Q^2(1 - x_i) \text{ and cyclic permutations.} \tag{14.65}$$

These are related by (14.60), which becomes

$$x_1 + x_2 + x_3 = 2. \tag{14.66}$$

An event with a given value of the set $x_i$ can then be plotted as a point in an equilateral triangle of height 1, as shown in figure 14.19. In order to find the limits of the allowed physical region in this $x_i$ space, and because it will be useful subsequently, we now transform from the overall three-body CMS to the CMS of 2 and 3 (figure 14.20). If $\tilde{\theta}$ is the angle between 1 and 3 in this system, then (problem 14.6)

$$\begin{aligned} x_2 &= (1 - x_1/2) + (x_1/2)\cos\tilde{\theta} \\ x_3 &= (1 - x_1/2) - (x_1/2)\cos\tilde{\theta}. \end{aligned} \tag{14.67}$$

The limits of the physical region are then clearly $\cos\tilde{\theta} = \pm 1$, which correspond to $x_2 = 1$ and $x_3 = 1$. By symmetry, we see that the entire perimeter of the triangle in figure 14.19 is the required boundary: physical events fall anywhere inside the triangle. (This is the massless limit of the classic Dalitz plot, first introduced by

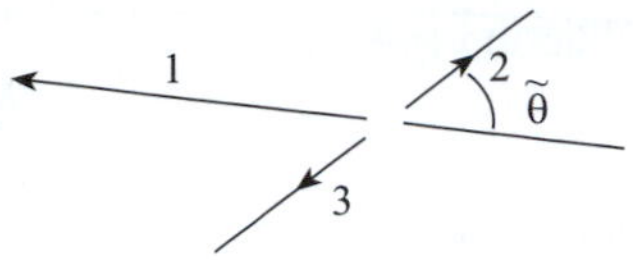


**Figure 14.20.** Definition of $\tilde{\theta}$.

Dalitz (1953) for the analysis of $K \to 3\pi$.) Lines of constant $\tilde{\theta}$ are shown in figure 14.19.

Now consider the distribution provided by the QCD bremsstrahlung process, equation (14.62), which can be written equivalently as

$$\frac{1}{\sigma_{\text{pt}}} \frac{\mathrm{d}^2\sigma}{\mathrm{d}x_1\mathrm{d}x_2} = \frac{2\alpha_s}{3\pi} \left( \frac{x_1^2 + x_2^2}{(1 - x_1)(1 - x_2)} \right) \tag{14.68}$$

where $\sigma_{\text{pt}}$ is the point-like $e^+e^- \to$ hadrons total cross-section of (9.100), and a factor of four has been introduced from the colour sum (problem 14.4). The factor in large parentheses is (14.63) written in terms of the $x_i$ (problem 14.7). The most striking feature of (14.68) is that it is *infinite* as $x_1$ or $x_2$, or both, tend to 1!

This is a quite different infinity from the ones encountered in the loop integrals of chapters 10 and 11. No integral is involved here—the tree amplitude itself becomes singular on the phase space boundary. We can trace the origin of the infinity back to the denominator factors $(p_1 \cdot p_3)^{-1} \sim (1 - x_2)^{-1}$ and $(p_2 \cdot p_3)^{-1} \sim (1 - x_1)^{-1}$ in (14.54). These become zero in two distinct configurations of the gluon momentum:

$$\text{(a)} \quad p_3 \propto p_1 \text{ or } p_3 \propto p_2 \text{ (using } p_i^2 = 0) \tag{14.69}$$

$$\text{(b)} \quad p_3 \to 0 \tag{14.70}$$

which are easily interpreted physically. Condition (a) corresponds to a situation in which the 4-momentum of the gluon is parallel to that of either the quark or the anti-quark: this is called a 'collinear divergence' and the configuration is pictured in figure 14.21($a$). If we restore the quark masses, $p_1^2 = m_1^2 \neq 0$ and $p_2^2 = m_2^2 \neq 0$, then the factor $(2p_1 \cdot p_3)^{-1}$, for example, becomes $((p_1 + p_3)^2 - m_1^2)^{-1}$ which only vanishes as $p_3 \to 0$, which is condition (b). The divergence of type (a) is therefore also termed a 'mass singularity', as it would be absent if the quarks had mass. Condition (b) corresponds to the emission of a very 'soft' gluon (figure 14.21($b$)) and is called a 'soft divergence'. In contrast to this, the gluon momentum $p_3$ in type (a) does *not* have to be vanishingly small.

It is apparent from these figures that in either of these two cases the observed final-state hadrons, after the fragmentation process, will in fact resemble a *two*-jet configuration. Such events will be found in the regions $x_1 \approx 1$ and/or $x_2 \approx 1$ of the kinematical plot shown in figure 14.19, which correspond to strips adjacent

**Figure 14.21.** Gluon configurations leading to divergences of equation (14.68): (*a*) gluon emitted approximately collinear with quark (or anti-quark): (*b*) soft gluon emission. The events are viewed in the overall CMS.

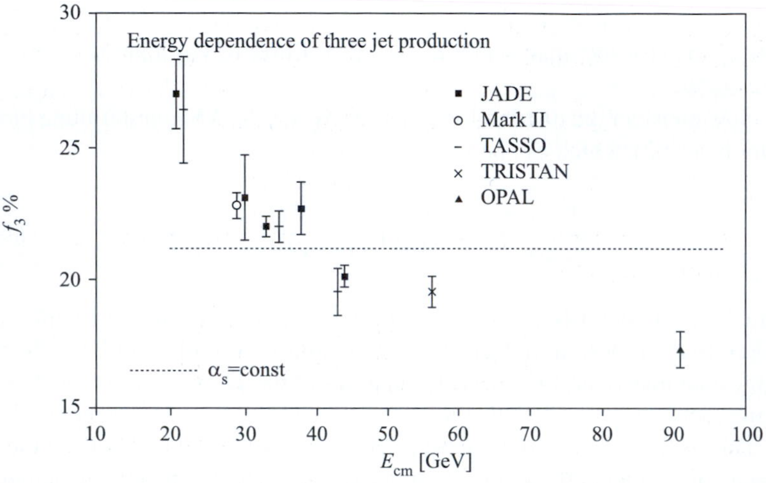


**Figure 14.22.** A compilation of three-jet fractions at different $e^+e^-$ annihilation energies. Adapted from Akrawy *et al* (OPAL) (1990); figure from Ellis *et al* (1996).

to two of the boundaries of the triangle, and to regions near the vertices of the triangle. Events inside the rounded triangular region should be mostly three-jet events. To isolate them, we must keep away from the boundaries of the triangle. The quantitative separation between two- and three-jet events is done by means of a *jet measure*, which needs to be defined in such a way as to be free of soft and collinear divergences. For example, Sterman and Weinberg (1977) defined two-jet events to be those in which all but a fraction $\epsilon$ of the total available energy is contained in two cones of half-angle $\delta$. The two-jet cross-section is then obtained by integrating the right-hand side of (14.68) over the relevant range of $x_1$ and $x_2$.

Assuming such a separation of three- and two-jet events can be done satisfactorily, their ratio carries important information—namely, it should be proportional to $\alpha_s$! This follows simply from the extra factor of $g_s$ associated with the gluon emissions in figure 14.15. Glossing over a number of technicalities (for which the reader is referred to Ellis *et al* (1996, section 3.3)), we show in figure 14.22 a compilation of data on the fraction of three-jet events at different $e^+e^-$ annihilation energies. The most remarkable feature of this figure is, of course, that this fraction—and, hence, $\alpha_s$—*changes with energy, decreasing as the energy increases*. This is, in fact, direct evidence for asymptotic freedom.

It is now time to start our discussion of the theoretical basis for this fundamental property.

## Problems

### 14.1

(a) Show that the anti-symmetric 3q combination of equation (14.2) is invariant under the transformation (14.14) for each colour wavefunction.

(b) Suppose that $p_\alpha$ and $q_\alpha$ stand for two $\mathrm{SU}(3)_\mathrm{c}$ colour wavefunctions, transforming under an infinitesimal $\mathrm{SU}(3)_\mathrm{c}$ transformation via

$$p' = (1 + \mathrm{i}\boldsymbol{\eta} \cdot \boldsymbol{\lambda}/2)p$$

and similarly for $q$. Consider the anti-symmetric combination of their components, given by

$$\begin{pmatrix} p_2q_3 - p_3q_2 \\ p_3q_1 - p_1q_3 \\ p_1q_2 - p_2q_1 \end{pmatrix} \equiv \begin{pmatrix} Q_1 \\ Q_2 \\ Q_3 \end{pmatrix};$$

that is, $Q_\alpha = \epsilon_{\alpha\beta\gamma} p_\beta q_\gamma$. Check that the three components $Q_\alpha$ transform as a $\mathbf{3}_\mathrm{c}^*$, in the particular case for which only the parameters $\eta_1, \eta_2, \eta_3$ and $\eta_8$ are non-zero. [*Note*: you will need the explicit forms of the $\boldsymbol{\lambda}$ matrices (appendix M); you need to verify the transformation law

$$Q' = (1 - \mathrm{i}\boldsymbol{\eta} \cdot \boldsymbol{\lambda}^*/2)Q.]$$

**14.2** Verify that the Lorentz-invariant 'contraction' $\epsilon_{\mu\nu\rho\sigma}\hat{F}^{\mu\nu}\hat{F}^{\rho\sigma}$ of two U(1) (Maxwell) field strength tensors is proportional to $\boldsymbol{E} \cdot \boldsymbol{B}$.

**14.3** Verify that the cross-section for the exchange of a single massless scalar gluon contains no '$1/\hat{t}^2$' factor.

**14.4** This problem is concerned with the evaluation of the 'colour factor' needed for equation (14.68). The 'colour wavefunction' part of the amplitude (14.54) is

$$a_c(c_3)\chi^\dagger(c_1)\frac{\lambda_c}{2}\chi(c_2) \tag{14.71}$$

where $c_1$, $c_2$ and $c_3$ label the colour degree of freedom of the quark, anti-quark and gluon respectively, and a sum on the repeated index $c$ is understood as usual. The $\chi$'s are the colour wavefunctions of the quark and anti-quark and are represented by three-component column vectors: a convenient choice is

$$\chi(r) = \begin{pmatrix} 1 \\ 0 \\ 0 \end{pmatrix} \qquad \chi(b) = \begin{pmatrix} 0 \\ 1 \\ 0 \end{pmatrix} \qquad \chi(g) = \begin{pmatrix} 0 \\ 0 \\ 1 \end{pmatrix}$$

by analogy with the spin wavefunctions of SU(2). The cross-section is obtained by forming the modulus squared of (14.71) and summing over the colour labels $c_i$:

$$\sum_{c_1,c_2,c_3} a_c(c_3)\chi_r^\dagger(c_1)\frac{(\lambda_c)_{rs}}{2}\chi_s(c_2)\chi_l^\dagger(c_2)\frac{(\lambda_d)_{lm}}{2}\chi_m(c_1)a_d^*(c_3) \tag{14.72}$$

where summation is understood on the matrix indices on the $\chi$'s and $\lambda$'s, which have been indicated explicitly. In this form the expression is very similar to the *spin* summations considered in chapter 8 (cf equation (8.60)). We proceed to convert (14.72) to a trace and to evaluate it as follows:

(i) Show that

$$\sum_{c_2} \chi_s(c_2)\chi_l^\dagger(c_2) = \delta_{sl}.$$

(ii) Assuming the analogous result

$$\sum_{c_3} a_c(c_3)a_d^*(c_3) = \delta_{cd}$$

show that (14.72) becomes

$$\frac{1}{4}\sum_{c=1}^{8} \mathrm{Tr}(\lambda_c)^2.$$

(iii) Using the $\lambda$'s given in appendix M, section M.4.5, show that

$$\sum_{c=1}^{8} \mathrm{Tr}(\lambda_c)^2 = 16.$$

and hence that the colour factor for (14.68) is four.

**14.5** Verify equation (14.55).

**14.6** Verify equation (14.67).

**14.7** Verify that expression (14.63) becomes the factor in large parentheses in equation (14.68), when expressed in terms of the $x_i$'s.

# 15

# QCD II: ASYMPTOTIC FREEDOM, THE RENORMALIZATION GROUP AND SCALING VIOLATIONS IN DEEP INELASTIC SCATTERING

## 15.1 QCD corrections to the parton model prediction for $\sigma(e^+e^- \to \text{hadrons})$

We begin by considering QCD corrections (at the one-loop, $O(\alpha_s)$, level) to the simple parton model prediction for the total $e^+e^-$ annihilation cross-section into hadrons (see figure 14.1). The parton model graph, shown again in figure 15.1 (assuming we are far from the $Z^0$ peak), has amplitude $F_\gamma$, say. The $O(\alpha_s)$ QCD corrections to $F_\gamma$ are shown in figure 15.2: we denote the amplitude for the sum of these processes by $F_{g,v}$, where 'v' stands for 'virtual', since these involve the emission and then reabsorption of gluons. The total cross-section from these contributions is thus proportional to $|F_\gamma + F_{g,v}|^2$—and this leads to a problem. The gluon loops of figure 15.2 contain, of course, the usual ('ultraviolet') divergences at large momenta. But they turn out *also* to diverge as the (virtual) gluon momenta approach zero. Such 'soft' divergences are usually called 'infrared' when they occur in loops—and they are *not* cured by renormalization, which is relevant to the high-energy (ultraviolet) divergence of Feynman integrals. Renormalization has nothing to offer the infrared problem. We ran into exactly the same trouble in chapter 11 for the case of the analogous QED corrections, of course, but did not give any details there of how the problem is solved. Now we need to be more explicit.

In fact, the gluon loops of figure 15.2 would also, *in the limit of zero quark mass*, exhibit further (non-ultraviolet) divergences, arising from 'collinear' configurations of the quarks and gluons in the loops. This is, after all, not unexpected: the gluon momenta in the loops run over all possible values, including those which gave trouble in the *real* gluon emission processes of figure 14.17, discussed in the last section of the previous chapter. Indeed, it is the latter processes which hold the key to dealing with these troublesome soft and collinear divergences of figure 15.2. The cure lies in a careful analysis of what is actually meant by the total annihilation cross-section to $q\bar{q}$. The point is that an outgoing quark (or anti-quark) cannot be distinguished, kinematically, from one which is accompanied by a soft or collinear gluon—just as, in the appropriate kinematic regions, there are ambiguities between two-jet and three-jet events, as

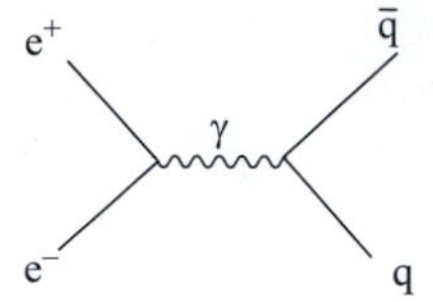


**Figure 15.1.** One-photon annihilation amplitude in $e^+e^- \to \bar{q}q$.

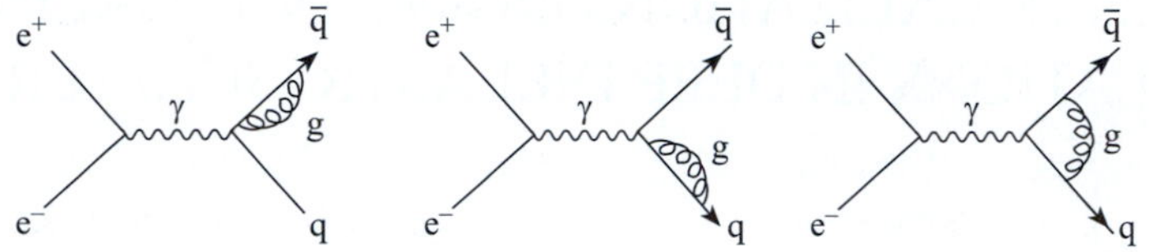


**Figure 15.2.** Virtual gluon corrections to figure 15.1.

we saw in section 14.4. Thus, to $|F_\gamma + F_{g,v}|^2$ should also be added the contribution to the total cross-section due to production of soft (and collinear) gluons in these dangerous kinematical regions. This will entail integrating (14.68) precisely over some area (call it $\eta$) of figure 14.19 close to the triangular boundary, as defined by a jet measure of some kind. This leads to a cross-section for the production of *real* soft and collinear gluons which is given by

$$\sigma_{g,r} = \sigma_{pt} \frac{2\alpha_s}{3\pi} \iint_\eta \frac{x_1^2 + x_2^2}{(1-x_1)(1-x_2)} \, dx_1 \, dx_2. \tag{15.1}$$

Clearly, as the region $\eta$ (parametrized in some way) tends to zero, (15.1) will diverge due to the vanishing denominators (soft and collinear divergences).

This refinement hardly seems to have helped: we now have two lots of 'infrared' type divergences to worry about, one in $|F_\gamma + F_{g,v}|^2$ and one in $\sigma_{g,r}$. Yet now comes the miracle. Note that three terms appear when squaring out $|F_\gamma + F_{g,v}|^2$: one of order $\alpha^2$ (from $|F_\gamma|^2$), another of order $\alpha^2\alpha_s^2$ (from $|F_{g,v}|^2$) and an *interference* term of order $\alpha^2\alpha_s$, *which is the same order as (15.1)*. Thus at order $\alpha^2\alpha_s$, (15.1) must be added precisely to this interference term—and the wonderful fact is that their divergences *cancel*. The complete cross-section at order $\alpha^2\alpha_s$ is found to be (see, for example, Pennington (1983) or Ellis *et al* (1996))

$$\sigma = \sigma_{pt}(1 + \alpha_s/\pi). \tag{15.2}$$

The remarkable cancellation of the soft and collinear divergences between the real and virtual emission processes is actually a general result. The Bloch–Nordsieck (1937) theorem states that 'soft' singularities cancel between real and virtual processes when one adds up all final states which are indistinguishable by virtue of the energy resolution of the apparatus. A theorem due to Kinoshita (1962) and Lee and Nauenberg (1964) states, roughly speaking, that mass

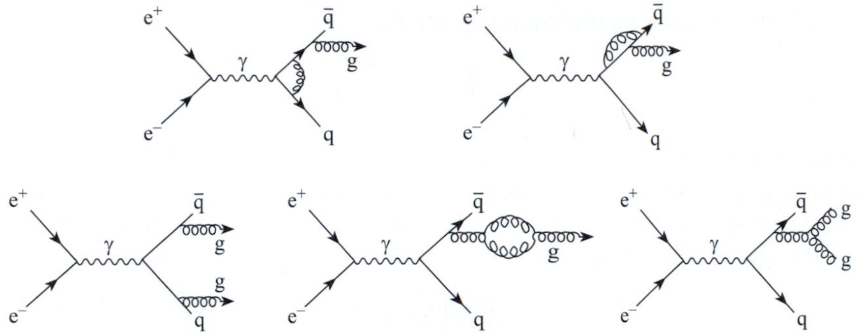


**Figure 15.3.** Some higher-order processes contributing to $e^+e^- \to$ hadrons at the parton level.

singularities are absent if one adds up all indistinguishable mass-degenerate states. If an 'inclusive' final state is considered (as in $e^+e^- \to$ hadrons), then only mass singularities from initial lines will remain. In the case of $e^+e^-$ annihilation, these do not arise since the gluon cannot be attached to the $e^+$ or $e^-$ lines. But in the case of deep inelastic scattering, when effectively a quark or anti-quark appears in the initial state, such uncancelled mass singularities will occur, as we shall see in section 15.7.1. They are of great importance physically, being essentially the origin of *scaling violations* in deep inelastic scattering.

We return to (15.2). At first sight, this result might appear satisfactory. Comparison with the data shown in figure 14.1 would suggest that $\alpha_s \sim 0.5$ or less,[1] so that (assuming the expansion parameter is $\alpha_s/\pi$) the implied perturbation series in powers of $\alpha_s$ would seem to be rapidly convergent. However, this is an illusion, which is dispelled as soon as we go to the next order in $\alpha_s$ (i.e. to the order $\alpha^2\alpha_s^2$ in the cross-section). Some typical graphs contributing to this order of the cross-section are shown in figure 15.3 (note that, as with the $O(\alpha^2\alpha_s)$ terms, some graphs will contribute via their modulus squared and some via interference terms). The result was obtained numerically by Dine and Sapirstein (1979) and analytically by Chetyrkin *et al* (1979) and by Celmaster and Gonsalves (1980). For our present purposes, the crucial feature of the answer is the appearance of a term

$$\sigma_{\text{pt}} \cdot \left[ -b\frac{\alpha_s^2}{\pi} \ln\left(\frac{s}{\mu^2}\right)\right] \tag{15.3}$$

where $\mu$ is a mass scale (about which we shall shortly have a lot more to say, but which for the moment may be thought of as related in some way to an average

[1] A more precise extraction of $\alpha_s$ can be made from the value of $R$ at the $Z^0$ peak, see Ellis *et al* (1996). The value of $\alpha_s$ at the $Z^0$ peak is approximately equal to 0.12.

quark mass), and the coefficient $b$ is given by

$$b = \left(\frac{33 - 2N_f}{12\pi}\right) \tag{15.4}$$

where $N_f$ is the number of 'active' flavours (e.g. $N_f = 5$ above the $b\bar{b}$ threshold). In (15.3), $s$ is the square of the total $e^+e^-$ energy (which we called $Q^2$ in section 14.4). The term (15.3) raises the following problem. The ratio between it and the $O(\alpha^2\alpha_s)$ term is clearly

$$-b\alpha_s \ln(s/\mu^2). \tag{15.5}$$

If we take $N_f = 5$, $\alpha_s \approx 0.4$, $\mu \sim 1$ GeV and $s \sim (10 \text{ GeV})^2$, (15.5) is of order 1 and can, in no sense, be regarded as a small perturbation.

Suppose that, nevertheless, we consider the sum of (15.2) and (15.3), which is

$$\sigma_{pt}\left[1 + \frac{\alpha_s}{\pi}\{1 - b\alpha_s \ln(s/\mu^2)\}\right]. \tag{15.6}$$

This suggests that one effect, at least, of these higher-order corrections is to convert $\alpha_s$ to an $s$-dependent quantity, namely $\alpha_s\{1 - b\alpha_s \ln(s/\mu^2)\}$. We have seen something very like this before, in equation (11.55), for the case of QED. There is, however, one remarkable difference: here the coefficient of the ln is *negative*, whereas that in (11.55) is positive. Apart from this (vital!) difference, however, we can reasonably begin to think in terms of an effective '$s$-dependent strong coupling constant $\alpha_s$'.

Pressing on with the next order $(\alpha^2\alpha_s^3)$ terms, we encounter a term (Samuel and Surguladze 1991, Gorishny *et al* 1991)

$$\sigma_{pt} \cdot \left[\alpha_s b \ln\left(\frac{s}{\mu^2}\right)\right]^2 \frac{\alpha_s}{\pi}, \tag{15.7}$$

and the ratio between this and (15.3) is precisely (15.5) once again! We are now strongly inclined to suspect that we are seeing, in this class of terms, an expansion of the form $(1+x)^{-1} = 1 - x + x^2 - x^3 \cdots$. If true, this would imply that all terms of the form (15.3) and (15.7) and higher, *sum up* to give (cf (11.61))

$$\sigma_{pt}\left[1 + \frac{\alpha_s/\pi}{1 + \alpha_s b \ln(s/\mu^2)}\right]. \tag{15.8}$$

The 're-summation' effected by (15.8) has a remarkable effect: the 'dangerous' large logarithms in (15.3) and (15.7) are now effectively in the *denominator* (cf (11.56)) and their effect is such as to *reduce* the effective value of $\alpha_s$ as $s$ increases—exactly the property of *asymptotic freedom*.

We hasten to say that of course this is not how the property was discovered! The foregoing remarks leave many questions unanswered—for example, are we guaranteed that still higher-order terms will indeed continue to contain pieces corresponding to the expression (15.8)? And what exactly is the mass parameter $\mu$? To address these questions we need to take a substantial detour.

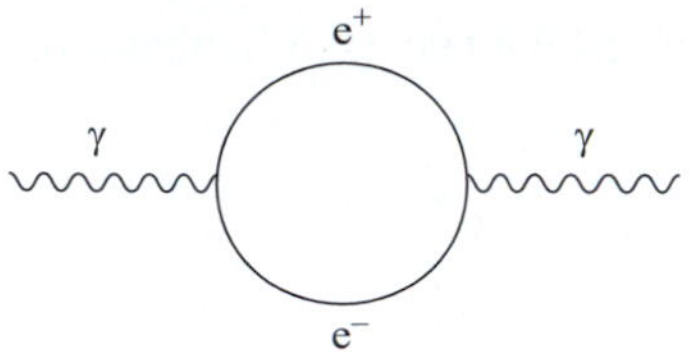


**Figure 15.4.** One-loop vacuum polarization contribution to $Z_3$.

## 15.2 The renormalization group and related ideas

### 15.2.1 Where do the large logs come from?

We have taken the title of this section from that of section 18.1 in Weinberg (1996), which we have found very illuminating, and to which we refer for a more detailed discussion.

As we have just mentioned, the phenomenon of 'large logarithms' arises also in the simpler case of QED. There, however, the factor corresponding to $\alpha_s b \sim \frac{1}{4}$ is $\alpha/3\pi \sim 10^{-3}$, so that it is only at quite unrealistically enormous $|q^2|$ values that the corresponding factor $(\alpha/3\pi)\ln(|q^2|/m_e^2)$ (where $m_e$ is the electron mass) becomes of order unity. Nevertheless, the origin of the logarithmic term is essentially the same in both cases and the technicalities are much simpler for QED (no photon self-interactions, no ghosts). We shall, therefore, forget about QCD for a while and concentrate on QED. Indeed, the discussion of renormalization of QED given in chapter 11 will be sufficient to answer the question in the title of this subsection.

For the answer does, in fact, fundamentally have to do with renormalization. Let us go back to the renormalization of the charge in QED. We learned in chapter 11 that the renormalized charge $e$ was given in terms of the 'bare' charge $e_0$ by the relation $e = e_0(Z_2/Z_1)Z_3^{1/2}$ (see (11.6)), where, in fact, due to the Ward identity $Z_1$ and $Z_2$ are equal (section 11.6), so that only $Z_3^{1/2}$ is needed. To order $e^2$ in renormalized perturbation theory, including only the $e^+e^-$ loop of figure 15.4, $Z_3$ is given by (cf (11.30))

$$Z_3^{[2]} = 1 + \Pi_\gamma^{[2]}(0) \tag{15.9}$$

where, from (11.23) and (11.24),

$$\Pi_\gamma^{[2]}(q^2) = 8e^2\mathrm{i}\int_0^1 \mathrm{d}x \int \frac{\mathrm{d}^4k'}{(2\pi)^4}\frac{x(1-x)}{(k'^2 - \Delta_\gamma + \mathrm{i}\epsilon)^2} \tag{15.10}$$

and $\Delta_\gamma = m_e^2 - x(1-x)q^2$ with $q^2 < 0$. We regularize the $k'$ integral by a cut-off

$\Lambda$, as explained in sections 10.3.1 and 10.3.2, obtaining (problem 15.1)

$$\Pi_{\gamma}^{[2]}(q^2) = -\frac{e^2}{\pi^2}\int_0^1 \mathrm{d}x\, x(1-x)\left\{\ln\left(\frac{\Lambda+\sqrt{\Lambda^2+\Delta_\gamma}}{\Delta_\gamma^{\frac{1}{2}}}\right) - \frac{\Lambda}{(\Lambda^2+\Delta_\gamma)^{\frac{1}{2}}}\right\}. \tag{15.11}$$

Setting $q^2 = 0$ and retaining the dominant $\ln\Lambda$ term, we find that

$$(Z_3^{[2]})^{1/2} = 1 - \left(\frac{\alpha}{3\pi}\right)\ln(\Lambda/m_{\mathrm{e}}). \tag{15.12}$$

It is not a coincidence that the coefficient $\alpha/3\pi$ of the ultraviolet divergence is also the coefficient of the $\ln(|q^2|/m_{\mathrm{e}}^2)$ term in (11.54)–(11.56): we need to understand why.

We first recall how (11.54) was arrived at. It refers to the *renormalized* self-energy part, which is defined by the 'subtracted' form

$$\bar{\Pi}_{\gamma}^{[2]}(q^2) = \Pi_{\gamma}^{[2]}(q^2) - \Pi_{\gamma}^{[2]}(0). \tag{15.13}$$

In the process of subtraction, the dependence on the cut-off $\Lambda$ disappears and we are left with

$$\bar{\Pi}_{\gamma}^{[2]}(q^2) = -\frac{2\alpha}{\pi}\int_0^1 \mathrm{d}x\, x(1-x)\ln\left[\frac{m_{\mathrm{e}}^2}{m_{\mathrm{e}}^2 - q^2x(1-x)}\right] \tag{15.14}$$

as in (11.36). For large values of $|q^2|$, this leads to the 'large log' term $(\alpha/3\pi)\ln(|q^2|/m_{\mathrm{e}}^2)$. Now, in order to form such a term, it is obviously not possible to have just '$\ln|q^2|$' appearing: the argument of the logarithm must be dimensionless so that some mass scale must be present to which $|q^2|$ can be compared. In the present case, that mass scale is evidently $m_{\mathrm{e}}$, which entered via the quantity $\Pi_{\gamma}^{[2]}(0)$ or, equivalently, via the renormalization constant $Z_3^{[2]}$ (cf (15.12)). This is the beginning of the answer to our questions.

Why is it $m_{\mathrm{e}}$ that enters into $\Pi_{\gamma}^{[2]}(0)$ or $Z_3$? Part of the answer—once again—is of course that a '$\ln\Lambda$' cannot appear in that form but must be '$\ln(\Lambda/\text{some mass})$'. So we must enquire: what determines the 'some mass'? With this question we have reached the heart of the problem (for the moment). The answer is, in fact, not immediately obvious: it lies in the *prescription used to define the renormalized coupling constant*—this prescription, whatever it is, determines $Z_3$.

The value (15.9) (or (11.30)) was determined from the requirement that the $O(e^2)$ corrected photon propagator (in the $\xi = 1$ gauge) had the simple form $-\mathrm{i}g_{\mu\nu}/q^2$ as $q^2 \to 0$; that is, as the photon goes on-shell. Now, this is a perfectly 'natural' definition of the renormalized charge—but it is by no means forced upon us. In fact, the appearance of a singularity in $Z_3^{[2]}$ as $m_{\mathrm{e}} \to 0$ suggests that it is inappropriate to the case in which fermion masses are neglected. We could, in principle, choose a different value of $q^2$, say $q^2 = -\mu^2$, at which to 'subtract'. Certainly the difference between $\Pi_\gamma^{[2]}(q^2 = 0)$ and $\Pi_\gamma^{[2]}(q^2 = -\mu^2)$ is finite as $\Lambda \to \infty$, so such a redefinition of 'the' renormalized charge would only amount to a finite shift. Nevertheless, even a finite shift is alarming to those accustomed to a certain 'sanctity' in the value $\alpha = \frac{1}{137}$! We have to concede, however, that if the point of renormalization is to render amplitudes finite by taking certain constants from experiment, then any choice of such constants should be as good as any other—for example, the 'charge' defined at $q^2 = -\mu^2$ rather than at $q^2 = 0$.

Thus, there is, actually, a considerable *arbitrariness* in the way renormalization can be done—a fact to which we did not draw attention in our earlier discussions in chapters 10 and 11. Nevertheless, it must somehow be the case that, despite this arbitrariness, the *physical results remain the same*. We shall come back to this important point shortly.

### 15.2.2 Changing the renormalization scale

The recognition that the *renormalization scale* ($-\mu^2$ in this case) is arbitrary suggests a way in which we might exploit the situation so as to avoid large '$\ln(|q^2|/m_e^2)$' terms: we renormalize at a *large* value of $\mu^2$! Consider what happens if we define a new $Z_3^{[2]}$ by

$$Z_3^{[2]}(\mu) = 1 + \Pi_\gamma^{[2]}(q^2 = -\mu^2). \tag{15.15}$$

Then for $\mu^2 \gg m_e^2$ but $\mu^2 \ll \Lambda^2$, we have

$$(Z_3^{[2]}(\mu))^{\frac{1}{2}} = 1 - \left(\frac{\alpha}{3\pi}\right)\ln(\Lambda/\mu) \tag{15.16}$$

and a new renormalized self-energy

$$\begin{aligned}\Pi_\gamma^{[2]}(q^2, \mu) &= \Pi_\gamma^{[2]}(q^2) - \Pi_\gamma^{[2]}(q^2 = -\mu^2) \\ &= -\frac{e^2}{2\pi^2}\int_0^1 \mathrm{d}x\, x(1-x)\ln\left[\frac{m_e^2 + \mu^2 x(1-x)}{m_e^2 - q^2 x(1-x)}\right].\end{aligned} \tag{15.17}$$

For $\mu^2$ and $-q^2$ both $\gg m_e^2$, the logarithm is now $\ln(|q^2|/\mu^2)$ which is small when $|q^2|$ is of order $\mu^2$. It seems, therefore, that with this different renormalization prescription we have 'tamed' the large logarithms.

However, we have forgotten that, for consistency, the '$e$' we should now be using is the one defined, in terms of $e_0$, via

$$e_\mu = (Z_3^{[2]}(\mu))^{\frac{1}{2}} e_0 = \left(1 - \frac{\alpha}{3\pi}\ln(\Lambda/\mu)\right) e_0 \tag{15.18}$$

rather than

$$e = (Z_3^{[2]})^{\frac{1}{2}} e_0 = \left(1 - \frac{\alpha}{3\pi}\ln(\Lambda/m_e)\right) e_0, \tag{15.19}$$

working always to one-loop order with an $e^+e^-$ loop. The relation between $e_\mu$ and $e$ is then

$$e_\mu = \frac{(1 - \frac{\alpha}{3\pi}\ln(\Lambda/\mu))}{(1 - \frac{\alpha}{3\pi}\ln(\Lambda/m_e))} e \approx \left(1 + \frac{\alpha}{3\pi}\ln(\mu/m_e)\right) e \tag{15.20}$$

to leading order in $\alpha$. Equation (15.20) indeed represents, as anticipated, a finite shift from '$e$' to '$e_\mu$' but the problem with it is that a 'large log' has resurfaced in the form of $\ln(\mu/m_e)$ (remember that our idea was to take $\mu^2 \gg m_e^2$). Although the numerical coefficient of the log in (15.20) is certainly small, a similar procedure applied to QCD will involve the larger coefficient $b\alpha_s$ as in (15.6) and the correction analogous to (15.20) will be of order 1, invalidating the approach.

We have to be more subtle. Instead of making one jump from $m_e^2$ to a large value $\mu^2$, we need to proceed in stages. We can calculate $e_\mu$ from $e$ as long as $\mu$ is not too different from $m_e$. Then we can proceed to $e_{\mu'}$ for $\mu'$ not too different from $\mu$, and so on. Rather than thinking of such a process in discrete stages $m_e \to \mu \to \mu' \to \cdots$, it is more convenient to consider infinitesimal steps—that is, we regard $e_{\mu'}$ at the scale $\mu'$ as being a continuous function of $e_\mu$ at scale $\mu$ and of whatever other dimensionless variables exist in the problem (since the $e$'s are themselves dimensionless). In the present case, these other variables are $\mu'/\mu$ and $m_e/\mu$, so that $e_{\mu'}$ must have the form

$$e_{\mu'} = E(e_\mu, \mu'/\mu, m_e/\mu). \tag{15.21}$$

Differentiating (15.21) with respect to $\mu'$ and letting $\mu' = \mu$, we obtain

$$\mu \frac{\mathrm{d}e_\mu}{\mathrm{d}\mu} = \beta(e_\mu, m_e/\mu) \tag{15.22}$$

where

$$\beta(e_\mu, m_e/\mu) = \left[\frac{\partial}{\partial z} E(e_\mu, z, m/\mu)\right]_{z=1}. \tag{15.23}$$

For $\mu \gg m_e$, equation (15.22) reduces to

$$\mu \frac{\mathrm{d}e_\mu}{\mathrm{d}\mu} = \beta(e_\mu, 0) \equiv \beta(e_\mu) \tag{15.24}$$

which is a form of the *Callan–Symanzik equation* (Callan 1970, Symanzik 1970): it governs the change of the coupling constant $e_\mu$ as the renormalization scale $\mu$ changes.

To this one-loop order, it is easy to calculate the crucial quantity $\beta(e_\mu)$. Returning to (15.18), we may write the bare coupling $e_0$ as

$$\begin{aligned} e_0 &= e_\mu \left(1 - \frac{\alpha}{3\pi}\ln(\Lambda/\mu)\right)^{-1} \\ &\approx e_\mu \left(1 + \frac{\alpha}{3\pi}\ln(\Lambda/\mu)\right) \\ &\approx e_\mu \left(1 + \frac{\alpha_\mu}{3\pi}\ln(\Lambda/\mu)\right) \end{aligned} \tag{15.25}$$

where the last step follows from the fact that $e$ and $e_\mu$ differ by $O(e^3)$, which would be a higher-order correction to (15.25). Now the unrenormalized coupling is certainly independent of $\mu$. Hence, differentiating (15.25) with respect to $\mu$ at fixed $e_0$, we find

$$\left.\frac{\mathrm{d}e_\mu}{\mathrm{d}\mu}\right|_{e_0} - \frac{e_\mu \alpha_\mu}{3\pi\mu} - \ln(\Lambda/\mu)\cdot\frac{e_\mu^2}{4\pi^2}\left.\frac{\mathrm{d}e_\mu}{\mathrm{d}\mu}\right|_{e_0} = 0. \tag{15.26}$$

Working to order $e_\mu^3$, we can drop the last term in (15.26), obtaining finally (to one-loop order)

$$\mu\left.\frac{\mathrm{d}e_\mu}{\mathrm{d}\mu}\right|_{e_0} = \frac{e_\mu^3}{12\pi^2}\left(\equiv \beta^{[2]}(e_\mu)\right). \tag{15.27}$$

We can now integrate equation (15.27) to obtain $e_\mu$ at an arbitrary scale $\mu$, in terms of its value at some scale $\mu = M$, chosen in practice large enough so that for variable scales $\mu$ greater than $M$ we can neglect $m_\mathrm{e}$ compared with $\mu$, but small enough so that $\ln(M/m_\mathrm{e})$ terms do not invalidate the perturbation theory calculation of $e_M$ from $e$. The solution of (15.27) is then (problem 15.2)

$$\ln(\mu/M) = 6\pi^2\left(\frac{1}{e_M^2} - \frac{1}{e_\mu^2}\right) \tag{15.28}$$

or, equivalently,

$$e_\mu^2 = \frac{e_M^2}{1 - \frac{e_M^2}{12\pi^2}\ln(\mu^2/M^2)} \tag{15.29}$$

which is

$$\alpha_\mu = \frac{\alpha_M}{1 - \frac{\alpha_M}{3\pi}\ln\left(\mu^2/M^2\right)} \tag{15.30}$$

where $\alpha = e^2/4\pi$. The crucial point is that the 'large log' is now in the *denominator* (and has coefficient $\alpha_M/3\pi$!). We note that the general solution of (15.24) may be written as

$$\mu = M\exp\int_{e_M}^{e_\mu}\frac{\mathrm{d}e}{\beta(e)}. \tag{15.31}$$

We have made progress in understanding how the coupling changes as the renormalization scale changes and in how 'large logarithmic' change as in (15.20) can be brought under control via (15.30). The final piece in the puzzle is to understand how this can help us with the large $-q^2$ behaviour of our cross-section, the problem from which we originally started.

### 15.2.3 The renormalization group equation and large $-q^2$ behaviour in QED

To see the connection we need to implement the fundamental requirement, stated at the end of section 15.2.1, that predictions for physically measurable quantities must *not* depend on the renormalization scale $\mu$. Consider, for example, our annihilation cross-section $\sigma$ for $e^+e^- \to$ hadrons, pretending that the one-loop corrections we are interested in are those due to QED rather than QCD. We need to work in the space-like region, so as to be consistent with all the foregoing discussion. To make this clear, we shall now denote the 4-momentum of the virtual photon by $q$ rather than $Q$ and take $q^2 < 0$ as in sections 15.2.1 and 15.2.2. Bearing in mind the way we used the 'dimensionless-ness' of the $e$'s in (15.21), let us focus on the dimensionless ratio $\sigma/\sigma_{\text{pt}} \equiv S$. Neglecting all masses, $S$ can only be a function of the dimensionless ratio $|q^2|/\mu^2$ and of $e_\mu$:

$$S = S(|q^2|/\mu^2, e_\mu). \tag{15.32}$$

But $S$ must ultimately have no $\mu$ dependence. It follows that *the $\mu^2$ dependence arising via the $|q^2|/\mu^2$ argument must cancel that associated with $e_\mu$*. This is why the $\mu^2$-dependence of $e_\mu$ controls the $|q^2|$ dependence of $S$, and hence of $\sigma$. In symbols, this condition is represented by the equation

$$\left(\left.\frac{\partial}{\partial\mu}\right|_{e_\mu} + \left.\frac{\mathrm{d}e_\mu}{\mathrm{d}\mu}\right|_{e_0}\frac{\partial}{\partial e_\mu}\right) S(|q^2|/\mu^2, e_\mu) = 0 \tag{15.33}$$

or

$$\left(\mu\left.\frac{\partial}{\partial\mu}\right|_{e_\mu} + \beta(e_\mu)\frac{\partial}{\partial e_\mu}\right) S(|q|^2/\mu^2, e_\mu) = 0. \tag{15.34}$$

Equation (15.34) is referred to as 'the renormalization group equation (RGE) for $S$'. The terminology goes back to Stueckelberg and Peterman (1953), who were the first to discuss the freedom associated with the choice of renormalization scale. The 'group' connotation is a trifle obscure—but all it really amounts to is the idea that if we do one infinitesimal shift in $\mu^2$ and then another, the result will be a third such shift; in other words, it is a kind of 'translation group'. It was, however, Gell-Mann and Low (1954) who realized how equation (15.34) could be used to calculate the large $|q^2|$ behaviour of $S$, as we now explain.

It is convenient to work in terms of $\mu^2$ and $\alpha$ rather than $\mu$ and $e$. Equation (15.34) is then

$$\left(\mu^2\left.\frac{\partial}{\partial\mu^2}\right|_{\alpha_\mu} + \beta(\alpha_\mu)\frac{\partial}{\partial\alpha_\mu}\right) S(|q^2|/\mu^2, \alpha_\mu) = 0, \tag{15.35}$$

where $\beta(\alpha_\mu)$ is defined by

$$\beta(\alpha_\mu) \equiv \mu^2\left.\frac{\partial\alpha_\mu}{\partial\mu^2}\right|_{e_0}. \tag{15.36}$$

From (15.36) and (15.27), we deduce that, to the one-loop order to which we are working,

$$\beta^{[2]}(\alpha_\mu) = \frac{e_\mu}{4\pi}\beta^{[2]}(e_\mu) = \frac{\alpha_\mu^2}{3\pi}. \tag{15.37}$$

Now introduce the important variable

$$t = \ln(|q^2|/\mu^2), \tag{15.38}$$

so that $|q^2|/\mu^2 = \mathrm{e}^t$. Equation (15.35) then becomes

$$\left[-\frac{\partial}{\partial t} + \beta(\alpha_\mu)\frac{\partial}{\partial \alpha_\mu}\right] S(\mathrm{e}^t, \alpha_\mu) = 0. \tag{15.39}$$

This is a first order differential equation which can be solved by implicitly defining a new function—the *running coupling* $\alpha(|q^2|)$—as follows (compare (15.31):

$$t = \int_{\alpha_\mu}^{\alpha(|q^2|)} \frac{\mathrm{d}\alpha}{\beta(\alpha)}. \tag{15.40}$$

To see how this helps, we have to recall how to differentiate an integral with respect to one of its limits—or, more generally, the formula

$$\frac{\partial}{\partial a}\int^{f(a)} g(x)\,\mathrm{d}x = g(f(a))\frac{\partial f}{\partial a}. \tag{15.41}$$

First, let us differentiate (15.40) with respect to $t$ at fixed $\alpha_\mu$: we obtain

$$1 = \frac{1}{\beta(\alpha(|q^2|))}\frac{\partial \alpha(|q^2|)}{\partial t}. \tag{15.42}$$

Next, differentiate (15.40) with respect to $\alpha_\mu$ at fixed $t$ (note that $\alpha(|q^2|)$ will depend on $\mu$ and hence on $\alpha_\mu$): we obtain

$$0 = \frac{\partial \alpha(|q^2|)}{\partial \alpha_\mu}\frac{1}{\beta(\alpha(|q^2|))} - \frac{1}{\beta(\alpha_\mu)} \tag{15.43}$$

the minus sign coming from the fact that $\alpha_\mu$ is the lower limit in (15.40). From (15.42) and (15.43), we find the result

$$\left[-\frac{\partial}{\partial t} + \beta(\alpha_\mu)\frac{\partial}{\partial \alpha_\mu}\right] \alpha(|q^2|) = 0. \tag{15.44}$$

It follows that $S(1, \alpha(|q^2|))$ is a solution of (15.39).

This is a remarkable result. It shows that all the dependence of $S$ on the (momentum)$^2$ variable $|q^2|$ enters through that of the running coupling $\alpha(|q^2|)$. Of course, this result is only valid in a regime of $-q^2$ which is much greater than all quantities with dimension (mass)$^2$—for example, the squares of all particle masses which do not appear in (15.32). This is why the technique applies only at 'high' $-q^2$. The result implies that if we can calculate $S(1, \alpha_\mu)$ (i.e. $S$ at the point $q^2 = -\mu^2$) at some definite order in perturbation theory, then replacing $\alpha_\mu$ by $\alpha(|q^2|)$ will allow us to predict the $q^2$-dependence (at large $-q^2$). All we need to do is solve (15.40). Indeed, for QED with one $e^+e^-$ loop, we have seen that $\beta^{[2]}(\alpha) = \alpha^2/3\pi$. Hence, integrating (15.40) we obtain

$$\alpha(|q^2|) = \frac{\alpha_\mu}{1 - \frac{\alpha_\mu}{3\pi} t} = \frac{\alpha_\mu}{1 - \frac{\alpha_\mu}{3\pi} \ln(|q^2|/\mu^2)}. \tag{15.45}$$

This is almost exactly the formula we proposed in (11.56), on plausibility grounds.[2]

Suppose now that the leading perturbative contribution to $S(1, \alpha_\mu)$ is $S_1\alpha_\mu$. Then the terms contained in $S(1, \alpha(|q^2|))$ in this approximation can be found by expanding (15.45) in powers of $\alpha_\mu$:

$$\begin{aligned} S(1, \alpha(|q^2|)) \approx S_1\alpha(|q^2|) &= S_1\alpha_\mu \left[1 - \frac{\alpha_\mu}{3\pi} t\right]^{-1} \\ &= S_1\alpha_\mu \left[1 + \frac{\alpha_\mu t}{3\pi} + \left(\frac{\alpha_\mu t}{3\pi}\right)^2 + \cdots\right] \end{aligned} \tag{15.46}$$

where $t = \ln(|q^2|/\mu^2)$. The next higher-order calculation of $S(1, \alpha_\mu)$ would be $S_2\alpha_\mu^2$, say, which generates the terms

$$S_2\alpha^2(|q^2|) = S_2\alpha_\mu^2 \left[1 + \frac{2\alpha_\mu t}{3\pi} + \cdots\right]. \tag{15.47}$$

Comparing (15.46) and (15.47), we see that each power of the large log factor appearing in (15.47) comes with one more power of $\alpha_\mu$ than in (15.46). Provided $\alpha_\mu$ is small, then, the *leading* terms in $t, t^2, \ldots$ are contained in (15.46). It is in this sense that replacing $S(1, \alpha_\mu)$ by $S(1, \alpha(|q^2|))$ sums all 'leading log terms'.

In fact, of course, the one-loop (and higher) corrections to $S$ in which we are really interested are those due to QCD, rather than QED, corrections. But the logic is exactly the same. The leading ($O(\alpha_s)$) perturbative contribution to $S = \sigma/\sigma_{pt}$ at $q^2 = -\mu^2$ is given in (15.2) as $\alpha_s(\mu^2)/\pi$. It follows that the 'leading log corrections' at high $-q^2$ are summed up by replacing this expression by $\alpha_s(|q^2|)/\pi$, where the running $\alpha_s(|q^2|)$ is determined by solving (15.40) with the QCD analogue of (15.37)—to which we now turn.

## 15.3 Back to QCD: asymptotic freedom

The reader may have realized, some time back, that the quantity $b$ introduced in (15.4) must be precisely the coefficient of $\alpha_s^2$ in the one-loop contribution to the

[2] The difference has to do, of course, with the different renormalization prescriptions. Equation (11.56) is written in terms of an '$\alpha$' defined at $q^2 = 0$ and without neglect of $m_e$.

$\beta$-function of QCD defined by

$$\beta_{\rm s} = \mu^2 \frac{\partial \alpha_{\rm s}}{\partial \mu^2}\bigg|_{\text{fixed bare } \alpha_{\rm s}}; \tag{15.48}$$

that is to say,

$$\beta_{\rm s}^{[2]} = -b\alpha_{\rm s}^2 \tag{15.49}$$

with

$$b = \frac{33 - 2N_{\rm f}}{12\pi}. \tag{15.50}$$

For $N_{\rm f} \leq 16$, the quantity $b$ is *positive*, so that the sign of (15.49)) is opposite to that of the QED analogue, equation (15.37). Correspondingly, (15.45) is replaced by

$$\alpha_{\rm s}(|q^2|) = \frac{\alpha_{\rm s}(\mu^2)}{[1 + \alpha_{\rm s}(\mu^2) b \ln(|q^2|/\mu^2)]} \tag{15.51}$$

and then replacing $\alpha_{\rm s}$ in (15.2) by (15.51) leads to (15.8).[3]

Thus, in QCD, the strong coupling runs in the opposite way to QED, becoming smaller at large values of $|q^2|$ (or small distances)—the property of asymptotic freedom. The justly famous result (15.50) was first obtained by Politzer (1973), Gross and Wilczek (1973) and 't Hooft. 't Hooft's result, announced at a conference in Marseilles in 1972, was not published. The published calculation of Politzer and of Gross and Wilczek quickly attracted enormous interest, because it immediately explained the 'paradoxical situation' referred to at the end of section 14.2.3: how the successful parton model could be reconciled with the undoubtedly very strong binding forces between quarks. The resolution, we now understand, lies in quite subtle properties of renormalized quantum field theory, involving first the exposure of 'large logarithms', and then their re-summation in terms of the running coupling. Not only did the result (15.50) explain the success of the parton model: it also, we repeat, opened the prospect of performing reliable perturbative calculations in a *strongly* interacting theory, at least at high $|q|^2$. For example, at sufficiently high $|q|^2$, we can reliably compute the $\beta$ function in perturbation theory. The result of Politzer and of Gross and Wilczek led rapidly to the general acceptance of QCD as the theory of strong interactions, a conclusion reinforced by the demonstration by Coleman and Gross (1973) that no theory without Yang–Mills fields possessed the property of asymptotic freedom.

In section 11.5.3 we gave the conventional physical interpretation of the way in which the running of the QED coupling tends to *increase* its value at distances $|q|^{-1}$ short enough to probe inside the screening provided by $e^+e^-$

[3] Except that, in (15.51), $\alpha_{\rm s}$ is evaluated at large *space-like* values of its argument, whereas in (15.8), it is wanted at large *time-like* values. Readers troubled by this may consult Pennington (1983) section 2.3.2, or Peskin and Schroeder (1995) section 18.5. The difficulty is evaded in the approach of section 15.6 below.

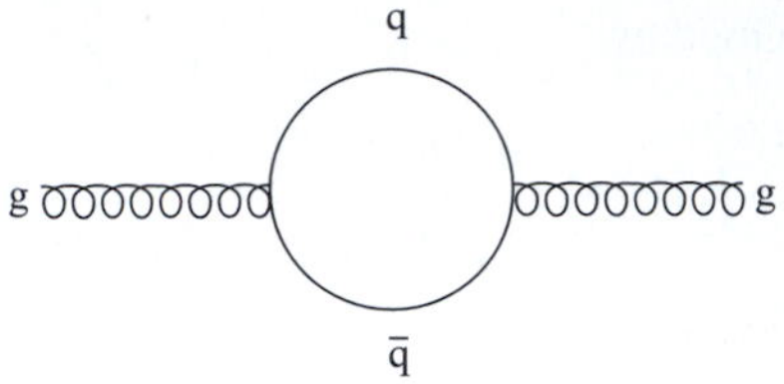


**Figure 15.5.** $q\bar{q}$ vacuum polarization correction to the gluon propagator.

pairs ($|q|^{-1} \ll m_e^{-1}$). This vacuum polarization screening effect is also present in (15.50) via the term $-2N_f/12\pi$, the value of which can be quite easily understood. It arises from the '$q\bar{q}$' vacuum polarization diagram of figure 15.5, which is precisely analogous to the $e^+e^-$ diagram used to calculate $\bar{\Pi}_\gamma^{[2]}(q^2)$ in QED. The only new feature in figure 15.5 is the presence of the $\frac{1}{2}\lambda$-matrices at each vertex. If '$a$' and '$b$' are the colour labels of the ingoing and outgoing gluons, the $\frac{1}{2}\lambda$-matrix factors must be

$$\sum_{\alpha,\beta=1}^{3} \left(\frac{1}{2}\lambda_a\right)_{\alpha\beta} \left(\frac{1}{2}\lambda_b\right)_{\beta\alpha} \tag{15.52}$$

since there are no free quark indices (of type $\alpha$, $\beta$) on the external legs of the diagram. It is simple to check that (15.52) has the value $\frac{1}{2}\delta_{ab}$ (this is, in fact, the way the $\lambda$'s are conventionally normalized). Hence, for one quark flavour we expect '$\alpha/3\pi$' to be replaced by '$\alpha_s/6\pi$', in agreement with the second term in (15.50).

The all-important, positive, first term must therefore be due to the gluons. The one-loop graphs contributing to the calculation of $b$ are shown in figure 15.6. They include figure 15.5, of course, but there are also, characteristically, graphs involving the gluon self-coupling which is absent in QED and also (in covariant gauges) ghost loops. We do not want to enter into the details of the calculation of $\beta(\alpha_s)$ here (they are given in Peskin and Schroeder 1995, chapter 16, for example) but it would be nice to have a simple intuitive picture of the 'anti-screening' result in terms of the gluon interactions, say. Unfortunately, no fully satisfactory simple explanation exists, though the reader may be interested to consult Hughes (1980, 1981) and Nielsen (1981) for a 'paramagnetic' type of explanation rather than a 'dielectric' one.

Returning to (15.51), we note that, despite appearances, it does not really involve two parameters—after all, (15.48) is only a first-order differential equation. By introducing

$$\ln \Lambda_{QCD}^2 = \ln \mu^2 - 1/(b\alpha_s(\mu^2)) \tag{15.53}$$

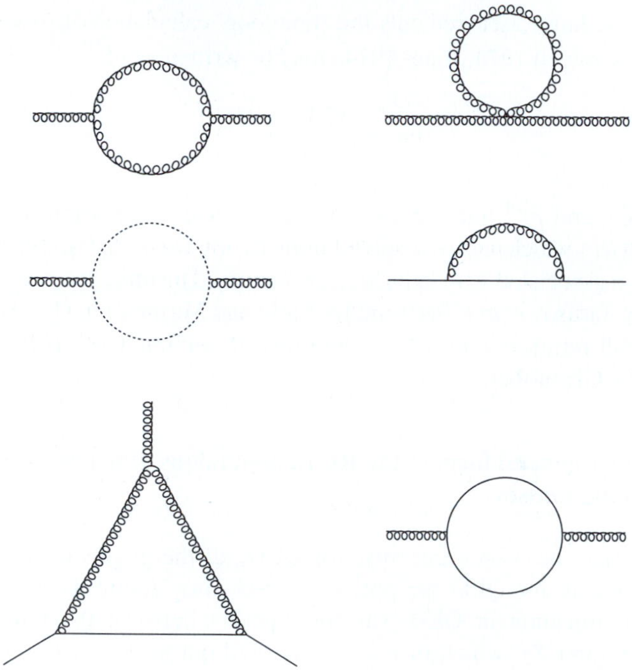

**Figure 15.6.** Graphs contributing to the one-loop $\beta$ function in QCD. A curly line represents a gluon, a broken line a ghost (see section 13.5.3) and a straight line a quark.

equation (15.51) can be rewritten (problem 15.3) as

$$\alpha_s(|q^2|) = \frac{1}{b \ln(|q^2|/\Lambda_{\mathrm{QCD}}^2)}. \tag{15.54}$$

Equation (15.54) is equivalent to (cf (15.40))

$$\ln\left(|q^2|/\Lambda_{\mathrm{QCD}}^2\right) = -\int_{\alpha_s(|q^2|)}^{\infty} \frac{d\alpha_s}{\beta_s^{[2]}} \tag{15.55}$$

with $\beta_s^{[2]} = -b\alpha_s^2$. $\Lambda_{\mathrm{QCD}}$ is, therefore, an integration constant representing the scale at which $\alpha_s$ would diverge to infinity (if we extended our calculation beyond its perturbative domain of validity). More usefully, $\Lambda_{\mathrm{QCD}}$ is a measure of the scale at which $\alpha_s$ really does become 'strong'. The extraction of a precise value of $\Lambda_{\mathrm{QCD}}$ is a complicated matter, as we shall briefly indicate in section 15.5, but a typical value is in the region of 200 MeV. Note that this is a distance scale of order (200 MeV)$^{-1}$ $\sim$ 1 fm, just about the size of a hadron—a satisfactory connection. Significantly, while perturbative QED is characterized by a dimensionless parameter $\alpha$, perturbative QCD requires a mass parameter ($\mu^2$ or $\Lambda_{\mathrm{QCD}}$).

So far we have discussed only the 'one-loop' calculation of $\beta(\alpha_s)$. The two-loop result (Caswell 1974, Jones 1974) may be written as $\beta_s^{[3]} = -bb'\alpha_s^3$, where

$$bb' = \frac{153 - 19N_f}{24\pi^2}. \tag{15.56}$$

Inserting $\beta_s^{[2]}$ and $\beta_s^{[3]}$ into (15.48) gives a corrected expression for $\alpha_s(|q^2|)$ in terms of $\alpha_s(\mu^2)$ which has to be solved numerically for $\alpha_s(|q^2|)$. Typically, the $b'$ coefficient is associated with $\ln\ln(|q^2|/\mu^2)$ terms. The three-loop result has been obtained by Tarasov *et al* (1980) and by Larin and Vermaseren (1993).

We shall return to $\sigma(e^+e^- \to \text{hadrons})$ in section 15.6. First we want to explore the RGE further.

## 15.4 A more general form of the RGE: anomalous dimensions and running masses

The reader may be wondering why, for QCD, all the graphs of figure 15.6 are needed, whereas for QED we got away with only figure 11.5. The reason for the simplification in QED was the equality between the renormalization constants $Z_1$ and $Z_2$, which therefore cancelled out in the relation between the renormalized and bare charges $e$ and $e_0$, as briefly stated before equation (15.9) (this equality was discussed in section 11.6). We recall that $Z_2$ is the field strength renormalization factor for the charged fermion in QED and $Z_1$ is the vertex part renormalization constant: their relation to the counter terms was given in equation (11.7). For QCD, although gauge invariance does imply generalizations of the Ward identity used to prove $Z_1 = Z_2$ (Taylor 1971, Slavnov 1972), the consequence is no longer the simple relation '$Z_1 = Z_2$' in this case, due essentially to the ghost contributions. In order to see what change $Z_1 \neq Z_2$ would make, let us return to the one-loop calculation of $\beta$ for QED, pretending that $Z_1 \neq Z_2$. We have

$$e_0 = \frac{Z_1}{Z_2} Z_3^{-\frac{1}{2}} e_\mu \tag{15.57}$$

where, because we are renormalizing at scale $\mu$, all the $Z_i$'s depend on $\mu$ (as in (15.16)) but we shall now not indicate this explicitly. Taking logs and differentiating with respect to $\mu$ at constant $e_0$, we obtain

$$\mu\frac{\mathrm{d}}{\mathrm{d}\mu}\bigg|_{e_0} \ln Z_1 - \mu\frac{\mathrm{d}}{\mathrm{d}\mu}\bigg|_{e_0} \ln Z_2 - \frac{1}{2}\mu\frac{\mathrm{d}}{\mathrm{d}\mu}\bigg|_{e_0} \ln Z_3 + \frac{\mu}{e_\mu}\frac{\mathrm{d}e_\mu}{\mathrm{d}\mu}\bigg|_{e_0} = 0. \tag{15.58}$$

Hence,

$$\beta(e_\mu) \equiv \mu\frac{\mathrm{d}e_\mu}{\mathrm{d}\mu}\bigg|_{e_0} = e_\mu\gamma_3 + 2e_\mu\gamma_2 - e_\mu\mu\frac{\mathrm{d}}{\mathrm{d}\mu}\ln Z_1 \tag{15.59}$$

where

$$\gamma_2 \equiv \frac{1}{2}\mu \frac{\mathrm{d}}{\mathrm{d}\mu}\bigg|_{e_0} \ln Z_2 \qquad \gamma_3 = \frac{1}{2}\mu \frac{\mathrm{d}}{\mathrm{d}\mu}\bigg|_{e_0} \ln Z_3. \tag{15.60}$$

To leading order in $e_\mu$, the $\gamma_3$ term in (15.60) reproduces (15.27) when (15.16) is used for $Z_3$, the other two terms in (15.58) cancelling via $Z_1 = Z_2$. So if, as in the case of QCD, $Z_1$ is not equal to $Z_2$, we need to introduce the contributions from loops determining the fermion field strength renormalization factor, as well as those related to the vertex parts (together with appropriate ghost loops), in addition to the vacuum polarization loop associated with $Z_3$.

Quantities such as $\gamma_2$ and $\gamma_3$ have an interesting and important significance, which we shall illustrate in the case of $\gamma_2$ for QED. $Z_2$ enters into the relation between the propagator of the bare fermion $\langle\Omega|T(\bar{\hat{\psi}}_0(x)\hat{\psi}_0(0))|\Omega\rangle$ and the renormalized one, via (cf (11.2))

$$\langle\Omega|T(\bar{\hat{\psi}}(x)\hat{\psi}(0)|\Omega\rangle = \frac{1}{Z_2}\langle\Omega|T(\bar{\hat{\psi}}_0(x)\hat{\psi}_0(0))|\Omega\rangle \tag{15.61}$$

where (cf section 10.1.3) $|\Omega\rangle$ is the vacuum of the interacting theory. The Fourier transform of (15.61) is, of course, the Feynman propagator:

$$\tilde{S}'_{\mathrm{F}}(q^2) = \int \mathrm{d}^4 x e^{\mathrm{i}q\cdot x}\langle\Omega|T(\bar{\hat{\psi}}(x)\hat{\psi}(0))|\Omega\rangle. \tag{15.62}$$

Suppose we now ask: what is the large $-q^2$ behaviour of (15.62) for space-like $q^2$, with $-q^2 \gg m^2$ where $m$ is the fermion mass? This sounds very similar to the question answered in 15.2.3 for the quantity $S(|q^2|/\mu^2, e_\mu)$. However, the latter was dimensionless, whereas (recalling that $\hat{\psi}$ has mass dimension $\frac{3}{2}$) $\tilde{S}'_{\mathrm{F}}(q^2)$ has dimension $M^{-1}$. This dimensionality is just what a propagator of the free-field form $i/(\not{q} - m)$ would provide.

Accordingly, we extract this $(\not{q})^{-1}$ factor (compare $\sigma/\sigma_{\mathrm{pt}}$) and consider the dimensionless ratio $\tilde{R}'_{\mathrm{F}}(|q^2|/\mu^2, \alpha_\mu) = \not{q}\tilde{S}'_{\mathrm{F}}(q^2)$. We might guess that, just as for $S(|q^2|/\mu^2, \alpha_\mu)$, to get the leading large $|q^2|$ behaviour we will need to calculate $\tilde{R}'_{\mathrm{F}}$ to some order in $\alpha_\mu$ and then replace $\alpha_\mu$ by $\alpha(|q^2|/\mu^2)$. But this is not quite all. The factor $Z_2$ in (15.61) will—as previously noted—depend on the renormalization scale $\mu$, just as $Z_3$ of (15.16) did. Thus, when we change $\mu$, the normalization of the $\hat{\psi}$'s will change via the $Z_2^{\frac{1}{2}}$ factors—by a finite amount here—and we must include this change when writing down the analogue of (15.34) for this case (i.e. the condition that the 'total change, on changing $\mu$, is zero'). The required equation is

$$\left[\mu^2 \frac{\partial}{\partial\mu^2}\bigg|_{\alpha_\mu} + \beta(\alpha_\mu)\frac{\partial}{\partial\alpha_\mu} + \gamma_2(\alpha_\mu)\right]\tilde{R}'_{\mathrm{F}}(|q^2|/\mu^2, \alpha_\mu) = 0. \tag{15.63}$$

The solution of (15.63) is somewhat more complicated than that of (15.34). We can gain insight into the essential difference caused by the presence of $\gamma_2$ by considering the special case $\beta(e_\mu) = 0$. In this case, we easily find

$$\tilde{R}'_{\mathrm{F}}(|q^2|/\mu^2, \alpha_\mu) \propto (\mu^2)^{-\gamma_2(\alpha_\mu)}. \tag{15.64}$$

But since $\tilde{R}'_{\mathrm{F}}$ can only depend on $\mu$ via $|q^2|/\mu^2$, we learn that if $\beta = 0$ then the large $|q^2|$ behaviour of $\tilde{R}'_{\mathrm{F}}$ is given by $(|q^2|/\mu^2)^{\frac{1}{2}\gamma_2}$—or, in other words, that at large $|q^2|$

$$\tilde{S}'_{\mathrm{F}}(|q^2|/\mu^2, \alpha_\mu) \propto \frac{1}{\not{q}}\left(\frac{|q^2|}{\mu^2}\right)^{\gamma_2(\alpha_\mu)}. \tag{15.65}$$

Thus, *at a zero of the $\beta$-function*, $\tilde{S}'_{\mathrm{F}}$ has an 'anomalous' power law dependence on $|q^2|$ (i.e. in addition to the obvious $\not{q}^{-1}$ factor), which is controlled by the parameter $\gamma_2$. The latter is called the 'anomalous dimension' of the fermion field, since its presence effectively means that the $|q^2|$ behaviour of $\tilde{S}'_{\mathrm{F}}$ is not determined by its 'normal' dimensionality $M^{-1}$. The behaviour (15.65) is often referred to as 'scaling with anomalous dimension', meaning that if we multiply $|q^2|$ by a scale factor $\lambda$, then $\tilde{S}'_{\mathrm{F}}$ is multiplied by $\lambda^{\gamma_2(e_\mu)-1}$ rather than just $\lambda^{-1}$. Anomalous dimensions turn out to play a vital role in the theory of critical phenomena—they are, in fact, closely related to 'critical exponents' (see section 16.6.3 and Peskin and Schroeder 1995, chapter 13). Scaling with anomalous dimensions is also exactly what occurs in deep inelastic scattering of leptons from nucleons, as we shall see in section 15.7.

The full solution of (15.63) for $\beta \neq 0$ is elegantly discussed in Coleman (1985, chapter 3); see also Peskin and Schroeder (1995) section 12.3. We quote it here:

$$\tilde{R}'_{\mathrm{F}}(|q^2|/\mu^2), \alpha_\mu) = \tilde{R}'_{\mathrm{F}}(1, \alpha(|q^2|/\mu^2)) \exp\left\{\int_0^t \mathrm{d}t' \gamma_2(\alpha(t'))\right\}. \tag{15.66}$$

The first factor is the expected one from section 15.2.3; the second results from the addition of the $\gamma_2$ term in (15.63). Suppose now that $\beta(\alpha)$ has a zero at some point $\alpha^*$, in the vicinity of which $\beta(\alpha) \approx -B(\alpha - \alpha^*)$ with $B > 0$. Then, near this point the evolution of $\alpha$ is given by (cf (15.40))

$$\ln(|q^2|/\mu^2) = \int_{\alpha_\mu}^{\alpha(|q^2|)} \frac{\mathrm{d}\alpha}{-B(\alpha - \alpha^*)} \tag{15.67}$$

which implies

$$\alpha(|q^2|) = \alpha^* + \text{constant} \times (\mu^2/|q^2|)^B. \tag{15.68}$$

Thus, asymptotically for large $|q^2|$, the coupling will evolve to the '*fixed point*' $\alpha^*$. In this case, at sufficiently large $-q^2$, the integral in (15.66) can be evaluated by setting $\alpha(t') = \alpha^*$, and $\tilde{R}'_{\mathrm{F}}$ will scale with an anomalous dimension $\gamma_2(\alpha^*)$ determined by the fixed point value of $\alpha$. The behaviour of such an $\alpha$ is shown

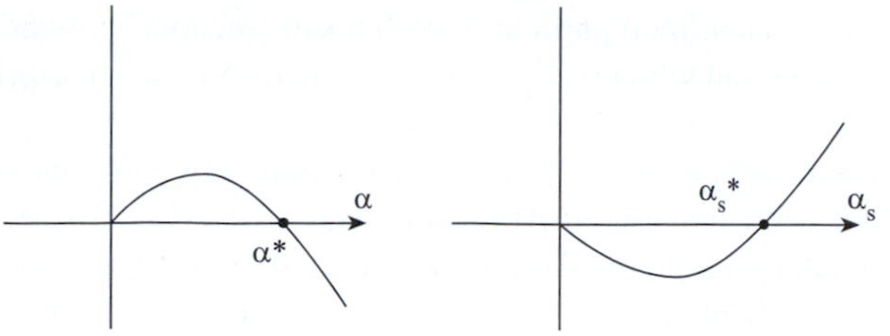


**Figure 15.7.** Possible behaviour of $\beta$ functions: (*a*) the slope is positive near the origin (as in QED) and negative near $\alpha = \alpha^*$; (*b*) the slope is negative at the origin (as in QCD) and positive near $\alpha_s = \alpha_s^*$.

in figure 15.7(*a*). We emphasize that there is no reason to believe that the QED $\beta$ function actually does behave like this.

The point $\alpha^*$ in figure 15.7(*a*) is called an ultraviolet-stable fixed point: $\alpha$ 'flows' towards it at large $|q^2|$. In the case of QCD, the $\beta$ function starts out negative, so that the corresponding behaviour (with a zero at a $\alpha_s^* \neq 0$) would look like that shown in figure 15.7(*b*). In this case, the reader can check (problem 15.4) that $\alpha_s^*$ is reached in the infrared limit $q^2 \to 0$, and so $\alpha_s^*$ is called an infrared-stable fixed point. Clearly it is the slope of $\beta$ near the fixed point that determines whether it is ultraviolet or infrared stable. This applies equally to a fixed point at the origin, so that QED is infrared stable at $\alpha = 0$ while QCD is ultraviolet stable at $\alpha_s = 0$.

We must now point out to the reader an error in the foregoing analysis in the case of a gauge theory. The quantity $Z_2$ is not gauge invariant in QED (or QCD), and so $\gamma_2$ depends on the choice of gauge. This is really no surprise because the full fermion propagator itself is not gauge invariant (the free-field propagator is gauge invariant, of course). What ultimately matters is that the complete physical amplitude for any process, at a given order of $\alpha$, be gauge invariant. Thus, the analysis given here really only applies—in this simple form—to non-gauge theories, such as the ABC model or to gauge-invariant quantities.

This is an appropriate point at which to consider the treatment of quark masses in the RGE-based approach. Up to now we have simply assumed that the relevant $|q^2|$ is very much greater than all quark masses, the latter therefore being neglected. While this may be adequate for the light quarks u, d, s, it seems surely a progressively worse assumption for c, b and t. However, in thinking about how to re-introduce the quark masses into our formalism, we are at once faced with a difficulty: how are they to be defined? For an unconfined particle such as a lepton, it seems natural to define 'the' mass as the position of the pole of the propagator (i.e. the 'on-shell' value $p^2 = m^2$), a definition we followed in chapters 10 and 11. Significantly, renormalization is required (in the shape of a mass counter-term) to achieve a pole at the 'right' physical mass $m$, in this sense. But this prescription cannot be used for a confined particle, which never 'escapes' beyond the range of the confining forces and whose propagator can, therefore, never approach the free form $\sim (\not{p} - m)^{-1}$.

Our present perspective on renormalization suggests an obvious way forward. Just as there was, in principle, no *necessity* to define the QED coupling parameter $e$ via an on-shell prescription, so here a mass parameter in the Lagrangian can be defined in any way we find convenient: all that is necessary is that it should be possible to determine its value from experiment. Effectively, we are regarding the '$m$' in a term such as $-m\bar{\hat{\psi}}(x)\hat{\psi}(x)$ as a 'coupling constant' having mass dimension 1 (and, after all, the ABC coupling itself had mass dimension 1). Incidentally, the operator $\bar{\hat{\psi}}(x)\hat{\psi}(x)$ *is* gauge invariant, as is any such *local* operator. Taking this point of view, it is clear that a renormalization scale will be involved in such a general definition of mass and we must expect to see our mass parameters 'evolve' with this scale, just as the gauge (or other) couplings do. In turn, this will get translated into a $|q^2|$-dependence of the mass parameters, just as for $\alpha(|q^2|)$ and $\alpha_s(|q^2|)$.

The RGE in such a scheme now takes the form

$$\left[\mu^2\frac{\partial}{\partial\mu^2} + \beta(\alpha_s)\frac{\partial}{\partial\alpha_s} + \sum_i \gamma_i(\alpha_s) + \gamma_m(\alpha_s)m\frac{\partial}{\partial m}\right]R(|q^2|/\mu^2, \alpha_s, m/|q|) = 0 \tag{15.69}$$

where the partial derivatives are taken at fixed values of the other two variables. Here the $\gamma_i$ are the anomalous dimensions relevant to the quantity $R$, and $\gamma_m$ is an analogous 'anomalous mass dimension', arising from finite shifts in the mass parameter when the scale $\mu^2$ is changed. Just as with the solution (15.66) of (15.63), the solution of (15.69) is given in terms of a 'running mass' $m(|q^2|)$. Formally, we can think of $\gamma_m$ in (15.69) as analogous to $\beta(\alpha_s)$ and $\ln m$ as analogous to $\alpha_s$. Then equation (15.42) for the running $\alpha_s$,

$$\frac{\partial\alpha_s(|q^2|)}{\partial t} = \beta(\alpha_s(|q^2|)) \tag{15.70}$$

where $t = \ln(|q^2|/\mu^2)$, becomes

$$\frac{\partial(\ln m(|q^2|))}{\partial t} = \gamma_m(\alpha_s(|q|^2)). \tag{15.71}$$

Equation (15.71) has the solution

$$m(|q^2|) = m(\mu^2)\exp\int_{\mu^2}^{|q^2|} \mathrm{d}\ln|q'^2|\gamma_m(\alpha_s(|q'^2|)). \tag{15.72}$$

To one-loop order in QCD, $\gamma_m(\alpha_s)$ turns out to be $-\frac{1}{\pi}\alpha_s$ (Peskin and Schroeder 1995, section 18.1). Inserting the one-loop solution for $\alpha_s$ in the form

(15.54), we find

$$m(|q^2|) = m(\mu^2)\left[\frac{\ln(\mu^2/\Lambda^2)}{\ln(|q^2|/\Lambda^2)}\right]^{1/\pi b} \tag{15.73}$$

where $(\pi b)^{-1} = 12/(33 - 2N_\mathrm{f})$. Thus the quark masses decrease logarithmically as $|q^2|$ increases, rather like $\alpha_\mathrm{s}(|q^2|)$. It follows that, in general, quark mass effects are suppressed both by explicit $m^2/|q^2|$ factors and by the logarithmic decrease given by (15.73). Further discussion of the treatment of quark masses is contained in Ellis *et al* (1996) section 2.4.

## 15.5 Some technicalities

We conclude our discussion of RGE's by commenting on a number of technical matters to which we should draw the reader's attention.

First, we have—for the sake of simplicity of exposition—conducted the entire discussion of renormalization effects in the framework of regularization by means of the ultraviolet cut-off $\Lambda$. However, we saw in section 11.3 that this was a potentially dangerous procedure in a gauge theory, since it could spoil gauge invariance. The latter is vital for two reasons: renormalizability depends upon it, and so does the elimination of unphysical states (i.e. the preservation of unitarity). While several gauge-invariant regularizations are available, the one now used most widely is 'dimensional regularization', due to 't Hooft and Veltman (1972). We describe this method briefly here: some more details are given in appendix N.

The basic idea is very simple (if, at first, rather strange). It is based on the observation that a typical logarithmically ultraviolet divergent one-loop diagram, such as the photon vacuum polarization with amplitude (see (11.23) and (11.24))

$$\Pi_\gamma^{[2]}(q^2) = 8e^2\mathrm{i}\int_0^1 \mathrm{d}x \int \frac{\mathrm{d}^4k'}{(2\pi)^4}\frac{x(1-x)}{(k'^2 - \Delta_\gamma + \mathrm{i}\epsilon)^2}, \tag{15.74}$$

would converge if the number of dimensions over which $k'$ was integrated were less than four.[4] Thus, if one can somehow calculate the integral as an analytic function of the dimensionality $d$ of spacetime, it will converge for $d < 4$ and have an identifiable singularity as $d$ approaches four, which the process of renormalization can remove.

The first step is to consider the $k'_0$ integral as a contour integral in the $k'_0$ plane, as we did in figure 10.8. This time, however, instead of evaluating it with a cut-off, we *rotate* the contour $C_\mathrm{R}$ in an anti-clockwise direction (thus avoiding the poles the location of which is determined by the '$+\mathrm{i}\epsilon$' term), so that it runs

[4] Note that dimensional regularization can equally well be used to deal with infrared divergences, by allowing $d$ to be greater than four. For example, the infrared divergence of section 15.1 can be handled this way.

along the imaginary axis: $k_0' \to \mathrm{i}k_4'$. Then (15.74) becomes

$$-8e^2 \int_0^1 \mathrm{d}x \int \frac{\mathrm{d}^4 k_{\mathrm{E}}'}{(2\pi)^4} \frac{x(1-x)}{(k_{\mathrm{E}}'^2 + \Delta_\gamma)^2} \tag{15.75}$$

where $k_{\mathrm{E}} = (k_4', \boldsymbol{k}')$ is the 'Euclidean' 4-momentum. Note that for $q^2 < 0$, the denominator is never zero, so there is no need to include any '$\mathrm{i}\epsilon$'. In dimensional regularization, one replaces the $k_{\mathrm{E}}'$-integral in (15.75) by

$$\int \frac{\mathrm{d}^d k_{\mathrm{E}}'}{(2\pi)^d} \frac{1}{(k_{\mathrm{E}}'^2 + \Delta_\gamma)^2} = \frac{1}{(4\pi)^{d/2}} \frac{\Gamma(2-d/2)}{\Gamma(2)} \left(\frac{1}{\Delta_\gamma}\right)^{2-d/2} \tag{15.76}$$

where $\Gamma$ is the gamma function (see, for example, Boas 1983, chapter 11). $\Gamma$ is related to the factorial function for integer values of its argument and satisfies

$$\Gamma(z+1) = z\Gamma(z) \qquad \Gamma(n) = (n-1)! \qquad \Gamma(1) = 1 \tag{15.77}$$

for general $z$ and integer $n$. It is clear from (15.77) that $\Gamma(z)$ has a pole at $z = 0$, so that (15.76) is indeed singular when $d = 4$. To isolate the singular behaviour we use the approximation

$$\Gamma(2-d/2) = \frac{2}{\epsilon} - \gamma + O(\epsilon) \tag{15.78}$$

where $\epsilon = 4 - d$ and $\gamma$ (not to be confused with the anomalous dimensions!) is the Euler–Mascheroni constant, having the value $\gamma \approx 0.5772$. Comparing (10.51) and (15.76), we can see that we may crudely identify '$\frac{1}{\epsilon} \sim \ln \Lambda$'.

Integrals such as (15.76) but with powers of $k_{\mathrm{E}}'$ in the numerator can be evaluated similarly (see appendix N). Using these results one finds that the non-gauge-invariant part of (11.18) does indeed cancel in this regularization procedure (problem 15.5).

We may expand the right-hand side of (15.76) in powers of $\epsilon$, using $x^\epsilon = e^{\epsilon \ln x} \approx 1 + \epsilon \ln x + \cdots$. We obtain the result (problem 15.6)

$$\frac{1}{(4\pi)^2}\left[\frac{2}{\epsilon} - \gamma + \ln 4\pi - \ln \Delta_\gamma + O(\epsilon)\right]. \tag{15.79}$$

The appearance of the dimensional quantity $\Delta_\gamma$ inside the logarithm is undesirable. We may rectify this by noting that although the fine structure constant $\alpha$ is dimensionless in four dimensions, the field dimension will change when we go to $d$ dimensions and, hence, so will that of the couplings if we want to keep the action dimensionless. Problem 15.7 shows that in $d$ dimensions the coupling $e$ has dimension $(\text{mass})^{\epsilon/2}$. It is natural, therefore, to rewrite the original $\alpha$ as $\alpha\mu^\epsilon$, where the new $\alpha$ is dimensionless in $d$ dimensions and $\mu$ is 'some mass'. In that case, the logarithm in (15.79) becomes $\ln(\Delta_\gamma/\mu^2)$. The reader will not need to be told that $\mu$ may be identified, most conveniently, with the renormalization scale introduced earlier.

The question now arises of how renormalization is to be done in this approach—and the answer is, just as before. We can, if we wish, define the renormalized $\bar{\Pi}_\gamma^{[2]}(q^2)$ by subtracting from $\Pi_\gamma^{[2]}(q^2)$ its value at $q^2 = 0$, as in (11.32). This will, in fact, give exactly the same result as (11.36), the $\gamma$ and $\ln 4\pi$ terms in (15.79) disappearing along with the $2/\epsilon$ singularity. However, we know very well by now that this is not the only possibility and that we can instead 'subtract' at a different point, say $q^2 = -\mu^2$. But in this dimensional regularization approach, a very simple prescription is most appealing: why not agree to define the renormalized $\Pi_\gamma^{[2]}$ by simply throwing away the singular term $2/\epsilon$ in (15.79)? This makes the remainder perfectly finite as $\epsilon \to 0$, though admittedly there are some odd-looking constants present. Such a procedure is called 'minimal subtraction', denoted by MS. Even better, from the point of view of simplicity, would be to get rid of the $-\gamma + \ln 4\pi \approx 1.95$ as well (this being a 'finite renormalization'), which is, after all, not that small numerically. This is called 'modified minimal subtraction' or $\overline{\text{MS}}$ ('em-ess-bar') (Bardeen *et al* 1978). This scheme tends to reduce loop corrections to their simplest form, but of course the resultant parameters may be less easily related to physically measurable quantities than in the 'on-shell' (i.e. at $q^2 = 0$) scheme. $\overline{\text{MS}}$ is the most widely used scheme as far as RGE-type applications are concerned.

It is now clear that, in addition to the (potential) dependence of calculated quantities on the renormalization scale $\mu^2$, there will also be a (potential) dependence on the renormalization scheme which is used. In fact, the full RGE equations of Stueckelberg and Peterman (1953) include variation due to the (suitably parametrized) renormalization scheme, and express the ultimate independence of physical qualities both of the choice of scale and of the choice of scheme.

We make two immediate points. First, the parameter $\Lambda_{\text{QCD}}$ introduced in (15.55) is scheme-dependent. The change from one scheme 'A' to another 'B' must involve a finite renormalization of the form (Ellis *et al* 1996, section 2.5)

$$\alpha_s^{\text{B}} = \alpha_s^{\text{A}}(1 + c_1\alpha_s^{\text{A}} + \cdots). \tag{15.80}$$

Note that this implies that the first two coefficients of the $\beta$ function are unchanged under this transformation, so that they are scheme-independent. From (15.55), the two corresponding values of $\Lambda_{\text{QCD}}$ are related by

$$\ln\left(\frac{\Lambda_{\text{B}}}{\Lambda_{\text{A}}}\right) = \frac{1}{2}\int_{\alpha_s^{\text{A}}(|q^2|)}^{\alpha_s^{\text{B}}(|q^2|)} \frac{\mathrm{d}x}{bx^2(1+\cdots)} \tag{15.81}$$

$$= \frac{c_1}{2b} \tag{15.82}$$

where we have taken $|q^2| \to \infty$ in (15.81) since the left-hand side is independent of $|q^2|$. Hence, the relationship between the $\Lambda_{\text{QCD}}$'s in different schemes is determined by the one-loop calculation which gives $c_1$ in (15.82). For example, changing from MS to $\overline{\rm MS}$ gives (problem 15.8)

$$\Lambda^2_{\overline{\rm MS}} = \Lambda^2_{\rm MS} \exp(\ln 4/\pi - \gamma). \tag{15.83}$$

For this and other reasons (Ellis *et al* 1996, section 2.5), it is more common to relate data to the value of $\alpha_s$ at a particular $|q^2|$ value, usually taken to be $M_Z^2$.

Second, we must stress that the $\mu^2$-independence of physical quantities only holds when they are evaluated *exactly*. As soon as the perturbative expansion is truncated, $\mu^2$-independence will break down. In general, it can be shown that changing the scale in something which has been calculated to $O(\alpha_s^n)$ induces changes which are $O(\alpha_s^{n+1})$. The more terms in the series one has, the less the effect of truncation should be—but, in practice, at any finite order, we may wonder if there is a 'best' renormalization prescription to use, which minimizes the various ambiguities. We shall not pursue this particular technicality any further here, referring the interested reader to Pennington (1983) section 4.2 or to Ellis *et al* (1996) section 3.1, for example.

## 15.6 $\sigma(e^+e^- \to$ hadrons) revisited

Armed with this new-found sophistication as regards renormalization matters, we may now return to the physical process which originally sparked this extensive detour. The higher-order (in $\alpha_s$) corrections to (15.2) are written as

$$\sigma = \sigma_{\rm pt}\left[1 + \frac{\alpha_s(\mu^2)}{\pi} + \sum_{n=2}^{\infty} C_n(s/\mu^2)\left(\frac{\alpha_s(\mu^2)}{\pi}\right)^n\right] \tag{15.84}$$

(see Ellis *et al* 1996, section 3.1, or the review of QCD by Hinchcliffe in Hagiwara *et al* 2002). The coefficient $C_2(1)$ was calculated by Dine and Sapirstein (1979), Chetyrkin *et al* (1979) and by Celmaster and Gonsalves (1980), and it has the value 1.411 for five flavours. The coefficient $C_3(1)$ was calculated by Samuel and Surguladze (1991) and by Gorishny *et al* (1991), and is equal to $-12.8$ for five flavours. The $\mu^2$-dependence of the coefficients is now fixed by the requirement that, order by order, the series (15.84) should be independent of the choice of scale $\mu^2$ (this is a 'direct' way of applying the RGE idea). Consider, for example, truncating at the $n = 2$ stage:

$$\sigma \approx \sigma_{\rm pt}\left(1 + \frac{\alpha_s(\mu^2)}{\pi} + C_2(s/\mu^2)(\alpha_s(\mu^2)/\pi)\right)^2. \tag{15.85}$$

Differentiating with respect to $\mu^2$ and setting the result to zero we obtain (problem 15.8)

$$\mu^2 \frac{{\rm d}C_2}{{\rm d}\mu^2} = -\frac{\pi\beta(\alpha_s(\mu^2))}{(\alpha_s(\mu^2))^2} \tag{15.86}$$

where an $O(\alpha_s^3)$ term has been dropped. Substituting the one-loop result (15.49)—as is consistent to this order—we find

$$C_2(s/\mu^2) = C_2(1) - \pi b \ln(s/\mu^2). \tag{15.87}$$

The second term on the right-hand side of (15.87) gives the contribution identified in (15.3).

In practice, corrections are necessary to account for effects due to the b and t quark masses (Chetyrkin and Kuhn 1993), and (at sufficiently high $\sqrt{s}$) for $Z^0$ effects. The current fitted value of $\alpha_s(M_Z^2)$ is 0.012 with an error of less than 5% (Hagiwara *et al* 2002).

The terms in (15.84) which involve logarithms are referred to as 'scaling violations' (compare the discussion of anomalous dimension in section (15.4)). The parton model cross-section $\sigma_{\rm pt}$ is proportional to $s^{-1}$ as could be predicted on simple dimensional grounds, if all masses are neglected. Thus, if this were the only contribution to $\sigma$, the latter would scale as $\lambda^{-2}$ when the momenta are all scaled by a factor $\lambda$. The logarithmic terms violate this simple (power law) scaling. We should also wonder whether something similar will happen to the *structure functions* introduced in section 9.1 and predicted to be scale-invariant functions in the free-parton model. This will be the topic of the final section of this chapter.

## 15.7 QCD corrections to the parton model predictions for deep inelastic scattering: scaling violations

As we saw in section 9.2, the parton model provides a simple intuitive explanation for the experimental observation that the nucleon structure functions in deep inelastic scattering depend, to a good first approximation, only on the dimensionless ratio $x = Q^2/2M\nu$, rather than on $Q^2$ and $\nu$ separately: this behaviour is referred to as 'scaling'. Here $M$ is the nucleon mass, and $Q^2$ and $\nu$ are defined in (9.7) and (9.8). In this section we shall show how QCD corrections to the simple parton model, calculated using RGE techniques, predict observable violations of scaling in deep inelastic scattering. As we shall see, comparison between the theoretical predictions and experimental measurements provides strong evidence for the correctness of QCD as the theory of nucleonic constituents.

### 15.7.1 Uncancelled mass singularities

The free-parton model amplitudes we considered in chapter 9 for deep inelastic lepton–nucleon scattering were of the form shown in figure 15.8 (cf figure 9.4). The obvious first QCD corrections will be due to real gluon emission by either the initial or final quark, as shown in figure 15.9, but to these we must add the one-loop virtual gluon processes of figure 15.10 in order (see later) to get rid of

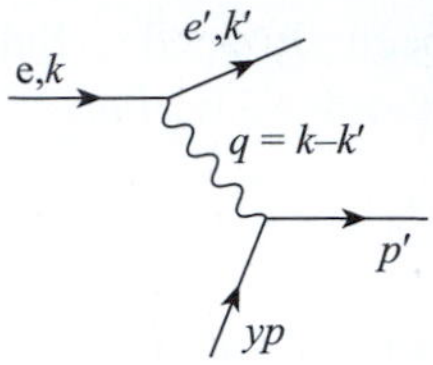


**Figure 15.8.** Electron–quark scattering via one-photon exchange.

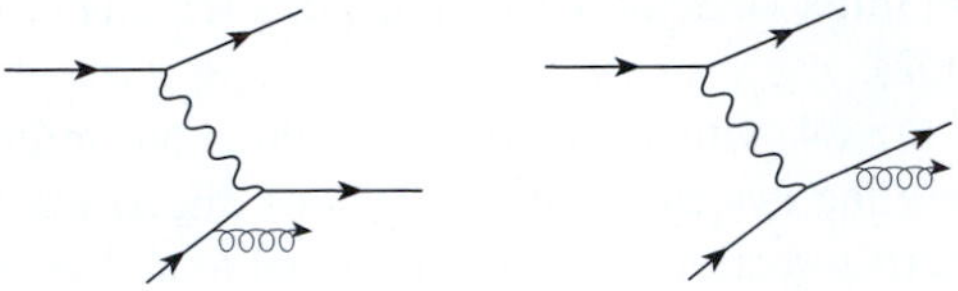

**Figure 15.9.** Electron–quark scattering with single-gluon emission

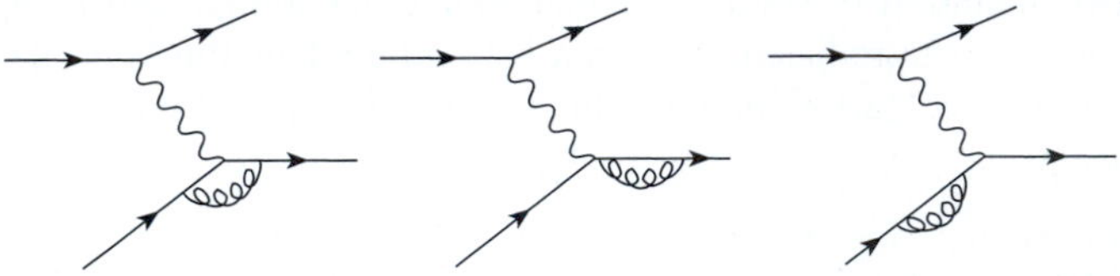

**Figure 15.10.** Virtual single–gluon corrections to figure 15.8.

infrared divergences similar to those encountered in section 15.1, and also the diagram of figure 15.11 corresponding to the presence of gluons in the nucleon. To simplify matters, we shall consider what is called a 'non-singlet structure function' $F_2^{\text{NS}}$, such as $F_2^{\text{ep}}$–$F_2^{\text{en}}$ in which the (flavour) singlet gluon contribution cancels out, leaving only the diagrams of figures 15.9 and 15.10.

We now want to perform, for these diagrams, calculations analogous to those of section 9.2, which enabled us to find the e–N structure functions $\nu W_2$ and $MW_1$ from the simple parton process of figure 15.8. There are two problems here: one is to find the parton level $W$'s corresponding to figure 15.9 (leaving aside figure 15.10 for the moment)—cf equations (9.29) and (9.30) in the case of the free-parton diagram in figure 15.8; the other is to relate these parton $W$'s to observed nucleon $W$'s via an integration over momentum fractions. In section 9.2 we solved the first problem by explicitly calculating the parton level $\mathrm{d}^2\sigma^i/\mathrm{d}Q^2\mathrm{d}\nu$ and picking off the associated $\nu W_2^i$, $W_1^i$. In principle, the same can be done here, starting from the five-fold differential cross-section for our $\mathrm{e}^- + \mathrm{q} \to \mathrm{e}^- + \mathrm{q} + \mathrm{g}$ process. However, a simpler—if somewhat heuristic—way is available. We note from (9.46) that, in general, $F_1 = MW_1$ is given by the transverse virtual photon

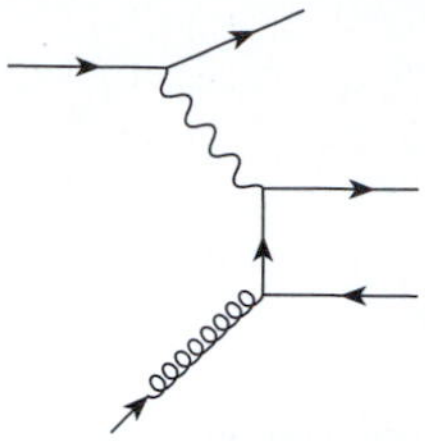

**Figure 15.11.** Electron–gluon scattering with q̄q production.

cross-section

$$W_1 = \sigma_T/(4\pi\alpha^2/K) = \tfrac{1}{2}\sum_{\lambda=\pm 1} \epsilon^*_\mu(\lambda)\epsilon_\nu(\lambda)W^{\mu\nu} \tag{15.88}$$

where $W^{\mu\nu}$ was defined in (9.3). Further, the Callan–Gross relation is still true (the photon only interacts with the charged partons, which are quarks with spin-$\frac{1}{2}$ and charge $e_i$), and so

$$F_2/x = 2F_1 = 2MW_1 = \sigma_T/(4\pi\alpha^2/2MK). \tag{15.89}$$

These formulae are valid for both parton and proton $W_1$'s and $W^{\mu\nu}$'s, with appropriate changes for parton masses $\hat{M}$. Hence, the parton level $2\hat{F}_1$ for figure 15.9 is just the transverse photon cross-section as calculated from the graphs of figure 15.12, divided by the factor $4\pi^2\alpha/2\hat{M}\hat{K}$, where, as usual, '^' denotes kinematic quantities in the corresponding parton process. This cross-section, however, is—apart from a colour factor—just the virtual Compton cross-section calculated in section 8.6. Also, taking the same (Hand) convention for the individual photon flux factors,

$$2\hat{M}\hat{K} = \hat{s}. \tag{15.90}$$

Thus, for the parton processes of figure 15.9,

$$\begin{aligned} 2\hat{F}_1 &= \hat{\sigma}_T/(4\pi^2\alpha/2\hat{M}\hat{K}) \\ &= \frac{\hat{s}}{4\pi^2\alpha}\int_{-1}^{1} \mathrm{d}\cos\theta \frac{4}{3}\frac{\pi e_i{}^2\alpha\alpha_s}{\hat{s}}\left(-\frac{\hat{t}}{\hat{s}} - \frac{\hat{s}}{\hat{t}} + \frac{2\hat{u}Q^2}{\hat{s}\hat{t}}\right) \end{aligned} \tag{15.91}$$

where, in going from (8.180) to (15.91), we have inserted a colour factor $\frac{4}{3}$ (problem 15.9), renamed the variables $t \rightarrow u, u \rightarrow t$ in accordance with figure 15.12, and replaced $\alpha^2$ by $e_i{}^2\alpha\alpha_s$.

Before proceeding with (15.91), it is helpful to consider the other part of the calculation—namely the relation between the nucleon $F_1$ and the parton $\hat{F}_1$. We mimic the discussion of section 9.2 but with one significant difference: the quark 'taken' from the proton has momentum fraction $y$ (momentum $yp$) but now

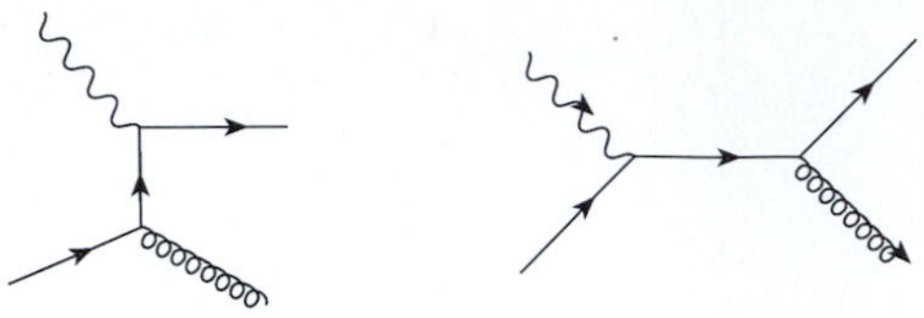

**Figure 15.12.** Virtual photon processes entering into figure 15.9.

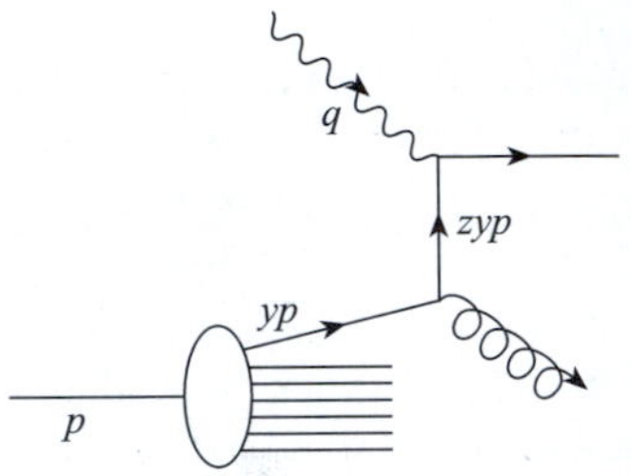


**Figure 15.13.** The first process of figure 15.12, viewed as a contribution to e$^-$–nucleon scattering.

its longitudinal momentum must be degraded in the final state due to the gluon bremsstrahlung process we are calculating. Let us call the quark momentum after gluon emission $zyp$ (figure 15.13). Then, assuming as in section 9.2 that it stays on-shell, we have

$$q^2 + 2zyq \cdot p = 0 \tag{15.92}$$

or

$$x = yz \qquad x = Q^2/2q \cdot p \qquad q^2 = -Q^2 \tag{15.93}$$

and we can write (cf (9.31))

$$\frac{F_2}{x} = 2F_1 = \sum_i \int_0^1 \mathrm{d}y\, f_i(y) \int_0^1 \mathrm{d}z\, 2\hat{F}_1^i \delta(x - yz) \tag{15.94}$$

where the $f_i(y)$ are the momentum distribution functions introduced in section 9.2 (we often call them $q(x)$ or $g(x)$ as the case may be) for parton type $i$, and the sum is over contributing partons. The reader may enjoy checking that (15.94) does reduce to (9.34) for free partons by showing that in that case $2\hat{F}_1^i = e_i^2\delta(1 - z)$ (see Halzen and Martin 1984, section 10.3 for help), so that $2F_1^{\text{free}} = \sum_i e_i^2 f_i(x)$.

To proceed further with the calculation (i.e. of (15.91) inserted into (15.94)), we need to look at the kinematics of the $\gamma$q $\to$ qg process, in the CMS. Referring to figure 15.14, we let $k$, $k'$ be the magnitudes of the CMS momenta $\boldsymbol{k}$, $\boldsymbol{k}'$. Then

$$\hat{s} = 4k'^2 = (yp + q)^2 = Q^2(1 - z)/z \qquad z = Q^2/(\hat{s} + Q^2)$$

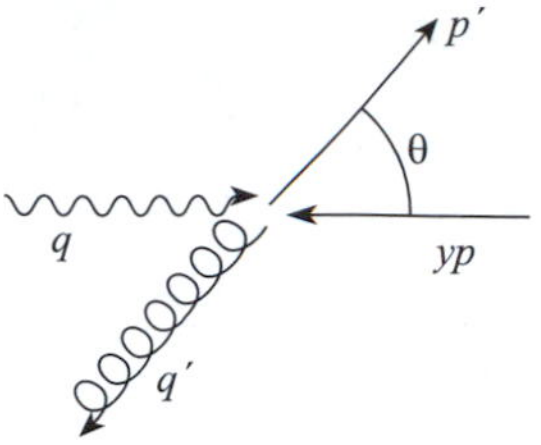


**Figure 15.14.** Kinematics for the parton process of figure 15.13.

$$\begin{aligned}\hat{t} &= (q-p')^2 = -2kk'(1-\cos\theta) = -Q^2(1-c)/2z \qquad c = \cos\theta \\ \hat{u} &= (q-q')^2 = -2kk'(1+\cos\theta) = -Q^2(1+c)/2z.\end{aligned} \tag{15.95}$$

We now note that in the integral (15.91) for $\hat{F}_1$, when we integrate over $c = \cos\theta$, we shall obtain an infinite result

$$\sim \int^1 \frac{\mathrm{d}c}{1-c} \tag{15.96}$$

associated with the vanishing of $\hat{t}$ in the 'forward' direction (i.e. when $q$ and $p'$ are parallel). This is a divergence of the 'collinear' type, in the terminology of section 14.4—or, as there, a 'mass singularity', occurring in the zero quark mass limit. If we simply replace the propagator factor $\hat{t}^{-1} = [(q-p')^2]^{-1}$ by $[(q-p')^2 - m^2]^{-1}$, where $m$ is a quark mass, then (15.96) becomes

$$\sim \int^1 \frac{\mathrm{d}c}{(1+2m^2z/Q^2) - c} \tag{15.97}$$

which will produce a factor of the form $\ln(Q^2/m^2)$ as $m^2 \to 0$. Thus $m$ regulates the divergence. As anticipated in section 15.1, we have here an uncancelled mass singularity and it *violates scaling*. This crucial physical result is present in the lowest-order QCD correction to the parton model, in this case. Such logarithmic violations of scaling are a characteristic feature of all QCD corrections to the free (scaling) parton model.

We may calculate the coefficient of the $\ln Q^2$ term by retaining in (15.91) only the terms proportional to $\hat{t}^{-1}$:

$$2\hat{F}_1^i \approx e_i^2 \int_{-1}^{1} \frac{\mathrm{d}c}{1-c} \left( \frac{\alpha_s}{2\pi} \frac{4}{3} \frac{1+z^2}{1-z} \right) \tag{15.98}$$

and so, for just one quark species, this QCD correction contributes (from (15.94)) a term

$$\frac{e_i^2 \alpha_s}{2\pi} \int_x^1 \frac{\mathrm{d}y}{y} q(y) \{ P_{qq}(x/y) \ln(Q^2/m^2) + C(x/y) \} \tag{15.99}$$

to $2F_1$, where

$$P_{qq}(z) = \frac{4}{3}\left(\frac{1+z^2}{1-z}\right) \tag{15.100}$$

is called a 'splitting function', and $C(x/y)$ has no mass singularity. The function $P_{qq}$ has an important physical interpretation: it is the probability that a quark, having radiated a gluon, is left with the fraction $z$ of its original momentum. Similar functions arise in QED in connection with what is called the 'equivalent photon approximation' (Weizsäcker 1934, Williams 1934, Chen and Zerwas 1975). The application of these techniques to QCD corrections to the free parton model is due to Altarelli and Parisi (1977) who thereby opened the way to the previous much simpler and more physical way of understanding scaling violations, which had previously been discussed mainly within the rather technical operator product formalism (Wilson 1969).

Our result so far is, therefore, that the 'free' quark distribution function $q(x)$, which depended only on the scaling variable $x$, becomes modified to

$$q(x) + \frac{\alpha_s}{2\pi}\int_x^1 \frac{dy}{y} q(y)\{P_{qq}(x/y)\ln(Q^2/m^2) + C(x/y)\} \tag{15.101}$$

due to lowest-order gluon radiation. Clearly, this corrected distribution function violates scaling because of the $\ln Q^2$ term but the result as it stands cannot represent a well-controlled approximation, since it diverges as $m^2 \to 0$. We must find some way of making sense, physically, of this uncancelled mass divergence in (15.101).

### 15.7.2 Factorization and the DGLAP equation

The key (following Ellis *et al* 1996, section 4.3.2) is to realize that when two partons are in the collinear configuration their relative momentum is very small, and hence the interaction between them is very strong, beyond the reach of a perturbative calculation. This suggests that we should absorb such uncalculable effects into a modified distribution function $q(x, \mu^2)$ given by

$$q(x,\mu^2) = q(x) + \frac{\alpha_s}{2\pi}\int_x^1 \frac{dy}{y} q(y) P_{qq}(x/y)\{\ln(\mu^2/m^2) + C(x/y)\} \tag{15.102}$$

which we have to take from experiment. Note that we have also absorbed the non-singular term $C(x/y)$ into $q(x, \mu^2)$. In terms of this quantity, then, we have

$$\begin{aligned} 2F_1(x, Q^2) &= e_i^2 \int_x^1 \frac{dy}{y} q(y,\mu^2)\left\{\delta(1-x/y) + \frac{\alpha_s}{2\pi}P_{qq}(x/y)\ln(Q^2/\mu^2)\right\} \\ &\equiv e_i^2 q(x, Q^2) \end{aligned} \tag{15.103}$$

to this order in $\alpha_s$, and for one quark type.

This procedure is, of course, very reminiscent of ultraviolet renormalization, in which ultraviolet divergences are controlled by similarly importing some quantities from experiment. In this example, we have essentially made use of the simple fact that

$$\ln(Q^2/m^2) = \ln(Q^2/\mu^2) + \ln(\mu^2/m^2). \tag{15.104}$$

The arbitrary scale $\mu$, which is analogous to a renormalization scale, is here referred to as a 'factorization scale'. It is the scale entering into the separation in (15.104) between one (uncalculable) factor which depends on the infrared parameter $m$ but not on $Q^2$ and the other (calculable) factor which depends on $Q^2$. The scale $\mu$ can be thought of as one which separates the perturbative short-distance physics from the non-perturbative long-distance physics. Thus, partons emitted at small transverse momenta $< \mu$ (i.e. approximately collinear processes) should be considered as part of the hadron structure and are absorbed into $q(x, \mu^2)$. Partons emitted at large transverse momenta contribute to the short-distance (calculable) part of the cross-section. Just as for the renormalization scale, the more terms that can be included in the perturbative contributions to the mass-singular terms (i.e. beyond (15.101)), the weaker the dependence on $\mu$ will be. In fact, for most purposes the factorization scale is chosen to be the same as the renormalization scale, as the notation has already implicitly assumed. We have demonstrated the possibility of factorization only to $O(\alpha_s)$ but proofs to all orders in perturbation theory exist: a review is provided in Collins and Soper (1987).

Different *factorization schemes* can also be employed, depending on how the non-singular part $C(x/y)$ is treated. In (15.102), as pointed out, we absorbed all of it into $q(x, \mu^2)$. This is why we obtained the simple result $2F_1(x, Q^2) = e_i^2 q(x, Q^2)$ in (15.103). This scheme is called the 'deep inelastic' scheme (DIS) (Altarelli *et al* 1978a, 1978b). A more common scheme is that in which the mass singularity is regulated in dimensional regularization and only the '$\ln 4\pi - \gamma$' bit, in addition to the singularity, is absorbed into the distribution. This is called the $\overline{\text{MS}}$ factorization scheme. In this case $2F_1(x, Q^2)$ will be given by an expression of the form

$$F_1(x, Q^2) = e_i^2 \int_x^1 \frac{\mathrm{d}y}{y} q(y, Q^2) \left\{\delta(1 - x/y) + \frac{\alpha_s}{2\pi} C_{\overline{\text{MS}}}(x/y)\right\} \tag{15.105}$$

to this order, where $C_{\overline{\text{MS}}}$ is a calculable function. Naturally, in analysing data the same factorization scheme must be employed consistently throughout.

Returning now to (15.103), the reader can guess what is coming next: we shall impose the condition that the physical quantity $F_1(x, Q^2)$ must be independent of the choice of factorization scale $\mu^2$. Differentiating (15.103) partially with respect to $\mu^2$, and setting the result to zero, we obtain (to order $\alpha_s$ on the right-hand side)

$$\mu^2 \frac{\partial q(x, \mu^2)}{\partial \mu^2} = \frac{\alpha_s(\mu^2)}{2\pi} \int_x^1 \frac{\mathrm{d}y}{y} P_{qq}(x/y) q(y, \mu^2). \tag{15.106}$$

This equation is the analogue of equation (15.36) describing the running of the coupling $\alpha_s$ with $\mu^2$, and is a fundamental equation in the theory of perturbative applications of QCD. It is called the DGLAP equation, after Dokshitzer (1977), Gribov and Lipatov (1972) and Altarelli and Parisi (1977) (it is also often referred to as the Altarelli–Parisi equation). The derivation here is not rigorous—for example we have assumed that it is correct to use $\alpha_s(\mu^2)$ on the right-hand side. A more sophisticated treatment (Georgi and Politzer 1974, Gross and Wilczek 1974) confirms the result and extends it to higher orders.

Equation (15.106) shows that, although perturbation theory cannot be used to calculate the distribution function $q(x, \mu^2)$ at any particular value $\mu^2 = \mu_0^2$, it can be used to predict how the distribution *changes* (or 'evolves') as $\mu^2$ varies. (We recall from (15.103) that $q(x, \mu_0^2)$ can be found experimentally via $q(x, \mu_0^2) = 2F_1(x, Q^2 = \mu_0^2)/e_i^2$.) Replacing $\mu^2$ by $Q^2$ then tells us how the structure function evolves with $Q^2$, via (15.103).

In general, the right-hand side of (15.106) will have to be supplemented by terms (calculable from figure 15.11) in which quarks are generated from the gluon distribution. The equations must then be closed by a corresponding one describing the evolution of the gluon distributions (Altarelli 1982). Such equations can be qualitatively understood as follows. The change in the distribution for a quark with momentum fraction $x$, which absorbs the virtual photon, is given by the integral over $y$ of the corresponding distribution for a quark with momentum fraction $y$, which radiated away (via a gluon) a fraction $x/y$ of its momentum with probability $(\alpha_s/2\pi)P_{qq}(x/y)$. This probability is high for large momentum fractions: high-momentum quarks lose momentum by radiating gluons. Thus, there is a predicted tendency for the distribution function $q(x, \mu^2)$ to get smaller at large $x$ as $\mu^2$ increases, and larger at small $x$ (due to the build-up of slower partons), while maintaining the integral of the distribution over $x$ as a constant. The effect is illustrated qualitatively in figure 15.15. In addition, the radiated gluons produce more $q\bar{q}$ pairs at small $x$. Thus, the nucleon may be pictured as having more and more constituents, all contributing to its total momentum, as its structure is probed on ever smaller distance (larger $\mu^2$) scales.

### 15.7.3 Comparison with experiment

Data on nucleon structure functions do indeed show such a trend. Figure 15.16 shows the $Q^2$-dependence of the proton structure function $F_2^p(x, Q^2) = \sum e_i^2 x f_i(x, Q^2)$ for various fixed $x$ values, as compiled by B Foster, A D Martin and M G Vincter for the 2002 Particle Data Group review (Hagiwara *et al* 2002). Clearly at larger $x$ ($x \geq 0.13$), the function gets smaller as $Q^2$ increases, while at smaller $x$ it increases.

Fits to the data have been made in various ways. One (theoretically convenient) way is to consider 'moments' (Mellin transforms) of the structure

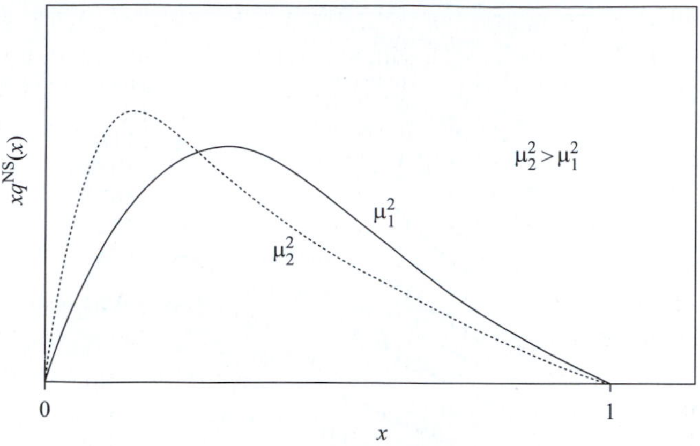


**Figure 15.15.** Evolution of the distribution function with $\mu^2$.

functions, defined by

$$M_q^n(t) = \int_0^1 dx x^{n-1} q(x,t) \tag{15.107}$$

where we have now introduced the variable $t = \ln \mu^2$. Taking moments of both sides of (15.106) and interchanging the order of the $x$ and $y$ integrations, we find that

$$\frac{dM_q^n(t)}{dt} = \frac{\alpha_s(t)}{2\pi} \int_0^1 dy\, y^{n-1} q(y,t) \int_0^y \frac{dx}{y} (x/y)^{n-1} P_{qq}(x/y). \tag{15.108}$$

Changing the variable to $z = x/y$ in the second integral and defining

$$\gamma_{qq}^n = 4 \int_0^1 dz z^{n-1} P_{qq}(z) \tag{15.109}$$

we obtain

$$\frac{dM_q^n(t)}{dt} = \frac{\alpha_s(t)}{8\pi} \gamma_{qq}^n M_q^n(t). \tag{15.110}$$

Thus the integral in (15.106)—which is of convolution type—has been reduced to product form by this transformation. Now we also know from (15.48) and (15.49) that

$$\frac{d\alpha_s}{dt} = -b\alpha_s^2 \tag{15.111}$$

with $b = (33 - 2N_f)/12\pi$ as usual, to this (one-loop) order. Thus (15.110) becomes

$$\frac{d \ln M_q^n}{d \ln \alpha_s} = -\frac{\gamma_{qq}^n}{8\pi b} \equiv -d_{qq}^n. \tag{15.112}$$

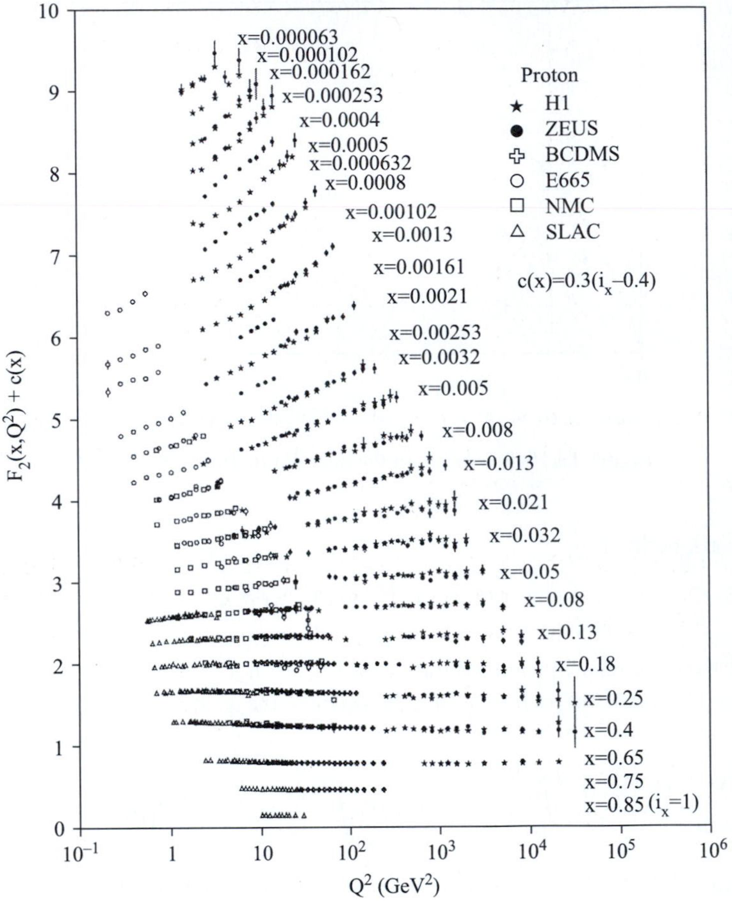


**Figure 15.16.** $Q^2$-dependence of the proton structure function $F_2^{\mathrm{p}}$ for various fixed $x$ values (Hagiwara *et al* 2002).

The solution to (15.112) is found to be

$$M_{\mathrm{q}}^n(t) = M_{\mathrm{q}}^n(t_0)\left(\frac{\alpha_{\mathrm{s}}(t_0)}{\alpha_{\mathrm{s}}(t)}\right)^{d_{\mathrm{qq}}^n}. \tag{15.113}$$

At this point we hit one more snag—but a familiar one. The function $P_{\mathrm{qq}}(z)$ of (15.100) is singular as $z \to 1$, in such a way as to make the integrals (15.109) for $\gamma_n$ diverge. This is clearly a standard infrared divergence (the quark momentum $yzp$ after gluon emission becomes equal to the quark momentum $yp$ before emission) and we expect that it can be cured by including the virtual gluon diagrams of figure 15.10, as indicated at the start of the section (and as was done analogously in the case of $\mathrm{e}^+\mathrm{e}^-$ annihilation). This has been verified explicitly by

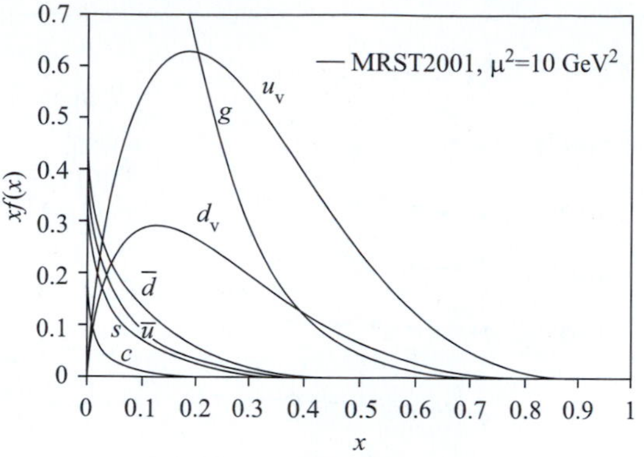


**Figure 15.17.** Distributions of $x$ times the unpolarized parton distributions $f(x, \mu^2)$ (where $f = u_\mathrm{V}, d_\mathrm{V}, \bar{u}, \bar{d}, s, c, g$) using the MRST2001 parametrization (Martin *et al* 2002) at a scale $\mu^2 = 10$ GeV$^2$ (from Hagiwara *et al* 2002).

Kim and Schilcher (1978) and by Altarelli *et al* (1978a, b, 1979). Alternatively, one can make the physical argument that the net number of quarks (i.e. the number of quarks minus the number of anti-quarks) of any flavour is conserved as $t$ varies:

$$\frac{\mathrm{d}}{\mathrm{d}t}\int_0^1 \mathrm{d}x\, q(x,t) = 0. \tag{15.114}$$

From (15.106) this implies

$$\int_0^1 \mathrm{d}x\, P_{\mathrm{qq}}^{+}(x,t) = 0 \tag{15.115}$$

where $P_{\mathrm{qq}}^{+}$ is the complete splitting function, including the effect of the gluon loops. This fixes the contribution of these loops (which only enter, in the leading log approximation, at $z \to 1$): for any function $f(z)$ regular as $z \to 1$, $P_{\mathrm{qq}}^{+}$ is defined by

$$\int_0^1 \mathrm{d}z\, f(z) P_{\mathrm{qq}}^{+}(z) = \int_0^1 \mathrm{d}z\, [f(z) - f(1)] P_{\mathrm{qq}}(z). \tag{15.116}$$

Applying this prescription to $\gamma_n$, we find (problem 15.10) that

$$\gamma_{\mathrm{qq}}^{n} = -\frac{8}{3}\left[1 - \frac{2}{n(n+1)} + 4\sum_{j=2}^{n}\frac{1}{j}\right] \tag{15.117}$$

and then

$$d_{\mathrm{qq}}^{n} = \frac{4}{33 - 2N_\mathrm{f}}\left[1 - \frac{2}{n(n+1)} + 4\sum_{j=2}^{n}\frac{1}{j}\right]. \tag{15.118}$$

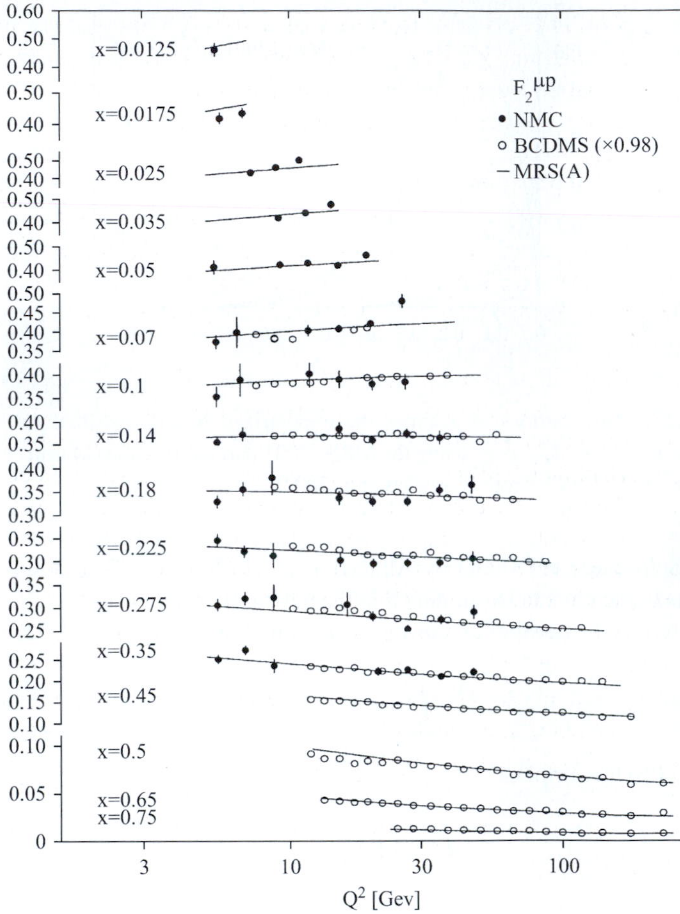


**Figure 15.18.** Data on the structure function $F_2$ in muon–proton deep inelastic scattering, from BCDMS (Benvenuti *et al* 1989) and NMC (Amaudruz *et al* 1992). The curves are QCD fits (Martin *et al* 1994) as described in the text.

We emphasize again that all the foregoing analysis is directly relevant only to distributions in which the flavour singlet gluon distributions do not contribute to the evolution equations. In the more general case, analogous splitting functions $P_{\mathrm{qg}}$, $P_{\mathrm{gq}}$ and $P_{\mathrm{gg}}$ will enter, folded appropriately with the gluon distribution function $g(x, t)$, together with the related quantities $\gamma^n_{\mathrm{qg}}$, $\gamma^n_{\mathrm{gq}}$ and $\gamma^n_{\mathrm{gg}}$. Equation (15.106) is then replaced by a $2 \times 2$ matrix equation for the evolution of the quark and gluon moments $M^n_{\mathrm{q}}$ and $M^n_{\mathrm{g}}$.

Returning to (15.113), one way of testing it is to plot the logarithm of one moment, $\ln M^n_{\mathrm{q}}$, *versus* the logarithm of another, $\ln M^m_{\mathrm{q}}$, for different $n, m$ values. These should give straight lines with slopes $(d^n_{\mathrm{qq}}/d^m_{\mathrm{qq}})$. Data support this prediction (Bosetti *et al* (1978)). However, since data do not exist for arbitrarily small $x$, the moments cannot be determined from the data without some additional assumptions. A more direct procedure, applicable to the non-singlet case too of course, is to choose a reference point $\mu_0^2$ and parametrize the parton distribution functions $f_i(x, t_0)$ in some way. These may then be evolved numerically, via the DGLAP equations, to the desired scale. Figure 15.17 shows a typical set of distributions at $\mu^2 = 10\ \text{GeV}^2$ (Martin *et al* 2002). A global numerical fit is then performed to determine the best values of the parameters, including $\Lambda_{\text{QCD}}$ which enters into $\alpha_s(t)$. An example of such a fit, due to Martin *et al* (1994), is shown in figure 15.18. For further details of QCD fits to deep inelastic data the reader is referred to Ellis *et al* (1996, chapter 4).

We conclude this chapter with two comments. First, the region of small $x$ (say $x \leq 10^{-2}$) requires special treatment and is the subject of ongoing theoretical and experimental interest (Ellis *et al* 1996 section 4.6). Second, the '$\gamma$' notation for the moments of the splitting functions (as in (15.109)) is not chosen accidentally. The same $\gamma$'s are indeed anomalous dimensions (cf section 15.4) of certain operators which appear in Wilson's 'operator product' approach to scaling violations, to which reference was made earlier (Wilson 1969). Readers keen to pursue this may consult Peskin and Schroeder (1995, chapter 18).

Finally, it is worth pausing to reflect on how far our understanding of *structure* has developed, via quantum field theory, from the simple 'fixed number of constituents' models which are useful in atomic and nuclear physics. When nucleons are probed on finer and finer scales, more and more partons (gluons, $q\bar{q}$ pairs) appear, in a way quantitatively predicted by QCD. The precise experimental confirmation of these predictions (and many others, as discussed by Ellis *et al* (1996)) constitutes a remarkable vote of confidence, by Nature, in relativistic quantum field theory.

## Problems

**15.1** Verify equation (15.11).

**15.2** Verify equation (15.28).

**15.3** Check that (15.51) can be rewritten as (15.54).

**15.4** Verify that for the type of behaviour of the $\beta$ function shown in figure 15.7(*b*), $\alpha_s^*$ is reached as $q^2 \to 0$.

**15.5** Verify using dimensional regularization that the non-gauge-invariant part of (11.18) cancels (see the text following equation (11.22)).

**15.6** Verify equation (15.79).

**15.7** Check that the electromagnetic charge $e$ has dimension $(\text{mass})^{\epsilon/2}$ in $d = 4 - \epsilon$ dimensions.

**15.8** Verify equation (15.83).

**15.9** Using the results of problem 14.6, explain why the colour factor for (15.91) is $\frac{4}{3}$.

**15.10** Verify equation (15.117).

# 16

# LATTICE FIELD THEORY AND THE RENORMALIZATION GROUP REVISITED

## 16.1 Introduction

Throughout this book, thus far, we have relied on perturbation theory as the calculational tool, justifying its use in the case of QCD by the smallness of the coupling constant at short distances: note, however, that this result itself required the summation of an infinite series of perturbative terms. As remarked at the end of section 15.3, the concomitant of asymptotic freedom is that $\alpha_s$ really does become strong at small $Q^2$ or at long distances of order $\Lambda_{QCD}^{-1} \sim 1$ fm. Here we have no prospect of getting useful results from perturbation theory: it is the *non-perturbative regime*. But this is precisely the regime in which quarks bind together to form hadrons. If QCD is indeed the true theory of the interaction between quarks, then it should be able to explain, ultimately, the vast amount of data that exists in low-energy hadronic physics. For example: what are the masses of mesons and baryons? Are there novel colourless states such as glueballs? Is $SU(2)_f$ or $SU(3)_f$ chiral symmetry spontaneously broken? What is the form of the effective interquark potential? What are the hadronic form factors, in electromagnetic (chapter 9) or weak (chapter 20) processes?

It is unlikely that answers to all these questions are going to be found by performing calculations analytically—as is possible, of course, for the tree diagrams of perturbation theory. Even in perturbation theory, however, one soon encounters integrals that have to be evaluated numerically. The standard way of doing this is to approximate the integral by some kind of discrete sum. Thus, a mesh of points (in general multi-dimensional) has to be chosen: perhaps it would make sense to formulate the theory on such a mesh—or 'lattice'—in the first place.

But there is a more fundamental point involved here. As we have seen several times in this book, one of the triumphs of theoretical physics over the past 50 years has been the success of renormalization techniques, first in 'taming' the ultraviolet divergences of quantum field theories and then in providing quantitative predictions for short-distance phenomena in QCD, via the RGE. But this immediately raises a question for any non-perturbative approach: how can we regulate the ultraviolet divergences and therefore define the theory, if we cannot get to grips with them via the specific divergent integrals supplied by perturbation theory? We need to be able to regulate the divergences in a way which does not rely on their appearance in the Feynman graphs of perturbation theory. As Wilson (1974, 1975) was the first to propose, one quite natural non-perturbative way of regulating ultraviolet divergences is to approximate continuous spacetime by a discrete lattice of points. Such a lattice will introduce a minimum distance—namely the lattice spacing '$a$' between neighbouring points. Since no two points can ever be closer than $a$, there is now a corresponding maximum momentum $\Lambda = \pi/a$ (see following equation (16.6)) in the lattice version of the theory. Thus the theory is automatically ultraviolet finite from the start, without presupposing the existence of any perturbative expansion; renormalization questions will, of course, enter when we consider the $a$ dependence of our parameters. As long as the lattice spacing is much smaller than the physical size of the hadrons one is studying, the lattice version of the theory should be a good approximation. Of course, Lorentz invariance is sacrificed in such an approach and replaced by some form of hypercubic symmetry: we must hope that for small enough $a$ this will not matter. We shall discuss how a simple field theory is 'discretized' in the next section: the following one will show how a gauge theory is discretized.

The discrete formulation of quantum field theory should be suitable for numerical computation. This at once seems to rule out any formalism based on non-commuting *operators*, since it is hard to see how they could be numerically simulated. Indeed, the same would be true of ordinary quantum mechanics. Fortunately, a formulation does exist which avoids operators: Feynman's *sum over paths* approach, which was briefly mentioned in section 5.2.2. This method is the essential starting point for the lattice approach to quantum field theory and it will be briefly introduced in section 16.4. The sum over paths approach does not involve quantum operators, but fermions still have to be accommodated somehow. The way this is done is briefly described in section 16.4 (see also appendix O).

It turns out that this formulation enables direct contact to be made between quantum field theory and *statistical mechanics*, as we shall discuss in section 16.5. This relationship has proved to be extremely fruitful, allowing physical insights and numerical techniques to pass from one subject to the other, in a way that has been very beneficial to both. In particular, the physics of renormalization and of the RGE is considerably illuminated from a lattice/statistical mechanics perspective, as we shall see in section 16.6. The chapter ends with some sample results obtained from lattice simulations of QCD.

## 16.2 Discretization

We start by considering a simple field theory involving a bosonic field $\phi$. Postponing until section 16.4 the question of exactly how we shall use it, we assume that we shall still want to formulate the theory in terms of an action of the form

$$S = \int \mathrm{d}^4x \, \mathcal{L}(\phi, \nabla\phi, \dot{\phi}). \tag{16.1}$$

It seems plausible that it might be advantageous to treat space and time as symmetrically as possible, from the start, by formulating the theory in 'Euclidean' space, instead of Minkowskian, by introducing $t = -\mathrm{i}\tau$: further motivation for doing this will be provided in section 16.4. In that case, the action (16.1) becomes

$$S \to -\mathrm{i}\int \mathrm{d}^3\boldsymbol{x}\,\mathrm{d}\tau\,\mathcal{L}\left(\phi, \boldsymbol{\nabla}\phi, \mathrm{i}\frac{\partial\phi}{\partial\tau}\right) \tag{16.2}$$

$$\equiv \mathrm{i}\int \mathrm{d}^3\boldsymbol{x}\,\mathrm{d}\tau\,\mathcal{L}_{\mathrm{E}} \equiv \mathrm{i}S_{\mathrm{E}}. \tag{16.3}$$

A typical free bosonic action is then

$$S_{\mathrm{E}}(\phi) = \tfrac{1}{2}\int \mathrm{d}^3\boldsymbol{x}\,\mathrm{d}\tau\,[(\partial_\tau\phi)^2 + (\boldsymbol{\nabla}\phi)^2 + m^2\phi^2]. \tag{16.4}$$

We now represent all of spacetime by a finite-volume 'hypercube'. For example, we may have $N_1$ lattice points along the $x$-axis, so that a field $\phi(x)$ is replaced by the $N_1$ numbers $\phi(n_1 a)$ with $n_1 = 0, 1, \ldots, N_1 - 1$. We write $L = N_1 a$ for the length of the cube side. In this notation, integrals and differentials are replaced by the finite difference expressions

$$\int \mathrm{d}x \to a\sum_{n_1} \qquad \frac{\partial\phi}{\partial x} \to \frac{1}{a}[\phi(n_1+1) - \phi(n_1)] \tag{16.5}$$

so that a typical integral (in one dimension) becomes

$$\int \mathrm{d}x\left(\frac{\partial\phi}{\partial x}\right)^2 \to a\sum_{n_1}\frac{1}{a^2}[\phi(n_1+1) - \phi(n_1)]^2. \tag{16.6}$$

As in all our previous work, we can alternatively consider a formulation in momentum space, which will also be discretized. It is convenient to impose periodic boundary conditions such that $\phi(x) = \phi(x+L)$. Then the allowed $k$-values may be taken to be $k_{\nu_1} = 2\pi\nu_1/L$ with $\nu_1 = -N_1/2+1, \ldots 0, \ldots N_1/2$ (we take $N_1$ to be even). It follows that the maximum allowed magnitude of the momentum is then $\pi/a$, indicating that $a^{-1}$ is (as anticipated) playing the role of our earlier momentum cut-off $\Lambda$. We then write

$$\phi(n_1) = \sum_{\nu_1}\frac{1}{(N_1 a)^{\frac{1}{2}}}\mathrm{e}^{\mathrm{i}2\pi\nu_1 n_1/N_1}\tilde{\phi}(\nu_1) \tag{16.7}$$

which has the inverse

$$\tilde{\phi}(\nu_1) = \left(\frac{a}{N_1}\right)^{\frac{1}{2}}\sum_{n_1}\mathrm{e}^{-\mathrm{i}2\pi\nu_1 n_1/N_1}\phi(n_1) \tag{16.8}$$

since (problem 16.1)

$$\frac{1}{N_1}\sum_{n_1=0}^{N_1-1} \mathrm{e}^{\mathrm{i}2\pi n_1(\nu_1-\nu_2)/N_1} = \delta_{\nu_1,\nu_2}. \tag{16.9}$$

Equation (16.9) is a discrete version of the $\delta$-function relation given in (E.25) of volume 1. A one-dimensional version of the mass term in (16.4) then becomes (problem 16.2)

$$\tfrac{1}{2}\int \mathrm{d}x\, m^2\phi(x)^2 \to \tfrac{1}{2}m^2\sum_{\nu_1}\tilde{\phi}(\nu_1)\tilde{\phi}(-\nu_1) \tag{16.10}$$

while

$$\frac{1}{2}\int \mathrm{d}x\left(\frac{\partial\phi}{\mathrm{d}x}\right)^2 \to \frac{2}{a^2}\sum_{\nu_1}\tilde{\phi}(\nu_1)\sin^2\left(\frac{\pi\nu_1}{N_1}\right)\tilde{\phi}(-\nu_1) \tag{16.11}$$

$$= \frac{1}{2a^2}\sum_{k_{\nu_1}}\tilde{\phi}(k_{\nu_1})4\sin^2\left(\frac{k_{\nu_1}a}{2}\right)\tilde{\phi}(-k_{\nu_1}). \tag{16.12}$$

Thus, a one-dimensional version of the free action (16.4) is

$$\frac{1}{2}\sum_{k_{\nu_1}}\tilde{\phi}(k_{\nu_1})\left[\frac{4\sin^2(k_{\nu_1}a/2)}{a^2}+m^2\right]\tilde{\phi}(-k_{\nu_1}). \tag{16.13}$$

In the continuum case, (16.13) would be replaced by

$$\frac{1}{2}\int\frac{\mathrm{d}k}{2\pi}\tilde{\phi}(k)[k^2+m^2]\tilde{\phi}(-k) \tag{16.14}$$

as usual, which implies that the propagator in the discrete case is proportional to

$$\left[\frac{4\sin^2(k_{\nu_1}a/2)}{a^2}+m^2\right]^{-1} \tag{16.15}$$

rather than to $[k^2+m^2]^{-1}$ (remember we are in one-dimensional Euclidean space). The two expressions do coincide in the continuum limit $a \to 0$. The manipulations we have been going through will be recognized by readers familiar with the theory of lattice vibrations and phonons.

Following the same procedure for fermion field leads, however, to difficulties. First note that the Euclidean Dirac matrices $\gamma_\mu^{\mathrm{E}}$ are related to the usual Minkowski ones $\gamma_\mu^{\mathrm{M}}$ by $\gamma_{1,2,3}^{\mathrm{E}} \equiv -\mathrm{i}\gamma_{1,2,3}^{\mathrm{M}}$, $\gamma_4^{\mathrm{E}} \equiv -\mathrm{i}\gamma_4^{\mathrm{M}} \equiv \gamma_0^{\mathrm{M}}$. They satisfy $\{\gamma_\mu^{\mathrm{E}},\gamma_\nu^{\mathrm{E}}\} = 2\delta_{\mu\nu}$ for $\mu = 1,2,3,4$. The Euclidean Dirac Lagrangian is then $\bar{\psi}(x)[\gamma_\mu^{\mathrm{E}}\partial_\mu + m]\psi(x)$, which should be written now in Hermitean form

$$m\bar{\psi}(x)\psi(x) + \tfrac{1}{2}\{\bar{\psi}(x)\gamma_\mu^{\mathrm{E}}\partial_\mu\psi(x) - (\partial_\mu\bar{\psi}(x))\gamma_\mu^{\mathrm{E}}\psi(x)\}. \tag{16.16}$$

The corresponding 'one-dimensional' discretized action is then

$$a\sum_{n_1} m\bar{\psi}(n_1)\psi(n_1) + \frac{1}{2}\left\{\sum_{n_1}\bar{\psi}(n_1)\gamma_1^{\rm E}\left[\frac{\psi(n_1+1)-\psi(n_1)}{a}\right]\right.$$
$$\left. - \sum_{n_1}\left(\frac{\bar{\psi}(n_1+1)-\bar{\psi}(n_1)}{a}\right)\gamma_1^{\rm E}\psi(n_1)\right\} \tag{16.17}$$

$$= a\sum_{n_1}\left\{m\bar{\psi}(n_1)\psi(n_1) + \frac{1}{2a^2}[\bar{\psi}(n_1)\gamma_1^{\rm E}\psi(n_1+1)\right.$$
$$\left. - \bar{\psi}(n_1+1)\gamma_1^{\rm E}\psi(n_1)]\right\}. \tag{16.18}$$

In momentum space this becomes (problem 16.3)

$$\sum_{k_{\nu_1}}\bar{\tilde{\psi}}(k_{\nu_1})\left[\mathrm{i}\gamma_1^{\rm E}\frac{\sin(k_{\nu_1}a)}{a} + m\right]\tilde{\psi}(-k_{\nu_1}) \tag{16.19}$$

and the inverse propagator is $[\mathrm{i}\gamma_1^{\rm E}\frac{\sin(k_{\nu_1}a)}{a} + m]$. Thus the propagator itself is

$$\left[m - \mathrm{i}\gamma_1^{\rm E}\frac{\sin(k_{\nu_1}a)}{a}\right]\Big/\left[m^2 + \frac{\sin^2(k_{\nu_1}a)}{a^2}\right]. \tag{16.20}$$

But here is the (first) problem with fermions: in addition to the correct continuum limit ($a \to 0$) found at $k_{\nu_1} \to 0$, an alternative finite $a \to 0$ limit is found at $k_{\nu_1} \to \pi/a$ (consider expanding $a^{-1}\sin[(\pi/a - \delta)a]$ for small $\delta$). Thus two modes survive as $a \to 0$, a phenomenon known as the 'fermion doubling problem' (actually in four dimensions there are 16 such corners of the hypercube). It is a consequence of the fact that the Dirac Lagrangian is linear in the derivatives.

Various solutions to this problem have been proposed (Wilson 1975, Susskind 1977, Banks *et al* 1976). Wilson (1974), for example, suggested adding a term of the form $\frac{1}{2}\bar{\psi}(n_1)[\psi(n_1+1) + \psi(n_1-1) - 2\psi(n_1)]$ to the Lagrangian, which changes our inverse propagator to

$$\left[\mathrm{i}\gamma_1^{\rm E}\frac{\sin(k_{\nu_1}a)}{a} + m\right] + \frac{1}{a}(1 - \cos(k_{\nu_1}a)). \tag{16.21}$$

By considering the expansion of the cosine near $k_{\nu_1} \approx 0$, it can be seen that the second term disappears in the continuum limit. However, for $k_{\nu_1} \approx \pi/a$ it gives a large term of order $1/a$, effectively banishing the 'doubled' state to a very high mass. Further discussion of this aspect of lattice fermions is contained in Montvay and Münster (1994).

A second problem concerning fermions is the more obvious one already alluded to: how are we to represent such entirely non-classical objects, which in particular obey the exclusion principle? We shall return to this question in section 16.4.

## 16.3 Gauge invariance on the lattice

Having explored the discretization of derivatives, it is now time to think about gauge invariance. In the usual (continuum) case, we saw in chapter 13 how this was implemented by replacing ordinary derivatives by *covariant derivatives*, the geometrical significance of which (in terms of parallel transport) was discussed in section 13.2 and 13.3. It is very instructive to see how the same ideas arise naturally in the lattice case.

We illustrate the idea in the simple case of the Abelian U(1) theory, QED. Consider, for example, a charged boson field $\phi(x)$, with charge $e$. To construct a gauge-invariant current, for example, we replaced $\phi^\dagger \partial_\mu \phi$ by $\phi^\dagger(\partial_\mu + \mathrm{i}eA_\mu)\phi$, so we ask: what is the discrete analogue of this? The term $\phi^\dagger(x)\frac{\partial}{\partial x}\phi(x)$ becomes, as we have seen,

$$\phi^\dagger(n_1)\frac{1}{a}[\phi(n_1+1) - \phi(n_1)a] \tag{16.22}$$

in one dimension. We do *not* expect (16.22) by itself to be gauge invariant and it is easy to check that it is not. Under a gauge transformation for the continuous case, we have

$$\phi(x) \to \mathrm{e}^{\mathrm{i}e\theta(x)}\phi(x),\ A(x) \to A(x) + \frac{\mathrm{d}\theta(x)}{\mathrm{d}x}; \tag{16.23}$$

then $\phi^\dagger(x)\phi(y)$ transforms by

$$\phi^\dagger(x)\phi(y) \to \mathrm{e}^{-\mathrm{i}e[\theta(x)-\theta(y)]}\phi^\dagger(x)\phi(y) \tag{16.24}$$

and is clearly not invariant. The essential reason is that this operator involves the fields at two *different* points—and so the term $\phi^\dagger(n_1)\phi(n_1+1)$ in (16.22) will not be gauge invariant either. Our discussion in chapter 13 should have prepared us for this: we are trying to compare two 'vectors' (here, fields) at two different points, when the 'coordinate axes' are changing as we move about. We need to parallel transport one field to the same point as the other, before they can be properly compared. The solution (13.62 ) shows us how to do this. Consider the quantity

$$\mathcal{O}(x, y) = \phi^\dagger(x)\exp\left[\mathrm{i}e\int_y^x A\,\mathrm{d}x'\right]\phi(y). \tag{16.25}$$

Under the gauge transformation (16.23), $\mathcal{O}(x, y)$ transforms by

$$\mathcal{O}(x, y) \to \phi^\dagger(x)\mathrm{e}^{-\mathrm{i}e\theta(x)}\exp^{\{\mathrm{i}e\int_y^x A\mathrm{d}x' + \mathrm{i}e[\theta(x)-\theta(y)]\}}\exp^{\mathrm{i}e\theta(y)}\phi(y) = \mathcal{O}(x, y) \tag{16.26}$$

and it is, therefore, gauge invariant. The familiar 'covariant derivative' rule can be recovered by letting $y = x + \mathrm{d}x$ for infinitesimal $\mathrm{d}x$, and by considering the gauge-invariant quantity

$$\lim_{\mathrm{d}x\to 0}\left[\frac{\mathcal{O}(x, x+\mathrm{d}x) - \mathcal{O}(x, x)}{\mathrm{d}x}\right]. \tag{16.27}$$

**Figure 16.1.** Link variable $U(n_2; n_1)$ in one dimension.

Evaluating (16.27), one finds (problem 16.4) the result

$$\phi^\dagger(x)\left(\frac{\mathrm{d}}{\mathrm{d}x} - \mathrm{i}eA\right)\phi(x) \tag{16.28}$$

$$\equiv \phi^\dagger(x) D_x \phi(x) \tag{16.29}$$

with the usual definition of the covariant derivative. In the discrete case, we merely keep the finite version of (16.25), and replace $\phi^\dagger(n_1)\phi(n_1+1)$ in (16.22) by the gauge-invariant quantity

$$\phi^\dagger(n_1) U(n_1, n_1+1) \phi(n_1+1) \tag{16.30}$$

where the *link variable* $U$ is defined by

$$U(n_1, n_1+1) = \exp\left[\mathrm{i}e \int_{(n_1+1)a}^{n_1 a} A\,\mathrm{d}x'\right]$$
$$\to \exp[-\mathrm{i}eA(n_1)a] \tag{16.31}$$

in the small $a$ limit. The generalization to more dimensions is straightforward. In the non-Abelian SU(2) or SU(3) case, '$eA$' in (16.31) is replaced by $gt^a A^a(n_1)$ where the $t$'s are the appropriate matrices, as in the continuum form of the covariant derivative. A link variable $U(n_2, n_1)$ may be drawn as in figure 16.1. Note that the order of the arguments is significant: $U(n_2, n_1) = U^{-1}(n_1, n_2)$ from (16.31), which is why the link carries an arrow.

Thus gauge-invariant discretized derivatives of charged fields can be constructed. What about the Maxwell action for the U(1) gauge field? This does not exist in only one dimension ($\partial_\mu A_\nu - \partial_\nu A_\mu$ cannot be formed) so let us move into two. Again, our discussion of the geometrical significance of $F_{\mu\nu}$ as a curvature (see section 13.3) guides us to the answer. Consider the product $U_\square$ of link variables around a square path (figure 16.2) of side $a$ (reading from the right):

$$U_\square = U(n_x, n_y; n_x, n_{y+1}) U(n_x, n_{y+1}; n_{x+1}, n_{y+1})$$
$$\times U(n_{x+1}, n_{y+1}; n_{x+1}, n_y) U(n_{x+1}, n_y; n_x, n_y). \tag{16.32}$$

It is straightforward to verify, first, that $U_\square$ is gauge invariant. Under a gauge transformation, the link $U(n_{x+1}, n_y; n_x, n_y)$, for example, transforms by a factor (cf equation (16.26))

$$\exp\{\mathrm{i}e[\theta(n_{x+1}, n_y) - \theta(n_x, n_y)]\} \tag{16.33}$$

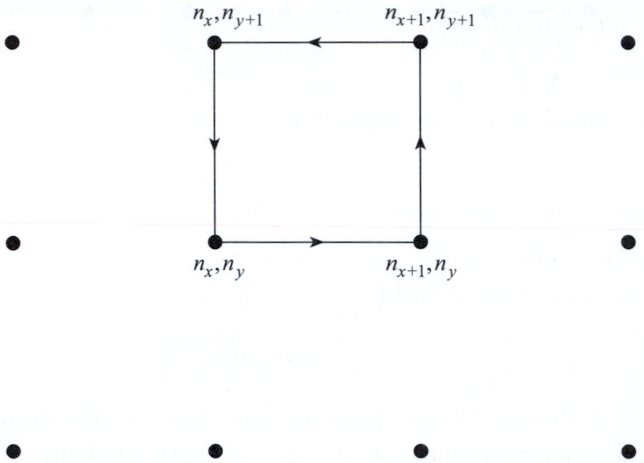


**Figure 16.2.** A simple plaquette in two dimensions.

and similarly for the three other links in $U_\Box$. In this Abelian case, the exponentials contain no matrices and the accumulated phase factors cancel out, verifying the gauge invariance. Next, let us see how to recover the Maxwell action. Adding the exponentials again, we can write

$$U_\Box \equiv \exp\{-ieaA_y(n_x, n_y) - ieaA_x(n_x, n_y+1) + ieaA_y(n_x+1, n_y) + ieaA_x(n_x, n_y)\} \tag{16.34}$$

$$= \exp\left\{-iea^2\left[\frac{A_x(n_x, n_y+1) - A_x(n_x, n_y)}{a}\right] + iea^2\left[\frac{A_y(n_x+1, n_y) - A_y(n_x, n_y)}{a}\right]\right\} \tag{16.35}$$

$$= \exp\left\{+iea^2\left(\frac{\partial A_y}{\partial x} - \frac{\partial A_x}{\partial y}\right)\right\} \tag{16.36}$$

using the derivative definition of (16.5). For small '$a$' we may expand the exponential in (16.36). We also take the real part to remove the imaginary terms, leading to

$$\sum_\Box (1 - \mathrm{Re}\, U_\Box) \to \tfrac{1}{2} \sum_\Box e^2 a^4 (F_{xy})^2 \tag{16.37}$$

where

$$F_{xy} = \frac{\partial A_y}{\partial x} - \frac{\partial A_x}{\partial y}$$

as usual. To relate this to the continuum limit we must note that we sum over each such *plaquette* with only one definite orientation, so that the sum over plaquettes is equivalent to half of the entire sum. Thus

$$\sum_{\square}(1 - \text{Re}\, U_{\square}) \to \tfrac{1}{4} \sum_{n_1, n_2} e^2 a^4 F_{xy}^2$$
$$\to e^2 a^2 \iint \tfrac{1}{4} F_{xy}^2 \,\mathrm{d}x\,\mathrm{d}y. \tag{16.38}$$

(Note that in two dimensions '$e$' has dimensions of mass.) In four dimensions similar manipulations lead to the form

$$S_{\mathrm{E}} = \frac{1}{e^2} \sum_{\square}(1 - \text{Re}\, U_{\square}) \to \frac{1}{4} \int \mathrm{d}^3\boldsymbol{x}\,\mathrm{d}\tau\, F_{\mu\nu}^2 \tag{16.39}$$

for the lattice action, as required. In the non-Abelian case, as noted earlier, '$eA$' is replaced by '$g\boldsymbol{t} \cdot \boldsymbol{A}$'; for SU(3), the analogue of the left-hand side of (16.37) is

$$S_g = \frac{6}{g^2} \sum_{\square} \text{Tr}\left(1 - \frac{1}{3} \text{Re}\, U_{\square}\right) \tag{16.40}$$

where the trace is over the SU(3) matrices.

## 16.4 Representation of quantum amplitudes

So we have a naturally gauge-invariant 'classical' field theory defined on a lattice, with a suitable continuum limit. (Actually, the $a \to 0$ limit of the quantum theory is, as we shall see in section 16.7, more subtle than the naive replacements (16.5) because of renormalization issues, as should be no surprise to the reader by now.) However, we have not yet considered how we are going to turn this classical lattice theory into a quantum one. The fact that the calculations are mostly going to have to be done numerically seems at once to require a formulation that avoids non-commuting operators. This is precisely what is provided by Feynman's *sum over paths* formulation of quantum mechanics (Feynman and Hibbs 1965) and of quantum field theory and it is, therefore, an essential element in the lattice approach to quantum field theory. In this section we give a brief introduction to this formalism.

In section 5.2.2, we stated that in this approach the amplitude for a quantum system, described by a Lagrangian $L$ depending on one degree of freedom $q(t)$, to pass from a state in which $q = q^{\mathrm{i}}$ at $t = t_{\mathrm{i}}$ to a state in which $q = q^{\mathrm{f}}$ at time $t = t_{\mathrm{f}}$, is proportional to (with $\hbar = 1$)

$$\sum_{\text{all paths } q(t)} \exp\left(\mathrm{i} \int_{t_{\mathrm{i}}}^{t_{\mathrm{f}}} L(q(t), \dot{q}(t))\,\mathrm{d}t\right) \tag{16.41}$$

where $q(t_{\mathrm{i}}) = q^{\mathrm{i}}$, and $q(t_{\mathrm{f}}) = q^{\mathrm{f}}$. We shall now provide some justification for this assertion.

We begin by recalling how, in ordinary quantum mechanics, state vectors and observables are related in the Schrödinger and Heisenberg pictures (see appendix I of volume 1). Let $\hat{q}$ be the canonical coordinate operator in the Schrödinger picture, with an associated complete set of eigenvectors $|q\rangle$ such that

$$\hat{q}|q\rangle = q|q\rangle. \tag{16.42}$$

The corresponding Heisenberg operator $\hat{q}_{\mathrm{H}}(t)$is defined by

$$\hat{q}_{\mathrm{H}}(t) = \mathrm{e}^{\mathrm{i}\hat{H}(t-t_0)}\hat{q}\mathrm{e}^{-\mathrm{i}\hat{H}(t-t_0)} \tag{16.43}$$

where $\hat{H}$ is the Hamiltonian and $t_0$ is the (arbitrary) time at which the two pictures coincide. Now define the Heisenberg picture state $|q_t\rangle_{\mathrm{H}}$ by

$$|q_t\rangle_{\mathrm{H}} = \mathrm{e}^{\mathrm{i}\hat{H}(t-t_0)}|q\rangle. \tag{16.44}$$

We then easily obtain from (16.42)–(16.44) the result

$$\hat{q}_{\mathrm{H}}(t)|q_t\rangle_{\mathrm{H}} = q|q_t\rangle_{\mathrm{H}} \tag{16.45}$$

which shows that $|q_t\rangle_{\mathrm{H}}$ is the (Heisenberg picture) state which at time $t$ is an eigenstate of $\hat{q}_{\mathrm{H}}(t)$ with eigenvalue $q$. Consider now the quantity

$$_{\mathrm{H}}\langle q^{\mathrm{f}}_{t_{\mathrm{f}}}|q^{\mathrm{i}}_{t_{\mathrm{i}}}\rangle_{\mathrm{H}} \tag{16.46}$$

which is, indeed, the amplitude for the system described by $\hat{H}$ to go from $q^{\mathrm{i}}$ at $t_{\mathrm{i}}$ to $q^{\mathrm{f}}$ at $t_{\mathrm{f}}$. Using (16.44), we can write

$$_{\mathrm{H}}\langle q^{\mathrm{f}}_{t_{\mathrm{f}}}|q^{\mathrm{i}}_{t_{\mathrm{i}}}\rangle_{\mathrm{H}} = \langle q^{\mathrm{f}}|\mathrm{e}^{-\mathrm{i}\hat{H}(t_{\mathrm{f}}-t_{\mathrm{i}})}|q^{\mathrm{i}}\rangle. \tag{16.47}$$

We want to understand how (16.47) can be represented as (16.41).

We shall demonstrate the connection explicitly for the special case of a free particle, for which

$$\hat{H} = \frac{\hat{p}^2}{2m}. \tag{16.48}$$

For this case, we can evaluate (16.47) directly as follows. Inserting a complete set of momentum eigenstates, we obtain[1]

$$\begin{aligned}\langle q^{\mathrm{f}}|\mathrm{e}^{-\mathrm{i}\hat{H}(t_{\mathrm{f}}-t_{\mathrm{i}})}|q^{\mathrm{i}}\rangle &= \int_{-\infty}^{\infty} \langle q^{\mathrm{f}}|p\rangle\langle p|\mathrm{e}^{-\mathrm{i}\hat{H}(t_{\mathrm{f}}-t_{\mathrm{i}})}|q^{\mathrm{i}}\rangle\,\mathrm{d}p \\ &= \frac{1}{2\pi}\int_{-\infty}^{\infty} \mathrm{e}^{\mathrm{i}pq^{\mathrm{f}}}\mathrm{e}^{-\mathrm{i}p^2(t_{\mathrm{f}}-t_{\mathrm{i}})/2m}\mathrm{e}^{-\mathrm{i}pq^{\mathrm{i}}}\,\mathrm{d}p \\ &= \frac{1}{2\pi}\int_{-\infty}^{\infty} \exp\left\{-\mathrm{i}\left[\frac{p^2(t_{\mathrm{f}}-t_{\mathrm{i}})}{2m} - p(q^{\mathrm{f}}-q^{\mathrm{i}})\right]\right\}\mathrm{d}p.\end{aligned} \tag{16.49}$$

[1] Remember that $\langle q|p\rangle$ is the $q$-space wavefunction of a state with definite momentum $p$ and is, therefore, a plane wave; we are using the normalization of equation (E.25) in volume 1.

To evaluate the integral, we complete the square via the steps

$$\begin{aligned}\frac{p^2(t_\mathrm{f}-t_\mathrm{i})}{2m} - p(q^\mathrm{f}-q^\mathrm{i}) &= \left(\frac{t_\mathrm{f}-t_\mathrm{i}}{2m}\right)\left[p^2 - \frac{2mp(q^\mathrm{f}-q^\mathrm{i})}{t_\mathrm{f}-t_\mathrm{i}}\right] \\ &= \left(\frac{t_\mathrm{f}-t_\mathrm{i}}{2m}\right)\left\{\left[p - \frac{m(q^\mathrm{f}-q^\mathrm{i})}{t_\mathrm{f}-t_\mathrm{i}}\right]^2 - \frac{m^2(q^\mathrm{f}-q^\mathrm{i})^2}{(t_\mathrm{f}-t_\mathrm{i})^2}\right\} \\ &= \left(\frac{t_\mathrm{f}-t_\mathrm{i}}{2m}\right)p'^2 - \frac{m(q^\mathrm{f}-q^\mathrm{i})^2}{2(t_\mathrm{f}-t_\mathrm{i})} \end{aligned} \tag{16.50}$$

where

$$p' = p - \frac{m(q^\mathrm{f}-q^\mathrm{i})}{t_\mathrm{f}-t_\mathrm{i}}. \tag{16.51}$$

We then shift the integration variable in (16.49) to $p'$ and obtain

$$\langle q^\mathrm{f}|\mathrm{e}^{-\mathrm{i}\hat{H}(t_\mathrm{f}-t_\mathrm{i})}|q^\mathrm{i}\rangle = \frac{1}{2\pi}\exp\left[\mathrm{i}\frac{m(q^\mathrm{f}-q^\mathrm{i})^2}{2(t_\mathrm{f}-t_\mathrm{i})}\right]\int_{-\infty}^{\infty}\mathrm{d}p'\exp\left[-\frac{\mathrm{i}(t_\mathrm{f}-t_\mathrm{i})p'^2}{2m}\right]. \tag{16.52}$$

As it stands, the integral in (16.52) is not well defined, being rapidly oscillatory for large $p'$. However, it is at this point that the motivation for passing to 'Euclidean' spacetime arises. If we make the replacement $t \to -\mathrm{i}\tau$, (16.52) becomes

$$\langle q^\mathrm{f}|\mathrm{e}^{-\hat{H}(\tau_\mathrm{f}-\tau_\mathrm{i})}|q^\mathrm{i}\rangle = \frac{1}{2\pi}\exp\left[-\frac{m(q^\mathrm{f}-q^\mathrm{i})^2}{2(\tau_\mathrm{f}-\tau_\mathrm{i})}\right]\int_{-\infty}^{\infty}\mathrm{d}p'\exp\left[-\frac{(\tau_\mathrm{f}-\tau_\mathrm{i})p'^2}{2m}\right] \tag{16.53}$$

and the integral is a simple convergent Gaussian. Using the result

$$\int_{-\infty}^{\infty}\mathrm{d}\xi\,\mathrm{e}^{-b\xi^2} = \sqrt{\frac{\pi}{b}} \tag{16.54}$$

we finally obtain

$$\langle q^\mathrm{f}|\mathrm{e}^{-\hat{H}(\tau_\mathrm{f}-\tau_\mathrm{i})}|q^\mathrm{i}\rangle = \left[\frac{m}{2\pi(\tau_\mathrm{f}-\tau_\mathrm{i})}\right]^{\frac{1}{2}}\exp\left[-\frac{m(q^\mathrm{f}-q^\mathrm{i})^2}{2(\tau_\mathrm{f}-\tau_\mathrm{i})}\right]. \tag{16.55}$$

We must now understand how the result (16.55) can be represented in the form (16.41). In Euclidean space, (16.41) is

$$\sum_{\text{paths}}\exp\left(-\int_{\tau_\mathrm{i}}^{\tau_\mathrm{f}}\frac{1}{2}m\left(\frac{\mathrm{d}q}{\mathrm{d}\tau}\right)^2\mathrm{d}\tau\right) \tag{16.56}$$

in the free-particle case. We interpret the $\tau$ integral in terms of a discretization procedure, similar to that introduced in section 16.2. We split the interval $\tau_\mathrm{f}-\tau_\mathrm{i}$

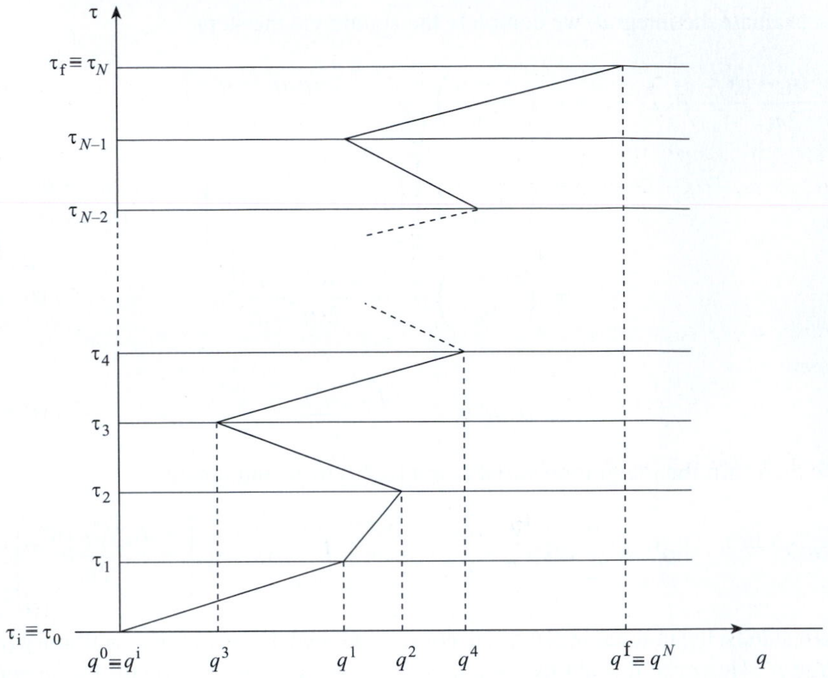


**Figure 16.3.** A 'path' from $q^0 \equiv q^{\mathrm{i}}$ at $\tau_{\mathrm{i}}$ to $q^N \equiv q^{\mathrm{f}}$ at $\tau_{\mathrm{f}}$, via the intermediate positions $q^1, q^2, \ldots, q^{N-1}$ at $\tau_1, \tau_2, \ldots, \tau_{N-1}$.

into $N$ segments each of size $\epsilon$, as shown in figure 16.3. The $\tau$-integral in (16.56) becomes the sum

$$m \sum_{j=1}^{N} \frac{(q^j - q^{j-1})^2}{2\epsilon} \tag{16.57}$$

and the 'sum over paths', in going from $q^0 \equiv q^{\mathrm{i}}$ at $\tau_{\mathrm{i}}$ to $q^N \equiv q^{\mathrm{f}}$ at $\tau_{\mathrm{f}}$, is now interpreted as a multiple integral over all the intermediate positions $q^1, q^2, \ldots, q^{N-1}$ which paths can pass through at 'times' $\tau_1, \tau_2, \ldots, \tau_{N-1}$:

$$\frac{1}{A(\epsilon)} \int\!\!\int \ldots \int \exp\left[-m \sum_{j=1}^{N} \frac{(q^j - q^{j-1})^2}{2\epsilon}\right] \frac{\mathrm{d}q^1}{A(\epsilon)} \frac{\mathrm{d}q^2}{A(\epsilon)} \cdots \frac{\mathrm{d}q^{N-1}}{A(\epsilon)} \tag{16.58}$$

where $A(\epsilon)$ is a normalizing factor, depending on $\epsilon$, which is to be determined.

The integrals in (16.58) are all of Gaussian form and since the integral of a Gaussian is again a Gaussian (cf the manipulations leading from (16.49) to (16.52), but without the 'i' in the exponents), we may perform all the integrations analytically. We follow the method of Feynman and Hibbs (1965, section 3.1).

Consider the integral over $q^1$:

$$I^1 \equiv \int \exp\left\{-\frac{m}{2\epsilon}[(q^2 - q^1)^2 + (q^1 - q^{\mathrm{i}})^2]\right\} \mathrm{d}q^1. \tag{16.59}$$

This can be evaluated by completing the square, shifting the integration variable and using (16.54) to obtain (problem 16.5)

$$I^1 = \left(\frac{\pi\epsilon}{m}\right)^{\frac{1}{2}} \exp\left[\frac{-m}{4\epsilon}(q^2 - q^{\mathrm{i}})^2\right]. \tag{16.60}$$

Now the procedure may be repeated for the $q^2$ integral

$$I^2 \equiv \int \exp\left\{-\frac{m}{4\epsilon}(q^2 - q^{\mathrm{i}})^2 - \frac{m}{2\epsilon}(q^3 - q^2)^2\right\} \mathrm{d}q^2 \tag{16.61}$$

which yields (problem 16.5)

$$I^2 = \left(\frac{4\pi\epsilon}{3m}\right)^{\frac{1}{2}} \exp\left[\frac{-m}{6\epsilon}(q^3 - q^{\mathrm{i}})^2\right]. \tag{16.62}$$

As far as the exponential factors in (16.55) in (16.56) are concerned, the pattern is now clear: after $n-1$ steps we shall have an exponential factor

$$\exp[-m(q^n - q^{\mathrm{i}})^2/(2n\epsilon)]. \tag{16.63}$$

Hence, after $N-1$ steps we shall have a factor

$$\exp[-m(q^{\mathrm{f}} - q^{\mathrm{i}})^2/2(\tau_{\mathrm{f}} - \tau_{\mathrm{i}})] \tag{16.64}$$

remembering that $q^N \equiv q^{\mathrm{f}}$ and that $\tau_{\mathrm{f}} - \tau_{\mathrm{i}} = N\epsilon$. So we have recovered the correct exponential factor of (16.55), and all that remains is to choose $A(\epsilon)$ in (16.58) so as to produce the same normalization as (16.55).

The required $A(\epsilon)$ is

$$A(\epsilon) = \sqrt{\frac{2\pi\epsilon}{m}} \tag{16.65}$$

as we now verify. For the first ($q^1$) integration, the formula (16.58) contains two factors of $A^{-1}(\epsilon)$, so that the result (16.60) becomes

$$\begin{aligned} \frac{1}{[A(\epsilon)]^2} I^1 &= \frac{m}{2\pi\epsilon}\left(\frac{\pi\epsilon}{m}\right)^{\frac{1}{2}} \exp\left[-\frac{m}{4\epsilon}(q^2 - q^{\mathrm{i}})^2\right] \\ &= \left(\frac{m}{2\pi 2\epsilon}\right)^{\frac{1}{2}} \exp\left[-\frac{m}{4\epsilon}(q^2 - q^{\mathrm{i}})^2\right]. \end{aligned} \tag{16.66}$$

For the second ($q^2$) integration, the accumulated constant factor is

$$\frac{1}{A(\epsilon)}\left(\frac{m}{2\pi 2\epsilon}\right)^{\frac{1}{2}}\left(\frac{4\pi\epsilon}{3m}\right)^{\frac{1}{2}} = \left(\frac{m}{2\pi 3\epsilon}\right)^{\frac{1}{2}}. \tag{16.67}$$

Proceeding in this way, one can convince oneself that after $N - 1$ steps, the accumulated constant is

$$\left(\frac{m}{2\pi N\epsilon}\right)^{\frac{1}{2}} = \left[\frac{m}{2\pi(\tau_{\mathrm{f}} - \tau_{\mathrm{i}})}\right]^{\frac{1}{2}} \tag{16.68}$$

as in (16.55).

The equivalence of (16.55) and (16.56) (in the sense $\epsilon \to 0$) is, therefore, established for the free-particle case. More general cases are discussed in Feynman and Hibbs (1965, chapter 5) and in Peskin and Schroeder (1995, chapter 9). The conventional notation for the path-integral amplitude is

$$\langle q^{\mathrm{f}}|\mathrm{e}^{-\hat{H}(\tau_{\mathrm{f}}-\tau_{\mathrm{i}})}|q^{\mathrm{i}}\rangle = \int \mathcal{D}q(\tau)\mathrm{e}^{-\int_{\tau_{\mathrm{i}}}^{\tau_{\mathrm{f}}} L\,\mathrm{d}\tau} \tag{16.69}$$

where the right-hand side of (16.69) is interpreted in the sense of (16.58).

We now proceed to discuss further aspects of the path-integral formulation. Consider the (Euclideanized) amplitude $\langle q^{\mathrm{f}}|e^{-\hat{H}(\tau_{\mathrm{f}}-\tau_{\mathrm{i}})}|q^{\mathrm{i}}\rangle$ and insert a complete set of energy eigenstates $|n\rangle$ such that $\hat{H}|n\rangle = E_n|n\rangle$:

$$\langle q^{\mathrm{f}}|\mathrm{e}^{-\hat{H}(\tau_{\mathrm{f}}-\tau_{\mathrm{i}})}|q^{\mathrm{i}}\rangle = \sum_n \langle q^{\mathrm{f}}|n\rangle\langle n|q^{\mathrm{i}}\rangle\mathrm{e}^{-E_n(\tau_{\mathrm{f}}-\tau_{\mathrm{i}})}. \tag{16.70}$$

Equation (16.70) shows that if we take the limits $\tau_{\mathrm{i}} \to -\infty$, $\tau_{\mathrm{f}} \to \infty$, then the state of lowest energy $E_0$ (the ground state) provides the dominant contribution. Thus, in this limit, our amplitude will represent the process in which the system begins in its ground state $|\Omega\rangle$ at $\tau_{\mathrm{i}} \to -\infty$, with $q = q^{\mathrm{i}}$, and ends in $|\Omega\rangle$ at $\tau_{\mathrm{f}} \to \infty$, with $q = q^{\mathrm{f}}$.

How do we represent propagators in this formalism? Consider the expression (somewhat analogous to a field theory propagator)

$$G_{\mathrm{fi}}(t_a, t_b) \equiv \langle q^{\mathrm{f}}_{t_{\mathrm{f}}}|T\{\hat{q}_{\mathrm{H}}(t_a)\hat{q}_{\mathrm{H}}(t_b)\}|q^{\mathrm{i}}_{t_{\mathrm{i}}}\rangle \tag{16.71}$$

where $T$ is the usual time-ordering operator. Using (16.43) and (16.44), (16.71) can be written, for $t_b > t_a$, as

$$G_{\mathrm{fi}}(t_a, t_b) = \langle q^{\mathrm{f}}|\mathrm{e}^{-\mathrm{i}\hat{H}(t_{\mathrm{f}}-t_b)}\hat{q}\mathrm{e}^{-\mathrm{i}\hat{H}(t_b-t_a)}\hat{q}\mathrm{e}^{-\mathrm{i}\hat{H}(t_a-t_{\mathrm{i}})}|q^{\mathrm{i}}\rangle. \tag{16.72}$$

Inserting a complete set of states and Euclideanizing, (16.72) becomes

$$\begin{aligned} G_{\mathrm{fi}}(t_a, t_b) = \int \mathrm{d}q^a\,\mathrm{d}q^b\,q^a q^b \langle q^{\mathrm{f}}|\mathrm{e}^{-\hat{H}(\tau_{\mathrm{f}}-\tau_b)}|q^b\rangle \\ \times \langle q^b|\mathrm{e}^{-\hat{H}(\tau_b-\tau_a)}|q^a\rangle\langle q^a|\mathrm{e}^{-\hat{H}(\tau_a-\tau_{\mathrm{i}})}|q^{\mathrm{i}}\rangle. \end{aligned} \tag{16.73}$$

Now, each of the three matrix elements has a discretized representation of the form (16.55) with, say, $N_1 - 1$ variables in the interval $(\tau_a, \tau_{\mathrm{i}})$, $N_2 - 1$ in $(\tau_b, \tau_a)$

and $N_3 - 1$ in $(\tau_{\rm f}, \tau_b)$. Each such representation carries one 'surplus' factor of $[A(\epsilon)]^{-1}$, making an overall factor of $[A(\epsilon)]^{-3}$. Two of these factors can be associated with the $\mathrm{d}q^a\,\mathrm{d}q^b$ integration in (16.73), so that we have a total of $N_1 + N_2 + N_3 - 1$ properly normalized integrations and one 'surplus' factor $[A(\epsilon)]^{-1}$ as in (16.58). If we now identify $q(\tau_a) \equiv q^a$, $q(\tau_n) \equiv q^b$, it follows that (16.73) is simply

$$\int \mathcal{D}q(\tau)q(\tau_a)q(\tau_b)\mathrm{e}^{-\int_{\tau_{\rm i}}^{\tau_{\rm f}} L\,\mathrm{d}\tau}. \tag{16.74}$$

In obtaining (16.74), we took the case $\tau_b > \tau_a$. Suppose alternatively that $\tau_a > \tau_b$. Then the order of $\tau_a$ and $\tau_b$ inside the interval $(\tau_{\rm i}, \tau_{\rm f})$ is simply reversed but since $q^a$ and $q^b$ in (16.73), or $q(\tau_a)$ and $q(\tau_b)$ in (16.74), are ordinary (commuting) numbers, the formula (16.74) is unaltered and it does actually represent the matrix element (16.71) of the time-ordered product.

The generalizations of these results to the field theory case are intuitively clear. For example, in the case of a single scalar field $\phi(\mathbf{x})$, we expect the analogue of (16.74) to be (cf (16.4))

$$\int \mathcal{D}\phi(x)\,\phi(x_a)\phi(x_b)\exp\left[-\int_{\tau_{\rm i}}^{\tau_{\rm f}} \mathcal{L}_{\rm E}(\phi, \nabla\phi, \partial_\tau\phi)\,\mathrm{d}^4 x_{\rm E}\right] \tag{16.75}$$

where

$$\mathrm{d}^4 x_{\rm E} = \mathrm{d}^3\boldsymbol{x}\,\mathrm{d}\tau \tag{16.76}$$

and the boundary conditions are given by $\phi(\mathbf{x}, \tau_{\rm i}) = \phi^{\rm i}(x)$, $\phi(\mathbf{x}, \tau_{\rm f}) = \phi^{\rm f}(x)$, $\phi(\mathbf{x}, \tau_a) = \phi^a(x)$ and $\phi(\mathbf{x}, \tau_b) = \phi^b(x)$, say. In (16.75), we have to understand that a *four*-dimensional discretization of Euclidean spacetime is implied, the fields being Fourier-analysed by four-dimensional generalizations of expressions such as (16.7). Just as in (16.71)-(16.74), (16.75) is equal to

$$\langle\phi^{\rm f}(x)|\mathrm{e}^{-\hat{H}\tau_{\rm f}}T\{\hat{\phi}_{\rm H}(x_a)\hat{\phi}_{\rm H}(x_b)\}\mathrm{e}^{\hat{H}\tau_{\rm i}}|\phi^{\rm i}(x)\rangle. \tag{16.77}$$

Taking the limits $\tau_{\rm i} \to -\infty$, $\tau_{\rm f} \to \infty$ will project out the configuration of lowest energy, as discussed after (16.70), which in this case is the (interacting) vacuum state $|\Omega\rangle$. Thus, in this limit, the surviving part of (16.77) is

$$\langle\phi^{\rm f}(x)|\Omega\rangle\mathrm{e}^{-E_\Omega\tau}\langle\Omega|T\{\hat{\phi}_{\rm H}(x_a)\hat{\phi}_{\rm H}(x_b)\}|\Omega\rangle\mathrm{e}^{-E_\Omega\tau}\langle\Omega|\phi^{\rm i}(x)\rangle \tag{16.78}$$

with $\tau \to \infty$. The exponential and overlap factors can be removed by dividing by the same quantity as (16.77) but without the additional fields $\phi(x_a)$ and $\phi(x_b)$. In this way, we obtain the formula for the *field theory propagator* in four-dimensional Euclidean space:

$$\langle\Omega|T\{\hat{\phi}_{\rm H}(x_a)\hat{\phi}_{\rm H}(x_b)\}|\Omega\rangle = \lim_{\tau\to\infty}\frac{\int \mathcal{D}\phi\phi(x_a)\phi(x_b)\exp[-\int_{-\tau}^{\tau}\mathcal{L}_{\rm E}\,\mathrm{d}^4x_{\rm E}]}{\int \mathcal{D}\phi\,\exp[-\int_{-\tau}^{\tau}\mathcal{L}_{\rm E}\,\mathrm{d}^4x_{\rm E}]}. \tag{16.79}$$

Vacuum expectation values (vevs) of time-ordered products of more fields will simply have more factors of $\phi$ on both sides.

Perturbation theory can be developed in this formalism also. Suppose $\mathcal{L}_E = \mathcal{L}_E^0 + \mathcal{L}_E^{\text{int}}$, where $\mathcal{L}_E^0$ describes a free scalar field and $\mathcal{L}_E^{\text{int}}$ is an interaction, for example $\lambda\phi^4$. Then, assuming $\lambda$ is small, the exponential in (16.79) can be expressed as

$$\exp\left[-\int d^4x_E\,(\mathcal{L}_E^0 + \mathcal{L}_E^{\text{int}})\right] = \left(\exp - \int d^4x_E\,\mathcal{L}_E^0\right)\left(1 - \lambda\int d^4x_E\,\phi^4 + \cdots\right) \tag{16.80}$$

and both numerator and denominator of (16.79) may be expressed as vevs of products of free fields. Compact techniques exist for analysing this formulation of perturbation theory (Ryder 1996, Peskin and Schroeder 1995) and one finds exactly the same 'Feynman rules' as in the canonical (operator) approach.

In the case of gauge theories, we can easily imagine a formula similar to (16.79) for the gauge-field propagator, in which the integral is carried out over all gauge fields $A_\mu(x)$ (in the U(1) case, for example). But we already know from chapter 7 (or from chapter 13 in the non-Abelian case) that, in the continuum limit, we shall not be able to construct a well-defined perturbation theory in this way, since the gauge-field propagator will not exist unless we 'fix the gauge' by imposing some constraint, such as the Lorentz gauge condition. Such constraints can be imposed on the corresponding path integral and, indeed, this was the route followed by Faddeev and Popov (1967) in first obtaining the Feynman rules for non-Abelian gauge theories, as mentioned in section 13.5.3.

In the discrete case, the appropriate integration variables are the link variables $\mathrm{U}(l_i)$ where $l_i$ is the $i^{\text{th}}$ link. They are elements of the relevant gauge group—for example $\mathrm{U}(n_1, n_1 + 1)$ of (16.3.1) is an element of U(1). In the case of the unitary groups, such elements typically have the form (cf (12.35)) $\sim$ exp (i Hermitean matrix), where the 'Hermitean matrix' can be parametrized in some convenient way—for example, as in (12.31) for SU(2). In all these cases, the variables in the parametrization of U vary over some bounded domain (they are essentially 'angle-type' variables, as in the simple U(1) case), and so, with a finite number of lattice points, the integral over the link variables is well-defined without gauge-fixing. The integration measure for the link variables can be chosen so as to be gauge invariant and, hence, provided the action is gauge invariant, the formalism provides well-defined expressions, independently of perturbation theory, for vevs of gauge-invariant quantities.

There remains one more conceptual problem to be addressed in this approach—namely, how are we to deal with fermions? It seems that we must introduce new variables which, though not quantum field operators, must nevertheless *anti*-commute with each other. Such 'classical' anti-commuting variables are called *Grassmann variables*, and are briefly described in appendix O. Further details are contained in Ryder (1996) and in Peskin and Schroeder (1995). For our purposes, the important point is that the fermion Lagrangian is *bilinear* in the (Grassmann) fermion fields $\psi$, the fermionic action having the form

$$S_\psi = \int \mathrm{d}^4 x_\mathrm{E}\, \bar{\psi} M(U) \psi \tag{16.81}$$

where $M$ is a matrix representing the Dirac operator $\mathrm{i}\not{D} - m_q$ in its discretized and Euclideanized form. This means that in a typical fermionic amplitude of the form (cf the denominator of (16.79))

$$Z_\psi = \int \mathcal{D}\bar{\psi}\mathcal{D}\psi \exp[-S_\psi] \tag{16.82}$$

one has essentially an integral of Gaussian type (albeit with Grassmann variables), which can actually be performed analytically.[2] The result is simply $\det[M(U)]$, the determinant of this matrix. The problem is that $M$ is a very large matrix indeed, if we want anything like a reasonably sized lattice (say 20 lattice spacings in each of the four dimensions); moreover, the gauge field degrees of freedom must be itemized (via the link variables $U$) at each site. At the time of writing, computers are just about reaching the stage of being able to calculate such a vast determinant numerically, but hitherto most calculations have been done in the *quenched approximation*, setting the determinant equal to a constant independent of the link variables $U$. This is equivalent to the neglect of closed fermion loops in a Feynman graph approach. In the quenched approximation, the expectation value of any operator $\mathcal{O}(U)$ is just

$$\langle \mathcal{O}(U) \rangle = \frac{\int \mathcal{D}U \mathcal{O}(U) \exp[-S_g(U)]}{\int \mathcal{D}U \exp[-S_g(U)]} \tag{16.83}$$

where $S_g(U)$ is the gauge action (16.40).

## 16.5 Connection with statistical mechanics

Not the least advantage of the path integral formulation of quantum field theory (especially in its lattice form) is that it enables a highly suggestive connection to be set up between quantum field theory and statistical mechanics. We introduce this connection now, by way of a preliminary to the discussion of renormalization in the following section.

The connection is made via the fundamental quantity of equilibrium statistical mechanics, the *partition function* $Z$ defined by

$$Z = \sum_{\text{configurations}} \exp\left(-\frac{H}{k_\mathrm{B} T}\right) \tag{16.84}$$

which is simply the 'sum over states' (or configurations) of the relevant degrees of freedom, with the Boltzmann weighting factor. $H$ is the classical Hamiltonian

[2] See appendix O.

evaluated for each configuration. Consider, for comparison, the denominator in (16.79), namely

$$Z_\phi = \int \mathcal{D}\phi \exp(-S_{\mathrm{E}}) \tag{16.85}$$

where

$$S_{\mathrm{E}} = \int \mathrm{d}^4 x_{\mathrm{E}}\, \mathcal{L}_{\mathrm{E}} = \int \mathrm{d}^4 x_{\mathrm{E}} \{\tfrac{1}{2}(\partial_\tau \phi)^2 + \tfrac{1}{2}(\boldsymbol{\nabla}\phi)^2 + \tfrac{1}{2}m^2\phi^2 + \lambda\phi^4\} \tag{16.86}$$

in the case of a single scalar field with mass $m$ and self-interaction $\lambda\phi^4$. The Euclideanized Lagrangian density $\mathcal{L}_{\mathrm{E}}$ is like an energy density: it is bounded from below and increases when the field has large magnitude or has large gradients in $\tau$ or $\boldsymbol{x}$. The factor $\exp(-S_{\mathrm{E}})$ is then a sensible statistical weight for the fluctuations in $\phi$, and $Z_\phi$ may be interpreted as the partition function for a system described by the field degree of freedom $\phi$ but, of course, in *four* 'spatial' dimensions.

The parallel becomes perhaps even stronger when we discretize spacetime. In an Ising model (see the following section), the Hamiltonian has the form

$$H = -J \sum_n s_n s_{n+1} \tag{16.87}$$

where $J$ is a constant and the sum is over lattice sites $n$; the system variables taking the values $\pm 1$. When (16.87) is inserted into (16.84), we arrive at something very reminiscent of the $\phi(n_1)\phi(n_1+1)$ term in (16.6). Naturally, the effective 'Hamiltonian' is not quite the same—though we may note that Wilson (1971b) argued that in the case of a $\phi^4$ interaction the parameters can be chosen so as to make the values $\phi = \pm 1$ the most heavily weighted in $S_{\mathrm{E}}$. Statistical mechanics does, of course, deal in three spatial dimensions, not the four of our Euclideanized spacetime. Nevertheless, it is remarkable that quantum field theory in three spatial dimensions appears to have such a close relationship to equilibrium statistical mechanics in four spatial dimensions.

One insight we may draw from this connection is that, in the case of pure gauge actions (16.39) or (16.40), the gauge coupling is seen to be analogous to an inverse temperature, by comparison with (16.84). For example, in (16.40) $6/g^2$ would play the rôle of $1/k_{\mathrm{B}}T$. One is led to wonder whether something like *transitions between different 'phases'* exist, as coupling constants (or other parameters) vary—and, indeed, such changes of 'phase' can occur.

A second point is somewhat related to this. In statistical mechanics, an important quantity is the *correlation length* $\xi$, which for a spin system may be defined via the *spin–spin correlation function*

$$G(\boldsymbol{x}) = \langle s(\boldsymbol{x})s(\boldsymbol{0})\rangle = \sum_{\text{all } s(\boldsymbol{x})} s(\boldsymbol{x})s(\boldsymbol{0})\mathrm{e}^{-H/k_{\mathrm{B}}T} \tag{16.88}$$

where we are once more reverting to a continuous $\boldsymbol{x}$ variable. For large $|\boldsymbol{x}|$, this takes the form

$$G(\boldsymbol{x}) \propto \frac{1}{|\boldsymbol{x}|} \exp\left(\frac{-|\boldsymbol{x}|}{\xi(T)}\right). \tag{16.89}$$

The Fourier transform of this (in the continuum limit) is

$$\tilde{G}(\boldsymbol{k}^2) \propto (\boldsymbol{k}^2 + \xi^{-2}(T))^{-1} \tag{16.90}$$

as we learned in section 2.3. Comparing (16.88) with (16.79), it is clear that (16.88) is proportional to the propagator (or Green function) for the field $s(\boldsymbol{x})$: (16.90) then shows that $\xi^{-1}(T)$ is playing the role of a mass term $m$. Now, near a critical point for a statistical system, correlations exist over very large scales $\xi$ compared to the inter-atomic spacing $a$; in fact, at the critical point $\xi(T_{\rm c}) \sim L$, where $L$ is the size of the system. In the quantum field theory, as indicated earlier, we may regard $a^{-1}$ as playing a role analogous to a momentum cut-off $\Lambda$, so the regime $\xi \gg a$ is equivalent to $m \ll \Lambda$, as was indeed always our assumption. Thus studying a quantum field theory this way is analogous to studying a four-dimensional statistical system near a critical point. This shows rather clearly why it is not going to be easy: correlations over all scales will have to be included. At this point, we are naturally led to the consideration of *renormalization* in the lattice formulation.

## 16.6 Renormalization and the renormalization group on the lattice

### 16.6.1 Introduction

In the continuum formulation which we have used elsewhere in this book, fluctuations over short distances of order $\Lambda^{-1}$ generally lead to divergences in the limit $\Lambda \to \infty$, which are controlled (in a renormalizable theory) by the procedure of renormalization. Such divergent fluctuations turn out, in fact, to affect a renormalizable theory only through the values of some of its parameters and, if these parameters are taken from experiment, all other quantities become finite, even as $\Lambda \to \infty$. This latter assertion is not easy to prove and, indeed, is quite surprising. However, this is by no means all there is to renormalization theory: we have seen the power of 'renormalization group' ideas in making testable predictions for QCD. Nevertheless, the methods of chapter 15 were rather formal and the reader may well feel the need of a more physical picture of what is going on. Such a picture was provided by Wilson (1971a) (see also Wilson and Kogut 1974), using the 'lattice + path integral' approach. Another important advantage of this formalism is, therefore, precisely the way in which, thanks to Wilson's work, it provides access to a more intuitive way of understanding renormalization theory. The aim of this section is to give a brief introduction to Wilson's ideas, so as to illuminate the formal treatment of the previous chapter.

In the 'lattice + path integral' approach to quantum field theory, the degrees of freedom involved are the values of the field(s) at each lattice site, as we have seen. Quantum amplitudes are formed by integrating suitable quantities over all values of these degrees of freedom, as in (16.79) for example. From this point of view, it should be possible to examine specifically how the 'short distance' or 'high momentum' degrees of freedom affect the result. In fact, the idea suggests

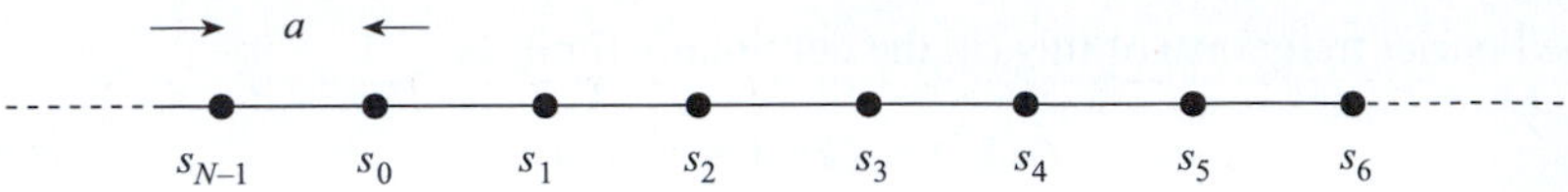


**Figure 16.4.** A portion of the one-dimensional lattice of spins in the Ising model.

itself that we might be able to perform explicitly the integration (or summation) over those degrees of freedom located near the cut-off $\Lambda$ in momentum space, or separated by only a lattice site or two in coordinate space. If we can do this, the result may be compared with the theory as originally formulated to see how this 'integration over short-distance degrees of freedom' affects the physical predictions of the theory. Having done this once, we can imagine doing it again—and indeed *iterating* the process, until eventually we arrive at some kind of 'effective theory' describing physics in terms of 'long-distance' degrees of freedom.

There are several aspects of such a programme which invite comment. First, the process of 'integrating out' short-distance degrees of freedom will obviously *reduce* the number of effective degrees of freedom, which is necessarily very large in the case $\xi \gg a$, as previously envisaged. Thus, it must be a step in the right direction. Second, this sketch of the 'integrating out' procedure suggests that, at any given stage of the integration, we shall be considering the system as described by parameters (including masses and couplings) *appropriate to that scale*, which is of course strongly reminiscent of RGE ideas. And third, we may perhaps anticipate that the result of this 'integrating out' will be not only to render the parameters of the theory scale-dependent but also, in general, to introduce new kinds of *effective interactions* into the theory. We now consider some simple examples which we hope will illustrate these points.

### 16.6.2 The one-dimensional Ising model

Consider first a simple one-dimensional Ising model with Hamiltonian (16.87) and partition function

$$Z = \sum_{\{s_n\}} \exp\left[K \sum_{n=0}^{N-1} s_n s_{n+1}\right] \tag{16.91}$$

where $K = J/(k_{\mathrm{B}}T) > 0$. In (16.91) all the $s_n$ variables take the values $\pm 1$ and the 'sum over $\{s_n\}$' means that all possible configurations of the $N$ variables $s_0, s_1, s_2, \ldots, s_{N-1}$ are to be included. The spin $s_n$ is located at the lattice site $na$ and we shall (implicitly) be assuming the periodic boundary condition $s_n = s_{N+n}$. Figure 16.4 shows a portion of the one-dimensional lattice with the spins on the sites, each site being separated by the lattice constant $a$. Thus, for this portion we

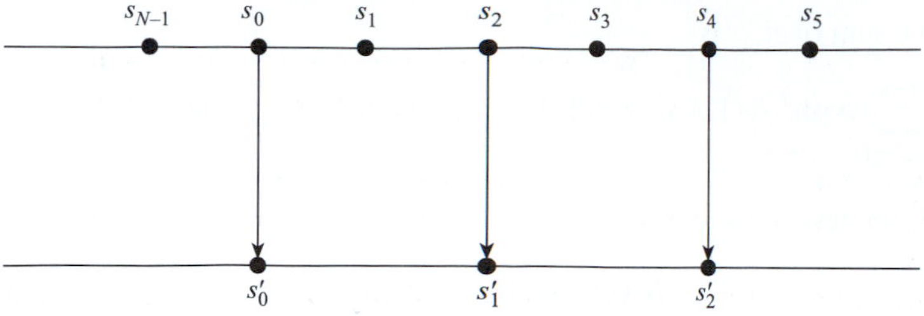


**Figure 16.5.** A 'coarsening' transformation applied to the lattice portion shown in figure 16.4. The new (primed) spin variables are situated twice as far apart as the original (unprimed) ones.

are evaluating

$$\sum_{s_{N-1},s_0,s_1,s_2,s_3,s_4} \exp[K(s_{N-1}s_0 + s_0s_1 + s_1s_2 + s_2s_3 + s_3s_4)]\,. \tag{16.92}$$

Now suppose we want to describe the system in terms of a 'coarser' lattice, with lattice spacing $2a$ and corresponding new spin variables $s'_n$. There are many ways we could choose to describe the $s'_n$ but here we shall only consider a very simple one (Kadanoff 1977) in which each $s'_n$ is simply identified with the $s_n$ at the corresponding site (see figure 16.5). For the portion of the lattice under consideration, then, (16.92) becomes

$$\sum_{s_{N-1},s'_0,s_1,s'_1,s_3,s'_2} \exp[K(s_{N-1}s'_0 + s'_0s_1 + s_1s'_1 + s'_1s_3 + s_3s'_2)]. \tag{16.93}$$

If we can now perform the sums over $s_1$ and $s_3$ in (16.93), we shall end up (for this portion) with an expression involving the 'effective' spin variables $s'_0$, $s'_1$ and $s'_2$, situated twice as far apart as the original ones and, therefore, providing a more 'coarse grained' description of the system. Summing over $s_1$ and $s_3$ corresponds to 'integrating out' two short-distance degrees of freedom as discussed earlier.

In fact, these sums are easy to do. Consider the quantity $\exp(Ks'_0s_1)$, expanded as a power series:

$$\exp(Ks'_0s_1) = 1 + Ks'_0s_1 + \frac{K^2}{2!} + \frac{K^3}{3!}(s'_0s_1) + \cdots \tag{16.94}$$

where we have used $(s'_0s_1)^2 = 1$. It follows that

$$\exp(Ks'_0s_1) = \cosh K(1 + s'_0s_1 \tanh K) \tag{16.95}$$

and similarly

$$\exp(Ks_1s'_1) = \cosh K(1 + s_1s'_1 \tanh K). \tag{16.96}$$

Thus the sum over $s_1$ is

$$\sum_{s_1=\pm 1} \cosh^2 K(1 + s_0' s_1 \tanh K + s_1 s_1' \tanh K + s_0' s_1' \tanh^2 K). \tag{16.97}$$

Clearly, the terms linear in $s_1$ vanish after summing and the $s_1$ sum becomes just

$$2\cosh^2 K(1 + s_0' s_1' \tanh^2 K). \tag{16.98}$$

Remarkably, (16.98) contains a new 'nearest-neighbour' interaction, $s_0' s_1'$, just like the original one in (16.91) but with an *altered coupling* (and a different spin-independent piece). In fact, we can write (16.98) in the standard form

$$\exp[g_1(K) + K' s_0' s_1'] \tag{16.99}$$

and then use (16.95) to set

$$\tanh K' = \tanh^2 K \tag{16.100}$$

and identify

$$g_1(K) = \ln\left(\frac{2\cosh^2 K}{\cosh K'}\right). \tag{16.101}$$

Exactly the same steps can be followed through for the sum on $s_3$ in (16.93)—and indeed for *all* the sums over the 'integrated out' spins. The upshot is that, apart from the accumulated spin-independent part, the new partition function, defined on a lattice of size $2a$, has the same form as the old one but with a new coupling $K'$ related to the old one $K$ by (16.100).

Equation (16.100) is an example of a *renormalization transformation*: the number of degrees of freedom has been halved, the lattice spacing has doubled and the coupling $K$ has been renormalized to $K'$.

It is clear that we could apply the same procedure to the new Hamiltonian, introducing a coupling $K''$ which is related to $K'$, and thence to $K$, by

$$\tanh K'' = (\tanh K')^2 = (\tanh K)^4. \tag{16.102}$$

This is equivalent to *iterating* the renormalization transformation; after $n$ iterations, the effective lattice constant is $2^n a$ and the effective coupling is given by

$$\tanh K^{(n)} = (\tanh K)^n. \tag{16.103}$$

The successive values $K', K'', \ldots$ of the coupling under these iterations can be regarded as a '*flow*' in the (one-dimensional) space of $K$-values: a *renormalization flow*.

Of particular interest is a point (or points) $K^*$ such that

$$\tanh K^* = \tanh^2 K^*. \tag{16.104}$$

**Figure 16.6.** 'Renormalization flow': the arrows show the direction of flow of the coupling $K$ as the lattice constant is increased. The starred values are fixed points.

This is called a *fixed point* of the renormalization tranformation. At such a point in $K$-space, changing the scale by a factor of 2 (or $2^n$ for that matter) will make no difference, which means that the system must be, in some sense, ordered. Remembering that $K = J/(k_B T)$, we see that $K = K^*$ when the temperature is 'tuned' to the value $T = T^* = J/(k_B K^*)$. Such a $T^*$ would be the temperature of a *critical point* for the thermodynamics of the system, corresponding to the onset of ordering. In the present case, the only fixed points are $K^* = \infty$ and $K^* = 0$. Thus there is no critical point at a non-zero $T^*$ and, hence, no transition to an ordered phase. However, we may describe the behaviour as $T \to 0$ as 'quasi-critical'. For large $K$, we may use

$$\tanh K \simeq 1 - 2e^{-2K} \tag{16.105}$$

to write (16.104) as

$$K^{(n)} = K - \tfrac{1}{2}\ln n \tag{16.106}$$

which shows that $K^n$ changes only very slowly (logarithmically) under iterations when in the vicinity of a very large value of $K$, so that this is 'almost' a fixed point.

We may represent the flow of $K$ under the renormalization transformation (16.103) as in figure 16.6. Note that the flow is away from the quasi-fixed point at $K^* = \infty (T = 0)$ and towards the (non-interacting) fixed point at $K^* = 0$.

Another way of looking for a critical (or fixed) point would be to calculate the correlation length $\xi(T)$ introduced in (16.89) and (16.90). At a critical point, $\xi \sim L$ (the system size), which goes to infinity in the thermodynamic limit. In the present model, with the Hamiltonian (16.87), we may calculate $\xi(T)$ exactly (problem 16.6) and find that

$$\xi(T) = \frac{-a}{\ln \tanh K(a)}. \tag{16.107}$$

Equation (16.107) confirms that there is no finite temperature $T$ at which $\xi \to \infty$ but as $T \to 0$ we do have $\xi \to \infty$ and the system is ordered. More precisely, from (16.107) we find

$$\xi(T) \simeq \frac{a}{2} e^{2K} \qquad \text{for } T \to 0 \tag{16.108}$$

or, in terms of the equivalent mass parameter $m(T) = \xi^{-1}(T)$ introduced after (16.90),

$$m(T) \simeq \frac{2}{a} e^{-2K}. \tag{16.109}$$

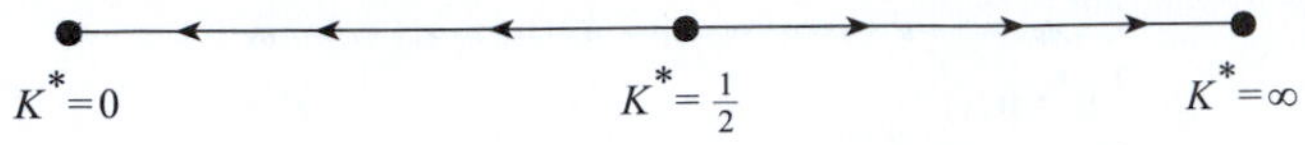


**Figure 16.7.** The renormalization flow for the transformation (16.113).

The expression (16.109) becomes very small at small $T$ but never vanishes at a finite value of $T$. In this sense, $T = 0$ is a kind of 'asymptotically reachable' fixed point.

It is interesting to consider the effect of the renormalization transformation (16.103) on $\xi$. Let us denote by $\xi_n(T)$ the effective correlation length after $n$ iterations, so that

$$\xi_n(T) = -\frac{a}{\ln \tanh K^{(n)}}. \tag{16.110}$$

From (16.103), it then easily follows that

$$\xi_n(T) = \frac{1}{n}\xi(T). \tag{16.111}$$

Equation (16.111) confirms what we might have expected: the correlation length is measured in units of the only available length unit, $a$, and when this increases to $na$ after $n$ iterations, $\xi$ must decrease by $n^{-1}$ so as to maintain the same physical distance. In particular, $\xi \to 0$ as $n \to \infty$.

### 16.6.3 Further developments and some connections with particle physics

A renormalization transformation which has a fixed point at a finite (neither zero nor infinite) value of the coupling is clearly of greater interest, since this will correspond to a critical point at a finite temperature. A simple such example given by Kadanoff (1977) is the transformation

$$K' = \tfrac{1}{2}(2K)^2 \tag{16.112}$$

for a doubling of the effective lattice size, or

$$K^{(n)} = \tfrac{1}{2}(2K)^n \tag{16.113}$$

for $n$ such iterations. The model leading to (16.113) involves fermions in one dimension but the details are irrelevant to our purpose here. The renormalization transformation (16.113) has three fixed points: $K^* = 0$, $K^* = \infty$ and the finite point $K^* = \frac{1}{2}$. The renormalization flow is shown in figure 16.7.

The striking feature of this flow is that the motion is always away from the finite fixed point, under successive iterations. This may be understood by recalling that at the fixed point (which is a critical point for the statistical system) the correlation length $\xi$ must be infinite (as $L \to \infty$). As we iterate away from

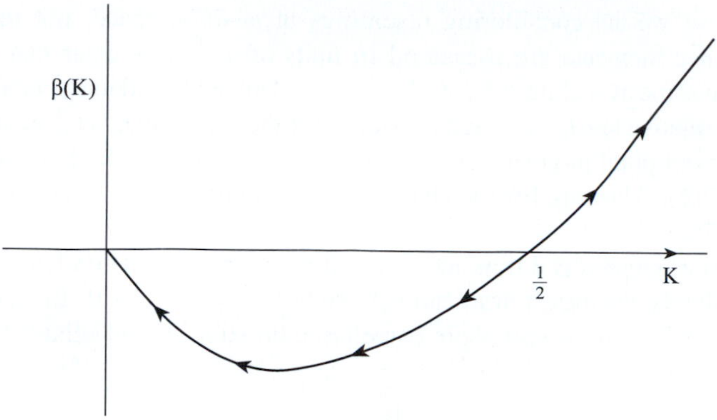


**Figure 16.8.** The $\beta$-function of (16.117)—the arrows indicate increasing $f$.

this point, $\xi$ must decrease according to (16.111), and therefore we must leave the fixed (or critical) point. For this model, $\xi$ is given by Kadanoff (1977) as

$$\xi = \frac{a}{|\ln 2K|} \tag{16.114}$$

which indeed goes to infinity at $K = \frac{1}{2}$.

Let us now begin to think about how all this may relate to the treatment of the renormalization group in particle physics, as given in the previous chapter. First, we need to consider a continuous change of scale, say by a factor of $f$. In the present model, the transformation (16.113) then becomes

$$K(fa) = \tfrac{1}{2}(2K(a))^f. \tag{16.115}$$

Differentiating (16.115) with respect to $f$, we find that

$$f\frac{\mathrm{d}K(fa)}{\mathrm{d}f} = K(fa)\ln[2K(fa)]. \tag{16.116}$$

We may reasonably call (16.116) a renormalization group equation, describing the 'running' of $K(fa)$ with the scale $f$, analogous to the RGE's for $\alpha$ and $\alpha_s$ considered in chapter 15. In this case, the $\beta$-function is

$$\beta(K) = K\ln(2K), \tag{16.117}$$

which is sketched in figure 16.8. The zero of $\beta$ is indeed at the fixed (critical) point $K = \frac{1}{2}$ and this is an infrared unstable fixed point, the flow being away from it as $f$ increases.

The foregoing is exactly analogous to the discussion in section 15.4—see, in particular, figure 15.7 and the related discussion. Note, however, that in the present case we are considering rescalings in *position* space, not momentum space. Since momenta are measured in units of $a^{-1}$, it is clear that scaling $a$ by $f$ is the same as scaling $k$ by $f^{-1} = t$, say. This will produce a change in sign in $\mathrm{d}K/\mathrm{d}t$ relative to $\mathrm{d}K/\mathrm{d}f$, and accounts for the fact that $K = \frac{1}{2}$ is an infrared unstable fixed point in figure 16.8, while $\alpha_{\mathrm{s}}^*$ is an infrared stable fixed point in figure 15.7(*b*). Allowing for the change in sign, figure 16.8 is quite analogous to figure 15.7(*a*).

We have emphasized that, at a critical point, the correlation length $\xi \to \infty$ or, equivalently, the mass parameter (cf (16.90)) $m = \xi^{-1} \to 0$. In this case, the Fourier transform of the spin–spin correlation function should behave as

$$\tilde{G}(\mathbf{k}^2) \propto \frac{1}{\mathbf{k}^2}. \tag{16.118}$$

This is indeed the $\mathbf{k}^2$-dependence of the propagator of a free, massless scalar particle, but—as we learned for the fermion propagator in section 15.4—it is no longer true in an interacting theory. In the interacting case, (16.118) generally becomes modified to

$$\tilde{G}(\mathbf{k}^2) \propto \frac{1}{(\mathbf{k}^2)^{1-\frac{\eta}{2}}} \tag{16.119}$$

or, equivalently,

$$G(\mathbf{x}) \propto \frac{1}{|\mathbf{x}|^{1+\eta}} \tag{16.120}$$

in three spatial dimensions, and in the continuum limit. Thus, at a critical point, the spin–spin correlation function exhibits scaling under the transformation $\mathbf{x}' = f\mathbf{x}$ but it is not free-field scaling. Comparing (16.119) with (15.65), we see that $\eta/2$ is precisely the *anomalous dimension* of the field $s(\mathbf{x})$, so—just as in section 15.4—we have an example of scaling with anomalous dimension. In the statistical mechanics case, $\eta$ is a *critical exponent*, one of a number of such quantities characterizing the critical behaviour of a system. In general, $\eta$ will depend on the coupling constant $\eta(K)$: at a non-trivial fixed point, $\eta$ will be evaluated at the fixed point value $K^*$, $\eta(K^*)$. Enormous progress was made in the theory of critical phenomena when the powerful methods of quantum field theory were applied to calculate critical exponents (see, for example, Peskin and Schroeder (1995, chapter 13), and Binney *et al* (1992)).

In our discussion so far, we have only considered simple models with just one 'coupling constant', so that diagrams of renormalization flow were one-dimensional. Generally, of course, Hamiltonians will consist of several terms and the behaviour of all their coefficients will need to be considered under a renormalization transformation. The general analysis of renormalization flow in multi-dimensional coupling space was given by Wegner (1972). In simple terms, the coefficients show one of three types of behaviour under renormalization transformations such that $a \to fa$, characterized by their behaviour in the vicinity of a fixed point:

(i) the difference from the fixed point value grows as $f$ increases, so that the system moves away from the fixed point (as in the single-coupling examples considered earlier);
(ii) the difference decreases as $f$ increases, so the system moves towards the fixed point; and
(iii) there is no change in the value of the coupling as $f$ changes.

The corresponding coefficients are called, respectively, (i) *relevant*, (ii) *irrelevant* and (iii) *marginal* couplings—the terminology is also frequently applied to the operators in the Hamiltonians themselves. The intuitive meaning of 'irrelevant' is clear enough: the system will head towards a fixed point as $f \to \infty$ whatever the initial values of the irrelevant couplings. The critical behaviour of the system will, therefore, be independent of the number and type of all irrelevant couplings and will be determined by the relatively few (in general) marginal and relevant couplings. Thus, *all* systems which flow close to the fixed point will display the same critical exponents determined by the dynamics of these few couplings. This explains the property of *universality* observed in the physics of phase transitions, whereby many apparently quite different physical systems are described (in the vicinity of their critical points) by the same critical exponents.

Additional terms in the Hamiltonian are, in fact, generally introduced following a renormalization transformation. A simple mathematical analogue may illustrate the point. Consider the expression

$$Z_{xy} = \int_{-\infty}^{\infty} \mathrm{d}x \int_{-\infty}^{\infty} \mathrm{d}y \, \exp[-(x^2 + y^2 + \lambda x^4 + \lambda x^2 y^2)] \tag{16.121}$$

which may be regarded as the 'partition function' for a system with two variables $x$ and $y$, the action being similar to that of two scalar fields with quartic couplings. Suppose we want to 'integrate out' the variable $y$. We write

$$\begin{aligned} Z_{xy} &= \int_{-\infty}^{\infty} \mathrm{d}x \, \exp[-(x^2 + \lambda x^4)] \int_{-\infty}^{\infty} \mathrm{d}y \, \exp[-(y^2 + \lambda x^2 y^2)] \\ &= \int_{-\infty}^{\infty} \mathrm{d}x \, \exp[-(x^2 + \lambda x^4)] \left( \frac{\pi}{1 + \lambda x^2} \right)^{\frac{1}{2}} . \end{aligned} \tag{16.122}$$

Assuming that $\lambda$ is small and may be treated perturbatively, (16.122) may be expanded as

$$Z_{xy} \simeq \pi^{\frac{1}{2}} \int_{-\infty}^{\infty} \mathrm{d}x \, \exp[-(x^2 + \lambda x^4)][1 - \tfrac{1}{2}\lambda x^2 + \tfrac{3}{8}\lambda^2 x^4 - \tfrac{5}{16}\lambda^3 x^6 + \cdots]. \tag{16.123}$$

The terms in the series expansion may be regarded as arising from an expansion of the exponential

$$\exp[-\tfrac{1}{2}\lambda x^2 - \tfrac{1}{4}\lambda^2 x^4 + \tfrac{3}{16}\lambda^3 x^6 - \cdots] \tag{16.124}$$

to order $\lambda^3$, so that to this order, (16.123) may be written as

$$Z_{xy} \simeq \pi^{\frac{1}{2}} \int_{-\infty}^{\infty} \mathrm{d}x \; \exp\{-[x^2(1+\tfrac{1}{2}\lambda)+x^4(\lambda-\tfrac{1}{4}\lambda^2)+\tfrac{3}{16}\lambda^3x^6+\cdots]\}. \quad (16.125)$$

The result (16.125) may be interpreted as (a) a renormalization of the 'mass' term $x^2$ so that its coefficient changes from 1 to $1+\frac{1}{2}\lambda$; (b) a renormalization of the $x^4$ term so that the coupling $\lambda$ changes to $\lambda - \frac{1}{4}\lambda^2$; and (c) the generation of a 'new interaction' of sixth order in the degree of freedom $x$. We may think of this new interaction (in more realistic quantum field cases) as supplying the effective interaction between the '$x$-fields' that was previously mediated by the '$y$-fields'.

In the quantum field case, we may expect that renormalization transformations associated with $a \to fa$, and iterations thereof, will in general lead to an effective theory involving all possible couplings allowed by whatever symmetries are assumed to be relevant. Thus, if we start with a typical '$\phi^4$' scalar theory as given by (16.86), we shall expect to generate all possible couplings involving $\phi$ and its derivatives. At first sight, this may seem disturbing: after all, the original theory (in four dimensions) is a renormalizable one, but an interaction such as $A\phi^6$ is *not* renormalizable according to the criterion given in section 11.8 ( in four dimensions $\phi$ has mass dimension unity, so that $A$ must have mass dimension $-2$). It is, however, essential to remember that in this 'Wilsonian' approach to renormalization, summations over momenta appearing in loops do not, after one iteration $a \to fa$, run up to the original cut-off value $\pi/a$, but only up to the lower cut-off $\pi/fa$. The additional interactions compensate for this change.

In fact, we shall now see how the coefficients of non-renormalizable interactions correspond precisely to *irrelevant* couplings in Wilson's approach, so that their effect becomes negligible as we iterate to scales much larger than $a$. We consider continuous changes of scale characterized by a factor $f$, and we discuss a theory with only a single scalar field $\phi$ for simplicity. Imagine, therefore, that we have integrated out, in (16.85), those components of $\phi(\mathbf{x})$ with $a < |\mathbf{x}| < fa$. We will be left with a functional integral of the form (16.85) but with $\phi(\mathbf{x})$ restricted to $|\mathbf{x}| > fa$, and with additional interaction terms in the action. In order to interpret the result in Wilson's terms, we must rewrite it so that it has the same general form as the original $Z_\phi$ of (16.85). A simple way to do this is to rescale distances by

$$\mathbf{x}' = \frac{\mathbf{x}}{f} \quad (16.126)$$

so that the functional integral is now over $\phi(\mathbf{x}')$ with $|\mathbf{x}'| > a$, as in (16.85). We now *define* the fixed point of the renormalization transformation to be that in which all the terms in the action are *zero*, except the 'kinetic' piece; this is the 'free-field' fixed point. Thus, we require the kinetic action to be unchanged:

$$\int \mathrm{d}^4x_{\mathrm{E}}\,(\partial_\mu\phi)^2 = \int \mathrm{d}^4x'_{\mathrm{E}}\,(\partial'_\mu\phi')^2$$

$$= \int \frac{1}{f^2} \mathrm{d}^4 x_{\mathrm{E}} \, (\partial_\mu \phi')^2 \tag{16.127}$$

from which it follows that $\phi' = f\phi$. Consider now a term of the form $A\phi^6$:

$$A \int \mathrm{d}^4 x_{\mathrm{E}} \, \phi^6 = \frac{A}{f^2} \int \mathrm{d}^4 x'_{\mathrm{E}} \, \phi'^6. \tag{16.128}$$

(16.128) shows that the 'new' $A'$ is related to the old one by $A' = A/f^2$ and in particular that, as $f$ increases, $A'$ decreases and is therefore an *irrelevant* coupling, tending to zero as we reach large scales. But such an interaction is precisely a non-renormalizable one (in four dimensions), according to the criterion of section 11.8. The mass dimension of $\phi$ is unity, and hence that of $A$ must be $-2$ so that the action is dimensionless: couplings with negative mass dimensions correspond to non-renormalizable interactions. The reader may verify the generality of this result for any interaction with $p$ powers of $\phi$ and $q$ derivatives of $\phi$.

However, the mass term $m^2\phi^2$ behaves differently:

$$m^2 \int \mathrm{d}^4 x_{\mathrm{E}} \, \phi^2 = m^2 f^2 \int \mathrm{d}^4 x'_{\mathrm{E}} \, \phi'^2 \tag{16.129}$$

showing that $m'^2 = m^2 f^2$ and the 'coupling' $m^2$ is *relevant*, since it grows with $f^2$. Such a term has positive mass dimension and corresponds to a 'super-renormalizable' interaction. Finally, the $\lambda\phi^4$ interaction transforms as

$$\lambda \int \mathrm{d}^4 x_{\mathrm{E}} \, \phi^4 = \lambda \int \mathrm{d}^4 x'_{\mathrm{E}} \, \phi'^4 \tag{16.130}$$

and so $\lambda' = \lambda$. The coupling is *marginal*, which may correspond (though not necessarily) to a renormalizable interaction. To find whether such couplings increase or decrease with $f$, we have to include higher-order loop corrections. The foregoing analysis in terms of the suppression of non-renormalizable interactions by powers of $f^{-1}$ parallels precisely the similar one in section 11.8. We saw that such terms were suppressed at low energies by factors of $E/\Lambda$, where $\Lambda$ is the cut-off scale beyond which the theory is supposed to fail on physical grounds (e.g. $\Lambda$ might be the Planck mass). The result is that as we renormalize, in Wilson's sense, down to much lower energy scales, the non-renormalizable terms disappear and we are left with an effective renormalizable theory. This is the field theory analogue of 'universality'.

One further word should be said about terms such as '$m^2\phi^2$' (which arise in the Higgs sector of the Standard Model, for instance). As we have seen, $m^2$ scales by $m'^2 = m^2 f^2$, which is a rapid growth with $f$. If we imagine starting at a very high scale, such as $10^{15}$ TeV and flowing down to 1 TeV, then the 'initial' value of $m$ will have to be very finely 'tuned' in order to end up with a mass of order 1 TeV. Thus, in this picture, it seems unnatural to have scalar particles with masses much less than the physical cut-off scale. We shall return to this problem in section 22.10.1.

We have strayed considerably from the initial purpose of this chapter—but we hope that the reader will agree that the extra insight gained into the physical meaning of renormalization has made the detour worthwhile. We now finally return to lattice QCD, with a brief survey of some of the results obtained numerically.

## 16.7 Numerical calculations

Even in the quenched approximation (section 16.4), computing demands are formidable. The lattice must be large enough so that the spatial dimension $R$ of the object we wish to describe—the Compton wavelength of a quark or the size of a hadron—fits comfortably inside it, otherwise the result will be subject to 'finite-size effects' as the hypercube side length $L$ is varied. We also need $R \gg a$ or else the granularity of the lattice resolution will become apparent. Further, as indicated earlier, we expect the mass $m$ (which is of order $R^{-1}$) to be very much less than $a^{-1}$. Thus, ideally, we need

$$a \ll R \sim 1/m \ll L = Na \tag{16.131}$$

so that $N$ must be large. Actual calculations are done by evaluating quantities such as (16.79) by 'Monte Carlo' methods (similar to the method which can be employed to evaluate multi-dimensional integrals).

Ignoring any statistical inaccuracy, the results will depend on the parameters $g_{\rm L}$ and $N$, where $g_{\rm L}$ is the bare lattice gauge coupling (we assume for simplicity that the quarks are massless). Despite the fact that $g_{\rm L}$ is dimensionless, we shall now see that its value actually controls the physical size of the lattice spacing, $a$, as a result of renormalization effects. The computed mass of a hadron $M$, say, must be related to the only quantity with mass dimension, $a^{-1}$, by a relation of the form

$$M = \frac{1}{a} f(g_{\rm L}). \tag{16.132}$$

Thus, in approaching the continuum limit $a \to 0$, we shall also have to change $g_{\rm L}$ suitably, so as to ensure that $M$ remains finite. This is, of course, quite analogous to saying that, in a renormalizable theory, the bare parameters of the theory depend on the momentum cut-off $\Lambda$ in such a way that, as $\Lambda \to \infty$, finite values are obtained for the corresponding physical parameters (see the last paragraph of section 10.1.2, for example). In practice, however, the extent to which the lattice '$a$' can really be taken to be very small is severely limited by the computational resources available—that is, essentially, by the number of mesh points $N$. Quenched calculations now use four-dimensional cubes with $N = 64$, for example. If we were to think of an $a$ of order 0.01 fm, so that $L \sim 0.64$ fm, the masses $m$ which could be simulated would be limited (from (16.131)) by

$m \gg 300$ MeV, which is a severe restriction against light quarks or hadrons. A more reasonable value might be $a \sim 0.1$ fm with $L \sim 6.4$ fm, which would allow the pion to be reached but note that, in this case, the equivalent momentum cut-off is $\hbar/a \sim 6$ GeV, which seems low and certainly rules out simulations with very massive quarks, in view of the left-hand inequality in (16.131). At all events, it is clear that the lattice cut-off is, in practice, a long way from the 'in principle' $\Lambda \to \infty$ situation.

Equation (16.132) should, therefore, really read as

$$M = \frac{1}{a} f\left(g_{\rm L}(a)\right). \tag{16.133}$$

As $a \to 0$, $M$ should be finite and independent of $a$. However, we know that the behaviour of $g_{\rm L}(a)$ at small scales is, in fact, calculable in perturbation theory, thanks to the asymptotic freedom of QCD. This will allow us to determine the form of $f(g_{\rm L})$, up to a constant, and lead to an interesting prediction for $M$ (equations (16.139)–(16.140)).

Differentiating (16.133), we find that

$$0 = \frac{\mathrm{d}M}{\mathrm{d}a} = -\frac{1}{a^2} f\left(g_{\rm L}(a)\right) + \frac{1}{a}\frac{\mathrm{d}f}{\mathrm{d}g_{\rm L}}\frac{\mathrm{d}g_{\rm L}(a)}{\mathrm{d}a} \tag{16.134}$$

so that

$$\left(a\frac{\mathrm{d}g_{\rm L}(a)}{\mathrm{d}a}\right)\frac{\mathrm{d}f}{\mathrm{d}g_{\rm L}} = f\left(g_{\rm L}(a)\right). \tag{16.135}$$

Meanwhile, the scale dependence of $g_{\rm L}$ is given (to one-loop order) by

$$a\frac{\mathrm{d}g_{\rm L}(a)}{\mathrm{d}a} = \frac{b}{4\pi} g_{\rm L}^3(a) \tag{16.136}$$

where the sign is the opposite of (15.48) since $a \sim \mu^{-1}$ is the relevant scale parameter here (compare the comments after equation (16.117)). The integration of (16.136) requires, as usual, a dimensionful constant of integration (cf (15.54)):

$$\frac{g_{\rm L}^2(a)}{4\pi} = \frac{1}{b\ln(1/a^2\Lambda_{\rm L}^2)}. \tag{16.137}$$

Equation (16.137) shows that $g_{\rm L}(a)$ tends logarithmically to zero as $a \to 0$, as we expect from asymptotic freedom. $\Lambda_{\rm L}$ can be regarded as a lattice equivalent of the continuum $\Lambda_{\rm QCD}$, and it is defined by

$$\Lambda_{\rm L} \equiv \lim_{g_{\rm L}\to 0} \frac{1}{a}\exp\left(-\frac{2\pi}{b g_{\rm L}^2}\right). \tag{16.138}$$

Equation (16.138) may also be read as showing that the lattice spacing $a$ must go exponentially to zero as $g_{\rm L}$ tends to zero. Higher-order corrections can, of course, be included.

In a similar way, integrating (16.135) using (16.136) gives, in (16.132),

$$M = \text{constant} \times \left[\frac{1}{a}\exp\left(-\frac{2\pi}{bg_{\mathrm{L}}^2}\right)\right] \tag{16.139}$$

$$= \text{constant} \times \Lambda_{\mathrm{L}}. \tag{16.140}$$

Equation (16.139) is known as *asymptotic scaling*: it predicts how any physical mass, expressed in lattice units $a^{-1}$, should vary as a function of $g_{\mathrm{L}}$. The form (16.140) is remarkable, as it implies that all calculated masses must be proportional, in the continuum limit $a \to 0$, to the same universal scale factor $\Lambda_{\mathrm{L}}$.

How are masses calculated on the lattice? The principle is very similar to the way in which the ground state was selected out as $\tau_{\mathrm{i}} \to -\infty$, $\tau_{\mathrm{f}} \to +\infty$ in (16.70). Consider a correlation function for a scalar field, for simplicity:

$$C(\tau) = \langle\Omega|\phi(\mathbf{x}=0,\tau)\phi(0)|\Omega\rangle$$

$$= \sum_n |\langle\Omega|\phi(0)|n\rangle|^2\,\mathrm{e}^{-E_n\tau}. \tag{16.141}$$

As $\tau \to \infty$, the term with the minimum value of $E_n$, namely $E_n = M_\phi$, will survive: $M_\phi$ can be measured from a fit to the exponential fall-off as a function of $\tau$.

The behaviour predicted by (16.139) and (16.140) can be tested in actual calculations. A quantity such as the $\rho$ meson mass is calculated (via a correlation function of the form (16.141), the result being expressed in terms of a certain number of lattice units $a^{-1}$ at a certain value of $g_{\mathrm{L}}$. By comparison with the known $\rho$ mass, $a^{-1}$ can be converted to GeV. Then the calculation is repeated for a different $g_{\mathrm{L}}$ value and the new $a^{-1}$ (GeV) extracted. A plot of $\ln[a^{-1}(\mathrm{GeV})]$ *versus* $1/g_{\mathrm{L}}^2$ should then give a straight line with slope $2\pi/b$ and intercept $\ln \Lambda_{\mathrm{L}}$. Figure 16.9 shows such a plot, taken from Ellis *et al* (1996), from which it appears that the calculations are indeed being performed close to the continuum limit. The value of $\Lambda_{\mathrm{L}}$ has been adjusted to fit the numerical data and has the value $\Lambda_{\mathrm{L}} = 1.74$ MeV in this case. This may seem alarmingly far from the kind of value expected for $\Lambda_{\mathrm{QCD}}$, but we must remember that the renormalization schemes involved in the two cases are quite different. In fact, we may expect $\Lambda_{\mathrm{QCD}} \approx 50\Lambda_{\mathrm{L}}$ (Montvay and Munster 1994, section 5.1.6).

Having fixed the physical state of $a$ by the $\rho$ mass, one can then go on to make predictions for the other low-lying hadrons. As one example, we show in figure 16.10 a precision calculation by the CP-PACS collaboration (Aoki *et al* 2000) of the spectrum of light hadrons in quenched QCD. The $\rho$ and $\pi$ masses are missing since they were used to fix, respectively, the lattice spacing (as just noted) and the u, d masses. For each hadron, two lattice results are shown: filled circles correspond to fixing the strange quark mass by fitting the K mass, and open circles correspond to fitting the $\phi$ mass. The horizontal lines are the experimental

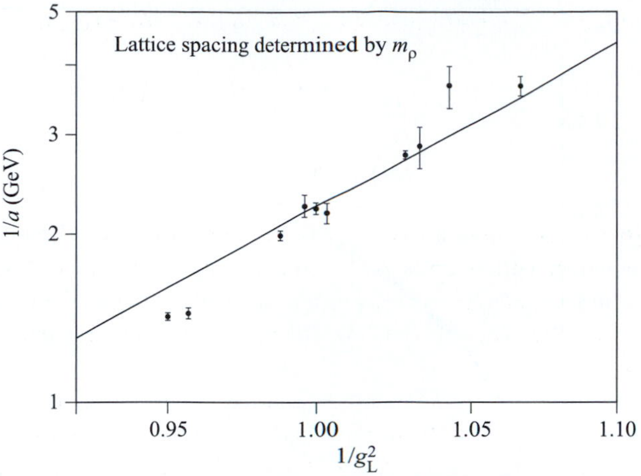


**Figure 16.9.** $\ln(a^{-1}$ in GeV) plotted against $1/g_L^2$, taken from Ellis *et al* (1996), as adapted from Allton (1995).

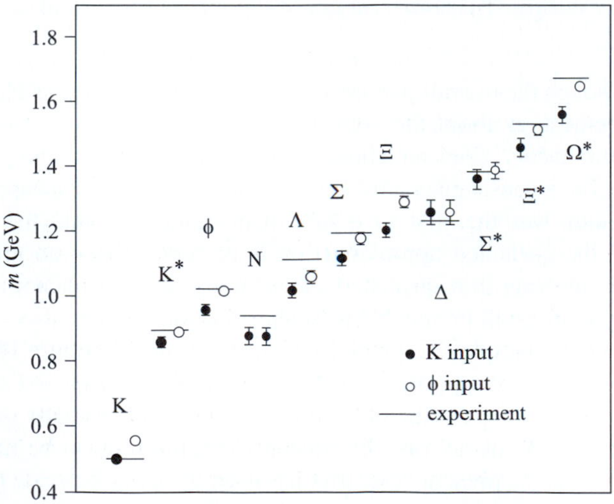


**Figure 16.10.** The mass spectrum of light mesons and baryons, containing u, d and s quarks, calculated in the quenched approximation (Aoki *et al* 2000): Filled circles are the results calculated by fixing the s quark mass to give the correct mass for the K meson; open circles are the results of fitting the s quark mass to the $\phi$ mass. The horizontal lines are the experimental values. The $\rho$ and $\pi$ masses are absent because they are used to fix the lattice spacing and the u, d masses.

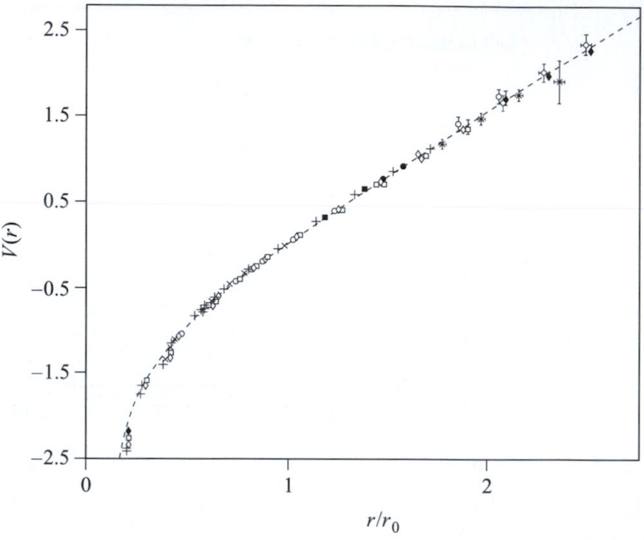


**Figure 16.11.** The static QCD potential expressed in units of $r_0$ (Allton *et al* 2002, UKQCD Collaboration). The broken curve is the functional form (16.145).

masses. Although the overall picture is quite impressive, there is clear evidence of a disagreement, at about the 10% level, between the pairs of results. This means that the 'theory' does not allow a consistent definition to be given of the s quark mass. The inconsistency must be attributed to the quenched approximation. This calculation was the first to establish quantitatively that the error to be expected of the quenched approximation is of order 10%—an encouragingly small value, implying that quenched calculations of other phenomenologically important quantities will be reliable to within that sort of error also.

As a second example of a precision result, we show in figure 16.11 a lattice calculation of the static $q\bar{q}$ potential (Allton *et al* 2002, UKQCD Collaboration) using two degenerate flavours of dynamical (i.e. unquenched) quarks[3] on a $16^3 \times 32$ lattice. As usual, one dimensionful quantity has to be fixed in order to set the scale. In the present case, this has been done via the scale parameter $r_0$ of Sommer (1994), defined by

$$r_0^2 \left.\frac{\mathrm{d}V}{\mathrm{d}r}\right|_{r=r_0} = 1.65. \tag{16.142}$$

Applying (16.142) to the Cornell (Eichten *et al* 1980) or Richardson (1979) phenomenological potentials gives $r_0 \simeq 0.49$ fm, conveniently in the range which is well determined by $c\bar{c}$ and $b\bar{b}$ data. The data are well described by

[3] Comparison with matched data in the quenched approximation revealed very little difference, in this case.

the expression

$$V(r) = V_0 + \sigma r - \frac{A}{r} \tag{16.143}$$

where, in accordance with (16.142),

$$\sigma = \frac{(1.65 - A)}{r_0^2} \tag{16.144}$$

and where $V_0$ has been chosen such that $V(r_0) = 0$. Thus, (16.143) becomes

$$r_0 V(r) = (1.65 - A)\left(\frac{r}{r_0} - 1\right) - A\left(\frac{r_0}{r} - 1\right). \tag{16.145}$$

Equation (16.143) is—up to a constant—exactly the functional form mentioned in chapter 2, equation (2.22). The quantity $\sqrt{\sigma}$ (there called $b$) is referred to as the 'string tension' and has a value of about 465 MeV in the present calculations. Phenomenological models suggest a value of around 440 MeV (Eichten *et al* 1980). The parameter $A$ is found to have a value of about 0.3. In lowest-order perturbation theory and in the continuum limit, $A$ would be given by one-gluon exchange as

$$A = \tfrac{4}{3}\alpha_s(\mu) \tag{16.146}$$

where $\mu$ is some energy scale. This would give $\alpha_s \simeq 0.22$, a reasonable value for $\mu \simeq 3$ GeV. Interestingly, the form (16.145) is predicted by the 'universal bosonic string model' (Lüscher *et al* 1980, Lüscher 1981), in which $A$ has the 'universal' value $\pi/12 \simeq 0.26$.

The existence of the linearly rising term with $\sigma > 0$ is a signal for confinement, since—if the potential maintained this form—it would cost an infinite amount of energy to separate a quark and an anti-quark. But at some point, enough energy will be stored in the 'string' to create a $q\bar{q}$ pair from the vacuum: the string then breaks and the two $q\bar{q}$ pairs form mesons. There is no evidence for string breaking in figure 16.11 but we must note that the largest distance probed is only about 1.3 fm.

Our third and last example of lattice QCD calculations concerns chiral symmetry breaking. We learned in section 12.3.2 that there is good evidence to believe that the hadron spectrum predicted by QCD should show signs of a symmetry (namely, chiral symmetry) which would be exact if the u and d quarks were massless, and which should survive as an approximate symmetry to the extent that $m_u$ and $m_d$ are small on hadronic scales. But we also noted that the most obvious signs of such a symmetry—parity doublets in the hadronic spectrum—are conspicuously absent. The resolution of this puzzle lies in the concept of 'spontaneous symmetry breaking', which forms the subject of the next part of this book (chapters 17–19). Nevertheless, we propose to include the topic at the present stage, since it is one on which significant progress has recently been made within the lattice approach. Besides, seeing the concept in action here will provide good motivation for the detailed study in part 7.

'Spontaneous symmetry breaking' is an essentially non-perturbative phenomenon and its possible occurrence in QCD is, therefore, a particularly suitable problem for investigation by lattice calculations. Unfortunately, it is difficult to construct a lattice theory with fermions—even massless ones, supposing we could somehow ignore the right-hand inequality in (16.131)—in such a way as not to violate chiral symmetry from the start. For example, the 'Wilson' (1974a) fermions mentioned earlier, while avoiding the fermion doubling problem, break chiral symmetry explicitly. This can easily be seen by noting (see (12.155) for example) that the crucial property required for chiral symmetry to hold is

$$\gamma_5 \not{D} + \not{D}\gamma_5 = 0 \tag{16.147}$$

where $\not{D}$ is the $\mathrm{SU}(3)_\mathrm{c}$-covariant Dirac derivative. But the modification to the derivative made in (16.21) contains a piece with no $\gamma$-matrix, which will not satisfy (16.147). Indeed, for a long time it was thought that, subject to quite mild assumptions, chiral symmetry simply could not be realized at non-zero lattice spacing: this is the content of the Nielsen–Ninomiya theorem (Nielsen and Ninomiya 1981a, b,c). Admittedly, terms responsible for the breaking will vanish in the continuum limit but it would be much better to start with an action that preserved chiral symmetry, or a suitable generalization of it, at finite lattice spacing, so that the effects attributable to spontaneous symmetry breaking can be studied at finite $a$, rather than having to be extracted only in the continuum limit. In particular, in view of the comments following (16.132), this would open up the possibility of being able to tackle light quarks and hadrons.

In the last few years a way has been found to formulate chiral gauge theories satisfactorily on the lattice at finite $a$. The key is to replace the condition (16.147) by the Ginsparg–Wilson (1982) relation

$$\gamma_5 \not{D} + \not{D}\gamma_5 = a\not{D}\gamma_5\not{D}. \tag{16.148}$$

This relation implies (Lüscher 1998) that the associated action has an exact symmetry, with infinitesimal variations proportional to

$$\delta\psi = \gamma_5(1 - \tfrac{1}{2}a\not{D})\psi \tag{16.149}$$

$$\delta\overline{\psi} = \overline{\psi}(1 - \tfrac{1}{2}a\not{D})\gamma_5. \tag{16.150}$$

The symmetry under (16.149)–(16.150), which reduces to (16.147) as $a \to 0$, provides a lattice theory with all the fundamental symmetry properties of continuum chiral gauge theories (Hasenfratz *et al* 1998). Finding an operator which satisfies (16.148) is, however, not so easy—but that problem has now been solved, indeed in three different ways: Kaplan's 'domain wall' fermions (Kaplan 1992); 'classically perfect fermions' (Hasenfratz and Niedermayer 1994); and 'overlap fermions' (Narayanan and Neuberger 1993a, b, 1994, 1995). All these approaches are being numerically implemented with very promising results.

As we shall see in chapter 17, a dramatic signal of the spontaneous breaking of a global symmetry is the appearance of a massless (Goldstone) boson: in the present (chiral) case, this is taken to be the pion. The physical pion is, of course, not massless and—as discussed in section 18.2—it is reasonable to suppose that this is because chiral symmetry is broken explicitly by (small) quark masses $m_{\rm u}$, $m_{\rm d}$ in the QCD Lagrangian. Indeed, we shall show in equation (18.66) that $m_\pi^2$ is proportional to $(m_{\rm u} + m_{\rm d})$. Here, therefore, is something that can be explicitly checked in a numerical simulation: the square of the pion mass should be proportional to the quark mass (taking $m_{\rm u} = m_{\rm d}$). Indeed, a simple model of chiral symmetry breaking implies the relation (Gasser and Leutwyler 1982)

$$m_\pi^2 = -\frac{(m_{\rm u} + m_{\rm d})}{2f_\pi^2}\langle\Omega|\bar{\hat{u}}\hat{u} + \bar{\hat{d}}\hat{d}|\Omega\rangle \tag{16.151}$$

where (cf (18.54)) $f_\pi \simeq 93$ MeV and $\langle\Omega|\bar{\hat{u}}\hat{u} + \bar{\hat{d}}\hat{d}|\Omega\rangle$ is the expectation value, in the physical vacuum, of the operator $\bar{\hat{u}}\hat{u} + \bar{\hat{d}}\hat{d}$. The crucial point here is that this vev remains non-zero even as the quark mass tends to zero—that is, as the 'explicit' chiral symmetry breaking is 'turned off'. The existence of such a non-zero vev for a field operator is a fundamental feature of spontaneous symmetry breaking, as we shall see in the following chapter. Let us note here, in particular, that the conventional definition of the vacuum $|\Omega\rangle$ that we have used hitherto would, of course, imply that the vev is zero. In section 18.1 we shall learn how, in the *Nambu vacuum* (see equation (18.11)), such a non-zero vev can arise.

A simple analogy may help us to see how such a non-zero vev may arise. Consider the quantity

$$Z_{\rm QCD} = \int \mathcal{D}U\,\mathcal{D}\psi\,\mathcal{D}\bar{\psi}\exp\left[-S_g - \int \bar{\psi}({\rm i}\not{D} - m)\psi\,{\rm d}^4x_{\rm E}\right] \tag{16.152}$$

where $S_g$ is the action for the gauge fields with link variables $U$ and only one fermion field $\psi$ of mass $m$ is treated. Then (cf (16.79)) it is clear that the vev of $\bar{\hat{\psi}}\hat{\psi}$ can be written as

$$\langle\Omega|\bar{\hat{\psi}}\hat{\psi}|\Omega\rangle = \frac{\partial}{\partial m}(\ln Z_{\rm QCD}) \tag{16.153}$$

and we are specifically thinking of the limit as $m \to 0$, taken *after* the infinite volume limit $a \to 0$ (see the penultimate paragraph of section 17.3.1).

Now consider an analogous problem in statistical mechanics, in which the degrees of freedom are spins which can interact with one another via a Hamiltonian $H_s$ and (via their associated magnetic moment $\mu$) with an external magnetic field $B$. The partition function is

$$Z_s = \sum \exp -(H_s - \mu s B)/k_{\rm B}T \tag{16.154}$$

where $s$ is the component of spin along the field $B$. The average value of $s$ is then given by

$$\langle s\rangle = \frac{k_{\rm B}T}{\mu}\frac{\partial \ln Z}{\partial B}. \tag{16.155}$$

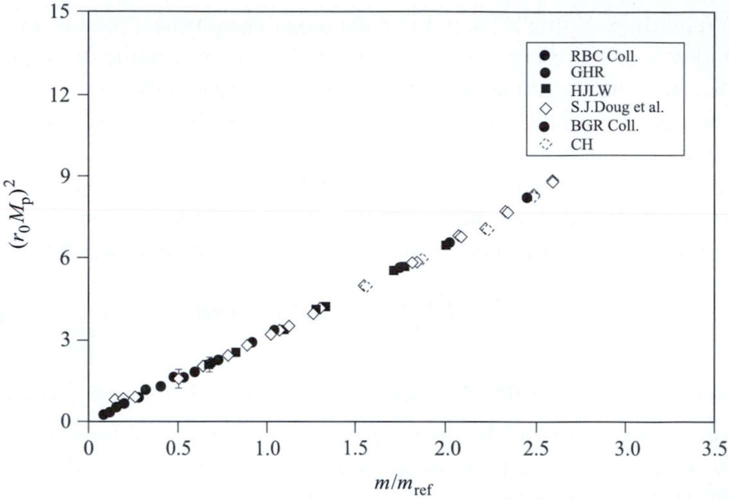


**Figure 16.12.** The pion mass squared in units of $r_0^{-2}$, *versus* the quark mass $m$, normalized at the reference point $m = m_{\text{ref}}$ (Giusti 2002).

In particular, we may regard a non-zero value of $\langle s \rangle$ as arising 'spontaneously' if it survives (in the thermodynamic limit $V \to \infty$) even as $B \to 0$. Such a non-zero $\langle s \rangle$ occurs in a ferromagnet below its transition temperature, where it is related to the 'internal field'. The limitations of this analogy will be discussed in section 17.3.1 but for the moment it does provide a useful physical picture in which '$\langle \Omega | \bar{\hat{\psi}} \hat{\psi} | \Omega \rangle$' can be regarded as an 'internal field' arising spontaneously in the ground state $|\Omega\rangle$, even when the 'external field' $m$ is reduced to zero. The existence of such a non-zero vacuum value, surviving in the symmetry limit $m \to 0$, is fundamental to the concept of spontaneous symmetry breaking. An even better analogy is to the 'condensates' relevant to superfluids and superconductors, which we shall study in chapter 17: the vev $\langle \Omega | \bar{\hat{u}} \hat{u} + \bar{\hat{d}} \hat{d} | \Omega \rangle$ is called the 'chiral condensate'.

Let us now see whether the prediction (16.151) is confirmed by numerical calculations in lattice QCD. Figure 16.12 shows a recent compilation (Giusti 2002) of calculations, done in the quenched approximation, of the pion mass squared (in units of $1/r_0^2$ where $r_0 = 0.5$ fm) *versus* quark mass $m$, normalized at the reference point $m = m_{\text{ref}}$ such that $m_\pi(m = m_{\text{ref}}) = \sqrt{2} m_{\text{K}}$ with $m_{\text{K}} = 495$ MeV (note that $2m_{\text{K}}^2 r_0^2 \simeq 3.16$). It is clear that the expectation $m_\pi^2 \propto m$ is generally well borne out. More sophisticated calculations predict corrections to (16.151), including terms of the form $m \ln m$ (Bernard and Golterman 1992, Sharpe 1992), which may be seen at the lowest $m$ values (Draper *et al* 2002). Note that the physical pion mass is at $(m_\pi r_0)^2 \simeq 0.13$ on figure 16.12: in physical units the lowest pion masses reached at present lie somewhat below 200 MeV and the lightest quark mass is about 15 MeV (Draper *et al* 2002). The actual physical values can be expected to be reached quite soon.

We have been able to give only a brief introduction into what is now, some 30 years after its initial inception by Wilson (1974), a highly mature field. A great deal of effort has gone into ingenious and subtle improvements to the lattice action, to the numerical algorithms and to the treatment of fermions—to name a few of the issues. Lattice QCD is now a major part of particle physics. From the perspective of this chapter and the previous one, we can confidently say that, both in the short-distance (perturbative) regime and in the long-distance (non-perturbative) regime, QCD is established as the correct theory of the strong interactions of quarks, beyond reasonable doubt.

## Problems

**16.1** Verify equation (16.9).

**16.2** Verify equation (16.10).

**16.3** Show that the momentum space version of (16.18) is (16.19).

**16.4** Use (16.25) in (16.27) to verify (16.28).

**16.5** Verify (16.60) and (16.62).

**16.6** In a modified one-dimensional Ising model, spin variables $s_n$ at sites labelled by $n = 1, 2, 3, \ldots, N$ take the values $s_n = \pm 1$ and the energy of each spin configuration is

$$E = -\sum_{n=1}^{N-1} J_n s_n s_{n+1}$$

where all the constants $J_n$ are positive. Show that the partition function $Z_N$ is given by

$$Z_N = 2\prod_{n=1}^{N-1}(2\cosh K_n)$$

where $K_n = J_n/k_{\mathrm{B}}T$. Hence, calculate the entropy for the particular case in which all the $J_n$'s are equal to $J$ and $N \gg 1$ and discuss the behaviour of the entropy in the limits $T \to \infty$ and $T \to 0$.

Let '$p$' denote a particular site such that $1 \ll p \ll N$. Show that the average value $\langle s_p s_{p+1}\rangle$ of the product $s_p s_{p+1}$ is given by

$$\langle s_p s_{p+1}\rangle = \frac{1}{Z_N}\frac{\partial Z_N}{\partial K_p}.$$

Show further that

$$\langle s_p s_{p+j}\rangle = \frac{1}{Z_N}\frac{\partial^j Z_N}{\partial K_p \partial K_{p+1}\ldots\partial K_{p+j}}.$$

Hence, show that, in the case $J_1 = J_2 = \cdots = J_N = J$,

$$\langle s_p s_{p+j} \rangle = \mathrm{e}^{-ja/\xi}$$

where

$$\xi = -a/[\ln(\tanh K)]$$

and $K = J/k_B T$. Discuss the physical meaning of $\xi$, considering the $T \to \infty$ and $T \to 0$ limits explicitly.

# PART 7

# SPONTANEOUSLY BROKEN SYMMETRY

# 17

# SPONTANEOUSLY BROKEN GLOBAL SYMMETRY

Previous chapters have introduced the non-Abelian symmetries SU(2) and SU(3) in both global and local forms, and we have seen how they may be applied to describe such typical physical phenomena as particle multiplets and massless gauge fields. Remarkably enough, however, these symmetries are also applied, in the Standard Model, in two cases where the physical phenomena appear to be very different. Consider the following two questions: (i) Why are there no signs in the baryonic spectrum, such as parity doublets in particular, of the global chiral symmetry introduced in section 12.3.2? (ii) How can weak interactions be described by a local non-Abelian gauge theory when we know the mediating gauge field quanta are not massless? The answers to these questions each involve the same fundamental idea, which is a crucial component of the Standard Model and perhaps also of theories which go beyond it. This is the idea that a symmetry can be 'spontaneously broken' or 'hidden'. By contrast, the symmetries considered hitherto may be termed 'manifest symmetries'.

The physical consequences of spontaneous symmetry breaking turn out to be rather different in the global and local cases. However, the essentials for a theoretical understanding of the phenomenon are contained in the simpler global case, which we consider in this chapter. The application to spontaneously broken chiral symmetry will be treated in chapter 18; spontaneously broken local symmetry will be discussed in chapter 19 and applied in chapter 22.

## 17.1 Introduction

We begin by considering, in response to question (i), what could go wrong with the argument for symmetry multiplets that we gave in chapter 12. To understand this, we must use the field theory formulation of section 12.3, in which the generators of the symmetry are Hermitian field operators and the states are created by operators acting on the vacuum. Thus, consider two states $|A\rangle$, $|B\rangle$:[1]

$$|A\rangle = \hat{\phi}_A^\dagger|0\rangle \qquad |B\rangle = \hat{\phi}_B^\dagger|0\rangle \tag{17.1}$$

[1] We now revert to the ordinary notation $|0\rangle$ for the vacuum state, rather than $|\Omega\rangle$, but it must be borne in mind that $|0\rangle$ is the full (interacting) vacuum.

where $\hat{\phi}_A^\dagger$ and $\hat{\phi}_B^\dagger$ are related to each other by (cf (12.100))

$$[\hat{Q}, \hat{\phi}_A^\dagger] = \hat{\phi}_B^\dagger \tag{17.2}$$

for some generator $\hat{Q}$ of a symmetry group, such that

$$[\hat{Q}, \hat{H}] = 0. \tag{17.3}$$

(17.2) is equivalent to

$$\hat{U}\hat{\phi}_A^\dagger\hat{U}^{-1} \approx \hat{\phi}_A^\dagger + \mathrm{i}\epsilon\hat{\phi}_B^\dagger \tag{17.4}$$

for an infinitesimal transformation $\hat{U} \approx 1 + \mathrm{i}\epsilon\hat{Q}$. Thus $\hat{\phi}_A^\dagger$ is 'rotated' into $\hat{\phi}_B^\dagger$ by $\hat{U}$, and the operators will create states related by the symmetry transformation. We want to see what assumptions are necessary to prove that

$$E_A = E_B \qquad \text{where } \hat{H}|A\rangle = E_A|A\rangle \text{ and } \hat{H}|B\rangle = E_B|B\rangle. \tag{17.5}$$

We have

$$E_B|B\rangle = \hat{H}|B\rangle = \hat{H}\hat{\phi}_B^\dagger|0\rangle = \hat{H}(\hat{Q}\hat{\phi}_A^\dagger - \hat{\phi}_A^\dagger\hat{Q})|0\rangle. \tag{17.6}$$

Now if

$$\hat{Q}|0\rangle = 0 \tag{17.7}$$

we can rewrite the right-hand side of (17.6) as

$$\begin{aligned} \hat{H}\hat{Q}\hat{\phi}_A^\dagger|0\rangle &= \hat{Q}\hat{H}\hat{\phi}_A^\dagger|0\rangle \qquad \text{using (17.3)} \\ &= \hat{Q}\hat{H}|A\rangle = E_A\hat{Q}|A\rangle \\ &= E_A\hat{Q}\hat{\phi}_A^\dagger|0\rangle = E_A(\hat{\phi}_B^\dagger + \hat{\phi}_A^\dagger\hat{Q})|0\rangle \qquad \text{using (17.2)} \\ &= E_A|B\rangle \qquad \text{if (17.7) holds;} \end{aligned} \tag{17.8}$$

whence, comparing (17.8) with (17.6), we see that

$$E_A = E_B \qquad \text{if (17.7) holds.} \tag{17.9}$$

Remembering that $\hat{U} = \exp(\mathrm{i}\alpha\hat{Q})$, we see that (17.7) is equivalent to

$$|0\rangle' \equiv \hat{U}|0\rangle = |0\rangle. \tag{17.10}$$

Thus, a multiplet structure will emerge provided that the vacuum is left invariant under the symmetry transformation. *The 'spontaneously broken symmetry' situation arises in the contrary case—that is, when the vacuum is not invariant under the symmetry,* which is to say when

$$\hat{Q}|0\rangle \neq 0. \tag{17.11}$$

In this case, the argument for the existence of symmetry multiplets breaks down and although the Hamiltonian or Lagrangian may exhibit a non-Abelian symmetry, this will not be manifested in the form of multiplets of mass-degenerate particles.

The preceding italicized sentence does correctly define what is meant by a spontaneously broken symmetry in field theory, but there is another way of thinking about it which is somewhat less abstract though also less rigorous. The basic condition is $\hat{Q}|0\rangle \neq 0$, and it seems tempting to infer that, in this case, the application of $\hat{Q}$ to the vacuum gives, not zero, but *another possible vacuum*, $|0\rangle'$. Thus we have the physically suggestive idea of 'degenerate vacua' (they must be degenerate since $[\hat{Q}, H] = 0$). We shall see in a moment why this notion, though intuitively helpful, is not rigorous.

It would seem, in any case, that the properties of the *vacuum* are all important, so we begin our discussion with a somewhat formal, but nonetheless fundamental, theorem about the quantum field vacuum.

## 17.2 The Fabri–Picasso theorem

Suppose that a given Lagrangian $\hat{\mathcal{L}}$ is invariant under some one-parameter continuous global internal symmetry with a conserved Noether current $\hat{j}^\mu$, such that $\partial_\mu \hat{j}^\mu = 0$. The associated 'charge' is the Hermitian operator $\hat{Q} = \int \hat{j}^0 \mathrm{d}^3\boldsymbol{x}$ and $\dot{\hat{Q}} = 0$. We have hitherto assumed that the transformations of such a U(1) group are representable in the space of physical states by unitary operations $\hat{U}(\lambda) = \exp \mathrm{i}\lambda\hat{Q}$ for arbitrary $\lambda$, with the vacuum invariant under $\hat{U}$, so that $\hat{Q}|0\rangle = 0$. Fabri and Picasso (1966) showed that there are actually *two* possibilities:

(a) $\hat{Q}|0\rangle = 0$ and $|0\rangle$ is an eigenstate of $\hat{Q}$ with eigenvalue 0, so that $|0\rangle$ is invariant under $\hat{U}$ (i.e. $\hat{U}|0\rangle = |0\rangle$); or

(b) $\hat{Q}|0\rangle$ does not exist in the space (its norm is infinite).

The statement (b) is technically more correct than the more intuitive statements '$\hat{Q}|0\rangle \neq 0$' or '$\hat{U}|0\rangle = |0\rangle'$', suggested before.

To prove this result, consider the vacuum matrix element $\langle 0|\hat{j}^0(x)\hat{Q}|0\rangle$. From translation invariance, implemented by the unitary operator[2] $\hat{U}(x) = \exp \mathrm{i}\hat{P}\cdot x$ (where $\hat{P}^\mu$ is the 4-momentum operator), we obtain

$$\begin{aligned}\langle 0|\hat{j}^0(x)\hat{Q}|0\rangle &= \langle 0|\mathrm{e}^{\mathrm{i}\hat{P}\cdot x}\hat{j}^0(0)\mathrm{e}^{-\mathrm{i}\hat{P}\cdot x}\hat{Q}|0\rangle \\ &= \langle 0|\mathrm{e}^{\mathrm{i}\hat{P}\cdot x}\hat{j}^0(0)\hat{Q}\mathrm{e}^{-\mathrm{i}\hat{P}\cdot x}|0\rangle\end{aligned}$$

where the second line follows from

$$[\hat{P}^\mu, \hat{Q}] = 0 \tag{17.12}$$

[2] If this seems unfamiliar, it may be regarded as the four-dimensional generalization of the transformation (I.7) in appendix I of volume 1, from Schrödinger picture operators at $t = 0$ to Heisenberg operators at $t \neq 0$.

since $\hat{Q}$ is an *internal* symmetry. But the vacuum is an eigenstate of $\hat{P}^\mu$ with eigenvalue zero, and so

$$\langle 0|\hat{j}^0(x)\hat{Q}|0\rangle = \langle 0|\hat{j}^0(0)\hat{Q}|0\rangle \tag{17.13}$$

which states that the matrix element we started from is, in fact, independent of $x$. Now consider the norm of $\hat{Q}|0\rangle$:

$$\langle 0|\hat{Q}\hat{Q}|0\rangle = \int \mathrm{d}^3\boldsymbol{x}\, \langle 0|\hat{j}^0(x)\hat{Q}|0\rangle \tag{17.14}$$

$$= \int \mathrm{d}^3\boldsymbol{x}\, \langle 0|\hat{j}^0(0)\hat{Q}|0\rangle \tag{17.15}$$

which must diverge in the infinite volume limit, unless $\hat{Q}|0\rangle = 0$. Thus, either $\hat{Q}|0\rangle = 0$ or $\hat{Q}|0\rangle$ has infinite norm. The foregoing can be easily generalized to non-Abelian symmetry operators $\hat{T}_i$.

Remarkably enough, the argument can also, in a sense, be reversed. Coleman (1986) proved that if an operator

$$\hat{Q}(t) = \int \mathrm{d}^3\boldsymbol{x}\, \hat{j}^0(x) \tag{17.16}$$

is the spatial integral of the $\mu = 0$ component of a 4-vector (but *not assumed* to be conserved) and if it annihilates the vacuum

$$\hat{Q}(t)|0\rangle = 0 \tag{17.17}$$

then in fact $\partial_\mu \hat{j}^\mu = 0$, $\hat{Q}$ is independent of $t$, and the symmetry is unitarily implementable by operators $\hat{U} = \exp(\mathrm{i}\lambda\hat{Q})$.

We might now simply proceed to the chiral symmetry application. We believe, however, that the concept of spontaneous symmetry breaking is so important to particle physics that a more extended discussion is amply justified. In particular, there are crucial insights to be gained by considering the analogous phenomenon in condensed matter physics. After a brief look at the ferromagnet, we shall describe the Bogoliubov model for the ground state of a superfluid, which provides an important physical example of a spontaneously broken global Abelian U(1) symmetry. We shall see that the excitations away from the ground state are *massless modes* and we shall learn, via Goldstone's theorem, that such modes are an inevitable result of spontaneously breaking a global symmetry. Next, we shall introduce the 'Goldstone model' which is the simplest example of a spontaneously broken global U(1) symmetry, involving just one complex scalar field. The generalization of this to the non-Abelian case will draw us in the direction of the Higgs sector of the Standard Model. Returning to condensed matter systems, we introduce the BCS ground state for a superconductor in a way which builds on the Bogoliubov model of a superfluid. We are then prepared for the application, in chapter 18, to spontaneous chiral symmetry breaking (question (i) of the first paragraph of this chapter), following Nambu's profound analogy with one aspect of superconductivity. In chapter 19 we shall see how a different aspect of superconductivity provides a model for the answer to question (ii).

## 17.3 Spontaneously broken symmetry in condensed matter physics

### 17.3.1 The ferromagnet

We have seen that everything depends on the properties of the vacuum state. An essential aid to understanding hidden symmetry in quantum field theory is provided by Nambu's (1960) remarkable insight that the *vacuum* state of a quantum field theory is analogous to the ground state of an interacting many-body system. It is the state of lowest energy—the equilibrium state, given the kinetic and potential energies as specified in the Hamiltonian. Now the ground state of a complicated system (for example, one involving interacting fields) may well have unsuspected properties—which may, indeed, be very hard to predict from the Hamiltonian. But we can postulate (even if we cannot yet prove) properties of the quantum field theory vacuum $|0\rangle$ which are analogous to those of the ground states of many physically interesting many-body systems—such as superfluids and superconductors, to name two with which we shall be principally concerned.

Now it is generally the case, in quantum mechanics, that the ground state of any system described by a Hamiltonian is non-degenerate. Sometimes we may meet systems in which apparently more than one state has the same lowest energy eigenvalue. Yet, in fact, none of these states will be the true ground state: tunnelling will take place between the various degenerate states, and the true ground state will turn out to be a unique linear superposition of them. This is, in fact, the only possibility for systems of finite spatial extent, though, in practice, a state which is not the true ground state may have an extremely long lifetime. However, in the case of fields (extending presumably throughout all space), the Fabri–Picasso theorem shows that there is an alternative possibility, which is often described as involving a 'degenerate ground state'—a term we shall now elucidate. In case (a) of the theorem, the ground state is unique. For, suppose that several ground states $|0, a\rangle$, $|0, b\rangle$, ... existed, with the symmetry unitarily implemented. Then one ground state will be related to another by

$$|0, a\rangle = \mathrm{e}^{\mathrm{i}\lambda\hat{Q}}|0, b\rangle \tag{17.18}$$

for some $\lambda$. However, in case (a) the charge annihilates a ground state, and so all of them are really identical. In case (b), however, we cannot write (17.18)—since $\hat{Q}|0\rangle$ does not exist—and we do have the possibility of many degenerate ground states. In simple models one can verify that these alternative ground states are all orthogonal to each other, in the infinite volume limit. And each member of every 'tower' of excited states, built on these alternative ground states, is also orthogonal to all the members of other towers. But any single tower must constitute a complete space of states. It follows that states in different towers belong to *different* complete spaces of states; that is, to different—and inequivalent—'worlds', each one built on one of the possible orthogonal ground states. In particular, they cannot be related by unitary transformations of the form $e^{i\lambda\hat{Q}}$.

At first sight, a familiar example of these ideas seems to be that of a ferromagnet below its Curie temperature $T_C$, so that the spins are fully aligned. Consider an 'ideal Heisenberg ferromagnet' with $N$ atoms each of spin-$\frac{1}{2}$, described by a Hamiltonian of Heisenberg exchange form $H_S = -J\sum \hat{\boldsymbol{S}}_i \cdot \hat{\boldsymbol{S}}_j$, where $i$ and $j$ label the atomic sites. This Hamiltonian is invariant under spatial rotations, since it only depends on the dot product of the spin operators. Such rotations are implemented by unitary operators $\exp(i\hat{\boldsymbol{S}}\cdot\boldsymbol{\alpha})$ where $\hat{\boldsymbol{S}} = \sum_i \hat{\boldsymbol{S}}_i$, and spins at different sites are assumed to commute. As usual with angular momentum in quantum mechanics, the eigenstates of $H_S$ are labelled by the eigenvalues of total squared spin, and of one component of spin, say of $\hat{S}_z = \sum_i \hat{S}_{iz}$. The quantum mechanical ground state of $H_S$ is an eigenstate with total spin quantum number $S = N/2$, and this state is $(2\cdot N/2+1) = (N+1)$-fold degenerate, allowing for all the possible eigenvalues $(N/2, N/2-1, \ldots, -N/2)$ of $\hat{S}_z$ for this value of $S$. We are free to choose any one of these degenerate states as 'the' ground state, say the state with eigenvalue $S_z = N/2$.

It is clear that the ground state is not invariant under the spin-rotation symmetry of $H_S$, which would require the eigenvalues $S = S_z = 0$. Furthermore, this ground state is degenerate. So two important features of what we have so far learned to expect of a spontaneously broken symmetry are present—namely, 'the ground state is not invariant under the symmetry of the Hamiltonian'; and 'the ground state is degenerate'. However, it has to be emphasized that this ferromagnetic ground state does, in fact, respect the symmetry of $H_S$ in the sense that it belongs to an irreducible representation of the symmetry group: the unusual feature is that it is not the 'trivial' (singlet) representation, as would be the case for an invariant ground state. The spontaneous symmetry breaking which is the true model for particle physics is that in which a many-body ground state is *not* an eigenstate (trivial or otherwise) of the symmetry operators of the Hamiltonian: rather it is a superposition of such eigenstates. We shall explore this for the superfluid and the superconductor in due course.

Nevertheless, there are some useful insights to be gained from the ferromagnet. First, consider two ground states differing by a spin rotation. In the first, the spins are all aligned along the 3-axis, say, and in the second along the axis $\hat{\boldsymbol{n}} = (0, \sin\alpha, \cos\alpha)$. Thus the first ground state is

$$\chi_0 = \begin{pmatrix} 1 \\ 0 \end{pmatrix}_1 \begin{pmatrix} 1 \\ 0 \end{pmatrix}_2 \cdots \begin{pmatrix} 1 \\ 0 \end{pmatrix}_N \qquad (N \text{ products}) \tag{17.19}$$

while the second is (cf (4.74))

$$\chi_0^{(\alpha)} = \begin{pmatrix} \cos\alpha/2 \\ \mathrm{i}\sin\alpha/2 \end{pmatrix}_1 \cdots \begin{pmatrix} \cos\alpha/2 \\ \mathrm{i}\sin\alpha/2 \end{pmatrix}_N . \tag{17.20}$$

The scalar product of (17.19) and (17.20) is $(\cos\alpha/2)^N$, which goes to zero as $N \to \infty$. Thus any two such 'rotated ground states' are, indeed, orthogonal in the infinite volume limit.

We may also enquire about the excited states built on one such ground state, say the one with $\hat{S}_z$ eigenvalue $N/2$. Suppose for simplicity that the magnet is one-dimensional (but the spins have all three components). Consider the state $\chi_n = \hat{S}_{n-}\chi_0$ where $\hat{S}_{n-}$ is the spin-lowering operator $\hat{S}_{n-} = (\hat{S}_{nx} - \mathrm{i}\hat{S}_{ny})$ at site $n$, such that

$$\hat{S}_{n-}\begin{pmatrix} 1 \\ 0 \end{pmatrix}_n = \begin{pmatrix} 0 \\ 1 \end{pmatrix}_n \tag{17.21}$$

so $\hat{S}_{n-}\chi_0$ differs from the ground state $\chi_0$ by having the spin at site $n$ flipped. The action of $\hat{H}_S$ on $\chi_n$ can be found by writing

$$\sum_{i\neq j} \hat{\boldsymbol{S}}_i \cdot \hat{\boldsymbol{S}}_j = \sum_{i\neq j} \tfrac{1}{2}(\hat{S}_{i-}\hat{S}_{j+} + \hat{S}_{j-}\hat{S}_{i+}) + \hat{S}_{iz}\hat{S}_{jz} \tag{17.22}$$

(remembering that spins on different sites commute), where $\hat{S}_{i+} = \hat{S}_{ix} + \mathrm{i}\hat{S}_{iy}$. Since all $\hat{S}_{i+}$ operators give zero on a spin 'up' state, the only non-zero contributions from the first (bracketed) term in (17.22) come from terms in which either $\hat{S}_{i+}$ or $\hat{S}_{j+}$ act on the 'down' spin at $n$, so as to restore it to 'up'. The 'partner' operator $\hat{S}_{i-}$ (or $\hat{S}_{j-}$) then simply lowers the spin at $i$ (or $j$), leading to the result

$$\sum_{i\neq j} \tfrac{1}{2}(\hat{S}_{i-}\hat{S}_{j+} + \hat{S}_{j-}\hat{S}_{i+})\chi_n = \sum_{i\neq n} \chi_i . \tag{17.23}$$

Thus the state $\chi_n$ is not an eigenstate of $\hat{H}_S$. However, a little more work shows that the superpostitions

$$\tilde{\chi}_q = \frac{1}{\sqrt{N}} \sum_n \mathrm{e}^{\mathrm{i}qna} \chi_n \tag{17.24}$$

*are* eigenstates. Here $q$ is one of the discretized wavenumbers produced by appropriate boundary conditions, as is usual in one-dimensional 'chain' problems (see section 16.2). The states (17.24) represent *spin waves* and they have the important feature that for low $q$ (long wavelength) their frequency $\omega$ tends to zero with $q$ (actually $\omega \propto q^2$). In this respect, therefore, they behave like massless particles when quantized—and this is another feature we should expect when a symmetry is spontaneously broken.

The ferromagnet gives us one more useful insight. We have been assuming that one particular ground state (e.g. the one with $S_z = N/2$) has been somehow 'chosen'. But what does the choosing? The answer to this is clear enough in the (perfectly realistic) case in which the Hamiltonian $\hat{H}_S$ is supplemented by a term $-g\mu B \sum_i \hat{S}_{iz}$, representing the effect of an applied field $B$ directed along the $z$-axis. This term will, indeed, ensure that the ground state is unique, and has $S_z = N/2$. Consider now the two limits $B \to 0$ and $N \to \infty$, both at finite temperature. When $B \to 0$ at finite $N$, the $N + 1$ different $S_z$ eigenstates become degenerate, and we have an ensemble in which each enters with an equal weight: there is, therefore, no loss of symmetry, even as $N \to \infty$ (but only *after* $B \to 0$). However, if $N \to \infty$ at finite $B \neq 0$, the single state with $S_z = N/2$ will be selected out as the unique ground state and this asymmetric situation will persist even in the limit $B \to 0$. In a (classical) mean-field theory approximation we suppose that an 'internal field' is 'spontaneously generated', which is aligned with the external $B$ and survives even as $B \to 0$, thus 'spontaneously' breaking the symmetry.

The ferromagnet, therefore, provides an easily pictured system exhibiting many of the features associated with spontaneous symmetry breaking; most importantly, it strongly suggests that what is really characteristic about the phenomenon is that it entails 'spontaneous ordering'.[3] Generally such ordering occurs below some characteristic 'critical temperature', $T_C$. The field which develops a non-zero equilibrium value below $T_C$ is called an 'order parameter'. This concept forms the basis of Landau's theory of second-order phase transitions (see, for example, chapter XIV of Landau and Lifshitz (1980)).

We now turn to an example much more closely analogous to the particle physics applications: the superfluid.

### 17.3.2 The Bogoliubov superfluid

Consider the non-relativistic Hamiltonian (in the Schrödinger picture)

$$\hat{H} = \frac{1}{2m}\int \mathrm{d}^3\boldsymbol{x}\, \nabla\hat{\phi}^\dagger \cdot \nabla\hat{\phi} + \tfrac{1}{2}\int\int \mathrm{d}^3\boldsymbol{x}\,\mathrm{d}^3\boldsymbol{y}\, v(|\boldsymbol{x}-\boldsymbol{y}|)\hat{\phi}^\dagger(\boldsymbol{x})\hat{\phi}^\dagger(\boldsymbol{y})\hat{\phi}(\boldsymbol{y})\hat{\phi}(\boldsymbol{x}) \qquad (17.25)$$

where $\hat{\phi}^\dagger(\boldsymbol{x})$ creates a boson of mass $m$ at position $\boldsymbol{x}$. This $\hat{H}$ describes identical bosons of mass $m$ interacting via a potential $v$, which is assumed to be weak (see, for example, Schiff 1968, section 55 or Parry 1973, chapter 1). We note at once that $\hat{H}$ is invariant under the global U(1) symmetry

$$\hat{\phi}(\boldsymbol{x}) \to \hat{\phi}'(\boldsymbol{x}) = \mathrm{e}^{-\mathrm{i}\alpha}\hat{\phi}(\boldsymbol{x}) \qquad (17.26)$$

the generator being the conserved number operator

$$\hat{N} = \int \hat{\phi}^\dagger\hat{\phi}\,\mathrm{d}^3\boldsymbol{x} \qquad (17.27)$$

[3] It is worth pausing to reflect on the idea that *ordering* is associated with *symmetry breaking*.

which obeys $[\hat{N}, \hat{H}] = 0$. Our ultimate concern will be with the way this symmetry is 'spontaneously broken' in the superfluid ground state. Naturally, since this is an Abelian, rather than a non-Abelian, symmetry the physics will not involve any (hidden) multiplet structure. But the nature of the 'symmetry breaking ground state' in this U(1) case (and in the BCS model of section 17.7) will serve as a physical model for non-Abelian cases also. We shall work always at zero temperature.

We begin by re-writing $\hat{H}$ in terms of mode creation and annihilation operators in the usual way. We expand $\hat{\phi}(\boldsymbol{x})$ as a superposition of solutions of the $v = 0$ problem, which are plane waves[4] quantized in a large cube of volume $\Omega$:

$$\hat{\phi}(\boldsymbol{x}) = \frac{1}{\Omega^{\frac{1}{2}}} \sum_{\boldsymbol{k}} \hat{a}_{\boldsymbol{k}} \mathrm{e}^{\mathrm{i}\boldsymbol{k}\cdot\boldsymbol{x}} \tag{17.28}$$

where $\hat{a}_{\boldsymbol{k}}|0\rangle = 0$, $\hat{a}^{\dagger}_{\boldsymbol{k}}|0\rangle$ is a one-particle state and $[\hat{a}_{\boldsymbol{k}}, \hat{a}^{\dagger}_{\boldsymbol{k}'}] = \delta_{\boldsymbol{k},\boldsymbol{k}'}$, with all other commutators vanishing. We impose periodic boundary conditions at the cube faces and the free particle energies are $\epsilon_k = \boldsymbol{k}^2/2m$. Inserting (17.28) into (17.25) leads (problem 17.1) to

$$\hat{H} = \sum_{\boldsymbol{k}} \epsilon_k \hat{a}^{\dagger}_{\boldsymbol{k}} a_{\boldsymbol{k}} + \frac{1}{2\Omega} \sum_{\Delta} \bar{v}(\boldsymbol{k}_1 - \boldsymbol{k}'_1) \hat{a}^{\dagger}_{\boldsymbol{k}_1} \hat{a}^{\dagger}_{\boldsymbol{k}_2} \hat{a}_{\boldsymbol{k}'_2} \hat{a}_{\boldsymbol{k}'_1} \Delta(\boldsymbol{k}_1 + \boldsymbol{k}_2 - \boldsymbol{k}'_1 - \boldsymbol{k}'_2) \tag{17.29}$$

where the sum is over all momenta $\boldsymbol{k}_1, \boldsymbol{k}_2, \boldsymbol{k}'_1, \boldsymbol{k}'_2$ subject to the conservation law imposed by the $\Delta$ function:

$$\Delta(\boldsymbol{k}) = 1 \qquad \text{if } \boldsymbol{k} = \boldsymbol{0} \tag{17.30}$$

$$= 0 \qquad \text{if } \boldsymbol{k} \neq \boldsymbol{0}. \tag{17.31}$$

The interaction term in (17.29) is easily visualized as in figure 17.1. A pair of particles in states $\boldsymbol{k}'_1, \boldsymbol{k}'_2$ is scattered (conserving momentum) to a pair in states $\boldsymbol{k}_1, \boldsymbol{k}_2$ via the Fourier transform of $v$:

$$\bar{v}(\boldsymbol{k}) = \int v(\boldsymbol{r}) \mathrm{e}^{\mathrm{i}\boldsymbol{k}\cdot\boldsymbol{r}} \, \mathrm{d}^3\boldsymbol{r}. \tag{17.32}$$

Now, below the superfluid transition temperature $T_S$, we expect from the statistical mechanics of Bose–Einstein condensation (Landau and Lifshitz 1980, section 62) that in the limit as $v \to 0$ the ground state has all the particles 'condensed' into the lowest energy state, which has $\boldsymbol{k} = \boldsymbol{0}$. Thus, in the limit $v \to 0$, the ground state will be proportional to

$$|N, 0\rangle = (\hat{a}^{\dagger}_0)^N |0\rangle. \tag{17.33}$$

When a weak repulsive $v$ is included, it is reasonable to hope that most of the particles remain in the condensate, only relatively few being excited to states with

[4] This is non-relativistic physics, so there is no anti-particle part.

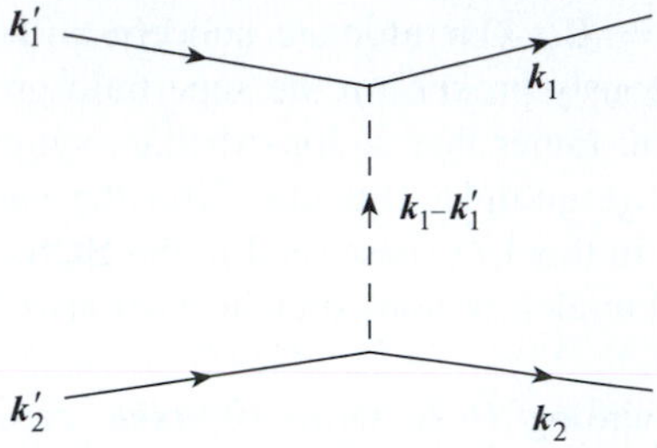


**Figure 17.1.** The interaction term in (17.29).

$\boldsymbol{k} \neq \boldsymbol{0}$. Let $N_0$ be the number of particles with $\boldsymbol{k} = \boldsymbol{0}$ where, by assumption, $N_0 \approx N$. We now consider the limit $N$ (and $N_0$) $\to \infty$ and $\Omega \to \infty$ such that the density $\rho = N/\Omega$ (and $\rho_0 = N_0/\Omega$) stays constant. Bogoliubov (1947) argued that, in this limit, we may effectively replace both $\hat{a}_0$ and $\hat{a}_0^\dagger$ in the second term in (17.29) by the number $N_0^{1/2}$. This amounts to saying that in the commutator

$$\frac{\hat{a}_0}{\Omega^{1/2}}\frac{\hat{a}_0^\dagger}{\Omega^{1/2}} - \frac{\hat{a}^\dagger}{\Omega^{1/2}}\frac{\hat{a}_0}{\Omega^{1/2}} = \frac{1}{\Omega} \tag{17.34}$$

the two terms on the left-hand side are each of order $N_0/\Omega$ and hence finite, while their difference may be neglected as $\Omega \to 0$. Replacing $\hat{a}_0$ and $\hat{a}_0^\dagger$ by $N_0^{1/2}$ leads (problem 17.2) to the following approximate form for $\hat{H}$:

$$\begin{aligned}\hat{H} \approx \hat{H}_{\mathrm{B}} \equiv {\sum_{\boldsymbol{k}}}' \hat{a}_{\boldsymbol{k}}^\dagger \hat{a}_{\boldsymbol{k}} E_k + \frac{1}{2}\frac{N^2}{\Omega}\bar{v}(0) \\ + \frac{1}{2}{\sum_{\boldsymbol{k}}}' \frac{N}{\Omega}\bar{v}(k)[\hat{a}_{\boldsymbol{k}}^\dagger \hat{a}_{-\boldsymbol{k}}^\dagger + \hat{a}_{\boldsymbol{k}}\hat{a}_{-\boldsymbol{k}}]\end{aligned} \tag{17.35}$$

where

$$E_k = \epsilon_k + \frac{N}{\Omega}\bar{v}(k) \tag{17.36}$$

primed summations do not include $\boldsymbol{k} = \boldsymbol{0}$ and terms which tend to zero as $\Omega \to \infty$ have been dropped (thus, $N_0$ has been replaced by $N$).

The most immediately striking feature of (17.35), as compared with $\hat{H}$ of (17.29), is that $\hat{H}_{\mathrm{B}}$ does not conserve the U(1) (number) symmetry (17.26) while $\hat{H}$ does: it is easy to see that for (17.26) to be a good symmetry, the number of $\hat{a}$'s must equal the number of $\hat{a}^\dagger$'s in every term. Thus, the ground state of $\hat{H}_{\mathrm{B}}$, $|\text{ground}\rangle_{\mathrm{B}}$, cannot be expected to be an eigenstate of the number operator. However, it is important to be clear that the number-non-conserving aspect of (17.35) is of a completely different kind, conceptually, from that which would be associated with a (hypothetical) '*explicit*' number violating term in the original

Hamiltonian—for example, the addition of a term of the form '$\hat{a}^\dagger \hat{a}\hat{a}$'. In arriving at (17.35), we have effectively replaced (17.28) by

$$\hat{\phi}_{\mathrm{B}}(\boldsymbol{x}) = \rho_0^{1/2} + \frac{1}{\Omega^{1/2}} \sum_{\boldsymbol{k}\neq \boldsymbol{0}} \hat{a}_{\boldsymbol{k}} \mathrm{e}^{\mathrm{i}\boldsymbol{k}\cdot\boldsymbol{x}} \tag{17.37}$$

where $\rho_0 = N_0/\Omega$, $N_0 \approx N$ and $N_0/\Omega$ remains finite as $\Omega \to \infty$. The limit is crucial here: it enables us to picture the condensate $N_0$ as providing an infinite reservoir of particles, with which excitations away from the ground state can exchange particle number. From this point of view, a number-non-conserving ground state may appear more reasonable. The ultimate test, of course, is whether such a state is a good approximation to the true ground state for a large but finite system.

What is $|\text{ground}\rangle_{\mathrm{B}}$? Remarkably, $\hat{H}_{\mathrm{B}}$ can be exactly diagonalized by means of the *Bogoliubov quasiparticle operators* (for $\boldsymbol{k} \neq \boldsymbol{0}$)

$$\hat{\alpha}_{\boldsymbol{k}} = f_k \hat{a}_{\boldsymbol{k}} + g_k \hat{a}^\dagger_{-\boldsymbol{k}} \qquad \hat{\alpha}^\dagger_{\boldsymbol{k}} = f_k \hat{a}^\dagger_{\boldsymbol{k}} + g_k \hat{a}^\dagger_{-\boldsymbol{k}} \tag{17.38}$$

where $f_k$ and $g_k$ are real functions of $k = |\boldsymbol{k}|$. We must again at once draw attention to the fact that this transformation does not respect the symmetry (17.26) either, since $\hat{a}_{\boldsymbol{k}} \to \mathrm{e}^{-\mathrm{i}\alpha}\hat{a}_{\boldsymbol{k}}$ while $\hat{a}^\dagger_{-\boldsymbol{k}} \to \mathrm{e}^{+\mathrm{i}\alpha}\hat{a}^\dagger_{-\boldsymbol{k}}$. In fact, the operators $\hat{\alpha}^\dagger_{\boldsymbol{k}}$ will turn out to be precisely *creation operators for quasiparticles* which exchange particle number with the ground state.

The commutator of $\hat{\alpha}_{\boldsymbol{k}}$ and $\hat{\alpha}^\dagger_{\boldsymbol{k}}$ is easily evaluated:

$$[\hat{\alpha}_{\boldsymbol{k}}, \hat{\alpha}^\dagger_{\boldsymbol{k}}] = f_k^2 - g_k^2 \tag{17.39}$$

while two $\hat{\alpha}$'s or two $\hat{\alpha}^\dagger$'s commute. We choose $f_k$ and $g_k$ such that $f_k^2 - g_k^2 = 1$, so that the $\hat{a}$'s and the $\hat{\alpha}$'s have the same (bosonic) commutation relations and the transformation (17.38) is then said to be 'canonical'. A convenient choice is $f_k = \cosh\theta_k$, $g_k = \sinh\theta_k$. We now assert that $\hat{H}_{\mathrm{B}}$ can be written in the form

$$\hat{H}_{\mathrm{B}} = \sum_{\boldsymbol{k}}{}' \omega_k \hat{\alpha}^\dagger_{\boldsymbol{k}} \hat{\alpha}_{\boldsymbol{k}} + \beta \tag{17.40}$$

for certain constants $\omega_k$ and $\beta$. Equation (17.40) implies, of course, that the eigenvalues of $\hat{H}_{\mathrm{B}}$ are $\beta + \sum_{\boldsymbol{k}}(n + 1/2)\omega_k$, and that $\hat{\alpha}^\dagger_{\boldsymbol{k}}$ acts as the creation operator for the quasiparticle of energy $\omega_k$, as just anticipated.

We verify (17.40) slightly indirectly. We note first that it implies that, for one particular mode operator $\hat{\alpha}^\dagger_{\boldsymbol{l}}$,

$$[\hat{H}_{\mathrm{B}}, \hat{\alpha}^\dagger_{\boldsymbol{l}}] = \omega_l \hat{\alpha}^\dagger_{\boldsymbol{l}}. \tag{17.41}$$

Substituting for $\hat{\alpha}^\dagger_{\boldsymbol{l}}$ from (17.38), we require

$$[\hat{H}_{\mathrm{B}}, \cosh\theta_l \hat{a}^\dagger_{\boldsymbol{l}} + \sinh\theta_l \hat{a}_{-\boldsymbol{l}}] = \omega_l(\cosh\theta_l\, \hat{a}^\dagger_{\boldsymbol{l}} + \sinh\theta_l \hat{a}_{-\boldsymbol{l}}) \tag{17.42}$$

which must hold as an identity in the $\hat{\alpha}$'s and $\hat{\alpha}^\dagger$'s. Using the expression (17.35) for $\hat{H}_{\mathrm{B}}$, and some patient work with the commutation relations (problem 17.3), one finds that

$$(\omega_l - E_l)\cosh\theta_l + \frac{N}{\Omega}\bar{v}(\boldsymbol{l})\sinh\theta_l = 0 \tag{17.43}$$

$$\frac{N}{\Omega}\bar{v}(\boldsymbol{l})\cosh\theta_l - (\omega_l + E_l)\sinh\theta_l = 0. \tag{17.44}$$

For consistency, therefore, we require

$$E_l^2 - \omega_l^2 - \left(\frac{N}{\Omega^2}\right)^2 (\bar{v}(\boldsymbol{l}))^2 = 0 \tag{17.45}$$

or (recalling the definitions of $E_l$ and $\epsilon_l$)

$$\omega_l = \left[\frac{\boldsymbol{l}^2}{2m}\left(\frac{\boldsymbol{l}^2}{2m} + 2\rho\bar{v}(\boldsymbol{l})\right)\right]^{1/2} \tag{17.46}$$

where $\rho = N/\Omega$. The value of $\tanh\theta_l$ is then determined via either (17.43) or (17.44).

Equation (17.46) is an important result, giving the frequency as a function of the momentum (or wavenumber); it is an example of a 'dispersion relation'. As long as $\bar{v}(\boldsymbol{l})$ is less singular than $\boldsymbol{l}^{-2}$ as $|\boldsymbol{l}| \to 0$, $\omega_l$ will tend to zero as $|\boldsymbol{l}| \to 0$ and we will have massless 'phonon-like' modes. In particular, if $\bar{v}(0) \neq 0$, the speed of sound will be $(\rho\bar{v}(0)/m)^{1/2}$. However, for large $|\boldsymbol{l}|$, $\omega_l$ behaves essentially like $\boldsymbol{l}^2/2m$ and the spectrum returns to the 'particle-like' one of massive bosons. Thus, (17.46) interpolates between phonon-like behaviour at small $|\boldsymbol{l}|$ and particle-like behaviour at large $|\boldsymbol{l}|$. Furthermore, we note that if, in fact, $\bar{v}(\boldsymbol{l}) \sim 1/\boldsymbol{l}^2$, then $\omega_l \to$ constant as $|\boldsymbol{l}| \to 0$ and the spectrum would *not* be that of a massless excitation. Indeed, if $\bar{v}(\boldsymbol{l}) \sim e^2/\boldsymbol{l}^2$, then $\omega_l \sim |e|(\rho/m)^{1/2}$ for small $|\boldsymbol{l}|$, which is just the 'plasma frequency'. Such a $\bar{v}$ is, of course, Colombic (the Fourier transform of $e^2/|\boldsymbol{x}|$), indicating that *in the case of such a long-range force the massless mode associated with spontaneous symmetry breaking acquires a mass.* This will be the topic of chapter 19.

Having discussed the spectrum of quasiparticle excitations, let us now concentrate on the ground state in this model. From (17.40), it is clear that it is defined as the state $|\text{ground}\rangle_{\mathrm{B}}$ such that

$$\hat{\alpha}_{\boldsymbol{k}}|\text{ground}\rangle_{\mathrm{B}} = 0 \qquad \text{for all } \boldsymbol{k} \neq \boldsymbol{0} \tag{17.47}$$

i.e. as the state with no non-zero-momentum quasiparticles in it. This is a complicated state in terms of the original $\hat{a}_{\boldsymbol{k}}$ and $\hat{a}_{\boldsymbol{k}}^\dagger$ operators, but we can give a formal expression for it, as follows. Since the $\hat{\alpha}$'s and $\hat{a}^\dagger$'s are related by a canonical transformation, there must exist a unitary operator $\hat{U}_{\mathrm{B}}$ such that

$$\hat{\alpha}_{\boldsymbol{k}} = \hat{U}_{\mathrm{B}}\hat{a}_{\boldsymbol{k}}\hat{U}_{\mathrm{B}}^{-1} \qquad \hat{a}_{\boldsymbol{k}} = \hat{U}_{\mathrm{B}}^{-1}\hat{\alpha}_{\boldsymbol{k}}\hat{U}_{\mathrm{B}}. \tag{17.48}$$

Now we know that $\hat{a}_{\boldsymbol{k}}|0\rangle = 0$. Hence it follows that

$$\hat{\alpha}_{\boldsymbol{k}}\hat{U}_{\mathrm{B}}|0\rangle = 0 \tag{17.49}$$

and we can identify $|\text{ground}\rangle_{\mathrm{B}}$ with $\hat{U}_{\mathrm{B}}|0\rangle$. In problem 17.4, $\hat{U}_{\mathrm{B}}$ is evaluated for an $\hat{H}_{\mathrm{B}}$ consisting of a single $\boldsymbol{k}$-mode only, in which case the operator effecting the transformation analogous to (17.48) is $\hat{U}_1 = \exp[\theta(\hat{a}\hat{a} - \hat{a}^\dagger\hat{a}^\dagger)/2]$ where $\theta$ replaces $\theta_k$ in this case. This generalizes (in the form of products of such operators) to the full $\hat{H}_{\mathrm{B}}$ case but we shall not need the detailed result: an analogous result for the BCS ground state is discussed more fully in section 17.7. The important point is the following. It is clear from expanding the exponentials that $\hat{U}_{\mathrm{B}}$ creates a state in which the number of $a$-quanta (i.e. the original bosons) *is not fixed*. Thus, unlike the simple non-interacting ground state $|N, 0\rangle$ of (17.33), $|\text{ground}\rangle_{\mathrm{B}} = \hat{U}_{\mathrm{B}}|0\rangle$ does not have a fixed number of particles in it: that is to say, it is *not* an eigenstate of the symmetry operator $\hat{N}$, as anticipated in the comment following (17.36). This is just the situation alluded to in the paragraph before equation (17.19), in our discussion of the ferromagnet.

Consider now the expectation value of $\hat{\phi}(\boldsymbol{x})$ in any state of definite particle number—that is, in an eigenstate of the symmetry operator $\hat{N}$; it is easy to see that this must vanish (there will always be a spare annihilation operator). However, this is *not* true of $\hat{\phi}_{\mathrm{B}}(\boldsymbol{x})$: for example, in the non-interacting ground state (17.33), we have

$$\langle N, 0|\hat{\phi}_{\mathrm{B}}(\boldsymbol{x})|N, 0\rangle = \rho_0^{1/2}. \tag{17.50}$$

Furthermore, using the inverse of (17.38)

$$\hat{a}_{\boldsymbol{k}} = \cosh\theta_k\hat{\alpha}_{\boldsymbol{k}} - \sinh\theta_k\hat{\alpha}^\dagger_{-\boldsymbol{k}} \tag{17.51}$$

together with (17.47), we find the similar result:

$${}_{\mathrm{B}}\langle\text{ground}|\hat{\phi}_{\mathrm{B}}(\boldsymbol{x})|\text{ground}\rangle_{\mathrm{B}} = \rho_0^{1/2}. \tag{17.52}$$

The question is now how to generalize (17.50) or (17.52) to the complete $\hat{\phi}(\boldsymbol{x})$ and the true ground state $|\text{ground}\rangle$, in the limit $N, \Omega \to \infty$ with fixed $N/\Omega$. We make the *assumption* that

$$\langle\text{ground}|\hat{\phi}(\boldsymbol{x})|\text{ground}\rangle \neq 0; \tag{17.53}$$

that is, we abstract from the Bogoliubov model the crucial feature that *the field acquires a non-zero expectation value in the true ground state*, in the infinite volume limit.

We are now at the heart of spontaneous symmetry breaking in field theory. Condition (17.53) has the form of an 'ordering' condition: it is analogous to the non-zero value of the total spin in the ferromagnetic case, but in (17.53)—we must again emphasize—$|\text{ground}\rangle$ is *not* an eigenstate of the symmetry operator $\hat{N}$. If it were, (17.53) would vanish, as we have just seen. Recalling the association 'quantum vacuum $\leftrightarrow$ many body ground state' we expect that the occurrence of a non-zero vacuum expectation value (vev) for an operator transforming non-trivially under a symmetry operator will be the key requirement for spontaneous symmetry breaking in field theory. In the next section we show how this requirement necessitates one (or more) massless modes, via Goldstone's theorem (1961).

Before leaving the superfluid, we examine (17.37) and (17.52) in another way, which is only rigorous for a finite system but is, nevertheless, very suggestive. Since the original $\hat{H}$ has a U(1) symmetry under which $\hat{\phi}$ transforms to $\hat{\phi}' = \exp(-\mathrm{i}\alpha)\hat{\phi}$, we should be at liberty to replace (17.37) by

$$\hat{\phi}'_{\mathrm{B}} = \mathrm{e}^{-\mathrm{i}\alpha}\rho_0^{1/2} + \frac{1}{\Omega^{1/2}}\sum_{k\neq 0}\hat{a}_{\boldsymbol{k}}\mathrm{e}^{-\mathrm{i}\alpha}\mathrm{e}^{\mathrm{i}\boldsymbol{k}\cdot\boldsymbol{x}}. \tag{17.54}$$

But in that case our condition (17.52) becomes

$$_{\mathrm{B}}\langle\text{ground}|\hat{\phi}'_{\mathrm{B}}|\text{ground}\rangle_{\mathrm{B}} = \mathrm{e}^{-\mathrm{i}\alpha}\,_{\mathrm{B}}\langle\text{ground}|\hat{\phi}_{\mathrm{B}}|\text{ground}\rangle_{\mathrm{B}}. \tag{17.55}$$

Now $\hat{\phi}'_{\mathrm{B}} = \hat{U}_\alpha\hat{\phi}_{\mathrm{B}}\hat{U}_\alpha^{-1}$ where $\hat{U}_\alpha = \exp(\mathrm{i}\alpha\hat{N})$. Hence (17.55) may be written as

$$_{\mathrm{B}}\langle\text{ground}|\hat{U}_\alpha\hat{\phi}_{\mathrm{B}}\hat{U}_\alpha^{-1}|\text{ground}\rangle_{\mathrm{B}} = \mathrm{e}^{-\mathrm{i}\alpha}\,_{\mathrm{B}}\langle\text{ground}|\hat{\phi}_{\mathrm{B}}|\text{ground}\rangle_{\mathrm{B}}. \tag{17.56}$$

If $|\text{ground}\rangle_{\mathrm{B}}$ were an eigenstate of $\hat{N}$ with eigenvalue $N$, say, then the $\hat{U}_\alpha$ factors in (17.56) would become just $\mathrm{e}^{\mathrm{i}\alpha N}\cdot\mathrm{e}^{-\mathrm{i}\alpha N}$ and would cancel out, leaving a contradiction. Instead, however, knowing that $|\text{ground}\rangle_{\mathrm{B}}$ is not an eigenstate of $\hat{N}$, we can regard $\hat{U}_\alpha^{-1}|\text{ground}\rangle_{\mathrm{B}}$ as an 'alternative ground state' $|\text{ground}, \alpha\rangle_{\mathrm{B}}$ such that

$$_{\mathrm{B}}\langle\text{ground}, \alpha|\hat{\phi}_{\mathrm{B}}|\text{ground}, \alpha\rangle_{\mathrm{B}} = \mathrm{e}^{-\mathrm{i}\alpha}\,_{\mathrm{B}}\langle\text{ground}|\hat{\phi}_{\mathrm{B}}|\text{ground}\rangle_{\mathrm{B}} \tag{17.57}$$

the original choice (17.52) corresponding to $\alpha = 0$. There are infinitely many such ground states since $\alpha$ is a continuous parameter. No physical consequence follows from choosing one rather than another but we do have to choose one, thus 'spontaneously' breaking the symmetry. In choosing say $\alpha = 0$, we are deciding (arbitrarily) to pick the ground state such that $_{\mathrm{B}}\langle\text{ground}|\hat{\phi}_{\mathrm{B}}|\text{ground}\rangle_{\mathrm{B}}$ is aligned in the 'real' direction. By hypothesis, a similar situation obtains for the true ground state. None of the states $|\text{ground}, \alpha\rangle$ is an eigenstate for $\hat{N}$: instead, they are certain coherent superpositions of states with different eigenvalues $N$, such that the expectation value of $\hat{\phi}_{\mathrm{B}}$ has a definite phase.

## 17.4 Goldstone's theorem

We return to quantum field theory proper and show, following Goldstone (1961) (see also Goldstone, Salam and Weinberg (1962)) how in case (b) of the Fabri–Picasso theorem massless particles will necessarily be present. Whether these particles will actually be observable depends, however, on whether the theory also contains gauge fields. In this chapter we are concerned solely with global symmetries, and gauge fields are absent: the local symmetry case is treated in chapter 19.

Suppose, then, that we have a Lagrangian $\hat{\mathcal{L}}$ with a continuous symmetry generated by a charge $\hat{Q}$, which is independent of time and is the space integral of the $\mu = 0$ component of a conserved Noether current:

$$\hat{Q} = \int \hat{j}_0(x)\,\mathrm{d}^3\boldsymbol{x}. \tag{17.58}$$

We consider the case in which the vacuum of this theory is not invariant, i.e. is not annihilated by $\hat{Q}$.

Suppose $\hat{\phi}(y)$ is some field operator which is not invariant under the continuous symmetry in question and consider the vacuum expectation value

$$\langle 0|[\hat{Q}, \hat{\phi}(y)]|0\rangle. \tag{17.59}$$

Just as in equation (17.13), translation invariance implies that this vev is, in fact, independent of $y$, and we may set $y = 0$. If $\hat{Q}$ were to annihilate $|0\rangle$, this would clearly vanish: we investigate the consequences of its *not* vanishing. Since $\hat{\phi}$ is not invariant under $\hat{Q}$, the commutator in (17.59) will give some other field, call it $\hat{\phi}'(0)$; thus, the hallmark of the hidden symmetry situation is the existence of some field (here $\hat{\phi}'(0)$) with *non-vanishing vacuum expectation value*, just as in (17.53).

From (17.58), we can write (17.59) as

$$\begin{aligned} 0 &\neq \langle 0|\hat{\phi}'(0)|0\rangle & (17.60)\\ &= \langle 0|\left[\int \mathrm{d}^3\boldsymbol{x}\ \hat{j}_0(x), \hat{\phi}(0)\right]|0\rangle. & (17.61)\end{aligned}$$

Since, by assumption, $\partial_\mu \hat{j}^\mu = 0$, we have, as usual,

$$\frac{\partial}{\partial x^0}\int \mathrm{d}^3\boldsymbol{x}\ \hat{j}_0(x) + \int \mathrm{d}^3\boldsymbol{x}\ \mathrm{div}\,\hat{\boldsymbol{j}}(x) = 0 \tag{17.62}$$

whence

$$\begin{aligned} \frac{\partial}{\partial x^0}\int \mathrm{d}^3\boldsymbol{x}\ \langle 0|[\hat{j}_0(x), \hat{\phi}(0)]|0\rangle &= -\int \mathrm{d}^3\boldsymbol{x}\ \langle 0|[\mathrm{div}\,\hat{\boldsymbol{j}}(x), \hat{\phi}(0)]|0\rangle & (17.63)\\ &= -\int \mathrm{d}\boldsymbol{S}\cdot\langle 0|[\hat{\boldsymbol{j}}(x), \hat{\phi}(0)]|0\rangle. & (17.64)\end{aligned}$$

If the surface integral vanishes in (17.64), (17.61) will be independent of $x_0$. The commutator in (17.64) involves local operators separated by a very large space-like interval and, therefore, the vanishing of (17.64) would seem to be unproblematic. Indeed so it is—with the exception of the case in which the symmetry is local and gauge fields are present. A detailed analysis of exactly how this changes the argument being presented here will take us too far afield at this point, and the reader is referred to Guralnik *et al* (1968) and Bernstein (1974). We shall treat the 'spontaneously broken' gauge theory case in chapter 19, but in less formal terms.

Let us now see how the independence of (17.61) on $x_0$ leads to the necessity for a massless particle in the spectrum. Inserting a complete set of states in (17.61), we obtain

$$0 \neq \int \mathrm{d}^3\boldsymbol{x} \sum_n \{\langle 0|\hat{j}_0(x)|n\rangle\langle n|\hat{\phi}(0)|0\rangle - \langle 0|\hat{\phi}(0)|n\rangle\langle n|\hat{j}_0(x)|0\rangle\} \tag{17.65}$$

$$= \int \mathrm{d}^3\boldsymbol{x} \sum_n \{\langle 0|\hat{j}_0(0)|n\rangle\langle n|\hat{\phi}(0)|0\rangle \mathrm{e}^{-\mathrm{i}p_n\cdot x} - \langle 0|\hat{\phi}(0)|n\rangle\langle n|\hat{j}_0(0)|0\rangle \mathrm{e}^{\mathrm{i}p_n\cdot x}\} \tag{17.66}$$

using translation invariance, with $p_n$ the 4-momentum eigenvalue of the state $|n\rangle$. Performing the spatial integral on the right-hand side, we find (omitting the irrelevant $(2\pi)^3$) that

$$0 \neq \sum_n \delta^3(\boldsymbol{p}_n)[\langle 0|\hat{j}_0(0)|n\rangle\langle n|\hat{\phi}(0)|0\rangle \mathrm{e}^{\mathrm{i}p_{n0}x_0} - \langle 0|\hat{\phi}(0)|n\rangle\langle n|\hat{j}_0(0)|0\rangle \mathrm{e}^{-\mathrm{i}p_{n0}x_0}]. \tag{17.67}$$

But this expression is independent of $x_0$. *Massive* states $|n\rangle$ will produce explicit $x_0$-dependent factors $\mathrm{e}^{\pm\mathrm{i}M_n x_0}$ ($p_{n0} \to M_n$ as the $\delta$-function constrains $\boldsymbol{p}_n = \boldsymbol{0}$); hence, the matrix elements of $\hat{j}_0$ between $|0\rangle$ and such a massive state must *vanish*, and such states contribute zero to (17.67). Equally, if we take $|n\rangle = |0\rangle$, (17.67) vanishes identically. But it has been assumed to be not zero. Hence, *some* state or states must exist among $|n\rangle$ such that $\langle 0|j_0|n\rangle \neq 0$ and yet (17.67) is independent of $x_0$. The only possibility is states whose energy $p_{n0}$ goes to zero as their 3-momentum does (from $\delta^3(\boldsymbol{p}_n)$). Such states are, of course, massless: they are called generically *Goldstone modes*. Thus, the existence of a non-vanishing vacuum expectation value for a field, in a theory with a continuous symmetry, appears to lead inevitably to the necessity of having a massless particle, or particles, in the theory. This is the Goldstone (1961) result.

The superfluid provided us with an explicit model exhibiting the crucial non-zero expectation value $\langle\text{ground}|\hat{\phi}|\text{ground}\rangle \neq 0$, in which the now expected massless mode emerged dynamically. We now discuss a simpler relativistic model, in which the symmetry breaking is brought about more 'by hand'—that is, by choosing a parameter in the Lagrangian appropriately. Although in a sense less 'dynamical' than the Bogoliubov superfluid (or the BCS superconductor, to be discussed shortly), this *Goldstone model* does provide a very simple example of the phenomenon of spontaneous symmetry breaking in field theory.

## 17.5 Spontaneously broken global U(1) symmetry: the Goldstone model

We consider, following Goldstone (1961), a complex scalar field $\hat{\phi}$ as in section 7.1, with

$$\hat{\phi} = \frac{1}{\sqrt{2}}(\hat{\phi}_1 - \mathrm{i}\hat{\phi}_2) \qquad \hat{\phi}^\dagger = \frac{1}{\sqrt{2}}(\hat{\phi}_1 + \mathrm{i}\hat{\phi}_2) \tag{17.68}$$

described by the Lagrangian

$$\hat{\mathcal{L}}_{\mathrm{G}} = (\partial_\mu \hat{\phi}^\dagger)(\partial^\mu \hat{\phi}) - \hat{V}(\hat{\phi}). \tag{17.69}$$

We begin by considering the 'normal' case in which the potential has the form

$$\hat{V} = \hat{V}_{\mathrm{S}} \equiv \tfrac{1}{4}\lambda(\hat{\phi}^\dagger \hat{\phi})^2 + \mu^2 \hat{\phi}^\dagger \hat{\phi} \tag{17.70}$$

with $\mu^2, \lambda > 0$. The Hamiltonian density is then

$$\hat{\mathcal{H}}_{\mathrm{G}} = \dot{\hat{\phi}}^\dagger \dot{\hat{\phi}} + \nabla \hat{\phi}^\dagger \cdot \nabla \hat{\phi} + \hat{V}(\hat{\phi}). \tag{17.71}$$

Clearly $\hat{\mathcal{L}}_{\mathrm{G}}$ is invariant under the global U(1) symmetry

$$\hat{\phi} \to \hat{\phi}' = \mathrm{e}^{-\mathrm{i}\alpha}\hat{\phi}, \tag{17.72}$$

the generator being $\hat{N}_\phi$ of (7.23). We shall see how this symmetry may be 'spontaneously broken'.

We know that everything depends on the nature of the ground state of this field system—that is, the vacuum of the quantum field theory. In general, it is a difficult, non-perturbative, problem to find the ground state (or a good approximation to it—witness the superfluid) but we can make some progress by first considering the theory *classically*. It is clear that the absolute minimum of the classical Hamiltonian $\mathcal{H}_{\mathrm{G}}$ is reached for

(i) $\phi =$ constant, which reduces the $\dot{\phi}$ and $\nabla\phi$ terms to zero; and
(ii) $\phi = \phi_0$, where $\phi_0$ is the minimum of the classical version of the potential, $V$.

For $V = V_{\mathrm{S}}$ as in (17.70) but without the hats and with $\lambda$ and $\mu^2$ both positive, the minimum of $V_{\mathrm{S}}$ is clearly at $\phi = 0$ and is unique. In the quantum theory, we expect to treat small oscillations of the field about this minimum as approximately harmonic, leading to the usual quantized modes. To implement this, we expand $\hat{\phi}$ about the classical minimum at $\phi = 0$, writing as usual

$$\hat{\phi} = \int \frac{\mathrm{d}^3 \boldsymbol{k}}{(2\pi)^3 \sqrt{2\omega}} [\hat{a}(k)\mathrm{e}^{-\mathrm{i}k\cdot x} + b^\dagger(k)\mathrm{e}^{\mathrm{i}k\cdot x}] \tag{17.73}$$

where the plane waves are solutions of the 'free' ($\lambda = 0$) problem. For $\lambda = 0$, the Lagrangian is simply

$$\hat{\mathcal{L}}_{\mathrm{free}} = \partial_\mu \hat{\phi}^\dagger \partial^\mu \hat{\phi} - \mu^2 \hat{\phi}^\dagger \hat{\phi} \tag{17.74}$$

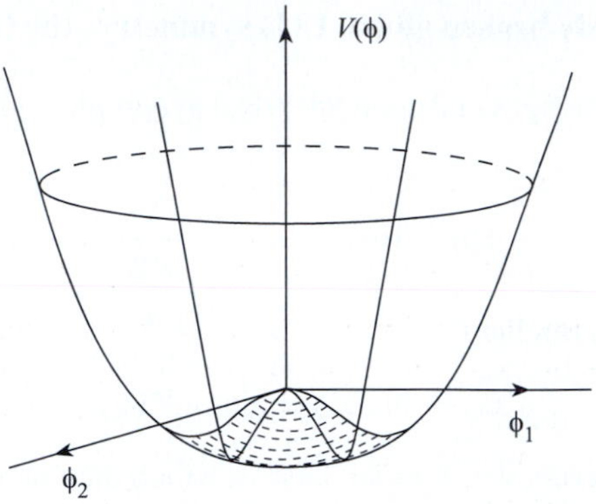


**Figure 17.2.** The classical potential $V_{\rm SB}$ of (17.77).

which represents a complex scalar field, consisting of two degrees of freedom, each with the same mass $\mu$ (see section 7.1). Thus, in (17.73), $\omega = (\boldsymbol{k}^2 + \mu^2)^{1/2}$ and the vacuum is defined by

$$\hat{a}(k)|0\rangle = \hat{b}(k)|0\rangle = 0 \tag{17.75}$$

and so clearly

$$\langle 0|\hat{\phi}|0\rangle = 0. \tag{17.76}$$

It seems reasonable to interpret quantum field average values as corresponding to classical field values and, on this interpretation, (17.76) is consistent with the fact that the classical minimum energy configuration has $\phi = 0$.

Consider now the case in which the classical minimum is *not* at $\phi = 0$. This can be achieved by altering the sign of $\mu^2$ in (17.70) 'by hand', so that the classical potential is now the 'symmetry breaking' one:

$$V = V_{\rm SB} \equiv \tfrac{1}{4}\lambda(\phi^\dagger\phi)^2 - \mu^2\phi^\dagger\phi. \tag{17.77}$$

This is sketched *versus* $\phi_1$ and $\phi_2$ in figure 17.2. This time, although the origin $\phi_1 = \phi_2 = 0$ is a stationary point, it is an (unstable) maximum rather than a minimum. The minimum of $V_{\rm SB}$ occurs when

$$(\phi^\dagger\phi) = \frac{2\mu^2}{\lambda} \tag{17.78}$$

or, alternatively, when

$$\phi_1^2 + \phi_2^2 = \frac{4\mu^2}{\lambda} \equiv v^2 \tag{17.79}$$

where

$$v = \frac{2|\mu|}{\lambda^{1/2}}. \tag{17.80}$$

The condition (17.79) can also be written as

$$|\phi| = v/\sqrt{2}. \tag{17.81}$$

To have a clearer picture, it is helpful to introduce the 'polar' variables $\rho(x)$ and $\theta(x)$ via

$$\phi(x) = (\rho(x)/\sqrt{2}) \exp(\mathrm{i}\theta(x)/v) \tag{17.82}$$

where, for convenience, the $v$ is inserted so that $\theta$ has the same dimension (mass) as $\rho$ and $\phi$. The minimum condition (17.81), therefore, represents the circle $\rho = v$: any point on this circle, at any value of $\theta$, represents a possible classical ground state—and it is clear that they are (infinitely) degenerate.

Before proceeding further, we briefly outline a condensed matter analogue of (17.77) and (17.81) which may help in understanding the change in sign of the parameter $\mu^2$. Consider the free energy $F$ of a ferromagnet as a function of the magnetization $\boldsymbol{M}$ at temperature $T$ and make an expansion of the form

$$F \approx F_0(T) + \mu^2(T)\boldsymbol{M}^2 + \frac{\lambda}{4}\boldsymbol{M}^4 + \cdots \tag{17.83}$$

valid for weak and slowly varying magnetization. If the parameter $\mu^2$ is positive, it is clear that $F$ has a simple 'bowl' shape as a function of $|\boldsymbol{M}|$, with a minimum at $|\boldsymbol{M}| = 0$. This is the case for $T$ greater than the ferromagnetic transition temperature $T_{\mathrm{C}}$. However, if one assumes that $\mu^2(T)$ changes sign at $T_{\mathrm{C}}$, becoming negative for $T < T_{\mathrm{C}}$, then $F$ will now resemble a vertical section of figure 17.2, the minimum being at $|\boldsymbol{M}| \neq 0$. Any direction of $\boldsymbol{M}$ is possible (only $|\boldsymbol{M}|$ is specified); but the system must choose one particular direction (e.g. via the influence of a very weak external field, as discussed in section 17.3.1) and when it does so the rotational invariance exhibited by $F$ of (17.83) is lost. This symmetry has been broken 'spontaneously'—though this is still only a classical analogue. Nevertheless, the model is essentially the Landau mean field theory of ferromagnetism, and it suggests that we should think of the 'symmetric' and 'broken symmetry' situations as different phases of the same system. It may also be the case in particle physics that parameters such as $\mu^2$ change sign as a function of $T$, or some other variable, thereby effectively precipitating a phase change.

If we maintain the idea that the vacuum expectation value of the quantum field should equal the ground-state value of the classical field, the vacuum in this $\mu^2 < 0$ case must, therefore, be $|0\rangle_{\mathrm{B}}$ such that ${}_{\mathrm{B}}\langle 0|\hat{\phi}|0\rangle_{\mathrm{B}}$ does not vanish, in contrast to (17.76). It is clear that this is exactly the situation met in the superfluid (but 'B' here will stand for 'broken symmetry') and is, moreover, the condition for the existence of massless (Goldstone) modes. Let us see how they emerge in this model.

In quantum field theory, particles are thought of as excitations from a ground state, which is the vacuum. Figure 17.2 strongly suggests that if we want a sensible quantum interpretation of a theory with the potential (17.77), we had better expand the fields about a point on the circle of minima, about which stable oscillations are likely, rather than about the obviously unstable point $\hat{\phi} = 0$. Let us pick the point $\rho = v$, $\theta = 0$ in the classical case. We might well guess that 'radial' oscillations in $\hat{\rho}$ would correspond to a conventional massive field (having a parabolic restoring potential), while 'angle' oscillations in $\hat{\theta}$—which pass through all the degenerate vacua—have no restoring force and are massless. Accordingly, we set

$$\hat{\phi}(x) = \frac{1}{\sqrt{2}}(v + \hat{h}(x)) \exp(-\mathrm{i}\hat{\theta}(x)/v) \tag{17.84}$$

and find (problem 17.5) that $\hat{\mathcal{L}}_{\mathrm{G}}$ (with $\hat{V} = \hat{V}_{\mathrm{SB}}$ of (17.77) with hats on) becomes

$$\hat{\mathcal{L}}_{\mathrm{G}} = \tfrac{1}{2}\partial_\mu \hat{h}\partial^\mu \hat{h} - \mu^2 \hat{h}^2 + \tfrac{1}{2}\partial_\mu \hat{\theta}\partial^\mu \hat{\theta} + \mu^4/\lambda + \cdots \tag{17.85}$$

the dots representing interaction terms which are cubic and quartic in $\hat{\theta}, \hat{h}$. Equation (17.85) shows that the particle spectrum in the 'spontaneously broken' case is dramatically different from that in the normal case: instead of two degrees of freedom with the same mass $\mu$, one (the $\theta$-mode) is massless and the other (the $h$-mode) has a mass of $\sqrt{2}\mu$. We expect the vacuum $|0\rangle_{\mathrm{B}}$ to be annihilated by the mode operators $\hat{a}_h$ and $\hat{a}_\theta$ for these fields. This implies, however, from (17.84) that

$$_{\mathrm{B}}\langle 0|\hat{\phi}|0\rangle_{\mathrm{B}} = v/\sqrt{2} \tag{17.86}$$

which is consistent with our interpretation of the vacuum expectation value (vev) as the classical minimum, and with the occurrence of massless modes. (The constant term in (17.85), which does not affect equations of motion, merely reflects the fact that the minimum value of $V_{\mathrm{SB}}$ is $-\mu^4/\lambda$.) The ansatz (17.84) and the non-zero vev (17.86) may be compared with (17.37) and (17.52), respectively, in the superfluid case.

Goldstone's model, then, contains much of the essence of spontaneous symmetry breaking in field theory: a non-zero vacuum value of a field which is not an invariant under the symmetry group, zero mass bosons and massive excitations in a direction in field space which is 'orthogonal' to the degenerate ground states. However, it has to be noted that the triggering mechanism for the symmetry breaking ($\mu^2 \to -\mu^2$) has to be put in by hand, in contrast to the—admittedly approximate but more 'dynamical'—Bogoliubov approach. The Goldstone model, in short, is essentially phenomenological.

As in the case of the superfluid, we may perfectly well choose a vacuum corresponding to a classical ground state with non-zero $\theta$, say $\theta = -\alpha$. Then

$$_{\mathrm{B}}\langle 0, \alpha|\hat{\phi}|0, \alpha\rangle_{\mathrm{B}} = \mathrm{e}^{-\mathrm{i}\alpha}\frac{v}{\sqrt{2}} \tag{17.87}$$

$$= \mathrm{e}^{-\mathrm{i}\alpha}\,{}_{\mathrm{B}}\langle 0|\hat{\phi}|0\rangle_{\mathrm{B}} \tag{17.88}$$

as in (17.57). But we know (see (7.27) and (7.28)) that

$$\mathrm{e}^{-\mathrm{i}\alpha}\hat{\phi} = \hat{\phi}' = \hat{U}_\alpha \hat{\phi} \hat{U}_\alpha^{-1} \tag{17.89}$$

where

$$\hat{U}_\alpha = e^{i\alpha\hat{N}_\phi}. \tag{17.90}$$

So (17.88) becomes

$$_B\langle 0, \alpha|\hat{\phi}|0, \alpha\rangle_B = {}_B\langle 0|\hat{U}_\alpha\hat{\phi}\hat{U}_\alpha^{-1}|0\rangle_B \tag{17.91}$$

and we may interpret $\hat{U}_\alpha^{-1}|0\rangle_B$ as the 'alternative vacuum' $|0, \alpha\rangle_B$ (this argument is, as usual, not valid in the infinite volume limit where $\hat{N}_\phi$ fails to exist).

It is interesting to find out what happens to the symmetry current corresponding to the invariance (17.72), in the 'broken symmetry' case. This current is given in (7.23) which we write again here in slightly different notation:

$$\hat{j}^\mu_\phi = i(\hat{\phi}^\dagger\partial^\mu\hat{\phi} - (\partial^\mu\hat{\phi})^\dagger\hat{\phi}) \tag{17.92}$$

normal ordering being understood. Written in terms of the $\hat{h}$ and $\hat{\theta}$ of (17.84), $\hat{j}^\mu_\phi$ becomes

$$\hat{j}^\mu_\phi = v\partial^\mu\hat{\theta} + 2\hat{h}\partial^\mu\hat{\theta} + \hat{h}^2\partial^\mu\hat{\theta}/v. \tag{17.93}$$

The term involving just the *single* field $\hat{\theta}$ is very remarkable: it tells us that there is a non-zero matrix element of the form

$$_B\langle 0|\hat{j}^\mu_\phi(x)|\theta, p\rangle = -ip^\mu v e^{-ip\cdot x} \tag{17.94}$$

where $|\theta, p\rangle$ stands for the state with one $\theta$-quantum (Goldstone boson), with momentum $p^\mu$. This is easily seen by writing the usual normal mode expansion for $\hat{\theta}$ and $\hat{h}$, and using the standard bosonic commutation relations for $\hat{a}_\theta(k)$, $\hat{a}^\dagger_\theta(k')$. In words, (17.94) asserts that, when the symmetry is spontaneously broken, the symmetry current connects the vacuum to a state with one Goldstone quantum, with an amplitude which is proportional to the *symmetry breaking vacuum expectation value* $v$. The matrix element (17.94), with $x = 0$, is precisely of the type that was shown to be non-zero in the proof of the Goldstone theorem, after (17.67). Note also that (17.94) is consistent with $\partial_\mu\hat{j}^\mu_\phi = 0$ only if $p^2 = 0$, as is required for the massless $\theta$.

We are now ready to generalize the Abelian U(1) model to the (global) non-Abelian case.

## 17.6 Spontaneously broken global non-Abelian symmetry

We can illustrate the essential features by considering a particular example, which, in fact, forms part of the Higgs sector of the Standard Model. We consider an SU(2) doublet but, this time, not of fermions as in section 12.3 but of bosons:

$$\hat{\phi} = \begin{pmatrix} \hat{\phi}^+ \\ \hat{\phi}^0 \end{pmatrix} \equiv \begin{pmatrix} \frac{1}{\sqrt{2}}(\hat{\phi}_1 + i\hat{\phi}_2) \\ \frac{1}{\sqrt{2}}(\hat{\phi}_3 + i\hat{\phi}_4) \end{pmatrix} \tag{17.95}$$

where the complex scalar field $\hat{\phi}^+$ destroys positively charged particles and creates negatively charged ones and the complex scalar field $\hat{\phi}^0$ destroys neutral particles and creates neutral anti-particles. As we shall see in a moment, the Lagrangian we shall use has an additional U(1) symmetry, so that the full symmetry is SU(2)×U(1). This U(1) symmetry leads to a conserved quantum number which we call $y$. We associate the physical charge $Q$ with the eigenvalue $t_3$ of the SU(2) generator $\hat{t}_3$, and with $y$, via

$$Q = e(t_3 + y/2) \tag{17.96}$$

so that $y(\phi^+) = 1 = y(\phi^0)$. Thus, $\phi^+$ and $\phi^0$ can be thought of as analogous to the hadronic iso-doublet ($K^+$, $K^0$).

The Lagrangian we choose is a simple generalization of (17.69) and (17.77):

$$\hat{\mathcal{L}}_\Phi = (\partial_\mu \hat{\phi}^\dagger)(\partial^\mu \hat{\phi}) + \mu^2 \hat{\phi}^\dagger \hat{\phi} - \frac{\lambda}{4}(\hat{\phi}^\dagger \hat{\phi})^2 \tag{17.97}$$

which has the 'spontaneous symmetry breaking' choice of sign for the parameter $\mu^2$. Plainly, for the 'normal' sign of $\mu^2$, in which '$+\mu^2\hat{\phi}^\dagger\hat{\phi}$' is replaced by '$-\mu^2\hat{\phi}^\dagger\hat{\phi}$', with $\mu^2$ positive in both cases, the free ($\lambda = 0$) part would describe a complex doublet, with four degrees of freedom, each with the same mass $\mu$. Let us see what happens in the broken symmetry case.

For the Lagrangian (17.97) with $\mu^2 > 0$, the minimum of the classical potential is at the point

$$(\phi^\dagger \phi)_{\min} = 2\mu^2/\lambda \equiv v^2/2. \tag{17.98}$$

As in the U(1) case, we interpret (17.98) as a condition on the vev of $\hat{\phi}^\dagger \hat{\phi}$,

$$\langle 0|\hat{\phi}^\dagger \hat{\phi}|0\rangle = v^2/2 \tag{17.99}$$

where now $|0\rangle$ is the symmetry-breaking ground state, *and the subscript 'B' is omitted*. Before proceeding, we note that (17.97) is invariant under global SU(2) transformations

$$\hat{\phi} \to \hat{\phi}' = \exp(-\mathrm{i}\boldsymbol{\alpha} \cdot \boldsymbol{\tau}/2)\hat{\phi} \tag{17.100}$$

but also under a separate global U(1) transformation

$$\hat{\phi} \to \hat{\phi}' = \exp(-\mathrm{i}\alpha)\hat{\phi} \tag{17.101}$$

where $\alpha$ is to be distinguished from $\boldsymbol{\alpha} \equiv (\alpha_1, \alpha_2, \alpha_3)$. The full symmetry is then referred to as SU(2)×U(1), which is the symmetry of the electroweak sector of the Standard Model, except that in that case it is a *local* symmetry.

As before, in order to get a sensible particle spectrum we must expand the fields $\hat{\phi}$ not about $\hat{\phi} = 0$ but about a point satisfying the stable ground state (vacuum) condition (17.98). That is, we need to define '$\langle 0|\hat{\phi}|0\rangle$' and expand about it, as in (17.84). In the present case, however, the situation is more complicated than (17.84) since the complex doublet (17.95) contains four real fields as indicated in (17.95), and (17.98) becomes

$$\langle 0|\hat{\phi}_1^2 + \hat{\phi}_2^2 + \hat{\phi}_3^2 + \hat{\phi}_4^2|0\rangle = v^2. \tag{17.102}$$

It is evident that we have a lot of freedom in choosing the $\langle 0|\hat{\phi}_i|0\rangle$ so that (17.102) holds, and it is not at first obvious what an appropriate generalization of (17.84) and (17.85) might be.

Furthermore, in this more complicated (non-Abelian) situation a qualitatively new feature can arise: it may happen that the chosen condition $\langle 0|\hat{\phi}_i|0\rangle \neq 0$ is *invariant* under some subset of the allowed symmetry transformations. This would effectively mean that this particular choice of the vacuum state respected that subset of symmetries, which would therefore not be 'spontaneously broken' after all. Since each broken symmetry is associated with a massless Goldstone boson, we would then get fewer of these bosons than expected. Just this happens (by design) in the present case.

Suppose, then, that we could choose the $\langle 0|\hat{\phi}_i|0\rangle$ so as to break this SU(2)×U(1) symmetry completely: we would then expect four massless fields. Actually, however, it is not possible to make such a choice. An analogy may make this point clearer. Suppose we were considering just SU(2) and the field '$\hat{\phi}$' was an SU(2)-triplet, $\hat{\boldsymbol{\phi}}$. Then we could always write $\langle 0|\hat{\boldsymbol{\phi}}|0\rangle = v\boldsymbol{n}$ where $\boldsymbol{n}$ is a unit vector; but this form is invariant under rotations about the $\boldsymbol{n}$-axis, irrespective of where that points. In the present case, by using the freedom of global SU(2)×U(1) phase changes, an arbitrary $\langle 0|\hat{\phi}|0\rangle$ can be brought to the form

$$\langle 0|\hat{\phi}|0\rangle = \begin{pmatrix} 0 \\ v/\sqrt{2} \end{pmatrix}. \tag{17.103}$$

In considering what symmetries are respected or broken by (17.103), it is easiest to look at infinitesimal transformations. It is then clear that the particular transformation

$$\delta\hat{\phi} = -\mathrm{i}\epsilon(1+\tau_3)\hat{\phi} \tag{17.104}$$

(which is a combination of (17.101) and the 'third component' of (17.100)) is still a symmetry of (17.103) since

$$(1+\tau_3)\begin{pmatrix} 0 \\ v/\sqrt{2} \end{pmatrix} = \begin{pmatrix} 0 \\ 0 \end{pmatrix} \tag{17.105}$$

so that

$$\langle 0|\phi|0\rangle = \langle 0|\phi + \delta\phi|0\rangle. \tag{17.106}$$

We say that 'the vacuum is invariant under (17.104)' and when we look at the spectrum of oscillations about that vacuum we expect to find only three massless bosons, not four.

Oscillations about (17.103) are conveniently parametrized by

$$\hat{\phi} = \exp(-\mathrm{i}(\hat{\boldsymbol{\theta}}(x) \cdot \boldsymbol{\tau}/2)v) \begin{pmatrix} 0 \\ \frac{1}{\sqrt{2}}(v + \hat{H}(x)) \end{pmatrix} \tag{17.107}$$

which is to be compared with (17.84). Inserting (17.107) into (17.97) (see problem 17.6), we find that no mass term is generated for the $\theta$ fields, while the $H$ field piece is

$$\hat{\mathcal{L}}_H = \tfrac{1}{2}\partial_\mu \hat{H} \partial^\mu \hat{H} - \mu^2 \hat{H}^2 + \text{interactions} \tag{17.108}$$

just as in (17.85), showing that $m_H = \sqrt{2}\mu$.

Let us now note carefully that whereas in the 'normal symmetry' case with the opposite sign for the $\mu^2$ term in (17.97), the free-particle spectrum consisted of a degenerate doublet of four degrees of freedom all with the same mass $\mu$, in the 'spontaneously broken' case, no such doublet structure is seen: instead, there is one massive scalar field and three massless scalar fields. The number of degrees of freedom is the same in each case, but the physical spectrum is completely different.

In the application of this to the electroweak sector of the Standard Model, the SU(2)×U(1) symmetry will be 'gauged' (i.e. made local), which is easily done by replacing the ordinary derivatives in (17.97) by suitable covariant ones. We shall see in chapter 19 that the result, with the choice (17.107), will be to end up with three *massive* gauge fields (those mediating the weak interactions) and one *massless* gauge field (the photon). We may summarize this (anticipated) result by saying, then, that when a spontaneously broken non-Abelian symmetry is gauged, those gauge fields corresponding to symmetries that are broken by the choice of $\langle 0|\hat{\phi}|0\rangle$ acquire a mass, while those that correspond to symmetries that are respected by $\langle 0|\hat{\phi}|0\rangle$ do not. Exactly how this happens will be the subject of chapter 19.

We end this chapter by considering a second important example of spontaneous symmetry breaking in condensed matter physics, as a preliminary to our discussion of chiral symmetry breaking in the following chapter.

## 17.7 The BCS superconducting ground state

We shall not attempt to provide a self-contained treatment of the Bardeen–Cooper–Schrieffer (1957)—or BCS—theory; rather, we wish simply to focus on one aspect of the theory, namely the occurrence of an *energy gap* separating the ground state from the lowest excited levels of the energy spectrum. The existence of such a gap is a fundamental ingredient of the theory of superconductivity; in the following chapter we shall see how Nambu (1960) interpreted a chiral symmetry breaking fermionic mass term as an analogous 'gap'. We emphasize at the outset that we shall here not treat electromagnetic interactions in the superconducting state, leaving that topic for chapter 19. Again, we work at zero temperature.

Our discussion will deliberately have some similarity to that of section 17.3.2. In the present case, of course, we shall be dealing with electrons—which are fermions—rather than the bosons of a superfluid. Nevertheless, we shall see that a similar kind of 'condensation' occurs in the superconductor too. Naturally, such a phenomenon can only occur for bosons. Thus, an essential element in the BCS theory is the identification of a mechanism whereby pairs of electrons become correlated, the behaviour of which may have some similarity to that of bosons. Now the Coulomb interaction between a pair of electrons is repulsive and it remains so despite the screening that occurs in a solid. But the positively charged ions do provide sources of attraction for the electrons, and may be used as intermediaries (via 'electron–phonon interactions') to promote an effective attraction between electrons in certain circumstances. At this point we recall the characteristic feature of a weakly interacting gas of electrons at zero temperature: thanks to the Exclusion Principle, the electrons populate single-particle energy levels up to some maximum energy $E_{\mathrm{F}}$ (the Fermi energy), whose value is fixed by the electron density. It turns out (see, for example, Kittel (1987) chapter 8) that electron–electron scattering, mediated by phonon exchange, leads to an effective attraction between two electrons whose energies $\epsilon_k$ lie in a thin band $E_{\mathrm{F}} - \omega_{\mathrm{D}} < \epsilon_k < E_{\mathrm{F}} + \omega_{\mathrm{D}}$ around $E_{\mathrm{F}}$, where $\omega_{\mathrm{D}}$ is the Debye frequency associated with lattice vibrations. Cooper (1956) was the first to observe that the Fermi 'sea' was unstable with respect to the formation of bound pairs, in the presence of an attractive interaction. What this means is that the energy of the system can be lowered by exciting a pair of electrons above $E_{\mathrm{F}}$, which then become bound to a state with a total energy less than $2E_{\mathrm{F}}$. This instability modifies the Fermi sea in a fundamental way: a sort of 'condensate' of pairs is created around the Fermi energy and we need a many-body formalism to handle the situation.

For simplicity, we shall consider pairs of equal and opposite momentum $\boldsymbol{k}$, so their total momentum is zero. It can also be argued that the effective attraction will be greater when the spins are anti-parallel but the spin will not be indicated explicitly in what follows: '$\boldsymbol{k}$' will stand for '$\boldsymbol{k}$ with spin up' and '$-\boldsymbol{k}$' for '$-\boldsymbol{k}$ with spin down'. With this by way of motivation, we thus arrive at the *BCS reduced Hamiltonian*

$$\hat{H}_{\mathrm{BCS}} = \sum_{\boldsymbol{k}} \epsilon_k \hat{c}^\dagger_{\boldsymbol{k}} \hat{c}_{\boldsymbol{k}} - V \sum_{\boldsymbol{k},\boldsymbol{k}'} \hat{c}^\dagger_{\boldsymbol{k}'} \hat{c}^\dagger_{-\boldsymbol{k}'} \hat{c}_{-\boldsymbol{k}} \hat{c}_{\boldsymbol{k}} \tag{17.109}$$

which is the starting point of our discussion. In (17.109), the $\hat{c}$'s are fermionic operators obeying the usual anti-commutation relations and the vacuum is such that $\hat{c}_{\boldsymbol{k}}|0\rangle = 0$. The sum is over states lying near $E_{\mathrm{F}}$, as before, and the single-particle energies $\epsilon_k$ are measured relative to $E_{\mathrm{F}}$. The constant $V$ (with the minus sign in front) represents a simplified form of the effective electron–electron attraction. Note that, in the non-interacting ($V = 0$) part, $\hat{c}^\dagger_{\boldsymbol{k}} \hat{c}_{\boldsymbol{k}}$ is the number operator for the electrons, which because of the Pauli principle has eigenvalues 0 or 1: this term is, of course, completely analogous to (7.50) and sums the single-particle energies $\epsilon_k$ for each occupied level.

We immediately note that $\hat{H}_{\rm BCS}$ is invariant under the global U(1) transformation

$$\hat{c}_{\boldsymbol{k}} \rightarrow \hat{c}'_{\boldsymbol{k}} = \mathrm{e}^{-\mathrm{i}\alpha}\hat{c}_{\boldsymbol{k}} \tag{17.110}$$

for all $\boldsymbol{k}$, which is equivalent to $\hat{\psi}'(\boldsymbol{x}) = \mathrm{e}^{-\mathrm{i}\alpha}\hat{\psi}(\boldsymbol{x})$ for the electron field operator at $\boldsymbol{x}$. Thus fermion number is conserved by $\hat{H}_{\rm BCS}$. However, just as for the superfluid, we shall see that the BCS ground state does not respect the symmetry.

We follow Bogoliubov (1958), Bogoliubov *et al* (1959) (see also Valatin (1958)) and make a canonical transformation on the operators $\hat{c}_{\boldsymbol{k}}$, $\hat{c}^{\dagger}_{-\boldsymbol{k}}$ similar to the one employed for the superfluid problem in (17.38), as motivated by the 'pair condensate' picture. We set

$$\begin{aligned} \hat{\beta}_{\boldsymbol{k}} &= u_k c_{\boldsymbol{k}} - v_k \hat{c}^{\dagger}_{-\boldsymbol{k}} \qquad & \beta^{\dagger}_{\boldsymbol{k}} &= u_k \hat{c}^{\dagger}_{\boldsymbol{k}} - v_k \hat{c}_{-\boldsymbol{k}} \\ \hat{\beta}_{-\boldsymbol{k}} &= u_k \hat{c}_{-\boldsymbol{k}} + v_k \hat{c}^{\dagger}_{\boldsymbol{k}} \qquad & \beta^{\dagger}_{-\boldsymbol{k}} &= u_k \hat{c}^{\dagger}_{-\boldsymbol{k}} + v_k \hat{c}_{\boldsymbol{k}} \end{aligned} \tag{17.111}$$

where $u_k$ and $v_k$ are real, depend only on $k = |\boldsymbol{k}|$ and are chosen so as to preserve *anti*-commutation relations for the $\beta$'s. This last condition implies (problem 17.7)

$$u_k^2 + v_k^2 = 1 \tag{17.112}$$

so that we may conveniently set

$$u_k = \cos\theta_k \qquad v_k = \sin\theta_k . \tag{17.113}$$

Just as in the superfluid case, the transformations (17.111) only make sense in the context of a number-non-conserving ground state, since they do not respect the symmetry (17.110). Although $\hat{H}_{\rm BCS}$ of (17.109) is number-conserving, we shall shortly make a crucial number-non-conserving approximation.

We seek a diagonalization of (17.109), analogous to (17.40), in terms of the mode operators $\hat{\beta}$ and $\hat{\beta}^{\dagger}$:

$$\hat{H}_{\rm BCS} = \sum_{\boldsymbol{k}} \omega_k (\hat{\beta}^{\dagger}_{\boldsymbol{k}}\hat{\beta}_{\boldsymbol{k}} + \hat{\beta}^{\dagger}_{-\boldsymbol{k}}\hat{\beta}_{-\boldsymbol{k}}) + \gamma \tag{17.114}$$

for certain constants $\omega_k$ and $\gamma$. It is easy to check (problem 17.8) that the form (17.114) implies

$$[\hat{H}_{\rm BCS}, \hat{\beta}^{\dagger}_{\boldsymbol{l}}] = \omega_l \hat{\beta}^{\dagger}_{\boldsymbol{l}} \tag{17.115}$$

as in (17.41), despite the fact that the operators obey *anti*commution relations. Equation (17.115) then implies that the $\omega_k$ are the energies of states created by the *quasiparticle operators* $\hat{\beta}^{\dagger}_{\boldsymbol{k}}$ and $\hat{\beta}^{\dagger}_{-\boldsymbol{k}}$, the ground state being defined by

$$\hat{\beta}_{\boldsymbol{k}}|\text{ground}\rangle_{\rm BCS} = \hat{\beta}_{-\boldsymbol{k}}|\text{ground}\rangle_{\rm BCS} = 0. \tag{17.116}$$

Substituting for $\hat{\beta}_l^\dagger$ in (17.115) from (17.111), we therefore require

$$[\hat{H}_{\rm BCS}, \cos\theta_l\,\hat{c}_l^\dagger - \sin\theta_l\,\hat{c}_{-l}] = \omega_l(\cos\theta_l\,\hat{c}_l^\dagger - \sin\theta_l\,\hat{c}_{-l}) \quad (17.117)$$

which must hold as an identity in the $\hat{c}_l$'s and $\hat{c}_l^\dagger$'s. Evaluating (17.117), one obtains (problem 17.9)

$$(\omega_l - \epsilon_l)\cos\theta_l - V\sin\theta_l \sum_k \hat{c}_{-k}\hat{c}_k = 0 \quad (17.118)$$

$$-V\cos\theta_l \sum_k \hat{c}_k^\dagger \hat{c}_{-k}^\dagger + (\omega_l + \epsilon_l)\sin\theta_l = 0. \quad (17.119)$$

It is at this point that we make the crucial 'condensate' assumption: we replace the *operator* expressions $\sum_k \hat{c}_{-k}\hat{c}_k$ and $\sum_k \hat{c}_k^\dagger\hat{c}_{-k}^\dagger$ by their average values, which are *assumed to be non-zero in the ground state*. Since these operators carry fermion number $\pm 2$, it is clear that this assumption is only valid if the ground state does not, in fact, have a definitive number of particles—just as in the superfluid case. We accordingly make the replacements

$$V\sum_k \hat{c}_{-k}\hat{c}_k \to V\,{}_{\rm BCS}\langle \text{ground}|\sum_k \hat{c}_{-k}\hat{c}_k|\text{ground}\rangle_{\rm BCS} \equiv \Delta \quad (17.120)$$

$$V\sum_k \hat{c}_k^\dagger\hat{c}_{-k}^\dagger \to V\,{}_{\rm BCS}\langle \text{ground}|\sum_k \hat{c}_k^\dagger\hat{c}_{-k}^\dagger|\text{ground}\rangle_{\rm BCS} \equiv \Delta^*. \quad (17.121)$$

In that case, equations (17.118) and (17.119) become

$$\omega_l\cos\theta_l = \epsilon_l\cos\theta_l + \Delta\sin\theta_l \quad (17.122)$$

$$\omega_l\sin\theta_l = -\epsilon_l\sin\theta_l + \Delta^*\cos\theta_l \quad (17.123)$$

which are consistent if

$$\omega_l = \pm[\epsilon_l^2 + |\Delta|^2]^{1/2}. \quad (17.124)$$

Equation (17.124) is the fundamental result at this stage. Recalling that $\epsilon_l$ is measured relative to $E_{\rm F}$, we see that it implies that all excited states are separated from $E_{\rm F}$ by a finite amount, namely $|\Delta|$.

In interpreting (17.124), we must however be careful to reckon energies for an excited state as relative to a BCS state having the same number of pairs, if we consider experimental probes which do not inject or remove electrons. Thus, considering a component of $|\text{ground}\rangle_{\rm BCS}$ with $N$ pairs, we may consider the excitation of two particles above a BCS state with $N-1$ pairs. The minimum energy for this to be possible is $2|\Delta|$. It is this quantity which is usually called the *energy gap*. Such an excited state is represented by $\beta_k^\dagger\beta_{-k}^\dagger|\text{ground}\rangle_{\rm BCS}$.

We shall need the expressions for $\cos\theta_l$ and $\sin\theta_l$ which may be obtained as follows. Squaring (17.122) and taking $\Delta$ now to be real and equal to $|\Delta|$, we obtain

$$|\Delta|^2(\cos^2\theta_l - \sin^2\theta_l) = 2\epsilon_l|\Delta|\cos\theta_l\sin\theta_l \quad (17.125)$$

which leads to

$$\tan 2\theta_l = |\Delta|/\epsilon_l \tag{17.126}$$

and then

$$\cos\theta_l = \left[\frac{1}{2}\left(1+\frac{\epsilon_l}{\omega_l}\right)\right]^{1/2} \qquad \sin\theta_l = \left[\frac{1}{2}\left(1-\frac{\epsilon_l}{\omega_l}\right)\right]^{1/2}. \tag{17.127}$$

All our experience to date indicates that the choice '$\Delta$ = real' amounts to a choice of phase for the ground-state value:

$$V\,{}_{\mathrm{BCS}}\langle\mathrm{ground}|\sum_{\boldsymbol{k}}\hat{c}_{-\boldsymbol{k}}c_{\boldsymbol{k}}|\mathrm{ground}\rangle_{\mathrm{BCS}} = |\Delta|. \tag{17.128}$$

By making use of the U(1) symmetry (17.110), other phases for $\Delta$ are equally possible.

The condition (17.128) has, of course, the by now anticipated form for a spontaneously broken U(1) symmetry, and we must therefore expect the occurrence of a massless mode. However, we may now recall that the electrons are charged, so that when electromagnetic interactions are included in the superconducting state, we have to allow the $\alpha$ in (17.110) to become a local function of $\boldsymbol{x}$. At the same time, the massless photon field will enter. Remarkably, we shall learn in chapter 19 that the expected massless (Goldstone) mode is, in this case, not observed: instead, that degree of freedom is incorporated into the gauge field, rendering it massive. As we shall see, this is the physics of the Meissner effect in a superconductor, and that of the 'Higgs mechanism' in the Standard Model.

An explicit formula for $\Delta$ can be found by using the definition (17.120), together with the expression for $\hat{c}_{\boldsymbol{k}}$ found by inverting (17.111):

$$\hat{c}_{\boldsymbol{k}} = \cos\theta_k\,\hat{\beta}_{\boldsymbol{k}} + \sin\theta_k\,\hat{\beta}^{\dagger}_{-\boldsymbol{k}}. \tag{17.129}$$

This gives, using (17.120) and (17.129),

$$\begin{aligned}|\Delta| &= V\,{}_{\mathrm{BCS}}\langle\mathrm{ground}|\sum_{\boldsymbol{k}}(\cos\theta_k\hat{\beta}_{-\boldsymbol{k}} + \sin\theta_k\,\hat{\beta}^{\dagger}_{\boldsymbol{k}})\\ &\quad\times(\cos\theta_k\,\hat{\beta}_{\boldsymbol{k}} + \sin\theta_k\,\hat{\beta}^{\dagger}_{-\boldsymbol{k}})|\mathrm{ground}\rangle_{\mathrm{BCS}}\\ &= V\,{}_{\mathrm{BCS}}\langle\mathrm{ground}|\sum_{\boldsymbol{k}}\cos\theta_k\sin\theta_k\hat{\beta}_{-\boldsymbol{k}}\hat{\beta}^{\dagger}_{-\boldsymbol{k}}|\mathrm{ground}\rangle_{\mathrm{BCS}}\\ &= V\sum_{k}\frac{|\Delta|}{2[\epsilon_k^2+|\Delta|^2]^{1/2}}.\end{aligned} \tag{17.130}$$

The sum in (17.130) is only over the small band $E_{\mathrm{F}} - \omega_{\mathrm{D}} < \epsilon_k < E_{\mathrm{F}} + \omega_{\mathrm{D}}$ over which the effective electron–electron attraction operates. Replacing the sum by an integral, we obtain the *gap equation*

$$1 = \frac{1}{2}V \cdot N_{\mathrm{F}} \int_{-\omega_{\mathrm{D}}}^{\omega_{\mathrm{D}}} \frac{\mathrm{d}\epsilon}{[\epsilon^2 + |\Delta|^2]^{\frac{1}{2}}}$$
$$= V N_{\mathrm{F}} \sinh^{-1}(\omega_{\mathrm{D}}/|\Delta|) \tag{17.131}$$

where $N_{\mathrm{F}}$ is the density of states at the Fermi level. Equation (17.131) yields

$$|\Delta| = \frac{\omega_{\mathrm{D}}}{\sinh(1/V N_{\mathrm{F}})} \approx 2\omega_{\mathrm{D}} \mathrm{e}^{-1/V N_{\mathrm{F}}} \tag{17.132}$$

for $V N_{\mathrm{F}} \ll 1$. This is the celebrated BCS solution for the gap parameter $|\Delta|$. Perhaps the most significant thing to note about it, for our purpose, is that the expression for $|\Delta|$ is not an analytic function of the dimensionless interaction parameter $V N_{\mathrm{F}}$ (it cannot be expanded as a power series in this quantity), and so no perturbative treatment starting from a normal ground state could reach this result. The estimate (17.132) is in reasonably good agreement with experiment, and may be refined.

The explicit form of the ground state in this model can be found by a method similar to the one indicated in section 17.3.2 for the superfluid. Since the transformation from the $\hat{c}$'s to the $\hat{\beta}$'s is canonical, there must exist a unitary operator which effects it via (compare (17.48))

$$\hat{U}_{\mathrm{BCS}}\, \hat{c}_{\boldsymbol{k}}\, \hat{U}_{\mathrm{BCS}}^{\dagger} = \hat{\beta}_{\boldsymbol{k}} \qquad \hat{U}_{\mathrm{BCS}}\, \hat{c}_{-\boldsymbol{k}}^{\dagger}\, \hat{U}_{\mathrm{BCS}}^{\dagger} = \hat{\beta}_{-\boldsymbol{k}}^{\dagger}. \tag{17.133}$$

The operator $\hat{U}_{\mathrm{BCS}}$ is (Blatt 1964, section V.4, Yosida 1958, and compare problem 17.4)

$$\hat{U}_{\mathrm{BCS}} = \prod_{\boldsymbol{k}} \exp[\theta_k (\hat{c}_{\boldsymbol{k}}^{\dagger} \hat{c}_{-\boldsymbol{k}}^{\dagger} - \hat{c}_{\boldsymbol{k}} \hat{c}_{-\boldsymbol{k}})]. \tag{17.134}$$

Then, since $\hat{c}_{\boldsymbol{k}}|0\rangle = 0$, we have

$$\hat{U}_{\mathrm{BCS}}^{\dagger} \hat{\beta}_{\boldsymbol{k}} \hat{U}_{\mathrm{BCS}}|0\rangle = 0 \tag{17.135}$$

showing that we may identify

$$|\mathrm{ground}\rangle_{\mathrm{BCS}} = \hat{U}_{\mathrm{BCS}}|0\rangle \tag{17.136}$$

via the condition (17.116). When the exponential in $\hat{U}_{\mathrm{BCS}}$ is expanded out and applied to the vacuum state $|0\rangle$, great simplifications occur. Consider the operator

$$\hat{s}_{\boldsymbol{k}} = \hat{c}_{\boldsymbol{k}}^{\dagger} \hat{c}_{-\boldsymbol{k}}^{\dagger} - \hat{c}_{\boldsymbol{k}} \hat{c}_{-\boldsymbol{k}}. \tag{17.137}$$

We have

$$\hat{s}_{\boldsymbol{k}}^2 = -\hat{c}_{\boldsymbol{k}}^{\dagger} \hat{c}_{-\boldsymbol{k}}^{\dagger} \hat{c}_{\boldsymbol{k}} \hat{c}_{-\boldsymbol{k}} - \hat{c}_{\boldsymbol{k}} \hat{c}_{-\boldsymbol{k}} \hat{c}_{\boldsymbol{k}}^{\dagger} \hat{c}_{-\boldsymbol{k}}^{\dagger} \tag{17.138}$$

so that $\hat{s}_{\boldsymbol{k}}^2|0\rangle = -|0\rangle$. It follows that

$$\begin{aligned}\exp(\theta_k \hat{s}_{\boldsymbol{k}})|0\rangle &= \left(1 + \theta_k \hat{s}_{\boldsymbol{k}} - \frac{\theta_k^2}{2} - \frac{\theta_k^3}{3}\hat{s}_{\boldsymbol{k}} \ldots\right)|0\rangle \\ &= (\cos\theta_k + \sin\theta_k\, \hat{s}_{\boldsymbol{k}})|0\rangle \\ &= (\cos\theta_k + \sin\theta_k\, \hat{c}_{\boldsymbol{k}}^\dagger \hat{c}_{-\boldsymbol{k}}^\dagger)|0\rangle \end{aligned} \tag{17.139}$$

and hence

$$|\text{ground}\rangle_{\text{BCS}} = \prod_{\boldsymbol{k}} (\cos\theta_k + \sin\theta_k\, \hat{c}_{\boldsymbol{k}}^\dagger \hat{c}_{-\boldsymbol{k}}^\dagger)|0\rangle. \tag{17.140}$$

As for the superfluid, (17.140) represents a coherent superposition of correlated pairs, with no restraint on the particle number.

We should emphasize that this is only the barest outline of a simple version of BCS theory, from which many subtleties have been omitted. Consider, for example, the binding energy $E_\text{b}$ of a pair, which to calculate one needs to evaluate the constant $\gamma$ in (17.114). To a good approximation, one finds (see, for example, Enz 1992) that $E_\text{b} \approx 3\Delta^2/E_\text{F}$. One can also calculate the approximate spatial extension of a pair, which is denoted by the *coherence length* $\xi$ and is of order $v_\text{F}/\pi\Delta$ where $k_\text{F} = mv_\text{F}$ is the Fermi momentum. If we compare $E_\text{b}$ to the Coulomb repulsion at a distance $\xi$, we find that

$$E_\text{b}/(\alpha/\xi) \sim a_0/\xi \tag{17.141}$$

where $a_0$ is the Bohr radius. Numerical values show that the right-hand side of (17.141), in conventional superconductors, is of order $10^{-3}$. Hence, the pairs are not really bound, only correlated, and as many as $10^6$ pairs may have their centres of mass within one coherence length of each other. Nevertheless, the simple theory presented here contains the essential features which underly all attempts to understand the dynamical occurrence of spontaneous symmetry breaking in fermionic systems.

We now proceed to an important application in particle physics.

## Problems

**17.1** Verify (17.29).

**17.2** Verify (17.35).

**17.3** Derive (17.43) and (17.44).

**17.4** Let

$$\hat{U}_\lambda = \exp[\tfrac{1}{2}\lambda\theta(\hat{a}^2 - \hat{a}^{\dagger 2}]$$

where $[\hat{a}, \hat{a}^\dagger] = 1$ and $\lambda$, $\theta$ are real parameters.

(a) Show that $\hat{U}_\lambda$ is unitary.

(b) Let

$$\hat{I}_\lambda = \hat{U}_\lambda \hat{a} \hat{U}_\lambda^{-1} \quad \text{and} \quad \hat{J}_\lambda = \hat{U}_\lambda \hat{a}^\dagger \hat{U}_\lambda^{-1}.$$

Show that

$$\frac{\mathrm{d}\hat{I}_\lambda}{\mathrm{d}\lambda} = \theta \hat{J}_\lambda$$

and that

$$\frac{\mathrm{d}^2 \hat{I}_\lambda}{\mathrm{d}\lambda^2} = \theta^2 \hat{I}_\lambda.$$

(c) Hence, show that

$$\hat{I}_\lambda = \cosh(\lambda\theta)\, \hat{a} + \sinh(\lambda\theta)\, \hat{a}^\dagger$$

and thus finally (compare (17.38) and (17.48)) that

$$\hat{U}_1 \hat{a} \hat{U}_1^{-1} = \cosh\theta\, \hat{a} + \sinh\theta\, \hat{a}^\dagger \equiv \hat{\alpha}$$

and

$$\hat{U}_1 \hat{a}^\dagger \hat{U}_1^{-1} = \sinh\theta\, \hat{a} + \cosh\theta\, \hat{a}^\dagger \equiv \hat{\alpha}^\dagger$$

where

$$\hat{U}_1 \equiv \hat{U}_{\lambda=1} = \exp[\tfrac{1}{2}\theta(\hat{a}^2 - \hat{a}^{\dagger 2})].$$

**17.5** Insert the ansatz (17.84) for $\hat{\phi}$ into $\hat{\mathcal{L}}_{\mathrm{G}}$ of (17.69) with $\hat{V} = \hat{V}_{\mathrm{SB}}$ of (17.77) and show that the result for the constant term and the quadratic terms in $\hat{h}$ and $\hat{\theta}$, is as given in (17.85).

**17.6** Verify that when (17.107) is inserted in (17.97), the terms quadratic in the fields $\hat{H}$ and $\hat{\theta}$ reveal that $\hat{\theta}$ is a massless field, while the quanta of the $\hat{H}$ field have mass $\sqrt{2}\mu$.

**17.7** Verify that the $\hat{\beta}$'s of (17.111) satisfy the required anti-commutation relations if (17.112) holds.

**17.8** Verify (17.115).

**17.9** Derive (17.118) and (17.119).

# 18

# CHIRAL SYMMETRY BREAKING

In section 12.3.2 we arrived at a puzzle: there seemed good reason to think that a world consisting of u and d quarks and their anti-particles, interacting via the colour gauge fields of QCD, should exhibit signs of the non-Abelian *chiral symmetry* $\mathrm{SU(2)_{f5}}$, which was exact in the massless limit $m_{\mathrm{u}}, m_{\mathrm{d}} \to 0$. But, as we showed, one of the simplest consequences of such a symmetry should be the existence of nucleon parity doublets, which are not observed. We can now resolve this puzzle by making the hypothesis (section 18.1) first articulated by Nambu (1960) and Nambu and Jona-Lasinio (1961a), that this chiral symmetry is *spontaneously broken* as a dynamical effect—presumably, from today's perspective, as a property of the QCD interactions, as discussed in section 18.2. If this is so, an immediate physical consequence should be the appearance of massless (Goldstone) bosons, one for every symmetry not respected by the vacuum. Indeed, returning to (12.169) which we repeat here for convenience,

$$\hat{T}_{+5}^{(\frac{1}{2})}|d\rangle = |\tilde{u}\rangle \tag{18.1}$$

we now interpret the state $|\tilde{u}\rangle$ (which is degenerate with $|d\rangle$) as $|d + \text{‘}\pi^+\text{’}\rangle$ where ‘$\pi^+$’ is a massless particle of positive charge but a *pseudo*scalar ($0^-$) rather than a scalar ($0^+$) since, as we saw, $|\tilde{u}\rangle$ has opposite parity to $|u\rangle$. In the same way, ‘$\pi^-$’ and ‘$\pi^0$’ will be associated with $\hat{T}_{-5}^{(\frac{1}{2})}$ and $\hat{T}_{35}^{(\frac{1}{2})}$. Of course, no such *massless* pseudoscalar particles are observed; but it is natural to hope that when the small up and down quark masses are included, the real pions ($\pi^+, \pi^-, \pi^0$) will emerge as ‘anomalously light’, rather than strictly massless. This is indeed how they do appear, particularly with respect to the octet of vector ($1^-$) mesons, which differ only in $\mathrm{q\bar{q}}$ spin alignment from the pseudoscalar ($0^-$) octet. As Nambu and Jona-Lasinio (1961a) stated, ‘It is perhaps not a coincidence that there exists such an entity [i.e. the Goldstone state(s)] in the form of the pion’.

If this was the only observable consequence of spontaneously breaking chiral symmetry, it would perhaps hardly be sufficient grounds for accepting the hypothesis. But there are two circumstances which greatly increase the phenomenological implications of the idea. First, the vector and axial vector symmetry currents $\hat{\boldsymbol{T}}^{(\frac{1}{2})\mu}$ and $\hat{\boldsymbol{T}}_5^{(\frac{1}{2})\mu}$ of the u–d strong interaction SU(2) symmetries (see (12.109) and (12.166)) happen to be the very same currents which enter into strangeness-conserving semileptonic weak interactions (such as

$n \to pe^-\bar{\nu}_e$ and $\pi^- \to \mu^-\bar{\nu}_\mu$), as we shall see in chapter 20. Thus, some remarkable connections between weak- and strong-interaction parameters can be established, such as the Goldberger–Treiman (1958) relation (see section 18.3) and the Adler–Weisberger (in Adler 1965) relation. Second, it turns out that the dynamics of the Goldstone modes, and their interactions with other hadrons such as nucleons, are strongly constrained by the underlying chiral symmetry of QCD; indeed, surprisingly detailed *effective theories* (see section 18.4) have been developed, which provide a very successful description of the low-energy dynamics of the hadronic degrees of freedom. Finally, we shall introduce the subject of chiral anomalies in section 18.5.

It would take us too far from our main focus on gauge theories to pursue these interesting avenues in detail. But we hope to convince the reader, in this chapter, that chiral symmetry breaking is an important part of the Standard Model, and to encourage further study of a subject which may at first sight appear somewhat peripheral to the Standard Model as conventionally understood.

## 18.1 The Nambu analogy

We recall from section 12.3.2 that for 'almost massless' fermions it is natural to use the representation (4.97) for the Dirac matrices, in terms of which the Dirac equation reads

$$E\phi = \boldsymbol{\sigma} \cdot \boldsymbol{p}\phi + m\chi \tag{18.2}$$

$$E\chi = -\boldsymbol{\sigma} \cdot \boldsymbol{p}\chi + m\phi. \tag{18.3}$$

Nambu (1960) and Nambu and Jona-Lasinio (1961a) pointed out a remarkable analogy between (18.2) and (18.3) and equations (17.122) and (17.123) which describe the elementary excitations in a superconductor (in the case $\Delta$ is real) and which we repeat here for convenience:

$$\omega_l \cos\theta_l = \epsilon_l \cos\theta_l + \Delta \sin\theta_l \tag{18.4}$$

$$\omega_l \sin\theta_l = -\epsilon_l \sin\theta_l + \Delta \cos\theta_l. \tag{18.5}$$

In (18.4) and (18.5), $\cos\theta_l$ and $\sin\theta_l$ are, respectively, the components of the electron destruction operator $\hat{c}_l$ and the electron creation operator $\hat{c}^\dagger_{-l}$ in the quasiparticle operator $\hat{\beta}_l$ (see (17.111)):

$$\hat{\beta}_l = \cos\theta_l\, \hat{c}_l - \sin\theta_l\, \hat{c}^\dagger_{-l}. \tag{18.6}$$

The superposition in $\hat{\beta}_l$ combines operators which transform differently under the U(1) (number) symmetry. The result of this spontaneous breaking of the U(1) symmetry is the creation of the gap $\Delta$ (or $2\Delta$ for a number-conserving excitation) and the appearance of a massless mode. If $\Delta$ vanishes, (17.126) implies that $\theta_l = 0$ and we revert to the symmetry-respecting operators $\hat{c}_l, \hat{c}^\dagger_{-l}$. Consider

**Figure 18.1.** The type of fermion–anti-fermion in the Nambu 'condensate'.

now (18.2) and (18.3). Here $\phi$ and $\chi$ are the components of definite chirality in the Dirac spinor $\omega$ (compare (12.150)), which is itself not a chirality eigenstate when $m \neq 0$. When $m$ vanishes, the Dirac equation for $\omega$ decouples into two separate ones for the chirality eigenstates $\phi_R \equiv \begin{pmatrix} \phi \\ 0 \end{pmatrix}$ and $\phi_L \equiv \begin{pmatrix} 0 \\ \chi \end{pmatrix}$. Nambu therefore made the following analogy:

$$\begin{aligned}
\text{superconducting gap parameter } \Delta &\leftrightarrow \text{Dirac mass } m \\
\text{quasiparticle excitation} &\leftrightarrow \text{massive Dirac particle} \\
\text{U(1) number symmetry} &\leftrightarrow \text{U(1)}_5 \text{ chirality symmetry} \\
\text{Goldstone mode} &\leftrightarrow \text{massless boson.}
\end{aligned}$$

In short, the mass of a Dirac particle arises from the (presumed) spontaneous breaking of a chiral (or $\gamma_5$) symmetry, and this will be accompanied by a massless boson.

Before proceeding, we should note that there are features of the analogy on both sides which need qualification. First, the particle symmetry we want to interpret this way is $\text{SU(2)}_{f5}$ not $\text{U(1)}_5$, so the appropriate generalization (Nambu and Jona-Lasinio 1961b) has to be understood. Second, we must again note that the BCS electrons are charged, so that in the real superconducting case we are dealing with a spontaneously broken *local* U(1) symmetry, not a global one. By contrast, the $\text{SU(2)}_{f5}$ chiral symmetry is not gauged.

As usual, the quantum field theory vacuum is analogous to the many-body ground state. According to Nambu's analogy, therefore, the vacuum for a massive Dirac particle is to be pictured as a condensate of correlated pairs of massive fermions. Since the vacuum carries neither linear nor angular momentum, the members of a pair must have equal and opposite spin: they therefore have the same helicity. However, since the vacuum does *not* violate fermion number conservation, one has to be a fermion and the other an anti-fermion. This means (recalling the discussion after (12.148)) that they have opposite chirality. Thus, a typical pair in the Nambu vacuum is as shown in figure 18.1. We may easily write down an expression for the Nambu vacuum, analogous to (17.140) for the BCS ground state. Consider solutions $\phi_+$ and $\chi_+$ of positive helicity in (18.2) and (18.3); then

$$E\phi_+ = |\boldsymbol{p}|\phi_+ + m\chi_+ \tag{18.7}$$
$$E\chi_+ = -|\boldsymbol{p}|\chi_+ + m\phi_+. \tag{18.8}$$

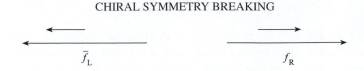

Comparing (18.7) and (18.8) with (18.4) and (18.5), we can read off the mixing coefficients $\cos\theta_p$ and $\sin\theta_p$ as (cf (17.127))

$$\cos\theta_p = \left[\frac{1}{2}\left(1 + \frac{|\boldsymbol{p}|}{E}\right)\right]^{1/2} \tag{18.9}$$

$$\sin\theta_p = \left[\frac{1}{2}\left(1 - \frac{|\boldsymbol{p}|}{E}\right]\right)^{1/2} \tag{18.10}$$

where $E = (m^2 + \boldsymbol{p}^2)^{1/2}$. The Nambu vacuum is then given by[1]

$$|0\rangle_{\mathrm{N}} = \prod_{\boldsymbol{p},s}(\cos\theta_p - \sin\theta_p \hat{c}_s^\dagger(\boldsymbol{p})\hat{d}_s^\dagger(-\boldsymbol{p}))|0\rangle_{m=0} \tag{18.11}$$

where $\hat{c}_s^\dagger$'s and $\hat{d}_s^\dagger$'s are the operators in *massless* Dirac fields. Depending on the sign of the helicity $s$, each pair in (18.11) carries $\pm 2$ units of chirality. We may check this by noting that in the mode expansion of the Dirac field $\hat{\psi}$, $\hat{c}_s(\boldsymbol{p})$ operators go with $u$-spinors for which the $\gamma_5$ eigenvalue equals the helicity, while $\hat{d}_s^\dagger(-\boldsymbol{p})$ operators accompany $v$-spinors for which the $\gamma_5$ eigenvalue equals minus the helicity. Thus, under a chiral transformation $\hat{\psi}' = \mathrm{e}^{-\mathrm{i}\beta\gamma_5}\hat{\psi}$, $\hat{c}_s \to \mathrm{e}^{-\mathrm{i}\beta s}\hat{c}_s$ and $\hat{d}_s^\dagger \to \mathrm{e}^{\mathrm{i}\beta s}\hat{d}_s^\dagger$, for a given $s$. Hence $\hat{c}_s^\dagger\hat{d}_s^\dagger$ acquires a factor $\mathrm{e}^{2\mathrm{i}\beta s}$. Thus the Nambu vacuum does not have a definite chirality and operators carrying non-zero chirality can have non-vanishing vacuum expectation values (vevs). A Dirac mass term $\bar{\hat{\psi}}\hat{\psi}$ is of just this kind, since under $\hat{\psi} = \mathrm{e}^{-\mathrm{i}\beta\gamma_5}\hat{\psi}$ we find that $\hat{\psi}^\dagger\gamma^0\hat{\psi} \to \hat{\psi}^\dagger\mathrm{e}^{\mathrm{i}\beta\gamma_5}\gamma^0\mathrm{e}^{-\mathrm{i}\beta\gamma_5}\hat{\psi} = \bar{\hat{\psi}}\mathrm{e}^{-2\mathrm{i}\beta\gamma_5}\hat{\psi}$. Thus, in analogy with (17.120), a Dirac mass is associated with a non-zero value for ${}_{\mathrm{N}}\langle 0|\bar{\hat{\psi}}\hat{\psi}|0\rangle_{\mathrm{N}}$.

In the original conception by Nambu and co-workers, the fermion under discussion was taken to be the nucleon, with '$m$' the (spontaneously generated) nucleon mass. The fermion–fermion interaction—necessarily invariant under chiral transformations—was taken to be of the four-fermion type. As we have seen in volume 1, this is actually a non-renormalizable theory but a physical cut-off was employed, somewhat analogous to the Fermi energy $E_{\mathrm{F}}$. Thus, the nucleon mass could not be dynamically predicted, unlike the analogous gap parameter $\Delta$ in BCS theory. Nevertheless, a gap equation similar to (17.131) could be formulated, and it was possible to show that when it had a non-trivial solution, a massless bound state automatically appeared in the $\mathrm{f\bar{f}}$ channel (Nambu and Jona-Lasinio 1961a). This work was generalized to the $\mathrm{SU(2)_{f5}}$ case by Nambu and Jona-Lasinio (1961b), who showed that if the chiral symmetry was broken explicitly by the introduction of a small nucleon mass ($\sim$ 5 MeV), then the Goldstone pions would have their observed non-zero (but small) mass. In addition, the Goldberger–Treiman (1958) relation was derived and a number of other applications were suggested. Subsequently, Nambu with other collaborators

[1] A different phase convention is used for $\hat{d}_s^\dagger(-\boldsymbol{p})$ as compared to that for $\hat{c}_{-\boldsymbol{k}}^\dagger$ in (17.111).

(Nambu and Lurie 1962, Nambu and Schrauner 1962) showed how the amplitudes for the emission of a single 'soft' (nearly massless, low momentum) pion could be calculated, for various processes. These developments culminated in the Adler–Weisberger relation (Adler 1965, Weisberger 1965) which involves *two* soft pions.

This work was all done in the absence of an agreed theory of the strong interactions (the N–J-L theory was an illustrative working model of dynamically-generated spontaneous symmetry breaking, but not a complete theory of strong interactions). QCD became widely accepted as that theory in around 1973. In this case, of course, the 'fermions in question' are quarks and the interactions between them are gluon exchanges, which conserve chirality as noted in section 12.3.2. The bulk of the quark masses inside bound states forming hadrons is then interpreted as being spontaneously generated, while a small explicit quark mass term in the Lagrangian is held to be responsible for the non-zero pion mass. Let us therefore now turn to two-flavour QCD.

### 18.1.1 Two flavour QCD and $\mathrm{SU(2)_{fL}} \times \mathrm{SU(2)_{fR}}$

Let us begin with the massless case, for which the fermionic part of the Lagrangian is

$$\hat{\mathcal{L}}_q = \bar{\hat{u}}\, \mathrm{i}\hat{\not{D}}\hat{u} + \bar{\hat{d}}\, \mathrm{i}\hat{\not{D}}\hat{d} \tag{18.12}$$

where $\hat{u}$ and $\hat{d}$ now stand for the field operators,

$$\hat{D}^{\mu} = \partial^{\mu} + \mathrm{i}g_{\mathrm{s}}\boldsymbol{\lambda}/2 \cdot \boldsymbol{A}^{\mu} \tag{18.13}$$

and the $\boldsymbol{\lambda}$ matrices act on the colour (r,b,g) degree of freedom of the u and d quarks. In addition to the local $\mathrm{SU(3)_c}$ symmetry, this Lagrangian is invariant under

(i) $\mathrm{U(1)_f}$ 'quark number' transformations

$$\hat{q} \to \mathrm{e}^{-\mathrm{i}\alpha}\hat{q} \tag{18.14}$$

(ii) $\mathrm{SU(2)_f}$ 'flavour isospin' transformations

$$\hat{q} \to \exp(-\mathrm{i}\boldsymbol{\alpha} \cdot \boldsymbol{\tau}/2)\,\hat{q} \tag{18.15}$$

(iii) $\mathrm{U(1)_{f5}}$ 'axial quark number' transformations

$$\hat{q} \to \mathrm{e}^{-\mathrm{i}\beta\gamma_5}\hat{q} \tag{18.16}$$

(iv) $\mathrm{SU(2)_{f5}}$ 'axial flavour isospin' transformations

$$\hat{q} \to \exp(-\mathrm{i}\boldsymbol{\beta} \cdot \boldsymbol{\tau}/2\gamma_5)\,\hat{q} \tag{18.17}$$

where

$$\hat{q} = \begin{pmatrix} \hat{u} \\ \hat{d} \end{pmatrix}. \tag{18.18}$$

Symmetry (i) is unbroken and its associated 'charge' operator (the quark number operator) commutes with all other symmetry operators, so it need not concern us further. Symmetry (ii) is the standard isospin symmetry of chapter 12, explicitly broken by the electromagnetic interactions (and by the difference in the masses $m_{\mathrm{u}}$ and $m_{\mathrm{d}}$, when included). Symmetry (iii) does not correspond to any known conservation law; however, there are not any near-massless isoscalar $0^-$ mesons, either, such as must be present if the symmetry is spontaneously broken. The $\eta$ meson is an isoscalar $0^-$ meson, but with a mass of 547 MeV it is considerably heavier than the pion. In fact, it can be understood as one of the Goldstone bosons associated with the spontaneous breaking of the larger group $\mathrm{SU(3)_{f5}}$, which includes the s quark (see, for example, Weinberg 1996, section 19.10). In that case, symmetry (iii) becomes extended to

$$\hat{u} \to \mathrm{e}^{-\mathrm{i}\beta\gamma_5}\hat{u} \qquad \hat{d} \to \mathrm{e}^{-\mathrm{i}\beta\gamma_5}\hat{d} \qquad \hat{s} \to \mathrm{e}^{-\mathrm{i}\beta\gamma_5}\hat{s} \tag{18.19}$$

but there is still a missing light isoscalar $0^-$ meson. It can be shown that its mass must be less than or equal to $\sqrt{3}\, m_\pi$ (Weinberg 1975); but no such particle exists. This is the well-known 'U(1) problem': it was resolved by 't Hooft (1976a, 1986), by showing that the inclusion of instanton configurations (Belavin *et al* 1975) in path integrals leads to violations of symmetry (iii)—see, for example, Weinberg (1996, section 23.5). Finally, symmetry (iv) is the one with which we are presently concerned.

The symmetry currents associated with (iv) are those already given in (12.166), but we give them again here in a slightly different notation which will be similar to the one used for weak interactions:

$$\hat{j}^\mu_{i,5} = \bar{\hat{q}}\gamma^\mu\gamma_5\frac{\tau_i}{2}\hat{q} \qquad i = 1, 2, 3. \tag{18.20}$$

Similarly, the currents associated with (ii) are

$$\hat{j}^\mu_i = \bar{\hat{q}}\gamma^\mu\frac{\tau_i}{2}\hat{q} \qquad i = 1, 2, 3. \tag{18.21}$$

The corresponding 'charges' are (compare (12.167))

$$\hat{Q}_{i,5} \equiv \int \hat{j}^0_{i,5}\,\mathrm{d}^3\boldsymbol{x} = \int \hat{q}^\dagger\gamma_5\frac{\tau_i}{2}\hat{q}\,\mathrm{d}^3\boldsymbol{x} \tag{18.22}$$

previously denoted by $\hat{T}^{(\frac{1}{2})}_{i,5}$ and (compare (12.101))

$$\hat{Q}_i = \int \hat{q}^\dagger\frac{\tau_i}{2}\hat{q}\,\mathrm{d}^3\boldsymbol{x} \tag{18.23}$$

previously denoted by $\hat{T}^{(\frac{1}{2})}_i$. As with all symmetries, it is interesting to discover the *algebra* of the generators, which are the six charges $\hat{Q}_i$, $\hat{Q}_{i,5}$ in this case.

Patient work with the anti-commutation relations for the operators in $\hat{q}(x)$ and $\hat{q}^\dagger(x)$ gives the results (problem 18.1)

$$[\hat{Q}_i, \hat{Q}_j] = \mathrm{i}\epsilon_{ijk}\hat{Q}_k \tag{18.24}$$
$$[\hat{Q}_i, \hat{Q}_{j,5}] = \mathrm{i}\epsilon_{ijk}\hat{Q}_{k,5} \tag{18.25}$$
$$[\hat{Q}_{i,5}, \hat{Q}_{j,5}] = \mathrm{i}\epsilon_{ijk}\hat{Q}_k. \tag{18.26}$$

Relation (18.24) has been seen before in (12.103) and simply states that the $\hat{Q}_i$'s obey an SU(2) algebra. A simple trick reduces the rather complicated algebra of (18.24)–(18.26) to something much simpler. Defining

$$\hat{Q}_{i,\mathrm{R}} = \tfrac{1}{2}(\hat{Q}_i + \hat{Q}_{i,5}) \qquad \hat{Q}_{i,\mathrm{L}} = \tfrac{1}{2}(\hat{Q}_i - \hat{Q}_{i,5}) \tag{18.27}$$

we find (problem 18.2)

$$[\hat{Q}_{i,\mathrm{R}}, \hat{Q}_{j,\mathrm{R}}] = \mathrm{i}\epsilon_{ijk}\hat{Q}_{k,\mathrm{R}} \tag{18.28}$$
$$[\hat{Q}_{i,\mathrm{L}}, \hat{Q}_{j,\mathrm{L}}] = \mathrm{i}\epsilon_{ijk}\hat{Q}_{k,\mathrm{L}} \tag{18.29}$$
$$[\hat{Q}_{i,\mathrm{R}}, \hat{Q}_{j,\mathrm{L}}] = 0. \tag{18.30}$$

The operators $\hat{Q}_{i,\mathrm{R}}, \hat{Q}_{i,\mathrm{L}}$ therefore behave like *two commuting (independent) angular momentum operators*, each obeying the algebra of SU(2). For this reason, the symmetry group of the combined symmetries (ii) and (iv) is called $\mathrm{SU(2)_{fL}} \times \mathrm{SU(2)_{fR}}$.

The decoupling effected by (18.27) has a simple interpretation. Referring to (18.22) and (18.23), we see that

$$\hat{Q}_{i,\mathrm{R}} = \int \hat{q}^\dagger \left(\frac{1+\gamma_5}{2}\right) \frac{\tau_i}{2} \hat{q} \, \mathrm{d}^3\boldsymbol{x} \tag{18.31}$$

and similarly for $\hat{Q}_{i,\mathrm{L}}$. But $((1 \pm \gamma_5)/2)$ are just the projection operators $P_{\mathrm{R,L}}$ introduced in section 12.3.2, which project out the chiral parts of any fermion field. Furthermore, it is easy to see that $P_\mathrm{R}^2 = P_\mathrm{R}$ and $P_\mathrm{L}^2 = P_\mathrm{L}$, so that $\hat{Q}_{i,\mathrm{R}}$ and $\hat{Q}_{i,\mathrm{L}}$ can also be written as

$$\hat{Q}_{i,\mathrm{R}} = \int \hat{q}_\mathrm{R}^\dagger \frac{\tau_i}{2} \hat{q}_\mathrm{R} \, \mathrm{d}^3\boldsymbol{x} \qquad \hat{Q}_{i,\mathrm{L}} = \int \hat{q}_\mathrm{L}^\dagger \frac{\tau_i}{2} \hat{q}_\mathrm{L} \, \mathrm{d}^3\boldsymbol{x} \tag{18.32}$$

where $\hat{q}_\mathrm{R} = ((1-\gamma_5)/2)q$, $\hat{q}_\mathrm{L} = ((1+\gamma_5)/2)\hat{q}$. In a similar way, the currents (18.20) and (18.21) can be written as

$$\hat{j}_i^\mu = \hat{j}_{i,\mathrm{R}}^\mu + \hat{j}_{i,\mathrm{L}}^\mu \qquad \hat{j}_{i,5}^\mu = \hat{j}_{i,\mathrm{R}}^\mu - \hat{j}_{i,\mathrm{L}}^\mu \tag{18.33}$$

where

$$\hat{j}_{i,\mathrm{R}} = \bar{\hat{q}}_\mathrm{R}\gamma^\mu \frac{\tau_i}{2} \hat{q}_\mathrm{R} \qquad \hat{j}_{i,\mathrm{L}}^\mu = \bar{\hat{q}}_\mathrm{L}\gamma^\mu \frac{\tau_i}{2} \hat{q}_\mathrm{L}. \tag{18.34}$$

Thus the $SU(2)_L$ and $SU(2)_R$ refer to the two chiral components of the fermion fields, which is why it is called *chiral symmetry*.

Under an infinitesimal SU(2) isospin transformation, $\hat{q}$ transforms by

$$\hat{q} \to \hat{q}' = (1 - \mathrm{i}\boldsymbol{\epsilon} \cdot \boldsymbol{\tau}/2)\hat{q} \tag{18.35}$$

while under an axial isospin transformation

$$\hat{q} \to \hat{q}' = (1 - \mathrm{i}\boldsymbol{\eta} \cdot \boldsymbol{\tau}/2\gamma_5)\hat{q}. \tag{18.36}$$

Multiplying (18.36) by $\gamma_5$ and adding the result to (18.35), we find that

$$\hat{q}'_{\mathrm{R}} = (1 - \mathrm{i}(\boldsymbol{\epsilon} + \boldsymbol{\eta}) \cdot \boldsymbol{\tau}/2)\hat{q}_{\mathrm{R}} \tag{18.37}$$

and similarly

$$\hat{q}'_{\mathrm{L}} = (1 - \mathrm{i}(\boldsymbol{\epsilon} - \boldsymbol{\eta}) \cdot \boldsymbol{\tau}/2)\hat{q}_{\mathrm{L}}. \tag{18.38}$$

Hence $\hat{q}_{\mathrm{R}}$ and $\hat{q}_{\mathrm{L}}$ transform quite independently,[2] which is why $[\hat{Q}_{i,\mathrm{R}}, \hat{Q}_{j,\mathrm{L}}] = 0$.

This formalism allows us to see immediately why (18.12) is chirally invariant: problem 18.3 verifies that $\hat{\mathcal{L}}_q$ can be written as

$$\hat{\mathcal{L}}_q = \bar{\hat{q}}_{\mathrm{R}}\mathrm{i}\hat{\not{D}}q_{\mathrm{R}} + \bar{\hat{q}}_{\mathrm{L}}\mathrm{i}\hat{\not{D}}\hat{q}_{\mathrm{L}} \tag{18.39}$$

which is plainly invariant under (18.37) and (18.38), since $\hat{D}$ is flavour-blind.

There is as yet no formal proof that this $SU(2)_L \times SU(2)_R$ chiral symmetry is spontaneously broken in QCD, though it can be argued that the larger symmetry $SU(3)_L \times SU(3)_R$—appropriate to three massless flavours—must be spontaneously broken (see Weinberg 1996, section 22.5). This is, of course, an issue that cannot be settled within perturbation theory (compare the comments after (17.132)). Numerical solutions of QCD on a lattice (see chapter 16) do provide strong evidence that quarks acquire significant dynamical ($SU(2)_{f5}$-breaking) mass.

Even granted that chiral symmetry is spontaneously broken in massless two-flavour QCD, how do we know that it breaks in such a way as to leave the isospin ('R+L') symmetry unbroken? A plausible answer can be given if we restore the quark mass terms via

$$\hat{\mathcal{L}}_m = m_{\mathrm{u}}\bar{\hat{u}}\hat{u} + m_{\mathrm{d}}\bar{\hat{d}}\hat{d} = \tfrac{1}{2}(m_{\mathrm{u}} + m_{\mathrm{d}})\bar{\hat{q}}\hat{q} + \tfrac{1}{2}(m_{\mathrm{u}} - m_{\mathrm{d}})\bar{\hat{q}}\tau_3\hat{q}. \tag{18.40}$$

Now

$$\bar{\hat{q}}\hat{q} = \bar{\hat{q}}_{\mathrm{L}}\hat{q}_{\mathrm{R}} + \bar{\hat{q}}_{\mathrm{R}}\hat{q}_{\mathrm{L}} \tag{18.41}$$

and

$$\bar{\hat{q}}\tau_3\hat{q} = \bar{\hat{q}}_{\mathrm{L}}\tau_3 q_{\mathrm{R}} + \bar{\hat{q}}_{\mathrm{R}}\tau_3\hat{q}_{\mathrm{L}}. \tag{18.42}$$

[2] We may set $\boldsymbol{\gamma} = \boldsymbol{\epsilon} + \boldsymbol{\eta}$ and $\boldsymbol{\delta} = \boldsymbol{\epsilon} - \boldsymbol{\eta}$.

Including these extra terms is somewhat analogous to switching on an external field in the ferromagnetic problem, which determines a preferred direction for the symmetry breaking. It is clear that neither of (18.41) and (18.42) preserves $SU(2)_L \times SU(2)_R$ since they treat the L and R parts differently. Indeed, from (18.37) and (18.38), we find that

$$\bar{\hat{q}}_L \hat{q}_R \to \bar{\hat{q}}'_L q'_R = \bar{\hat{q}}_L (1 + i(\boldsymbol{\epsilon} - \boldsymbol{\eta}) \cdot \boldsymbol{\tau}/2)(1 - i(\boldsymbol{\epsilon} + \boldsymbol{\eta}) \cdot \boldsymbol{\tau}/2)\hat{q}_R \quad (18.43)$$

$$= \bar{\hat{q}}_L \hat{q}_R - i\boldsymbol{\eta} \cdot \bar{\hat{q}}_L \boldsymbol{\tau} \hat{q}_R \quad (18.44)$$

and

$$\bar{\hat{q}}_R \hat{q}_L \to \bar{\hat{q}}_R \hat{q}_L + i\boldsymbol{\eta} \cdot \bar{\hat{q}}_R \boldsymbol{\tau} \hat{q}_L. \quad (18.45)$$

Equations (18.44) and (18.45) confirm that the term $\bar{\hat{q}}\hat{q}$ in (18.40) is invariant under the isospin part of $SU(2)_L \times SU(2)_R$ (since $\boldsymbol{\epsilon}$ is not involved) but not invariant under the axial isospin transformations parametrized by $\boldsymbol{\eta}$. The $\bar{\hat{q}}\tau_3\hat{q}$ term explicitly breaks the third component of isospin (resembling an electromagnetic effect) but its magnitude may be expected to be smaller than that of the $\bar{\hat{q}}\hat{q}$ term, being proportional to the difference of the masses, rather than their sum. This suggests that the vacuum will 'align' in such a way as to preserve isospin but break axial isospin.

## 18.2 Pion decay and the Goldberger–Treiman relation

We now discuss some of the rather surprising phenomenological implications of spontaneously broken chiral symmetry—specifically, the spontaneous breaking of the axial isospin symmetry. We start by ignoring any 'explicit' quark masses, so that the axial isospin current is conserved, $\partial_\mu \hat{j}^\mu_{i,5} = 0$. From sections 17.4 and 17.5 (suitably generalized) we know that this current has non-zero matrix elements between the vacuum and a 'Goldstone' state which, in our case, is the pion. We therefore set (cf (17.94))

$$\langle 0|\hat{j}^\mu_{i,5}(x)|\pi_j, p\rangle = ip^\mu f_\pi e^{-ip\cdot x}\delta_{ij} \quad (18.46)$$

where $f_\pi$ is a constant with dimensions of mass, which we expect to be related to a symmetry-breaking vev. The precise relation between $f_\pi$ and a vev depends on the dynamical theory (or model) being considered: for example, in the $\sigma$-model of section 18.3, in which $\hat{j}^\mu_{i,5}$ is given by (18.81) and the field $\sigma$ develops a vev given in (18.86), we find that $f_\pi = -v_\sigma$. Note that (18.46) is consistent with $\partial_\mu \hat{j}^\mu_{i,5} = 0$ if $p^2 = 0$, i.e. if the pion is massless.

We treat $f_\pi$ as a phenomenological parameter. Its value can be determined from the rate for the decay $\pi^+ \to \mu^+\nu_\mu$ by the following reasoning. In chapter 20 we shall learn that the effective weak Hamiltonian density for this low-energy *strangeness-non-changing semileptonic transition* is

$$\hat{\mathcal{H}}_W(x) = \frac{G_F}{\sqrt{2}} \cos\theta_C \bar{\hat{\psi}}_d(x)\gamma^\mu(1 - \gamma_5)\hat{\psi}_u(x)$$

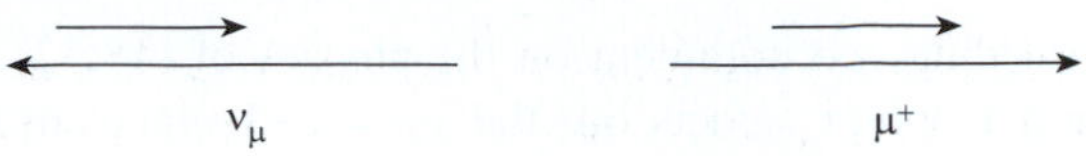


**Figure 18.2.** Helicities of *massless* leptons in $\pi^+ \to \mu^+\nu_\mu$ due to the 'V – A' interaction.

$$\times\,[\hat{\bar{\psi}}_{\nu_e}(x)\gamma_\mu(1-\gamma_5)\hat{\psi}_e(x)+\hat{\bar{\psi}}_{\nu_\mu}(x)\gamma_\mu(1-\gamma_5)\hat{\psi}_\mu(x)] \quad (18.47)$$

where $G_F$ is the Fermi constant and $\theta_C$ is the Cabibbo angle. Thus, the lowest-order contribution to the $S$-matrix is

$$\begin{aligned}
&-\mathrm{i}\langle\mu^+, p_1;\nu_\mu, p_2|\int \mathrm{d}^4x\,\hat{\mathcal{H}}_W(x)|\pi^+, p\rangle\\
&\quad=-\mathrm{i}\frac{G_F}{\sqrt{2}}\cos\theta_C\int \mathrm{d}^4x\,\langle\mu^+, p_1;\nu_\mu, p_2|\hat{\bar{\psi}}_{\nu_\mu}(x)\gamma_\mu(1-\gamma_5)\hat{\psi}_\mu(x)|0\rangle\\
&\quad\times\langle 0|\hat{\bar{\psi}}_d(x)\gamma^\mu(1-\gamma_5)\hat{\psi}_u(x)|\pi^+, p\rangle. \qquad (18.48)
\end{aligned}$$

The leptonic matrix element gives $\bar{u}_\nu(p_2)\gamma_\mu(1-\gamma_5)v_\mu(p_1)\mathrm{e}^{\mathrm{i}(p_1+p_2)\cdot x}$ in the usual way. For the pionic one, we note that

$$\hat{\bar{\psi}}_d(x)\gamma^\mu(1-\gamma_5)\hat{\psi}_u(x)=\hat{j}_1^\mu(x)-\mathrm{i}\hat{j}_2^\mu(x)-\hat{j}_{1,5}^\mu(x)+\mathrm{i}\hat{j}_{2,5}^\mu(x) \qquad (18.49)$$

from (18.20) and (18.21). Further, the currents $\hat{j}_i^\mu$ can have no matrix elements between the vacuum (which is a $0^+$ state) and the $\pi$ (which is $0^-$), by the following argument. From Lorentz invariance such a matrix element has to be a 4-vector. But since the initial and final parities are different, it would have to be an axial 4-vector.[3] However, the only 4-vector available is the pion's momentum $p^\mu$ which is an ordinary (not an axial) 4-vector. On the other hand, precisely for this reason the axial currents $\hat{j}_{i,5}^\mu$ do have a non-zero matrix element, as in (18.46). Noting that $|\pi^+\rangle=\frac{1}{\sqrt{2}}|\pi_1+\mathrm{i}\pi_2\rangle$, we find that

$$\begin{aligned}
\langle 0|\hat{\bar{\psi}}_d(x)\gamma^\mu(1-\gamma_5)\hat{\psi}_u(x)|\pi^+, p\rangle &= -\frac{\mathrm{i}}{\sqrt{2}}\langle 0|\hat{j}_{1,5}^\mu-\mathrm{i}\hat{j}_{2,5}|\pi_1+\mathrm{i}\pi_2\rangle \qquad (18.50)\\
&=\sqrt{2}p^\mu f_\pi \mathrm{e}^{-\mathrm{i}p\cdot x} \qquad (18.51)
\end{aligned}$$

so that (18.48) becomes

$$\mathrm{i}(2\pi)^4\delta^4(p_1+p_2-p)[-G_F\cos\theta_C\bar{u}_\nu(p_2)\gamma_\mu(1-\gamma_5)v(p_1)p^\mu f_\pi]. \qquad (18.52)$$

The quantity in brackets is, therefore, the invariant amplitude for the process, $\mathcal{M}$. Using $p=p_1+p_2$, we may replace $\not{p}$ in (18.52) by $m_\mu$, taking the neutrino to be massless.

[3] See page 284.

Before proceeding, we comment on the physics of (18.52). The $(1-\gamma_5)$ factor acting on a $v$ spinor selects out the $\gamma_5 = -1$ eigenvalue which, *if the muon was massless*, would correspond to positive helicity for the $\mu^+$ (compare the discussion in section 12.3.2). Likewise, taking the $(1-\gamma_5)$ through the $\gamma^0\gamma^\mu$ factor to act on $u_\nu^\dagger$, it selects the negative helicity neutrino state. Hence, the configuration is as shown in figure 18.2, so that the leptons carry off a net spin angular momentum. But this is forbidden, since the pion spin is zero. Hence, the amplitude vanishes for massless muons and neutrinos. Now the muon, at least, is not massless and some 'wrong' helicity is present in its wavefunction, in an amount proportional to $m_\mu$. This is why, as we have just remarked after (18.52), the amplitude is proportional to $m_\mu$. The rate is therefore proportional to $m_\mu^2$. This is a very important conclusion, because it implies that the rate to muons is $\sim (m_\mu/m_\mathrm{e})^2 \sim (400)^2$ times greater than the rate to electrons—a result which agrees with experiment, while grossly contradicting the naive expectation that the rate with the larger energy release should dominate. This, in fact, is one of the main indications for the 'vector–axial vector', or 'V – A', structure of (18.47), as we shall see in more detail in section 20.4.

Problem 18.4 shows that the rate computed from (18.52) is

$$\Gamma_{\pi\to\mu\nu} = \frac{G_\mathrm{F}^2 m_\mu^2 f_\pi^2 (m_\pi^2 - m_\mu^2)^2}{4\pi m_\pi^3}\cos^2\theta_\mathrm{C}. \tag{18.53}$$

Neglecting radiative corrections, this enables the value

$$f_\pi \simeq 93 \text{ MeV} \tag{18.54}$$

to be extracted.

Consider now another matrix element of $\hat{j}^\mu_{i,5}$, this time between nucleon states. Following an analysis similar to that in section 8.8 for the matrix elements of the electromagnetic current operator between nucleon states, we write

$$\begin{aligned}\langle \mathrm{N}, p'|\hat{j}^\mu_{i,5}(0)|\mathrm{N}, p\rangle \\ = \bar{u}(p')\left[\gamma^\mu\gamma_5 F_1^5(q^2) + \frac{\mathrm{i}\sigma^{\mu\nu}}{2M}q_\nu\gamma_5 F_2^5(q^2) + q^\mu\gamma_5 F_3^5(q^2)\right]\frac{\tau_i}{2}u(p)\end{aligned} \tag{18.55}$$

where the $F_i^5$'s are certain form factors, $M$ is the nucleon mass and $q = p - p'$. The spinors in (18.55) are understood to be written in flavour and Dirac space. Since (with massless quarks) $\hat{j}^\mu_{i,5}$ is conserved—that is $q_\mu\hat{j}^\mu_{i,5}(0) = 0$ (cf (8.99))—we find that

$$\begin{aligned}0 &= \bar{u}(p')[\not{q}\gamma_5 F_1^5(q^2) + q^2\gamma_5 F_3^5(q^2)]\frac{\tau_i}{2}u(p) \\ &= \bar{u}(p')[(\not{p} - \not{p}')\gamma_5 F_1^5(q^2) + q^2\gamma_5 F_3^5(q^2)]\frac{\tau_i}{2}u(p) \\ &= \bar{u}(p')[-2M\gamma_5 F_1^5(q^2) + q^2\gamma_5 F_3^5(q^2)]\frac{\tau_i}{2}u(p)\end{aligned} \tag{18.56}$$

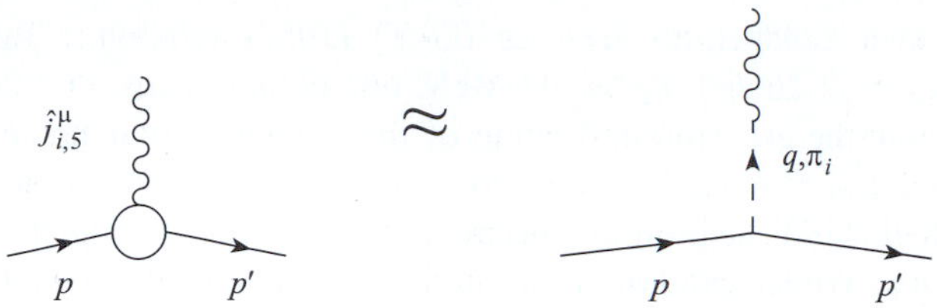


**Figure 18.3.** One pion intermediate state contribution to $F_3^5$.

using $\not p\gamma_5 = -\gamma_5\not p$ and the Dirac equations for $u(p), \bar{u}(p')$. Hence, the form factors $F_1^5$ and $F_3^5$ must satisfy

$$2MF_1^5(q^2) = q^2F_3^5(q^2). \tag{18.57}$$

Now the matrix element (18.55) enters into neutron $\beta$-decay (as does the matrix element of $\hat{j}_i^{\mu}(0)$). Here, $q^2 \simeq 0$ and (18.57) appears to predict, therefore, that either $M = 0$ (which is certainly not so) or $F_1^5(0) = 0$. But $F_1^5(0)$ can be measured in $\beta$ decay and is found to be approximately equal to 1.26: it is conventionally called $g_{\rm A}$. The only possible conclusion is that $F_3^5$ *must contain a part proportional to* $1/q^2$. Such a contribution can only arise from the propagator of a massless particle—which, of course, is the pion. This elegant physical argument, first given by Nambu (1960), sheds a revealing new light on the phenomenon of spontaneous symmetry breaking: the existence of the massless particle coupled to the symmetry current $\hat{j}^{\mu}_{i,5}$ 'saves' the conservation of the current.

We calculate the pion contribution to $F_3^5$ as follows. The process is pictured in figure 18.3. The pion-current matrix element is given by (18.46), and the (massless) propagator is $\mathrm{i}/q^2$. For the $\pi$–N vertex, the conventional Lagrangian is

$$\mathrm{i}g_{\pi\mathrm{NN}}\hat{\pi}_i\bar{\hat{N}}\gamma_5\tau_i\hat{N} \tag{18.58}$$

which is SU(2)$_{\rm f}$-invariant and parity-conserving since the pion field is a pseudoscalar and so is $\bar{N}\gamma_5N$. Putting these pieces together, the contribution of figure 18.3 to the current matrix element is

$$2g_{\pi\mathrm{NN}}\bar{u}(p')\gamma_5\frac{\tau_i}{2}u(p)\frac{\mathrm{i}}{q^2} - \mathrm{i}q^{\mu}f_{\pi} \tag{18.59}$$

and so

$$F_3^5(q^2) = \frac{1}{q^2}2g_{\pi\mathrm{NN}}f_{\pi} \tag{18.60}$$

from this contribution. Combining (18.57) with (18.60), we deduce

$$g_{\rm A} \equiv \lim_{q^2\to 0} F_1^5(q^2) = \frac{g_{\pi\mathrm{NN}}f_{\pi}}{M} \tag{18.61}$$

the well-known Goldberger–Treiman (G–T) (1958) relation. Taking $M = 939$ MeV, $g_{\mathrm{A}} = 1.26$ and $f_\pi = 93$ MeV, one obtains $g_{\pi\mathrm{NN}} \approx 12.7$, which is only 5% below the experimental value of this effective pion–nucleon coupling constant.

In the real world, the pion mass is not zero and neither are the 'explicit' quark masses $m_{\mathrm{u}}, m_{\mathrm{d}}$. With $m_{\mathrm{u}}$ and $m_{\mathrm{d}}$ reinstated, the equations of motion for the quark fields are

$$\mathrm{i}\hat{\not{D}}\hat{q} = \boldsymbol{m}\hat{q} \qquad -\mathrm{i}\hat{D}_\mu\hat{\bar{q}}\gamma^\mu = \hat{\bar{q}}\boldsymbol{m} \tag{18.62}$$

where

$$\boldsymbol{m} = \begin{pmatrix} m_{\mathrm{u}} & 0 \\ 0 & m_{\mathrm{d}} \end{pmatrix}. \tag{18.63}$$

We can re-calculate $\partial_\mu \hat{j}^\mu_{i,5}$ and find (problem 18.5) that

$$\partial_\mu \hat{j}^\mu_{i,5} = \mathrm{i}\hat{\bar{q}}\left\{\boldsymbol{m}, \frac{\tau_i}{2}\right\}\gamma_5\hat{q}. \tag{18.64}$$

Let us take the case $i = 1$, for example. Then

$$\{\boldsymbol{m}, \tau_1\} = (m_{\mathrm{u}} + m_{\mathrm{d}})\tau_1. \tag{18.65}$$

Now consider the matrix element

$$\partial_\mu \langle 0|\hat{j}^\mu_{i,5}(0)|\pi_1(p)\rangle = -p^2 f_\pi = \tfrac{1}{2}(m_{\mathrm{u}} + m_{\mathrm{d}})\langle 0|\mathrm{i}\,\hat{\bar{q}}\tau_1\gamma_5\hat{q}|\pi_1\rangle. \tag{18.66}$$

Since $p^2 = m_\pi^2$, we see that $m_\pi^2$ is proportional to the sum of quark masses and tends to zero as they do.

We can repeat the argument leading to the G–T relation but retaining $m_\pi^2 \neq 0$. Equation (18.46) tells us that $\partial_\mu \hat{j}^\mu_{i,5}/(m_\pi^2 f_\pi)$ behaves like a properly normalized pion field, at least when operating on a near mass-shell pion state. This means that the one-nucleon matrix element of $\partial_\mu \hat{j}^\mu_{i,5}$ is (cf (18.59))

$$2g_{\pi\mathrm{NN}}\bar{u}(p')\gamma_5\frac{\tau_i}{2}u(p)\frac{\mathrm{i}}{q^2 - m_\pi^2}m_\pi^2 f_\pi \tag{18.67}$$

while from (18.55) it is given by

$$\mathrm{i}\bar{u}(p')[-2M\gamma_5 F_1^5(q^2) + q^2\gamma_5 F_3^5(q^2)]\frac{\tau_i}{2}u(p). \tag{18.68}$$

Hence,

$$-2MF_1^5(q^2) + q^2 F_3^5(q^2) = \frac{2g_{\pi\mathrm{NN}}m_\pi^2 f_\pi}{q^2 - m_\pi^2}. \tag{18.69}$$

Also, in place of (18.60), we now have

$$F_3^5(q^2) = \frac{1}{q^2 - m_\pi^2}2g_{\pi\mathrm{NN}}f_\pi. \tag{18.70}$$

Equations (18.69) and (18.70) are consistent for $q^2 = m_\pi^2$ if

$$F_1^5(q^2 = m_\pi^2) = g_{\pi\mathrm{NN}} f_\pi / M. \tag{18.71}$$

$F_1^5(q^2)$ varies only slowly from $q^2 = 0$ to $q^2 = m_\pi^2$, since it contains no rapidly varying pion pole contribution, and so we recover the G–T relation again.

Amplitudes involving *two* 'Goldstone' pions can be calculated by an extension of these techniques. We refer the interested reader to Georgi (1984).

We now turn to another example of a phenomenological model exhibiting spontaneously broken axial isospin symmetry, this time realized in terms of hadronic (nucleon and pion) degrees of freedom, rather than quarks. It will be a somewhat more complicated model than that of section 17.4, though similar to it in that the spontaneous breaking is put in 'by hand' via a suitable potential. As we will see, it effectively embodies many of the preceding results.

## 18.3 The linear and nonlinear $\sigma$-models

The linear $\sigma$-model involves a massless fermion isodoublet $\hat{\psi}$ (which will be identified with the nucleon—its mass being generated spontaneously, as we shall see) and a massless pseudoscalar isotriplet $\hat{\boldsymbol{\pi}}$ (the pions). There is also a scalar field $\hat{\sigma}$ which is an isoscalar. The Lagrangian is (with $\lambda > 0$)

$$\begin{aligned}\hat{\mathcal{L}}_\sigma = {} & \bar{\hat{\psi}} \mathrm{i} \partial\!\!\!/ \hat{\psi} + \mathrm{i} g_{\pi\mathrm{NN}} \bar{\hat{\psi}} \boldsymbol{\tau} \gamma_5 \hat{\psi} \cdot \boldsymbol{\pi} + g_{\pi\mathrm{NN}} \bar{\hat{\psi}} \hat{\psi} \hat{\sigma} + \tfrac{1}{2} \partial_\mu \hat{\boldsymbol{\pi}} \cdot \partial^\mu \bar{\hat{\boldsymbol{\pi}}} \\ & + \tfrac{1}{2} \partial_\mu \hat{\sigma} \partial^\mu \hat{\sigma} - \tfrac{1}{2} \mu^2 (\hat{\sigma}^2 + \hat{\boldsymbol{\pi}}^2) - \tfrac{1}{4} \lambda (\hat{\sigma}^2 + \hat{\boldsymbol{\pi}}^2)^2. \end{aligned} \tag{18.72}$$

We have seen all the different parts of this before: the massless fermions, the $\boldsymbol{\pi}$–$\psi$ coupling as in (18.58), an analogous $\sigma$–$\psi$ coupling (here with the same coupling constant), and $\boldsymbol{\pi}$ and $\sigma$ fields with a symmetrical mass parameter $\mu^2$ and a quartic potential.

What are the global symmetries of (18.72)? We can at once infer that it is invariant under global isospin transformations if $\hat{\psi}$ is an isodoublet, $\hat{\boldsymbol{\pi}}$ an isotriplet and $\hat{\sigma}$ an isoscalar, transforming by

$$\hat{\psi} \to \hat{\psi} = (1 - \mathrm{i}\boldsymbol{\epsilon} \cdot \boldsymbol{\tau}/2)\hat{\psi} \tag{18.73}$$

$$\hat{\boldsymbol{\pi}} \to \hat{\boldsymbol{\pi}}' = \boldsymbol{\pi} + \boldsymbol{\epsilon} \times \hat{\boldsymbol{\pi}} \qquad \text{(cf (12.64))} \tag{18.74}$$

$$\hat{\sigma} \to \hat{\sigma}' = \hat{\sigma} \tag{18.75}$$

under an infinitesimal $\mathrm{SU}(2)_\mathrm{f}$ transformation. The associated symmetry current is

$$\hat{j}_i^{\mu\,(\sigma)} = \tfrac{1}{2} \bar{\hat{\psi}} \gamma^\mu \tau_i \hat{\psi} + (\hat{\boldsymbol{\pi}} \times \partial^\mu \hat{\boldsymbol{\pi}})_i. \tag{18.76}$$

The first term of (18.76) is as in (18.21) and the second is equivalent to (12.124). The corresponding charges

$$\hat{Q}_i^{(\sigma)} = \int \hat{j}_i^{0\,(\sigma)} \, \mathrm{d}^3 \boldsymbol{x} \tag{18.77}$$

are constants of the motion and obey the SU(2) algebra (18.24). Note that all these algebraic results hold independently of the specific realization of the operators in terms of the fields in the model under consideration (quarks on the one hand, pions and nucleons on the other).

However, (18.72) is also invariant under a further set of transformations (see problem 18.6), namely

$$\hat{\psi} \to \hat{\psi}' = (1 - \mathrm{i}\boldsymbol{\eta} \cdot \boldsymbol{\tau}/2\,\gamma_5)\hat{\psi} \tag{18.78}$$

$$\hat{\boldsymbol{\pi}} \to \hat{\boldsymbol{\pi}}' = \boldsymbol{\pi} + \boldsymbol{\eta}\sigma \tag{18.79}$$

$$\hat{\sigma} \to \hat{\sigma}' = \hat{\sigma} - \boldsymbol{\eta} \cdot \hat{\boldsymbol{\pi}} \tag{18.80}$$

where $\boldsymbol{\eta}$ is a second set of three infinitesimal parameters. Transformation (18.78) is the same as (18.36) and is, therefore, again an axial isospin transformation on the doublet $\hat{\psi}$. This suggests we call the second set of currents $\hat{j}^{\mu}_{i,5}$, where now

$$\hat{j}^{\mu}_{i,5} = \tfrac{1}{2}\bar{\hat{\psi}}\gamma^{\mu}\gamma_5\tau_i\hat{\psi} + (\hat{\sigma}\,\partial^{\mu}\pi_i - \hat{\pi}_i\partial^{\mu}\hat{\sigma}). \tag{18.81}$$

Again, the first term is as in (18.20) and there is a new 'mesonic' piece. It can easily be verified that $\hat{j}^{\mu}_{i,5}$ would not be conserved if we added an explicit mass for $\hat{\psi}$. Remarkably, the charges

$$\hat{Q}^{(\sigma)}_{i,5} = \int \hat{j}^{0\,(\sigma)}_{i,5}\,\mathrm{d}^3\boldsymbol{x} \tag{18.82}$$

and $\hat{Q}^{(\sigma)}_{i}$ satisfy relations of the form (18.24)–(18.26). Thus, once again, we have a model of interacting fields with a global $\mathrm{SU(2)_L \times SU(2)_R}$ symmetry.

As far as the fermion field $\hat{\psi}$ is concerned, we know that the 'L' and 'R' refer to the components of different chirality. But how can we understand this SU(2)×SU(2) structure for the meson fields, for which of course no $\gamma_5$ matrix can enter? Just as the algebra of SU(2) is the same as that of the generators of 3D rotations (section 12.1.1), so the algebra of SU(2)×SU(2) turns out to be the same as that of the generators of rotations in a *four*-dimensional Euclidean space—here, of course, an 'internal' space involving the field components (see appendix M, section M.4.3). If we label the four directions as 1, 2, 3 and 4, we can imagine rotations in the planes 12, 13 and 23 which would be 'spatial' rotations, leading to an SU(2) algebra. But there are also rotations in the 14, 24 and 34 directions, which are analogous to spacetime (velocity) transformations in special relativity. This makes six 'rotations' in all, which is the same number of generators as the three $\hat{Q}^{(\sigma)}_{i}$'s together with the three $\hat{Q}^{(\sigma)}_{i,5}$'s. Of course, this by itself by no means proves that the algebra of the generators of SO(4) (the special orthogonal group in 4D—i.e. the group of 4D rotations) is the same as SU(2)×SU(2). But the SO(4) symmetry of (18.72) is apparent, at least in the meson sector, if we regard $(\hat{\sigma}, \hat{\boldsymbol{\pi}})$ as being the four components of a '4-vector'. The transformations of SO(4) preserve the (length)$^2$ of 4-vectors—in this case, therefore, of $\sigma^2 + \hat{\boldsymbol{\pi}}^2$. Just this combination is visible in the potential terms of (18.72), and we can easily verify from (18.79) and (18.80) that, indeed, (to first order in $\boldsymbol{\eta}$) $\hat{\sigma}'^2 + \hat{\boldsymbol{\pi}}'^2 = \sigma^2 + \hat{\boldsymbol{\pi}}^2$ as required. Finally, note that (18.79) and (18.80) are analogous to an infinitesimal velocity transformation in relativity, with $\hat{\sigma} \to t$, $\hat{\boldsymbol{\pi}} \to \boldsymbol{x}$, and $t^2 - \boldsymbol{x}^2$ invariant rather than $\hat{\sigma}^2 + \hat{\boldsymbol{\pi}}^2$. So (18.74) and (18.75) tell us how $\hat{\boldsymbol{\pi}}$ and $\hat{\sigma}$ transform under ordinary $\mathrm{SU(2)_f}$, while (18.79) and (18.80) record their transformations under the other SU(2), which we may reasonably call $\mathrm{SU(2)_{f5}}$ in view of (18.78). Then the associated 'L' and 'R' generators can be found by writing $\hat{Q}^{(\sigma)}_{i,\mathrm{L}} = \frac{1}{2}(\hat{Q}^{(\sigma)}_i - \hat{Q}^{(\sigma)}_{i,5})$, $\hat{Q}^{(\sigma)}_{i,\mathrm{R}} = \frac{1}{2}[\hat{Q}^{(\sigma)}_i + \hat{Q}^{(\sigma)}_{i,5}]$ as before. These relations provide us with the 'L' and 'R' transformation law for the meson fields in this model.

Thus far we have supposed that the parameter $\mu^2$ in (18.72) is positive, representing a normal mass parameter. The symmetry is then unbroken, the ground state having $\langle 0|\hat{\sigma}|0\rangle = \langle 0|\hat{\boldsymbol{\pi}}|0\rangle = 0$. We now consider the symmetry breaking case $\mu^2 < 0$, as in (17.77) and (17.97). The classical potential in (18.72) now becomes

$$V(\boldsymbol{\pi}, \sigma) = -\tfrac{1}{2}\mu^2(\sigma^2 + \boldsymbol{\pi}^2) + \tfrac{1}{4}\lambda(\sigma^2 + \boldsymbol{\pi}^2)^2 \tag{18.83}$$

where the 'new' $\mu^2$ in (18.83) is positive. This potential has a minimum when

$$\boldsymbol{\pi}^2 + \sigma^2 = v_\sigma^2 \tag{18.84}$$

with

$$v_\sigma = (\mu^2/\lambda)^{\frac{1}{2}}. \tag{18.85}$$

(18.84) generalizes the circular minimum of figure 17.2 to the surface of a sphere in 4D; it is also, in fact, exactly analogous to (17.102)—a point to which we shall eventually return in chapter 22. As in these previous cases, we interpret (18.84) as $\langle 0|\hat{\boldsymbol{\pi}}^2 + \hat{\sigma}^2|0\rangle = v_\sigma^2$, and we need to choose one particular ground state before we can get a proper particle interpretation. We choose

$$\langle 0|\hat{\boldsymbol{\pi}}|0\rangle = 0 \qquad \langle 0|\hat{\sigma}|0\rangle = v_\sigma. \tag{18.86}$$

Then $\hat{\boldsymbol{\pi}}$ will have a standard expansion in terms of $\hat{a}$'s and $\hat{a}^\dagger$'s, while for $\hat{\sigma}$ we need to set

$$\hat{\sigma} = v_\sigma + \hat{\sigma}' \tag{18.87}$$

and then expand $\hat{\sigma}'$ in terms of creation and annihilation operators. Introducing (18.87) into (18.72), we find (problem 18.7) that $\hat{\mathcal{L}}_\sigma$ becomes

$$\begin{aligned}\hat{\mathcal{L}}_{\sigma'} = {} & \bar{\hat{\psi}}(\mathrm{i}\not{\partial} + g_{\pi\mathrm{NN}}v_\sigma)\hat{\psi} + \mathrm{i}g_{\pi\mathrm{NN}}\bar{\hat{\psi}}\boldsymbol{\tau}\gamma_5\hat{\psi}\cdot\hat{\boldsymbol{\pi}} + g_{\pi\mathrm{NN}}\bar{\hat{\psi}}\hat{\psi}\hat{\sigma}' \\ & + \tfrac{1}{2}\partial_\mu\hat{\boldsymbol{\pi}}\cdot\partial^\mu\hat{\boldsymbol{\pi}} + \tfrac{1}{2}\partial_\mu\hat{\sigma}'\partial^\mu\hat{\sigma}' + \mu^2\hat{\sigma}'^2 - \lambda v_\sigma\hat{\sigma}'(\hat{\sigma}'^2 + \hat{\boldsymbol{\pi}}^2) \\ & - \tfrac{1}{4}\lambda(\hat{\sigma}'^2 + \hat{\boldsymbol{\pi}}^2)^2 + \text{constant}.\end{aligned} \tag{18.88}$$

The spontaneous breaking typified by (18.87) has, therefore, induced the following results:

(i) the fermion has acquired a mass $-g_{\pi NN}v_\sigma$, proportional to the symmetry breaking parameter $v_\sigma$;
(ii) the $\hat{\boldsymbol{\pi}}$'s remain massless, being the three Goldstone modes 'perpendicular' to the one selected out by the symmetry-breaking condition (18.86);
(iii) the $\hat{\sigma}'$ field has a mass $\sqrt{2}\mu$, corresponding to oscillations in the 'radial' direction in figure 17.2;
(iv) there are new trilinear couplings between $\hat{\boldsymbol{\pi}}$ and $\sigma'$, proportional to $v_\sigma$; and
(v) the $\mathrm{SU(2)_f}$ symmetry of (18.72) is preserved, since the vacuum choice (18.86) respects it, but $\mathrm{SU(2)_{f5}}$ is spontaneously broken.

We may therefore regard $\hat{\mathcal{L}}_\sigma$ as some kind of 'effective Lagrangian' embodying the $\mathrm{SU(2)_L \times SU(2)_R}$ symmetry of QCD, broken spontaneously in just the same way as we assumed for QCD. The empirical consequences—massless pions, a massive nucleon—are the same but, of course, there is no real dynamical explanation of the symmetry breakdown here, just a 'by hand' choice of the sign of $\mu^2$ in (18.72).

This model can be easily modified to include a finite mass for the pions. In the QCD case, we saw that a quark mass term broke the full $\mathrm{SU(2) \times SU(2)}$ symmetry explicitly, while leaving $\mathrm{SU(2)_f}$ intact. In the same way, the addition of a term $+c\hat{\sigma}$ to $\hat{\mathcal{L}}_\sigma$ will have the same effect (again, it is analogous to an 'alignment field'). One quickly verifies (problem 18.8) that now $\partial_\mu \hat{j}^\mu_{i,5}$ is no longer zero but is given by

$$\partial_\mu \hat{j}^\mu_{i,5} = -c\hat{\pi}_i. \tag{18.89}$$

Thus, just as in the previous section, $\partial_\mu \hat{j}^\mu_{i,5}$ is proportional to the pion field. In fact, for consistency with (18.46), we should have

$$c = -m_\pi^2 f_\pi. \tag{18.90}$$

We can check (18.90) (at least to tree level in the interactions of $\hat{\mathcal{L}}_\sigma + c\hat{\sigma}$) as follows. With the explicit symmetry-breaking addition $c\hat{\sigma}$, the minimum (18.84) gets shifted to a new point such that $\langle 0|\hat{\sigma}|0\rangle$—which we still call $v_\sigma$—satisfies

$$v_\sigma(-\mu^2 + \lambda v_\sigma^2) = c. \tag{18.91}$$

Note that $v_\sigma$ returns to $(\mu^2/\lambda)^{1/2}$ for $c \to 0$. In addition, there is a pion mass term $-\frac{1}{2}(-\mu^2 + \lambda^2 v_\sigma^2)\hat{\boldsymbol{\pi}}^2$. Previously this was of course zero (from (18.85)), but now from (18.91) we can identify $m_\pi^2$ as $c/v_\sigma$. This will be consistent with (18.90) if our symmetry breaking parameter $v_\sigma$ is identified with $-f_\pi$. In that case, from item (i) we learn that the nucleon mass is $g_{\pi \mathrm{NN}} f_\pi$, and we recover a G–T type relation (at tree level) with $g_\mathrm{A} = 1$.

We can ask to what extent the rather simple Lagrangian (18.72) (with the $\mu^2$ term as in (18.83)) describes other features of low-energy interactions among pions and nucleons. The $\pi$–N interaction itself leads, of course, precisely to the 'one pion exchange' potential between two nucleons, as postulated by Yukawa, but now incorporating the conservation of isospin. Furthermore, this $\pi$–N theory is renormalizable, although in view of the magnitude of $g_{\pi \mathrm{NN}}$, perturbation theory is of little value. The main phenomenological problem with (18.72) is that there is no plausible candidate, among the observed mesons with masses below 1 GeV, for the '$\sigma$' ($0^+$) meson. We can actually get rid of $\sigma$ by supposing that its mass $\sqrt{2}\mu$ is very large indeed but in such a way that the ratio $v_\sigma = (\mu^2/\lambda)^{1/2}$ remains finite. This implies that $\lambda$ also becomes very large, rising as the square of the scaling of $\mu$. However, the value of the potential at the minimum is $-\frac{1}{4}\frac{\mu^4}{\lambda}$, which becomes very large and negative. Thus in terms of a picture such as figure 17.2, the potential has a very deep and narrow minimum and (assuming a semi-classical picture) the fields are effectively constrained to lie on the 'chiral circle'

$$\hat{\sigma}^2 + \hat{\boldsymbol{\pi}}^2 = f_\pi^2. \tag{18.92}$$

The result of imposing (18.92) is rather remarkable. Let us consider now just the meson sector. The potential terms disappear altogether from (18.72) and we are left with pions interacting via the interaction

$$\hat{\mathcal{L}}_{\mathrm{nl}\sigma\mathrm{m}} = \tfrac{1}{2}\partial_\mu\left(\sqrt{f_\pi^2 - \hat{\boldsymbol{\pi}}^2}\right)\partial^\mu\left(\sqrt{f_\pi^2 - \hat{\boldsymbol{\pi}}^2}\right) \tag{18.93}$$

$$= \frac{1}{2}\frac{(\hat{\boldsymbol{\pi}}\cdot\partial_\mu\hat{\boldsymbol{\pi}})(\hat{\boldsymbol{\pi}}\cdot\partial^\mu\hat{\boldsymbol{\pi}})}{(f_\pi^2 - \hat{\boldsymbol{\pi}}^2)}. \tag{18.94}$$

We interpret the denominator in (18.94) via its expansion

$$\hat{\mathcal{L}}_{\mathrm{nl}\sigma\mathrm{m}} = \frac{1}{2f_\pi^2}(\hat{\boldsymbol{\pi}}\cdot\partial\hat{\boldsymbol{\pi}})(\hat{\boldsymbol{\pi}}\cdot\partial\hat{\boldsymbol{\pi}})(1 - \hat{\boldsymbol{\pi}}^2/f_\pi^2)^{-1} \tag{18.95}$$

$$= \frac{1}{2f_\pi^2}(\hat{\boldsymbol{\pi}}\cdot\partial_\mu\hat{\boldsymbol{\pi}})(\hat{\boldsymbol{\pi}}\cdot\partial^\mu\hat{\boldsymbol{\pi}}) + O(\hat{\boldsymbol{\pi}}^6). \tag{18.96}$$

The first term in (18.96) describes a pion–pion scattering process of the form $\pi + \pi \to \pi + \pi$, for which it makes a quite specific prediction, since $f_\pi$ is known. The result was first given by Weinberg (1966), using a different technique, and is consistent with experiment (see Donoghue *et al* (1992) for a review).

Relation (18.96) invites a number of comments. First, we note that the effective interaction $\frac{1}{2f_\pi^2}(\boldsymbol{\pi}\cdot\partial\boldsymbol{\pi})^2$ is not renormalizable, since it has a coefficient with dimension $(\text{mass})^{-2}$. However, the discussion in section 11.8 showed that such a Lagrangian could still be useful, even in loops, provided one worked at energies below the scale set by the dimensional coupling. Here that scale is $f_\pi = 93$ MeV (or perhaps this multiplied by numbers such as $2\pi$, if we are lucky). At any rate, we expect the theory to work only for energies not too far from threshold. Nevertheless, it is striking that *symmetry conditions have determined the low-energy dynamics of the Goldstone modes.* This is a general feature and clearly a most important one. It opens up a large field of 'effective Lagrangians' for low-energy hadronic physics (Donoghue *et al* 1992).

Second, it is interesting that the effective $\pi$–$\pi$ interaction involves two *derivatives*, so that the corresponding Feynman amplitudes for $\pi(p_1) + \pi(p_2) \rightarrow \pi(p_1') + \pi(p_2')$ contain two powers of the momenta, which from Lorentz invariance must appear in the form of $s = (p_1 + p_2)^2$, $t = (p_1 - p_1')^2$ or $u = (p_1 - p_2')^2$, where $p_1 + p_2 = p_1' + p_2'$ expresses 4-momentum conservation. They therefore vanish (for $m_\pi^2 = 0$) at the points $s = 0$, $t = 0$ or $u = 0$. On the other hand, derivatives are absent from the meson sector of the 'spontaneously broken' model (18.88), suggesting a contradiction. In particular, the interaction $\frac{1}{4}\lambda(\hat{\boldsymbol{\pi}}^2)^2$ would seem to lead to a constant (non-vanishing) contribution as the momenta went to zero. However, there will also be contributions from $\sigma'$ exchange of the form $A/(q^2 - m_\sigma^2)$ where $m_\sigma^2 = 2\mu^2$ and $A$ is proportional to $\lambda^2 v_\sigma^2$. Expanding this in powers of $q^2/m_\sigma^2$, we find a leading term $-A/m_\sigma^2 \sim -\lambda^2 v_\sigma^2/m_\sigma^2 \sim -\lambda^2 v_\sigma^2/\mu^2 \sim -\lambda$. A proper calculation shows that indeed such terms exactly cancel the $\frac{1}{4}\lambda(\hat{\boldsymbol{\pi}}^2)^2$ ones (see, for example, Donoghue *et al* (1992)), leaving the leading contribution proportional to $q^2$ (where $q$ is any of the possible momentum transfers).

Despite the low-energy success of (18.96) (which is called the 'nonlinear $\sigma$-model' in this context), it fails to account for prominent phenomena as the energy scale rises into the 500–1000 MeV region. In particular, there is no sign of the $T = 1$, $J^P = 1^- \pi - \pi$ resonance called the $\rho$ (see section 9.5), with a mass of 770 MeV. The situation is no better with the 'linear $\sigma$-model' of (18.88). The importance of $\rho$-meson exchange in hadronic dynamics was first stressed by Sakurai (1960) in his 'vector meson dominance' theory. It is phenomenologically rather successful but it has not yet been possible to derive it directly from QCD. A combination of the low-energy Goldstone mode dynamics and the (at present) phenomenological $\rho$-meson contribution provides a good representation of non-strange mesonic dynamics below about 1 GeV.

## 18.4 Chiral anomalies

In all our discussions of symmetries so far—unbroken, approximate and spontaneously broken—there is one result on which we have relied and never queried. We refer to Noether's theorem, as discussed in section 12.3.1. This states that for every continuous symmetry of a Lagrangian, there is a corresponding conserved current. We demonstrated this result in some special cases, but we have now to point out that while it is undoubtedly valid at the level of the *classical* Lagrangian and field equations, we did not investigate whether quantum corrections might violate the classical conservation law. This can, in fact, happen and when it does the afflicted current (or its divergence) is said to be 'anomalous' or to contain an 'anomaly'. General analysis shows that anomalies occur in renormalizable theories of fermions coupled to both vector and axial vector currents. In particular, therefore, we may expect an anomaly when we introduce electromagnetism into our chiral models (such as the linear $\sigma$-model), since then

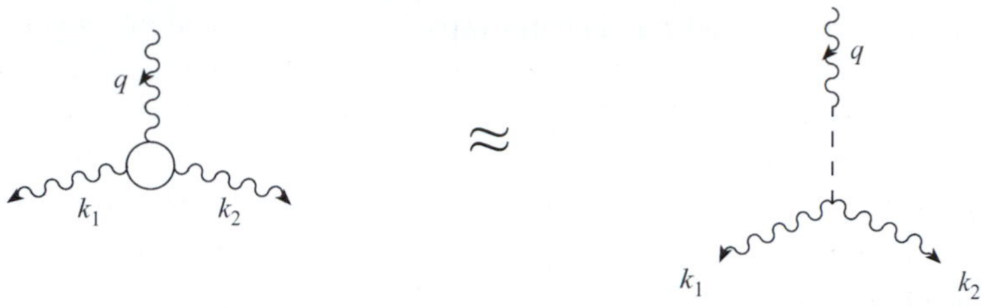


**Figure 18.4.** The amplitude considered in (18.97) and the one-pion intermediate state contribution to it.

both axial vector and vector symmetry currents will be involved. This is an example of a *chiral anomaly*, a typical illustration of which is provided by the calculation of the rate for $\pi^0 \to \gamma\gamma$, to be discussed shortly.

One way of understanding how anomalies arise is through consideration of the renormalization process, which is in general necessary once we get beyond the classical ('tree level') approximation. As we saw in volume 1, this will invariably entail some *regularization* of divergent integrals. But the specific example of the $O(e^2)$ photon self-energy studied in section 11.3 showed that a simple cut-off form of regularization already violated the current conservation (or gauge invariance) condition (11.21). In that case, it was possible to find alternative regularizations which respected electromagnetic current conservation and were satisfactory. Anomalies arise when *both* axial and vector symmetry currents are present, since it is not possible to find a regularization scheme which preserves both vector and axial vector current conservation (Adler 1970, Jackiw 1972, Adler and Bardeen 1969).

The need for particular care in the calculation of $\pi^0 \to \gamma\gamma$ was first recognized by Schwinger (1951), using a different approach. A full exposition of the anomaly in the axial vector current in spinor electrodynamics was first given by Adler (1969) and the occurrence of the anomaly in the $\sigma$-model (with electromagnetic interactions) was pointed out by Bell and Jackiw (1969). A more modern non-perturbative perspective is provided by Peskin and Schroeder (1995, chapter 19).

We shall not attempt an extended discussion of this technical subject. But we do want to alert the reader to the existence of these anomalies; to indicate how they arise in one simple model; and to explain why, in some cases, they are to be welcomed, while in others they must be eliminated.

We consider the classic case of $\pi^0 \to \gamma\gamma$, in the context of spontaneously broken global chiral symmetry with massless quarks and pions. The axial isospin current $\hat{j}^{\mu}_{i,5}(x)$ should then be conserved, but we shall see that this implies that the amplitude for $\pi^0 \to \gamma\gamma$ must vanish, as first pointed out by Veltman (1967) and Sutherland (1967). We begin by writing the matrix element of $\hat{j}^{\mu}_{3,5}(x)$ between the vacuum and a $2\gamma$ state, in momentum space, as

$$\int \mathrm{d}^4 x\, \mathrm{e}^{-\mathrm{i}q\cdot x} \langle \gamma, k_1, \epsilon_1; \gamma, k_2, \epsilon_2 | \hat{j}^{\mu}_{3,5}(x) | 0 \rangle$$
$$= (2\pi)^4 \delta^4(k_1 + k_2 - q) \epsilon^*_{1\nu}(k_1) \epsilon^*_{2\lambda}(k_2) \mathcal{M}^{\mu\nu\lambda}(k_1, k_2). \qquad (18.97)$$

As in figure 18.3, one contribution to $\mathcal{M}^{\mu\nu\lambda}$ has the form (constant$/q^2$) due to the massless $\pi^0$ propagator, shown in figure 18.4. This is because, once again, when chiral symmetry is spontaneously broken, the axial current connects the pion state to the vacuum, as described by the matrix element (18.46). The contribution of the process shown in figure 18.4 to $\mathcal{M}^{\mu\nu\lambda}$ is then

$$\mathrm{i}q^{\mu} f_{\pi} \frac{\mathrm{i}}{q^2} \mathrm{i}A\epsilon^{\nu\lambda\alpha\beta} k_{1\alpha} k_{2\beta} \qquad (18.98)$$

where we have parametrized the $\pi^0 \to \gamma\gamma$ amplitude as $A\epsilon^{\nu\lambda\alpha\beta}\epsilon^*_{1\nu}(k_1)\epsilon^*_{2\lambda}(k_2)$ $k_{1\alpha}k_{2\beta}$. Note that this automatically incorporates electromagnetic gauge invariance (the amplitude vanishes when the polarization vector of either photon is replaced by its 4-momentum, due to the anti-symmetry of the $\epsilon$ symbol), and it is symmetrical under interchange of the photon labels. Now consider replacing $\hat{j}^{\mu}_{3,5}(x)$ in (18.97) by $\partial_{\mu}\hat{j}^{\mu}_{3,5}(x)$, which should be zero. A partial integration in (18.97) then shows that this implies that

$$q_{\mu}\mathcal{M}^{\mu\nu\lambda} = 0 \qquad (18.99)$$

which with (18.98) implies that $A = 0$, and hence that $\pi^0 \to \gamma\gamma$ is forbidden. It is important to realize that all other contributions to $\mathcal{M}^{\mu\nu\lambda}$, apart from the $\pi^0$ one shown in figure 18.4, will *not* have the $1/q^2$ factor in (18.98) and will, therefore, give a vanishing contribution to $q_{\mu}\mathcal{M}^{\mu\nu\lambda}$ at $q^2 = 0$ which is the on-shell point for the (massless) pion.

It is, of course, true that $m_{\pi}^2 \neq 0$. But estimates (Adler 1969) of the consequent corrections suggest that the predicted rate of $\pi^0 \to \gamma\gamma$ for real $\pi^0$'s is far too small. Consequently, there is a problem for the hypothesis of spontaneously broken (approximate) chiral symmetry.

In such a situation, it is helpful to consider a detailed calculation performed within a specific model. In the present case, we want a model which exemplifies spontaneously broken chiral symmetry, so the Lagrangian $\hat{\mathcal{L}}_{\sigma}$ of (18.72) is an obvious choice, when enlarged to include electromagnetism in the usual gauge-invariant way. For our purposes, it will be sufficient to simplify (18.72) so as to include only one fermion of charge $e$ (the proton) and two mesons, the $\sigma$ and $\pi_3$. This was the model considered by Bell and Jackiw (1969) and also by Adler (1969). It is also effectively the model used long before, by Steinberger (1949), in the first calculation of the $\pi^0 \to \gamma\gamma$ rate. To order $\alpha$, there are two graphs to consider, shown in figure 18.5(*a*) and (*b*). It turns out that the fermion loop integral is actually convergent: details of its evaluation may be found in

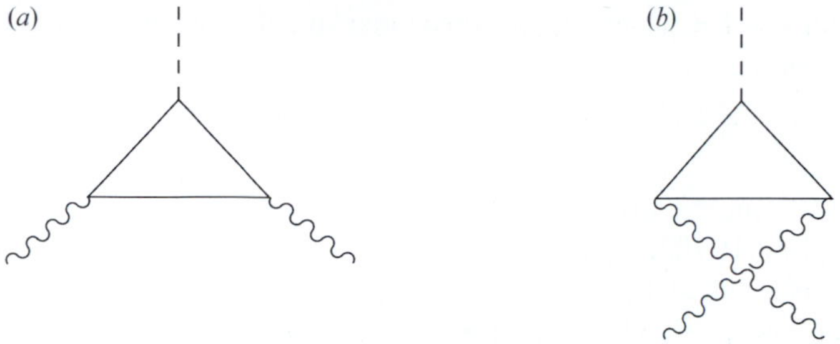


**Figure 18.5.** The two $O(\alpha)$ graphs contributing to $\pi^0 \to \gamma\gamma$ decay in the simplified version of (18.72).

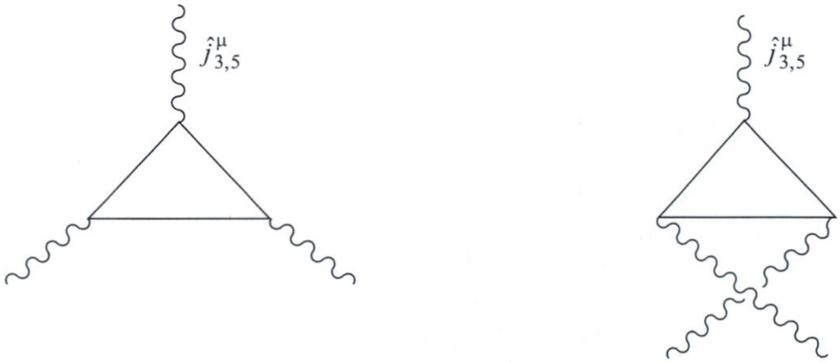


**Figure 18.6.** $O(\alpha)$ contributions to the matrix element in (18.97).

Itzykson and Zuber (1980, section 11.5.2). In the limit $q^2 \to 0$, the result is (using $g/m = f_\pi^{-1}$)

$$A = \frac{e^2}{4\pi^2 f_\pi} \tag{18.100}$$

where $A$ is the $\pi^0 \to \gamma\gamma$ amplitude introduced earlier. Problem 18.9 evaluates the $\pi^0 \to \gamma\gamma$ rate using (18.100) to give

$$\Gamma(\pi^0 \to 2\gamma) = \frac{\alpha^2}{64\pi^3} \frac{m_\pi^3}{f_\pi^2}. \tag{18.101}$$

(18.101) is in very good agreement with experiment.

In principle, various possibilities now exist. But a careful analysis of the 'triangle' graph contributions to the matrix element $\mathcal{M}^{\mu\nu\lambda}$ of (18.97), shown in figure 18.6, reveals that the fault lies in assuming that a regularization exists such that for these amplitudes the conservation equation $q_\mu \langle \gamma\gamma | \hat{j}^{\mu}_{3,5}(0)|0\rangle = 0$ can be maintained, at the same time as electromagnetic gauge invariance. In fact, no such regularization can be found. When the amplitudes of figure 18.6 are calculated using an (electromagnetic) gauge-invariant procedure, one finds a non-zero result for $q_\mu \langle \gamma\gamma | \hat{j}^{\mu}_{3,5}(0)|0\rangle$ (again the details are given in Itzykson and Zuber (1980)).

This implies that $\partial_\mu \hat{j}^\mu_{3,5}(x)$ is not zero after all, the calculation producing the specific value

$$\partial_\mu \hat{j}^\mu_{3,5}(x) = -\frac{e^2}{32\pi^2}\epsilon^{\alpha\nu\beta\lambda}\hat{F}_{\alpha\nu}\hat{F}_{\beta\lambda} \tag{18.102}$$

where the $\hat{F}$'s are the usual electromagnetic field strength tensors.

Equation (18.102) means that (18.99) is no longer valid, so that $A$ need no longer vanish: indeed, (18.102) predicts a definite value for $A$, so we need to see if it is consistent with (18.100). Taking the vacuum $\to 2\gamma$ matrix element of (18.102) produces (problem 18.10)

$$\mathrm{i}q_\mu \mathcal{M}^{\mu\nu\lambda} = \frac{e^2}{4\pi^2}\epsilon^{\alpha\nu\beta\lambda}k_{1\alpha}k_{2\mu} \tag{18.103}$$

which is indeed consistent with (18.97) and (18.100), after suitably interchanging the labels on the $\epsilon$ symbol.

Equation (18.102) is a typical example of 'an anomaly'—the violation, at the quantum level, of a symmetry of the classical Lagrangian. It might be thought that the result (18.102) is only valid to order $\alpha$ (though the $O(\alpha^2)$ correction would presumably be very small). But Adler and Bardeen (1969) showed that such 'triangle' loops give the *only* anomalous contributions to the $\hat{j}^\mu_{i,5} - \gamma - \gamma$ vertex, so that (18.102) is true to all orders in $\alpha$.

The triangles considered earlier actually used a fermion with integer charge (the proton). We clearly should use quarks, which carry fractional charge. In this case, the previous numerical value for $A$ is multiplied by the factor $\tau_3 Q^2$ for each contributing quark. For the u and d quarks of chiral SU(2)×SU(2), this gives 1/3. Consequently agreement with experiment is lost unless there exist three replicas of each quark, identical in their electromagnetic and SU(2)×SU(2) properties. Colour supplies just this degeneracy, and thus the $\pi^0 \to \gamma\gamma$ rate is important evidence for such a degree of freedom.

In the foregoing discussion, the axial isospin current was associated with a global symmetry: only the electromagnetic currents (in the case of $\pi^0 \to \gamma\gamma$) were associated with a local (gauged) symmetry and they remained conserved (anomaly free). If, however, we have an anomaly in a current associated with a local symmetry, we will have a serious problem. The whole rather elaborate construction of a quantum gauge field theory relies on current conservation equations such as (11.21) or (13.164) to eliminate unwanted gauge degrees of freedom and ensure unitarity of the $S$-matrix. So anomalies in currents coupled to gauge fields cannot be tolerated. As we shall see in chapter 20, and is already evident from (18.48), axial currents are indeed present in weak interactions and they are coupled to the $W^\pm, Z^0$ gauge fields. Hence, if this theory is to be satisfactory at the quantum level, all anomalies must somehow cancel away. That this is possible rests essentially on the observation that the anomaly (18.102) is independent of the mass of the circulating fermion. Thus cancellations are, in principle, possible between quark and lepton 'triangles' in the weak interaction case. Remarkably enough, complete cancellation of all anomalies does occur in the GSW theory (see Peskin and Schroeder 1995, section 20.2). Bouchiat *et al* (1972) were the first to point out that each generation of quarks and leptons will be separately anomaly free if the fractionally charged quarks come in three colours. Anomaly cancellation is a powerful constraint on possible theories ('t Hooft 1980, Weinberg 1996, section 22.5).

## Problems

**18.1** Verify (18.24)–(18.26).

**18.2** Verify (18.28)–(18.30).

**18.3** Show that $\mathcal{L}_q$ of (18.12) can be written as (18.39).

**18.4** Show that the rate for $\pi^+ \to \mu^+\nu_\mu$, calculated from the lowest-order matrix element (18.52), is given by (18.53).

**18.5** Verify (18.64).

**18.6** Show that (18.72) is invariant under the transformations (18.78)–(18.80).

**18.7** Show that after making the 'shift' (18.87), the Lagrangian (18.72) becomes (18.88).

**18.8** Show that when a term $c\hat{\sigma}$ is added to $\mathcal{L}_\sigma$ of (18.72), the divergence of the axial vector current is given by $\partial_\mu \hat{j}^\mu_{i,5}(x) = -c\hat{\pi}_i$.

**18.9** Verify (18.101), and calculate the $\pi^0$ lifetime in seconds.

**18.10** Verify (18.103).

# 19

# SPONTANEOUSLY BROKEN LOCAL SYMMETRY

In earlier parts of this book we have briefly indicated why we might want to search for a *gauge* theory of the weak interactions. The reasons include: (a) the goal of unification (e.g. with the U(1) gauge theory QED), as mentioned in section 2.5; and (b) certain 'universality' phenomena (to be discussed more fully in chapter 20), which are reminiscent of a similar situation in QED (see comment (ii) in section 3.6 and also section 11.6) and which are particularly characteristic of a non-Abelian gauge theory, as pointed out in section 13.1 after equation (13.44). However, we also know from section 2.5 that weak interactions are short-ranged, so that their mediating quanta must be massive. At first sight, this seems to rule out the possibility of a gauge theory of weak interactions, since a simple gauge boson mass violates gauge invariance, as we pointed out for the photon in section 11.4 and for non-Abelian gauge quanta in section 13.51, and we will review again in the following section. Nevertheless, there is a way of giving gauge field quanta a mass, which is by '*spontaneously breaking*' the gauge (i.e. local) symmetry. This is the topic of the present chapter. The detailed application to the electroweak theory will be made in chapter 22.

## 19.1 Massive and massless vector particles

Let us begin by noting an elementary (classical) argument for why a gauge field quantum cannot have mass. The electromagnetic potential satisfies the Maxwell equation (cf (3.21))

$$\Box A^{\nu} - \partial^{\nu}(\partial_{\mu}A^{\mu}) = j^{\nu}_{\text{em}} \tag{19.1}$$

which, as discussed in section 3.3, is invariant under the gauge transformation

$$A^{\mu} \to A'^{\mu} = A^{\mu} - \partial^{\mu}\chi. \tag{19.2}$$

However, if $A^{\mu}$ were to represent a *massive* field, the relevant wave equation would be

$$(\Box + M^2)A^{\nu} - \partial^{\nu}(\partial_{\mu}A^{\mu}) = j^{\nu}_{\text{em}}. \tag{19.3}$$

To get this, we have simply replaced the massless 'Klein–Gordon' operator $\Box$ by the corresponding massive one, $\Box + M^2$ (compare sections 4.1 and 5.3). Equation (19.3) is manifestly *not* invariant under (19.2) and it is precisely the mass term $M^2A^{\nu}$ that breaks the gauge invariance. The same conclusion follows

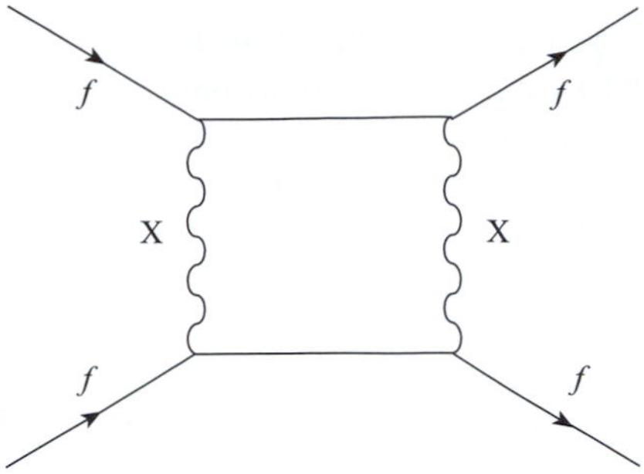


**Figure 19.1.** Fermion–fermion scattering via exchange of two X bosons.

in a Lagrangian treatment: to obtain (19.3) as the corresponding Euler–Lagrange equation, one adds a mass term $+\frac{1}{2}M^2 A_\mu A^\mu$ to the Lagrangian of (7.63) (see also sections 11.4 and 13.5.1) and this clearly violates invariance under (19.2). Similar reasoning holds for the non-Abelian case too. Perhaps, then, we must settle for a theory involving massive charged vector bosons, $W^\pm$ for example, without it being a gauge theory.

Such a theory is certainly possible but it will not be *renormalizable*, as we now discuss. Consider figure 19.1, which shows some kind of fermion–fermion scattering (we need not be more specific), proceeding in fourth-order perturbation theory via the exchange of two massive vector bosons, which we will call X-particles. To calculate this amplitude, we need the propagator for the X-particle, which can be found by following the 'heuristic' route outlined in section 7.3.2 for photons. We consider the momentum–space version of (19.3) for the corresponding $X^\nu$ field, but without the current on the right-hand side (so as to describe a free field):

$$[(-k^2 + M^2)g^{\nu\mu} + k^\nu k^\mu]\tilde{X}_\mu(k) = 0 \tag{19.4}$$

which should be compared with (7.87). Apart from the 'i$\epsilon$', the propagator should be proportional to the inverse of the quantity in the square brackets in (19.4). Problem 19.1 shows that, *unlike* the (massless) photon case, this inverse does exist and is given by

$$\frac{-g^{\mu\nu} + k^\mu k^\nu/M^2}{k^2 - M^2}. \tag{19.5}$$

A proper field-theoretic derivation would yield this result multiplied by an overall factor 'i' as usual and would also include the 'i$\epsilon$' via $k^2 - M^2 \to k^2 - M^2 + \mathrm{i}\epsilon$. We remark immediately that (19.5) gives nonsense in the limit $M \to 0$, thus indicating already that a massless vector particle seems to be a very different kind of thing from a massive one (we cannot just take the massless limit of the latter).

Now consider the loop integral in figure 19.1. At each vertex we will have a coupling constant $g$, associated with an interaction Lagrangian having the general form $g\hat{\bar{\psi}}\gamma_\mu\hat{\psi}\hat{X}^\mu$ (a $\gamma_\mu\gamma_5$ coupling could also be present but will not affect the argument). Just as in QED, this '$g$' is dimensionless but, as we warned the reader in section 11.8, this may not guarantee renormalizability and, indeed, this is a case where it does not. To get an idea of why this might be so, consider the leading divergent behaviour of figure 19.1. This will be associated with the $k^\mu k^\nu$ terms in the numerator of (19.5), so that the leading divergence is effectively

$$\sim \int \mathrm{d}^4k \left(\frac{k^\mu k^\nu}{k^2}\right)\left(\frac{k^\rho k^\sigma}{k^2}\right)\frac{1}{\not k}\frac{1}{\not k} \tag{19.6}$$

for high $k$-values (we are not troubling to get all the indices right, we are omitting the spinors altogether and we are looking only at the large-$k$ part of the propagators). Now the first two bracketed terms in (19.6) behave like a constant at large $k$, so that the divergence becomes

$$\sim \int \mathrm{d}^4k \frac{1}{\not k}\frac{1}{\not k} \tag{19.7}$$

which is quadratically divergent and, indeed, exactly what we would get in a 'four-fermion' theory—see (11.89) for example. This strongly suggests that the theory is non-renormalizable.

Where have these dangerous powers of $k$ in the numerator of (19.6) come from? The answer is simple and important. They come from the *longitudinal* polarization state of the massive X-particle, as we shall now explain. The free-particle wave equation is

$$(\Box + M^2)X^\nu - \partial^\nu(\partial_\mu X^\mu) = 0 \tag{19.8}$$

and plane-wave solutions have the form

$$X^\nu = \epsilon^\nu \mathrm{e}^{-\mathrm{i}k\cdot x}. \tag{19.9}$$

Hence, the polarization vectors $\epsilon^\nu$ satisfy the condition

$$(-k^2 + M^2)\epsilon^\nu + k^\nu k_\mu \epsilon^\mu = 0. \tag{19.10}$$

Taking the 'dot' product of (19.10) with $k_\nu$ leads to

$$M^2 k\cdot\epsilon = 0 \tag{19.11}$$

which implies (for $M^2 \neq 0$!)

$$k\cdot\epsilon = 0. \tag{19.12}$$

Equation (19.12) is a covariant condition, which has the effect of ensuring that there are just three independent polarization vectors, as we expect for a spin-1 particle. Let us take $k^\mu = (k^0, 0, 0, |\boldsymbol{k}|)$: then the $x$- and $y$-directions are 'transverse' while the $z$-direction is 'longitudinal'. Now, in the rest frame of the

X, such that $k_{\text{rest}} = (M, 0, 0, 0)$, (19.12) reduces to $\epsilon^0 = 0$ and we may choose three independent $\epsilon$'s as

$$\epsilon^\mu(k_{\text{rest}}, \lambda) = (0, \boldsymbol{\epsilon}(\lambda)) \tag{19.13}$$

with

$$\boldsymbol{\epsilon}(\lambda = \pm 1) = \mp 2^{-1/2}(1, \pm \mathrm{i}, 0) \tag{19.14}$$

$$\boldsymbol{\epsilon}(\lambda = 0) = (0, 0, 1). \tag{19.15}$$

The $\boldsymbol{\epsilon}$'s are 'orthonormalized' so that (cf (7.83))

$$\boldsymbol{\epsilon}(\lambda)^* \cdot \boldsymbol{\epsilon}(\lambda') = \delta_{\lambda\lambda'}. \tag{19.16}$$

These states have definite spin projection ($\lambda = \pm 1, 0$) along the $z$-axis. For the result in a general frame, we can Lorentz transform $\epsilon^\mu(k_{\text{rest}}, \lambda)$ as required. For example, in a frame such that $k^\mu = (k^0, 0, 0, |\boldsymbol{k}|)$, we find that

$$\epsilon^\mu(k, \lambda = \pm 1) = \epsilon^\mu(k_{\text{rest}}, \lambda = \pm 1) \tag{19.17}$$

as before, but the longitudinal polarization vector becomes (problem 19.2)

$$\epsilon^\mu(k, \lambda = 0) = M^{-1}(|\boldsymbol{k}|, 0, 0, k^0). \tag{19.18}$$

Note that $k \cdot \epsilon^\mu(k, \lambda = 0) = 0$ as required.

From (19.17) and (19.18), it is straightforward to verify the result (problem 19.3)

$$\sum_{\lambda=0,\pm 1} \epsilon^\mu(k, \lambda)\epsilon^{\nu *}(k, \lambda) = -g^{\mu\nu} + k^\mu k^\nu / M^2. \tag{19.19}$$

Consider now the propagator for a spin-$\frac{1}{2}$ particle, given in (7.60):

$$\frac{\mathrm{i}(\not{k} + m)}{k^2 - m^2 + \mathrm{i}\epsilon}. \tag{19.20}$$

Equation (7.61) shows that the factor in the numerator of (19.20) arises from the spin sum

$$\sum_s u_\alpha(k, s)\bar{u}_\beta(k, s) = (\not{k} + m)_{\alpha\beta}. \tag{19.21}$$

In just the same way, the massive spin-1 propagator is given by

$$\frac{\mathrm{i}[-g^{\mu\nu} + k^\mu k^\nu / M^2]}{k^2 - M^2 + \mathrm{i}\epsilon} \tag{19.22}$$

and the numerator in (19.22) arises from the spin sum (19.19). Thus, the dangerous factor $k^\mu k^\nu / M^2$ can be traced to the spin sum (19.19): in particular, at large values of $k$ the longitudinal state $\epsilon^{\mu}(k, \lambda = 0)$ is proportional to $k^{\mu}$, and this is the origin of the numerator factors $k^{\mu}k^{\nu}/M^2$ in (19.22).

We shall not give further details here (see also section 20.3) but merely state that theories with massive charged vector bosons are indeed non-renormalizable. Does this matter? In section 11.8 we explained why it is thought that the relevant theories at presently accessible energy scales should be renormalizable theories. Is there, then, any way of getting rid of the offending '$k^{\mu}k^{\nu}$' terms in the X-propagator, so as (perhaps) to render the theory renormalizable? Consider the photon propagator of chapter 7 repeated here:

$$\frac{\mathrm{i}[-g^{\mu\nu} + (1-\xi)k^{\mu}k^{\nu}/k^2]}{k^2 + \mathrm{i}\epsilon}. \tag{19.23}$$

This contains somewhat similar factors of $k^{\mu}k^{\nu}$ (admittedly divided by $k^2$ rather than $M^2$) but they are gauge-dependent and can, in fact, be 'gauged away' entirely, by choice of the gauge parameter $\xi$ (namely by taking $\xi = 1$). But, as we have seen, such 'gauging'—essentially the freedom to make gauge transformations—seems to be possible only in a massless vector theory.

A closely related point is that, as section 7.3.1 showed, free photons exist in only two polarization states (electromagnetic waves are purely transverse), instead of the three we might have expected for a vector (spin-1) particle—and as do indeed exist for massive vector particles. This gives another way of seeing in what way a massless vector particle is really very different from a massive one: the former has only two (spin) degrees of freedom, while the latter has three, and it is not at all clear how to 'lose' the offending longitudinal state smoothly (certainly not, as we have seen, by letting $M \to 0$ in (19.5)).

These considerations therefore suggest the following line of thought: is it possible somehow to create a theory involving massive vector bosons, in such a way that the dangerous $k^{\mu}k^{\nu}$ term can be 'gauged away', making the theory renormalizable? The answer is yes, via the idea of *spontaneous breaking* of gauge symmetry. This is the natural generalization of the spontaneous global symmetry breaking considered in chapter 17. By way of advance notice, the crucial formula is (19.75) for the propagator in such a theory, which is to be compared with (19.22).

The first serious challenge to the then widely held view that electromagnetic gauge invariance requires the photon to be massless was made by Schwinger (1962). Soon afterwards, Anderson (1963) pointed out that several situations in solid state physics could be interpreted in terms of an effectively massive electromagnetic field. He outlined a general framework for treating the phenomenon of the acquisition of mass by a gauge boson, and discussed its possible relevance to contemporary attempts (Sakurai 1960) to interpret the recently discovered vector mesons $(\rho, \omega, \phi, \ldots)$ as the gauge quanta associated with a local extension of hadronic flavour symmetry. From his discussion, it is clear that Anderson had his doubts about the hadronic application, precisely because, as he remarked, gauge bosons can only acquire a mass if the symmetry is spontaneously broken. This has the consequence, as we saw in chapter 17, that the multiplet structure ordinarily associated with a non-Abelian symmetry would be lost. But we know that flavour symmetry, even if admittedly not exact, certainly leads to identifiable multiplets, which are at least approximately degenerate in mass. It was Weinberg (1967) and Salam (1968) who made the correct application of these ideas to the generation of mass for the gauge quanta associated with the weak force. There is, however, nothing specifically relativistic about the basic mechanism involved, nor need we start with the non-Abelian case. In fact, the physics is well illustrated by the non-relativistic Abelian (i.e. electromagnetic) case—which is nothing but the physics of superconductivity. Our presentation is influenced by that of Anderson (1963).

## 19.2 The generation of 'photon mass' in a superconductor: the Meissner effect

In chapter 17, section 17.7, we gave a brief introduction to some aspects of the BCS theory of superconductivity. We were concerned mainly with the nature of the BCS ground state and with the non-perturbative origin of the energy gap for elementary excitations. In particular, as noted after (17.128), we omitted completely all electromagnetic couplings of the electrons in the 'microscopic' Hamiltonian. It is certainly possible to complete the BCS theory in this way, so as to include within the same formalism a treatment of electromagnetic effects (e.g. the Meissner effect) in a superconductor. We refer interested readers to the book by Schrieffer (1964, chapter 8). Instead, we shall follow a less 'microscopic' and somewhat more 'phenomenological' approach, which has a long history in theoretical studies of superconductivity and is, in some ways, actually closer (at least formally) to our eventual application in particle physics.

In section 17.3.1 we introduced the concept of an 'order parameter', a quantity which was a measure of the 'degree of ordering' of a system below some transition temperature. In the case of superconductivity, the order parameter (in this sense) is taken to be a complex scalar field $\psi$, as originally proposed by Ginzburg and Landau (1950), well before the appearance of BCS theory. Subsequently, Gorkov (1959) and others showed how the Ginzburg–Landau description could be derived from BCS theory, in certain domains of temperature and magnetic field. This work all relates to static phenomena. More recently, an analogous 'effective theory' for time-dependent phenomena (at zero temperature) has been derived from a BCS-type model (Aitchison *et al* 1995). For the moment, we shall follow a more qualitative approach.

The Ginzburg–Landau field $\psi$ is commonly referred to as the 'macroscopic wave function'. This terminology originates from the recognition that in the BCS ground state a macroscopic number of Cooper pairs have 'condensed' into the state of lowest energy, a situation similar to that in the Bogoliubov superfluid. Further, this state is highly *coherent*, all pairs having the same total momentum (namely zero, in the case of (17.140)). These considerations suggest that a successful phenomenology can be built by invoking the idea of a macroscopic wavefunction $\psi$, describing the condensate. Note that $\psi$ is a 'bosonic' quantity, referring essentially to *paired* electrons. Perhaps the single most important property of $\psi$ is that it is assumed to be normalized to the *total* density of Cooper pairs $n_c$ via the relation

$$|\psi|^2 = n_c = n_s/2 \tag{19.24}$$

where $n_s$ is the density of superconducting electrons. The quantities $n_c$ and $n_s$ will depend on temperature $T$, tending to zero as $T$ approaches the superconducting transition temperature $T_c$ from below. The precise connection between $\psi$ and the microscopic theory is indirect; in particular, $\psi$ has no knowledge of the coordinates of individual electron pairs. Nevertheless, as an 'empirical' order parameter, it may be thought of as in some way related to the ground-state 'pair' expectation value introduced in (17.121); in particular, the charge associated with $\psi$ is taken to be $-2e$ and the mass is $2m_e$.

The Ginzburg–Landau description proceeds by considering the quantum-mechanical electromagnetic current associated with $\psi$, in the presence of a static external electromagnetic field described by a vector potential $\boldsymbol{A}$. This current was considered in section 3.4 and is given by the gauge-invariant form of (A.7), namely

$$\boldsymbol{j}_{\mathrm{em}} = \frac{-2e}{4m_e \mathrm{i}}[\psi^*(\nabla + 2\mathrm{i}e\boldsymbol{A})\psi - \{(\nabla + 2\mathrm{i}e\boldsymbol{A})\psi\}^*\psi]. \tag{19.25}$$

Note that we have supplied an overall factor of $-2e$ to turn the Schrödinger 'number density' current into the appropriate electromagnetic current. Assuming now that, consistently with (19.24), $\psi$ is varying primarily through its *phase* degree of freedom $\phi$, rather than its modulus $|\psi|$, we can rewrite (19.25) as

$$\boldsymbol{j}_{\mathrm{em}} = -\frac{2e^2}{m_e}\left(\boldsymbol{A} + \frac{1}{2e}\nabla\phi\right)|\psi|^2 \tag{19.26}$$

where $\psi = \mathrm{e}^{\mathrm{i}\phi}|\psi|$. We easily verify that (19.26) is invariant under the gauge transformation (3.40), which can be written in this case as

$$\boldsymbol{A} \to \boldsymbol{A} + \nabla\chi \tag{19.27}$$

$$\phi \to \phi - 2e\chi. \tag{19.28}$$

We now replace $|\psi|^2$ in (19.26) by $n_s/2$ in accordance with (19.24) and take the curl of the resulting equation to obtain

$$\nabla \times \boldsymbol{j}_{\mathrm{em}} = -\left(\frac{e^2 n_s}{m_e}\right)\boldsymbol{B}. \tag{19.29}$$

Equation (19.29) is known as the London equation (London 1950) and is one of the fundamental phenomenological relations in superconductivity.

The significance of (19.29) emerges when we combine it with the (static) Maxwell equation

$$\nabla \times \boldsymbol{B} = \boldsymbol{j}_{\mathrm{em}}. \tag{19.30}$$

Taking the curl of (19.30) and using $\nabla \times (\nabla \times \boldsymbol{B}) = \nabla(\nabla \cdot \boldsymbol{B}) - \nabla^2 \boldsymbol{B}$ and $\nabla \cdot \boldsymbol{B} = 0$, we find that

$$\nabla^2 \boldsymbol{B} = \left(\frac{e^2 n_{\mathrm{s}}}{m_{\mathrm{e}}}\right) \boldsymbol{B}. \tag{19.31}$$

The variation of magnetic field described by (19.31) is a very characteristic one encountered in a number of contexts in condensed matter physics. First, we note that the quantity $(e^2 n_{\mathrm{s}}/m_{\mathrm{e}})$ must—in our units—have the dimensions of $(\text{length})^{-2}$, by comparison with the left-hand side of (19.31). Let us write

$$\left(\frac{e^2 n_{\mathrm{s}}}{m_{\mathrm{e}}}\right) = \frac{1}{\lambda^2}. \tag{19.32}$$

Next, consider for simplicity one-dimensional variation

$$\frac{\mathrm{d}^2 \boldsymbol{B}}{\mathrm{d}x^2} = \frac{1}{\lambda^2} \boldsymbol{B} \tag{19.33}$$

in the half-plane $x \geq 0$, say. Then the solutions of (19.33) have the form

$$\boldsymbol{B}(x) = \boldsymbol{B}_0 \exp -(x/\lambda) \tag{19.34}$$

the exponentially growing solution being rejected as unphysical. The field, therefore, penetrates only a distance of order $\lambda$ into the region $x \geq 0$. The range parameter $\lambda$ is called the *screening length*. This expresses the fact that, in a medium such that (19.29) holds, the magnetic field will be 'screened out' from penetrating further into the medium.

The physical origin of the screening is provided by Lenz's law: when a magnetic field is applied to a system of charged particles, induced EMF's are set up which accelerate the particles and the magnetic effect of the resulting currents tends to cancel (or screen) the applied field. On the atomic scale, this is the cause of atomic diamagnetism. Here the effect is occurring on a macroscopic scale (as mediated by the 'macroscopic wavefunction' $\psi$) and leads to the Meissner effect—the exclusion of flux from the interior of a superconductor. In this case, screening currents are set up within the superconductor, over distances of order $\lambda$ from the exterior boundary of the material. These exactly cancel—perfectly screen—the applied flux density in the interior. With $n_{\mathrm{s}} \sim 4 \times 10^{28}\ \mathrm{m}^{-3}$ (roughly one conduction electron per atom), we find that

$$\lambda = \left(\frac{m_{\mathrm{e}}}{n_{\mathrm{s}} e^2}\right)^{1/2} \approx 10^{-8}\ \mathrm{m} \tag{19.35}$$

which is the correct order of magnitude for the thickness of the surface layer within which screening currents flow, and over which the applied field falls to zero. As $T \to T_c$, $n_s \to 0$ and $\lambda$ becomes arbitrarily large, so that flux is no longer screened.

It is quite simple to interpret equation (19.31) in terms of an 'effective non-zero photon mass'. Consider the equation (19.8) for a free massive vector field. Taking the divergence via $\partial_\nu$ leads to

$$M^2 \partial_\nu X^\nu = 0 \tag{19.36}$$

(cf (19.11)) and so (19.8) can be written as

$$(\Box + M^2) X^\nu = 0 \tag{19.37}$$

which simply expresses the fact that each component of $X^\nu$ has mass $M$. Now consider the static version of (19.37), in the rest frame of the X-particle in which (see equation (19.13)) the $\nu = 0$ component vanishes. Equation (19.37) reduces to

$$\nabla^2 \boldsymbol{X} = M^2 \boldsymbol{X} \tag{19.38}$$

which is exactly the same in form as (19.31) (if $\boldsymbol{X}$ were the electromagnetic field $\boldsymbol{A}$, we could take the curl of (19.38) to obtain (19.31) via $\boldsymbol{B} = \nabla \times \boldsymbol{A}$). The connection is made precise by making the association

$$M^2 = \left(\frac{e^2 n_s}{m_e}\right) = \frac{1}{\lambda^2}. \tag{19.39}$$

Equation (19.39) shows very directly another way of understanding the 'screening length $\leftrightarrow$ photon mass' connection: in our units $\hbar = c = 1$, a mass has the dimension of an inverse length and so we naturally expect to be able to interpret $\lambda^{-1}$ as an equivalent mass (for the photon, in this case).

This treatment conveys much of the essential physics behind the phenomenon of 'photon mass generation' in a superconductor. In particular, it suggests rather strongly that a *second* field, in addition to the electromagnetic one, is an essential element in the story (here, it is the $\psi$ field). This provides a partial answer to the puzzle about the discontinuous change in the number of spin degrees of freedom in going from a massless to a massive gauge field: actually, some other field has to be supplied. Nevertheless, many questions remain unanswered so far. For example, how is all the foregoing related to what we learned in chapter 17 about spontaneous symmetry breaking? Where is the Goldstone mode? Is it really all gauge invariant? And what about Lorentz invariance? Can we provide a Lagrangian description of the phenomenon? The answers to these questions are mostly contained in the model to which we now turn, which is due to Higgs (1964) and is essentially the *local* version of the U(1) Goldstone model of section 17.5.

## 19.3 Spontaneously broken local U(1) symmetry: the Abelian Higgs model

This model is just $\hat{\mathcal{L}}_{\rm G}$ of (17.69) and (17.77), extended so as to be locally, rather than merely globally, U(1) invariant. Due originally to Higgs (1964), it provides a deservedly famous and beautifully simple model for investigating what happens when a *gauge* symmetry is spontaneously broken.

To make (17.69) locally U(1) invariant, we need only replace the $\partial$'s by $\hat{D}$'s according to the rule (7.120) and add the Maxwell piece. This produces

$$\hat{\mathcal{L}}_{\rm H} = [(\partial^\mu + {\rm i}q\hat{A}_\mu)\hat{\phi}]^\dagger[(\partial_\mu + {\rm i}q\hat{A}_\mu)\hat{\phi}] - \tfrac{1}{4}\hat{F}_{\mu\nu}\hat{F}^{\mu\nu} - \tfrac{1}{4}\lambda(\hat{\phi}^\dagger\hat{\phi})^2 + \mu^2(\hat{\phi}^\dagger\hat{\phi}). \tag{19.40}$$

(19.40) is invariant under the local version of (17.72), namely

$$\hat{\phi}(x) \to \hat{\phi}'(x) = {\rm e}^{-{\rm i}\hat{\alpha}(x)\hat{\phi}(x)} \tag{19.41}$$

when accompanied by the gauge transformation on the potentials

$$\hat{A}^\mu(x) \to \hat{A}'^\mu(x) = \hat{A}^\mu(x) + \frac{1}{q}\partial^\mu\hat{\alpha}(x). \tag{19.42}$$

Before proceeding any further, we note at once that this model contains four field degrees of freedom—two in the complex scalar Higgs field $\hat{\phi}$ and two in the massless gauge field $\hat{A}^\mu$.

We learned in section 17.5 that the form of the potential terms in (19.40) (specifically the $\mu^2$ one) does not lend itself to a natural particle interpretation, which only appears after making a 'shift to the classical minimum', as in (17.84). But there is a remarkable difference between the global and local cases. In the present (local) case, the phase of $\hat{\phi}$ is completely arbitrary, since any change in $\hat{\alpha}$ of (19.41) can be compensated by an appropriate transformation (19.42) on $\hat{A}^\mu$, leaving $\hat{\mathcal{L}}_{\rm H}$ the same as before. Thus, the field $\hat{\theta}$ in (17.84) can be 'gauged away' altogether, if we choose! But $\hat{\theta}$ was precisely the Goldstone field, in the global case. This must mean that there is somehow no longer any physical manifestation of the massless mode. This is the first unexpected result in the local case. We may also be reminded of our desire to 'gauge away' the longitudinal polarization states for a 'massive gauge' boson: we shall return to this later.

However, a degree of freedom (the Goldstone mode) cannot simply disappear. Somehow the system must keep track of the fact that we started with four degrees of freedom. To see what is going on, let us study the field equation for $\hat{A}^\nu$, namely

$$\Box\hat{A}^\nu - \partial^\nu(\partial_\mu\hat{A}^\mu) = \hat{j}^\nu_{\rm em} \tag{19.43}$$

where $\hat{j}^\nu_{\rm em}$ is the electromagnetic current contained in (19.40). This current can be obtained just as in (7.137) and is given by

$$\hat{j}^\nu_{\rm em} = {\rm i}q(\hat{\phi}^\dagger\partial^\nu\hat{\phi} - (\partial^\nu\hat{\phi}^\dagger)\hat{\phi}) - 2q^2\hat{A}^\nu\hat{\phi}^\dagger\hat{\phi}. \tag{19.44}$$

We now insert the field parametrization (cf (17.84))

$$\hat{\phi}(x) = \frac{1}{\sqrt{2}}(v + \hat{h}(x)) \exp(-\mathrm{i}\hat{\theta}(x)/v) \tag{19.45}$$

into (19.40) where $v/\sqrt{2} = 2^{1/2}|\mu|/\lambda^{\frac{1}{2}}$ is the position of the minimum of the classical potential as a function of $|\phi|$, as in (17.81). We obtain (problem 19.4)

$$\hat{j}^{\nu}_{\mathrm{em}} = -v^2 q^2 \left( \hat{A}^{\nu} - \frac{\partial^{\nu}\hat{\theta}}{vq} \right) + \text{terms quadratic and cubic in the fields.} \tag{19.46}$$

The linear part of the right-hand side of (19.46) is directly analogous to the non-relativistic current (19.26), interpreting $\psi$ as essentially playing the role of '$\hat{\phi}$' and $|\psi|^2$ the role of $v^2$. Retaining just the linear terms in (19.46) (the others would appear on the right-hand side of equation (19.47) following, where they would represent interactions) and placing this $\hat{j}^{\nu}_{\mathrm{em}}$ in (19.43), we obtain

$$\Box \hat{A}^{\nu} - \partial^{\nu}\partial_{\mu}\hat{A}^{\mu} = -v^2 q^2 \left( \hat{A}^{\nu} - \frac{\partial^{\nu}\theta}{vq} \right). \tag{19.47}$$

Now a gauge transformation on $\hat{A}^{\nu}$ has the form shown in (19.42), for arbitrary $\hat{\alpha}$. So we can certainly regard the whole expression $(\hat{A}^{\nu} - \partial^{\nu}\hat{\theta}/vq)$ as a perfectly acceptable gauge field. Let us define

$$\hat{A}'^{\nu} = \hat{A}^{\nu} - \frac{\partial^{\nu}\hat{\theta}}{vq}. \tag{19.48}$$

Then, since we know (or can easily verify) that the left-hand side of (19.47) is invariant under (19.42), the resulting equation for $\hat{A}'^{\nu}$ is

$$\Box \hat{A}'^{\nu} - \partial^{\nu}\partial_{\mu}\hat{A}'^{\mu} = -v^2 q^2 \hat{A}'^{\nu} \tag{19.49}$$

or

$$(\Box + v^2 q^2)\hat{A}'^{\nu} - \partial^{\nu}\partial_{\mu}\hat{A}'^{\mu} = 0. \tag{19.50}$$

But (19.50) is nothing but the equation (19.8) for a free massive vector field, with mass $M = vq$! This fundamental observation was first made, in the relativistic context, by Englert and Brout (1964), Higgs (1964) and Guralnik *et al* (1964); for a full account, see Higgs (1966).

The foregoing analysis shows us two things. First, the current (19.46) is indeed a relativistic analogue of (19.26), in that it provides a 'screening' (mass generation) effect on the gauge field. Second, equation (19.48) shows how the *phase* degree of freedom of the Higgs field $\hat{\phi}$ has been incorporated into a new gauge field $\hat{A}'^{\nu}$, which is massive and, therefore, has '*three*' spin degrees of freedom. In fact, we can go further. If we imagine plane-wave solutions for $\hat{A}'^{\nu}$, $\hat{A}^{\nu}$ and $\hat{\theta}$, we see that the $\partial^{\nu}\hat{\theta}/vq$ part of (19.48) will contribute something proportional to $k^\nu/M$ to the polarization vector of $A'^\nu$ (recall $M = vq$). But this is exactly the (large $k$) behaviour of the longitudinal polarization vector of a massive vector particle. We may therefore say that the massless gauge field $\hat{A}^\nu$ has 'swallowed' the Goldstone field $\hat{\theta}$ via (19.48) to make the massive vector field $\hat{A}'^\nu$. The Goldstone field disappears as a massless degree of freedom, and reappears, via its gradient, as the longitudinal part of the massive vector field. In this way the four degrees of freedom are all now safely accounted for: three are in the massive vector field $\hat{A}'^\nu$ and one is in the real scalar field $\hat{h}$ (to which we shall return shortly).

In this (relativistic) case, we know from Lorentz covariance that all the components (transverse and longitudinal) of the vector field must have the same mass and this has, of course, emerged automatically from our covariant treatment. But the transverse and longitudinal degrees of freedom respond differently in the non-relativistic (superconductor) case. There, the longitudinal part of $\boldsymbol{A}$ couples strongly to longitudinal excitations of the electrons: primarily, as Bardeen (1957) first recognized, to the collective density fluctuation mode of the electron system—that is, to plasma oscillations. This is a high-frequency mode and is essentially the one discussed in section 17.3.2, after equation (17.46). When this aspect of the dynamics of the electrons is included, a fully gauge-invariant description of the electromagnetic properties of superconductors, within the BCS theory, is obtained (Schreiffer 1964, chapter 8).

We return to equations (19.48)–(19.50). Taking the divergence of (19.50) leads, as we have seen, to the condition

$$\partial_\mu \hat{A}'^\mu = 0 \tag{19.51}$$

on $\hat{A}'^\mu$. It follows that in order to interpret the relation (19.48) as a gauge transformation on $\hat{A}^\nu$ we must, to be consistent with (19.51), regard $\hat{A}^\mu$ as being in a gauge specified by

$$\partial_\mu \hat{A}^\mu = \frac{1}{vq}\Box\hat{\theta} = \frac{1}{M}\Box\hat{\theta}. \tag{19.52}$$

In going from the situation described by $\hat{A}^\mu$ and $\hat{\theta}$ to one described by $\hat{A}'^\mu$ alone via (19.48), we have evidently chosen a gauge function (cf (19.42))

$$\hat{\alpha}(x) = -\hat{\theta}(x)/v. \tag{19.53}$$

Recalling then the form of the associated local phase change on $\hat{\phi}(x)$,

$$\hat{\phi}(x) \to \hat{\phi}'(x) = \mathrm{e}^{-\mathrm{i}\hat{\alpha}(x)}\hat{\phi}(x) \tag{19.54}$$

we see that the phase of $\hat{\phi}$ in (19.45) has been reduced to zero, in this choice of gauge. Thus it is indeed possible to 'gauge $\hat{\theta}$ away' in (19.45), but then the vector field we must use is $\hat{A}'^\mu$, satisfying the massive equation (19.50) (ignoring other interactions). In superconductivity, the choice of gauge which takes the macroscopic wavefunction to be real (i.e. $\phi = 0$ in (19.26)) is called the 'London gauge'. In the next section we shall discuss a subtlety in the argument which applies in the case of real superconductors, and which leads to the phenomenon of flux quantization.

The fact that this 'Higgs mechanism' leads to a massive vector field can be seen very economically by working in the particular gauge for which $\hat{\phi}$ is real and inserting the parametrization (cf (19.45))

$$\hat{\phi} = \frac{1}{\sqrt{2}}(v + \hat{h}) \tag{19.55}$$

into the Lagrangian $\hat{\mathcal{L}}_{\mathrm{H}}$. Retaining only the terms quadratic in the fields, one finds (problem 19.5) that

$$\begin{aligned} \hat{\mathcal{L}}_{\mathrm{H}}^{\mathrm{quad}} = & -\tfrac{1}{4}(\partial_\mu \hat{A}_\nu - \partial_\nu \hat{A}_\mu)(\partial^\mu \hat{A}^\nu - \partial^\nu \hat{A}^\mu) + \tfrac{1}{2}q^2 v^2 \hat{A}_\mu \hat{A}^\mu \\ & + \tfrac{1}{2}\partial_\mu \hat{h}\partial^\mu \hat{h} - \mu^2 \hat{h}^2. \end{aligned} \tag{19.56}$$

The first line of (19.56) is exactly the Lagrangian for a spin-1 field of mass $vq$—i.e. the Maxwell part with the addition of a mass term (note that the sign of the mass term is correct for the spatial (physical) degrees of freedom); and the second line is the Lagrangian of a scalar particle of mass $\sqrt{2}\mu$. The latter is the mass of excitations of the Higgs field $\hat{h}$ away from its vacuum value (compare the global U(1) case discussed in section 17.5). The necessity for the existence of one or more massive *scalar* particles ('Higgs bosons') when a gauge symmetry is spontaneously broken in this way was first pointed out by Higgs (1964).

We may now ask: what happens if we start with a certain phase $\hat{\theta}$ for $\hat{\phi}$ but do *not* make use of the gauge freedom in $\hat{A}^\nu$ to reduce $\hat{\theta}$ to zero? We shall see in section 19.5 that the equation of motion, and hence the propagator for the vector particle *depend on the choice of gauge*; furthermore, Feynman graphs involving quanta corresponding to the degree of freedom associated with the phase field $\hat{\theta}$ will have to be included for a consistent theory, even though this must be an unphysical degree of freedom, as follows from the fact that a gauge can be chosen in which this field vanishes. That the propagator is gauge dependent should, on reflection, come as a relief. After all, if the massive vector boson generated in this way were *simply* described by the wave equation (19.50), all the troubles with massive vector particles outlined in section 19.1 would be completely unresolved. As we shall see, a different choice of gauge from that which renders $\hat{\phi}$ real has precisely the effect of ameliorating the bad high-energy behaviour associated with (19.50). This is ultimately the reason for the following wonderful fact: *massive vector theories, in which the vector particles acquire mass through the spontaneous symmetry breaking mechanism, are renormalizable* ('t Hooft 1971b).

However, before discussing other gauges than the one in which $\hat{\phi}$ is given by (19.55), we first explore another interesting aspect of superconductivity.

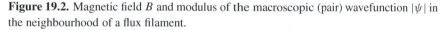

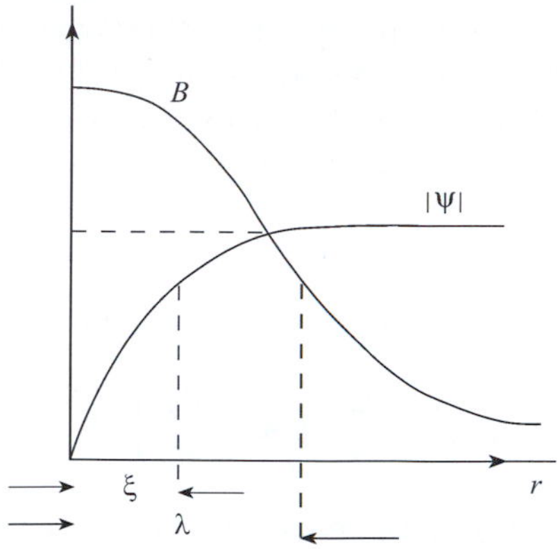


**Figure 19.2.** Magnetic field $B$ and modulus of the macroscopic (pair) wavefunction $|\psi|$ in the neighbourhood of a flux filament.

## 19.4 Flux quantization in a superconductor

Though a slight diversion, it is convenient to include a discussion of flux quantization at this point, while we have a number of relevant results assembled. Apart from its intrinsic interest, the phenomenon may also provide a useful physical model for the 'confining' property of QCD, already discussed in section 16.7.

Our discussion of superconductivity so far has dealt, in fact, with only one class of superconductors, called type I: these remain superconducting throughout the bulk of the material (exhibiting a complete Meissner effect), when an external magnetic field of less than a certain critical value is applied. There is a quite separate class—type-II superconductors—which allow partial entry of the external field, in the form of thin filaments of flux. Within each filament the field is high and the material is not superconducting. Outside the core of the filaments, the material is superconducting and the field dies off over the characteristic penetration length $\lambda$. Around each filament of magnetic flux, there circulates a vortex of screening current: the filaments are often called vortex lines. It is as if numerous thin cylinders, each enclosing flux, had been drilled in a block of type-I material, thereby producing a non-simply connected geometry.

In real superconductors, screening currents are associated with the macroscopic pair wavefunction (field) $\psi$. For type-II behaviour to be possible, $|\psi|$ must vanish at the centre of a flux filament and rise to the constant value appropriate to the superconducting state over a distance $\xi < \lambda$, where $\xi$ is the 'coherence length' of section 17.7. According to the Ginzburg–Landau (GL) theory, a more precise criterion is that type-II behaviour holds if $\xi < 2^{1/2}\lambda$: both $\xi$ and $\lambda$ are, of course, temperature-dependent. The behaviour of $|\psi|$ and $B$ in the vicinity of a flux filament is shown in figure 19.2. Thus, whereas for simple type-I superconductivity, $|\psi|$ is simply set equal to a constant, in the type-II case $|\psi|$ has the variation shown in this figure. Solutions of the coupled GL equations for $\boldsymbol{A}$ and $\psi$ can be obtained which exhibit this behaviour.

An important result is that the flux through a vortex line is quantized. To see this, we write

$$\psi = \mathrm{e}^{\mathrm{i}\phi}|\psi| \tag{19.57}$$

as before. The expression for the electromagnetic current is

$$\boldsymbol{j}_{\mathrm{em}} = -\frac{q^2}{m}\left(\boldsymbol{A} - \frac{\nabla\phi}{q}\right)|\psi|^2 \tag{19.58}$$

as in (19.26), but in (19.58) we are leaving the charge parameter $q$ undetermined for the moment: the mass parameter $m$ will be unimportant. Rearranging, we have

$$\boldsymbol{A} = -\frac{m}{q^2|\psi|^2}\boldsymbol{j}_{\mathrm{em}} + \frac{\nabla\phi}{q}. \tag{19.59}$$

Let us integrate equation (19.59) around any closed loop $C$ in the type-II superconductor, which encloses a flux (or vortex) line. Far enough away from the vortex, the screening currents $\boldsymbol{j}_{\mathrm{em}}$ will have dropped to zero, and hence

$$\oint_C \boldsymbol{A}\cdot\mathrm{d}\boldsymbol{s} = \frac{1}{q}\oint_C \nabla\phi\cdot\mathrm{d}\boldsymbol{s} = \frac{1}{q}[\phi]_C \tag{19.60}$$

where $[\phi]_C$ is the change in phase around $C$. If the wavefunction $\psi$ is single-valued, the change in phase $[\phi]_C$ for any closed path can only be zero or an integer multiple of $2\pi$. Transforming the left-hand side of (19.60) by Stoke's theorem, we obtain the result that the flux $\Phi$ through any surface spanning $C$ is quantized:

$$\Phi = \int \boldsymbol{B}\cdot\mathrm{d}\boldsymbol{S} = \frac{2\pi n}{q} = n\Phi_0 \tag{19.61}$$

where $\Phi_0 = 2\pi/q$ is the flux equation (or $2\pi\hbar/q$ in ordinary units). It is not entirely self-evident why $\psi$ should be single-valued, but experiments do indeed demonstrate the phenomenon of flux quantization, in units of $\Phi_0$ with $|q| = 2e$ (which may be interpreted as the charge on a Cooper pair, as usual). The phenomenon is seen in non-simply connected specimens of type-I superconductors (i.e. ones with holes in them, such as a ring), and in the flux filaments of type-II materials: in the latter case each filament carries a single flux quantum $\Phi_0$.

It is interesting to consider now a situation—so far entirely hypothetical—in which a magnetic monopole is placed in a superconductor. Dirac showed (1931) that, for consistency with quantum mechanics, the monopole strength $g_{\mathrm{m}}$ had to satisfy the 'Dirac quantization condition'

$$qg_{\mathrm{m}} = n/2 \tag{19.62}$$

where $q$ is any electronic charge and $n$ is an integer. It follows from (19.62) that the flux $4\pi g_{\mathrm{m}}$ out of any closed surface surrounding the monopole is quantized in units of $\Phi_0$. Hence, a flux filament in a superconductor can originate from, or be terminated by, a Dirac monopole (with the appropriate sign of $g_{\mathrm{m}}$), as was first pointed out by Nambu (1974).

This is the basic model which, in one way or another, underlies many theoretical attempts to understand confinement. The monopole–antimonopole pair in a type-II superconducting vacuum, joined by a quantized magnetic flux filament, provides a model of a meson. As the distance between the pair—the length of the filament—increases, so does the energy of the filament, at a rate proportional to its length, since the flux cannot spread out in directions transverse to the filament. This is exactly the kind of linearly rising potential energy required by hadron spectroscopy (see equation (16.143)). The configuration is stable because there is no way for the flux to leak away: it is a conserved quantized quantity.

For the eventual application to QCD, one will want (presumably) particles carrying non-zero values of the colour quantum numbers to be confined. These quantum numbers are the analogues of electric charge in the U(1) case, rather than of magnetic charge. We imagine, therefore, interchanging the roles of magnetism and electricity in all of the foregoing. Indeed, the Maxwell equations have such a symmetry when monopoles are present. The essential feature of the superconducting ground state was that it involved the coherent state formed by condensation of electrically charged bosonic fermion pairs. A vacuum which confined filaments of $\boldsymbol{E}$ rather than $\boldsymbol{B}$ may be formed as a coherent state of condensed magnetic monopoles (Mandelstam 1976, 't Hooft 1976b). These $\boldsymbol{E}$ filaments would then terminate on electric charges. Now magnetic monopoles do not occur naturally as solutions of QED: they would have to be introduced by hand. Remarkably enough, however, solutions of the magnetic monopole type do occur in the case of non-Abelian gauge field theories, whose symmetry is spontaneously broken to an electromagnetic $\mathrm{U(1)_{em}}$ gauge group. Just this circumstance can arise in a grand unified theory which contains $\mathrm{SU(3)_c}$ and a residual $\mathrm{U(1)_{em}}$. Incidentally, these monopole solutions provide an illuminating way of thinking about charge quantization: as Dirac (1931) pointed out, the existence of just one monopole implies, from his quantization condition (19.62), that charge is quantized.

When these ideas are applied to QCD, $\boldsymbol{E}$ and $\boldsymbol{B}$ must be understood as the appropriate colour fields (i.e. they carry an $\mathrm{SU(3)_c}$ index). The group structure of SU(3) is also quite different from that of U(1) models, and we do not want to be restricted just to static solutions (as in the GL theory, here used as an analogue). Whether in fact the real QCD vacuum (ground state) is formed as some such coherent plasma of monopoles, with confinement of electric charges and flux, is a subject of continuing research; other schemes are also possible. As so often stressed, the difficulty lies in the non-perturbative nature of the confinement problem.

## 19.5 't Hooft's gauges

We must now at last grasp the nettle and consider what happens if, in the parametrization

$$\hat{\phi} = |\hat{\phi}| \exp(\mathrm{i}\hat{\theta}(x)/v) \tag{19.63}$$

we do not choose the gauge (cf (19.52))

$$\partial_\mu \hat{A}^\mu = \Box\hat{\theta}/M. \tag{19.64}$$

This was the gauge that enabled us to transform away the phase degree of freedom and reduce the equation of motion for the electromagnetic field to that of a massive vector boson. Instead of using the modulus and phase as the two independent degrees of freedom for the complex Higgs field $\hat{\phi}$, we now choose to parametrize $\hat{\phi}$, quite generally, by the decomposition

$$\hat{\phi} = 2^{-1/2}[v + \hat{\chi}_1(x) + \mathrm{i}\hat{\chi}_2(x)] \tag{19.65}$$

where the vacuum values of $\hat{\chi}_1$ and $\hat{\chi}_2$ are zero. Substituting this form for $\hat{\phi}$ into the master equation for $\hat{A}^\nu$ (obtained from (19.43) and (19.44))

$$\Box\hat{A}^\nu - \partial^\nu(\partial_\mu\hat{A}^\mu) = \mathrm{i}q[\hat{\phi}^\dagger\partial^\nu\hat{\phi} - (\partial^\nu\hat{\phi})^\dagger\hat{\phi}] - 2q^2\hat{A}^\nu\hat{\phi}^\dagger\hat{\phi} \tag{19.66}$$

leads to the equation of motion

$$\begin{aligned}(\Box + M^2)\hat{A}^\nu - \partial^\nu(\partial_\mu\hat{A}^\mu) = & -M\partial^\nu\hat{\chi}_2 + q(\hat{\chi}_2\partial^\nu\hat{\chi}_1 - \hat{\chi}_1\partial^\nu\hat{\chi}_2) \\ & - q^2\hat{A}^\nu(\hat{\chi}_1^2 + 2v\hat{\chi}_1 + \hat{\chi}_2^2)\end{aligned} \tag{19.67}$$

with $M = qv$. At first sight this just looks like the equation of motion of an ordinary massive vector field $\hat{A}^\nu$ coupled to a rather complicated current. However, this certainly cannot be right, as we can see by a count of the degrees of freedom. In the previous gauge we had four degrees of freedom, counted either as two for the original massless $\hat{A}^\nu$ plus one each for $\hat{\theta}$ and $\hat{h}$, or as three for the massive $\hat{A}'^\nu$ and one for $\hat{h}$. If we take this new equation at face value, there seem to be three degrees of freedom for the massive field $\hat{A}^\nu$ and one for each of $\hat{\chi}_1$ and $\hat{\chi}_2$, making *five* in all. Actually, we know perfectly well that we can make use of the freedom gauge choice to set $\hat{\chi}_2$ to zero, say, reducing $\hat{\phi}$ to a real quantity and eliminating a spurious degree of freedom: we have then returned to the form (19.55). In terms of (19.67), the consequence of the unwanted degree of freedom is quite subtle, but it is basic to all gauge theories and already appeared in the photon case, in section 7.3.2. The difficulty arises when we try to calculate the propagator for $\hat{A}^\nu$ from equation (19.67).

The operator on the left-hand side can be simply inverted, as was done in section 19.1, to yield (apparently) the standard massive vector boson propagator

$$\mathrm{i}(-g^{\mu\nu} + k^\mu k^\nu/M^2)/(k^2 - M^2). \tag{19.68}$$

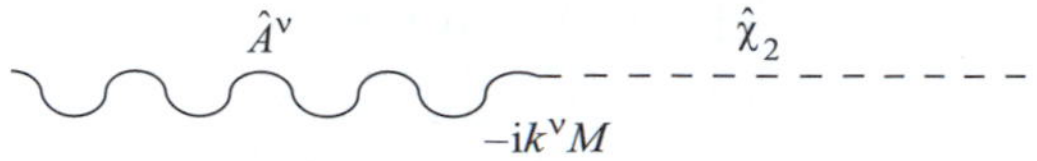


**Figure 19.3.** $\hat{A}^\nu$–$\hat{\chi}_2$ coupling.

**Figure 19.4.** Series for the full $\hat{A}^\nu$ propagator.

However, the current on the right-hand side is rather peculiar: instead of having only terms corresponding to $\hat{A}^\nu$ coupling to two or three particles, there is also a term involving only one field. This is the term $-M\partial^\nu\hat{\chi}_2$, which tells us that $\hat{A}^\nu$ actually couples directly to the scalar field $\chi_2$ via the gradient coupling $(-M\partial^\nu)$. In momentum space this corresponds to a coupling strength $-ik^\nu M$ and an associated vertex as shown in figure 19.3. Clearly, for a scalar particle, the momentum 4-vector is the only quantity that can couple to the vector index of the vector boson. The existence of this coupling shows that the propagators of $\hat{A}^\nu$ and $\hat{\chi}_2$ are necessarily mixed: the complete vector propagator must be calculated by summing the infinite series shown diagrammatically in figure 19.4. This complication is, of course, completely eliminated by the gauge choice $\hat{\chi}_2 = 0$. However, we are interested in pursuing the case $\hat{\chi}_2 \neq 0$.

In figure 19.4 the only unknown factor is the propagator for $\hat{\chi}_2$. This can be easily found by substituting (19.65) into $\hat{\mathcal{L}}_{\rm H}$ and examining the part which is quadratic in the $\hat{\chi}$'s: we find (problem 19.6) that

$$\hat{\mathcal{L}}_{\rm H} = \tfrac{1}{2}\partial_\mu\hat{\chi}_1\partial^\mu\hat{\chi}_1 + \tfrac{1}{2}\partial_\mu\hat{\chi}_2\partial^\mu\hat{\chi}_2 - \mu^2\hat{\chi}_1^2 + \text{cubic and quartic terms.} \tag{19.69}$$

Equation (19.69) confirms that $\hat{\chi}_1$ is a massive field with mass $\sqrt{2}\mu$ (like the $\hat{h}$ in (19.56)), while $\hat{\chi}_2$ is massless. The $\hat{\chi}_2$ propagator is, therefore, $\mathrm{i}/k^2$. Now that all the elements of the diagrams are known, we can formally sum the series by generalizing the well-known result ((cf 10.12) and (11.26))

$$(1-x)^{-1} = 1 + x + x^2 + x^3 + \cdots. \tag{19.70}$$

Diagrammatically, we rewrite the propagator of figure 19.4 as in figure 19.5 and perform the sum. Inserting the expressions for the propagators and vector–scalar coupling and keeping track of the indices, we finally arrive at the result (problem 19.7)

$$\mathrm{i}\left(\frac{-g^{\mu\lambda} + k^\mu k^\lambda/M^2}{k^2 - M^2}\right)(g^\nu_\lambda - k^\nu k_\lambda/k^2)^{-1} \tag{19.71}$$

for the full propagator. But the inverse required in (19.71) is precisely (with a lowered index) the one we needed for the photon propagator in (7.88)—and, as we saw there, it does not exist. At last the fact that we are dealing with a gauge theory has caught up with us!

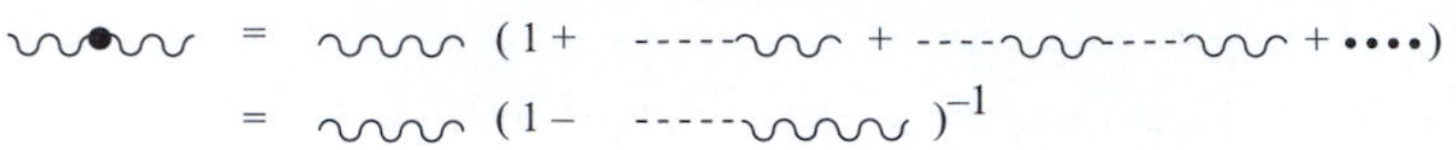

**Figure 19.5.** Formal summation of the series in figure 19.4.

As we saw in section 7.3.2, to obtain a well-defined gauge field propagator we need to *fix the gauge*. A clever way to do this in the present (spontaneously broken) case was suggested by 't Hooft (1971b). His proposal was to set

$$\partial_\mu \hat{A}^\mu = M\xi\hat{\chi}_2 \tag{19.72}$$

where $\xi$ is an arbitrary gauge parameter[1] (not to be confused with the superconducting coherence length). This condition is manifestly covariant and, moreover, it effectively reduces the degrees of freedom by one. Inserting (19.72) into (19.67), we obtain

$$(\Box + M^2)\hat{A}^\nu - \partial^\nu(\partial_\mu\hat{A}^\mu)(1 - 1/\xi) = q(\hat{\chi}_2\partial^\nu\hat{\chi}_1 - \hat{\chi}_1\partial^\nu\hat{\chi}_2) \tag{19.73}$$

$$- q^2\hat{A}^\nu(\hat{\chi}_1^2 + 2v\hat{\chi}_1 + \hat{\chi}_2^2). \tag{19.74}$$

The operator appearing on the left-hand side now *does* have an inverse (see problem 19.8) and yields the general form for the gauge boson propagator

$$\mathrm{i}\left[-g^{\mu\nu} + \frac{(1-\xi)k^\mu k^\nu}{k^2 - \xi M^2}\right](k^2 - M^2)^{-1}. \tag{19.75}$$

This propagator is very remarkable.[2] The standard massive vector boson propagator

$$\mathrm{i}(-g^{\mu\nu} + k^\mu k^\nu/M^2)(k^2 - M^2)^{-1} \tag{19.76}$$

is seen to correspond to the limit $\xi \to \infty$ and, in this gauge, the high-energy disease outlined in section 19.1 appears to threaten renormalizability (in fact, it can be shown that there is a consistent set of Feynman rules for this gauge and the theory is renormalizable thanks to many cancellations of divergences). For any finite $\xi$, however, the high-energy behaviour of the gauge boson propagator is actually $\sim 1/k^2$, which is as good as the *renormalizable* theory of QED (in Lorentz gauge). Note, however, that there seems to be another pole in the propagator (19.75) at $k^2 = \xi M^2$: this is surely unphysical since it depends on the arbitrary parameter $\xi$. A full treatment ('t Hooft 1971b) shows that this pole is always cancelled by an exactly similar pole in the propagator for the $\hat{\chi}_2$

[1] We shall not enter here into the full details of quantization in such a gauge: we shall effectively treat (19.72) as a classical field relation.

[2] A vector boson propagator of similar form was first introduced by Lee and Yang (1962) but their discussion was not within the framework of a spontaneously broken theory, so that Higgs particles were not present and the physical limit was obtained *only* as $\xi \to 0$.

field itself. These finite-$\xi$ gauges are called *R gauges* (since they are 'manifestly renormalizable') and typically involve unphysical Higgs fields such as $\hat{\chi}_2$. The infinite-$\xi$ gauge is known as the *U gauge* (U for unitary) since only physical particles appear in this gauge. For tree-diagram calculations, of course, it is easiest to use the U-gauge Feynman rules: the technical difficulties with this gauge choice only enter in loop calculations, for which the R-gauge choice is easier.

Note that in our master formula (19.75) for the gauge boson propagator the limit $M \to 0$ may be safely taken (compare the remarks about this limit for the 'naive' massive vector boson propagator in section 19.1). This yields the massless vector boson (photon) propagator in a general $\xi$-gauge, exactly as in equation (7.119) or (19.23).

We now proceed with the generalization of these ideas to the non-Abelian SU(2) case, which is the one relevant to the electroweak theory. The general non-Abelian case was treated by Kibble (1967).

## 19.6 Spontaneously broken local SU(2)×U(1) symmetry

We shall limit our discussion of the spontaneous breaking of a local non-Abelian symmetry to the particular case needed for the electroweak part of the Standard Model. This is, in fact, just the local version of the model studied in section 17.6. As noted there, the Lagrangian $\hat{\mathcal{L}}_\Phi$ of (17.97) is invariant under global SU(2) transformations of the form (17.100), and also global U(1) transformations (17.101). Thus, in the local version, we shall have to introduce three SU(2) gauge fields (as in section 13.1), which we call $\hat{W}_i^\mu(x)$ $(i = 1, 2, 3)$, and one U(1) gauge field $\hat{B}^\mu(x)$. We recall that the scalar field $\hat{\phi}$ is an SU(2)-doublet

$$\hat{\phi} = \begin{pmatrix} \hat{\phi}^+ \\ \hat{\phi}^0 \end{pmatrix} \tag{19.77}$$

so that the SU(2) covariant derivative acting on $\hat{\phi}$ is as given in (13.10), namely

$$\hat{D}^\mu = \partial^\mu + \mathrm{i}g\boldsymbol{\tau}\cdot\hat{\boldsymbol{W}}^\mu/2. \tag{19.78}$$

To this must be added the U(1) piece, which we write as $\mathrm{i}g'\hat{B}^\mu/2$, the $\frac{1}{2}$ being for later convenience. The Lagrangian is, therefore,

$$\hat{\mathcal{L}}_{\mathrm{G}\Phi} = (\hat{D}_\mu\hat{\phi})^\dagger(\hat{D}^\mu\hat{\phi}) + \mu^2\hat{\phi}^\dagger\hat{\phi} - \frac{\lambda}{4}(\hat{\phi}^\dagger\hat{\phi})^2 - \frac{1}{4}\hat{\boldsymbol{F}}_{\mu\nu}\cdot\hat{\boldsymbol{F}}^{\mu\nu} - \frac{1}{4}\hat{G}_{\mu\nu}\hat{G}^{\mu\nu} \tag{19.79}$$

where

$$\hat{D}^\mu\hat{\phi} = (\partial^\mu + \mathrm{i}g\boldsymbol{\tau}\cdot\hat{\boldsymbol{W}}^\mu/2 + \mathrm{i}g'\hat{B}^\mu/2)\hat{\phi} \tag{19.80}$$

$$\hat{\boldsymbol{F}}^{\mu\nu} = \partial^\mu\hat{\boldsymbol{W}}^\nu - \partial^\nu\hat{\boldsymbol{W}}^\mu - g\hat{\boldsymbol{W}}^\mu \times \hat{\boldsymbol{W}}^\nu \tag{19.81}$$

and

$$\hat{G}^{\mu\nu} = \partial^\mu \hat{B}^\nu - \partial^\nu \hat{B}^\mu. \tag{19.82}$$

We must now decide how to choose the non-zero vacuum expectation value that breaks this symmetry. The essential point for the electroweak application is that, after symmetry breaking, we should be left with three massive boson gauge bosons (which will be the $W^\pm$ and $Z^0$) and one massless gauge boson, the photon. We may reasonably guess that the massless boson will be associated with a symmetry that is *un*broken by the vacuum expectation value. Put differently, we certainly do not want a 'superconducting' massive photon to emerge from the theory in this case, as the physical vacuum is not an electromagnetic superconductor. This means that we do not want to give a vacuum value to a charged field (as is done in the BCS ground state). However, we do want it to behave as a 'weak' superconductor, generating mass for $W^\pm$ and $Z^0$. The choice suggested by Weinberg (1967)) was

$$\langle 0|\hat{\phi}|0\rangle = \begin{pmatrix} 0 \\ v/\sqrt{2} \end{pmatrix} \tag{19.83}$$

where $v/\sqrt{2} = \sqrt{2}\mu/\lambda^{1/2}$, which we have already considered in the global case in section 17.6. As pointed out there, (19.83) implies that the vacuum remains invariant under the combined transformation of 'U(1) + third component of SU(2) isospin'—that is, (19.83) implies

$$(\tfrac{1}{2} + t_3^{(\frac{1}{2})})\langle 0|\hat{\phi}|0\rangle = 0 \tag{19.84}$$

and hence

$$\langle 0|\hat{\phi}|0\rangle \to (\langle 0|\hat{\phi}|0\rangle)' = \exp[\mathrm{i}\alpha(\tfrac{1}{2} + t_3^{(1/2)})]\langle 0|\hat{\phi}|0\rangle = \langle 0|\hat{\phi}|0\rangle \tag{19.85}$$

where, as usual, $t_3^{(1/2)} = \tau_3/2$ (we are using lower case $t$ for isospin now, anticipating that it is the *weak*, rather than hadronic, isospin—see chapter 21).

We now need to consider oscillations about (19.83) in order to see the physical particle spectrum. As in (17.107), we parametrize these conveniently as

$$\hat{\phi} = \exp(-\mathrm{i}\hat{\boldsymbol{\theta}}(x)\cdot\boldsymbol{\tau}/2v)\begin{pmatrix} 0 \\ \frac{1}{\sqrt{2}}(v + \hat{H}(x)) \end{pmatrix} \tag{19.86}$$

(compare (19.45)). However, this time, in contrast to (17.107) but just as in (19.55), we can reduce the phase fields $\hat{\boldsymbol{\theta}}$ to zero by an appropriate gauge transformation, and it is simplest to examine the particle spectrum in this (*unitary*) gauge. Substituting

$$\hat{\phi} = \begin{pmatrix} 0 \\ \frac{1}{\sqrt{2}}(v + \hat{H}(x)) \end{pmatrix} \tag{19.87}$$

into (19.79) and retaining only terms which are second order in the fields (i.e. kinetic energies or mass terms), we find (problem 19.9) that

$$\begin{aligned}\hat{\mathcal{L}}^{\text{free}}_{\text{G}\Phi} &= \tfrac{1}{2}\partial_\mu \hat{H}\partial^\mu \hat{H} - \mu^2 \hat{H}^2 \\ &\quad - \tfrac{1}{4}(\partial_\mu \hat{W}_{1\nu} - \partial_\nu \hat{W}_{1\mu})(\partial^\mu \hat{W}_1^\nu - \partial^\nu \hat{W}_1^\mu) + \tfrac{1}{8}g^2 v^2 \hat{W}_{1\mu}\hat{W}_1^\mu \\ &\quad - \tfrac{1}{4}(\partial_\mu \hat{W}_{2\nu} - \partial_\nu \hat{W}_{2\mu})(\partial^\mu \hat{W}_2^\nu - \partial^\nu \hat{W}_2^\mu) + \tfrac{1}{8}g^2 v^2 \hat{W}_{2\mu}\hat{W}_2^\mu \\ &\quad - \tfrac{1}{4}(\partial_\mu \hat{W}_{3\nu} - \partial_\nu \hat{W}_{3\mu})(\partial^\mu \hat{W}_3^\nu - \partial^\nu \hat{W}_3^\mu) - \tfrac{1}{4}\hat{G}_{\mu\nu}\hat{G}^{\mu\nu} \\ &\quad + \tfrac{1}{8}v^2(g\hat{W}_{3\mu} - g'\hat{B}_\mu)(g\hat{W}_3^\mu - g'\hat{B}^\mu). \end{aligned} \tag{19.88}$$

The first line of (19.88) tells us that we have a scalar field of mass $\sqrt{2}\mu$ (the Higgs boson, again). The next two lines tell us that the components $\hat{W}_1$ and $\hat{W}_2$ of the triplet $(\hat{W}_1, \hat{W}_2, \hat{W}_3)$ acquire a mass (cf (19.56) in the U(1) case)

$$M_1 = M_2 = gv/2 \equiv M_{\text{W}}. \tag{19.89}$$

The last two lines show us that the fields $\hat{W}_3$ and $\hat{B}$ are mixed. But they can easily be unmixed by noting that the last term in (19.88) involves only the combination $g\hat{W}_3^\mu - g'\hat{B}^\mu$, which evidently acquires a mass. This suggests introducing the normalized linear combination

$$\hat{Z}^\mu = \cos\theta_{\text{W}}\hat{W}_3^\mu - \sin\theta_{\text{W}}\hat{B}^\mu \tag{19.90}$$

where

$$\cos\theta_{\text{W}} = g/(g^2 + g'^2)^{1/2} \qquad \sin\theta_{\text{W}} = g'/(g^2 + g'^2)^{1/2} \tag{19.91}$$

together with the orthogonal combination

$$\hat{A}^\mu = \sin\theta_{\text{W}}\hat{W}_3^\mu + \cos\theta_{\text{W}}\hat{B}^\mu. \tag{19.92}$$

We then find that the last two lines of (19.88) become

$$-\tfrac{1}{4}(\partial_\mu \hat{Z}_\nu - \partial_\nu \hat{Z}_\mu)(\partial_\mu \hat{Z}^\nu - \partial^\nu \hat{Z}^\mu) + \tfrac{1}{8}v^2(g^2 + g'^2)\hat{Z}_\mu \hat{Z}^\mu - \tfrac{1}{4}\hat{F}_{\mu\nu}\hat{F}^{\mu\nu} \tag{19.93}$$

where

$$\hat{F}_{\mu\nu} = \partial_\mu \hat{A}_\nu - \partial_\nu \hat{A}_\mu. \tag{19.94}$$

Thus

$$M_{\text{Z}} = \tfrac{1}{2}v(g^2 + g'^2)^{1/2} = M_{\text{W}}/\cos\theta_{\text{W}} \tag{19.95}$$

and

$$M_{\text{A}} = 0. \tag{19.96}$$

Counting degrees of freedom as in the local U(1) case, we originally had 12 in (19.79)—three massless $\hat{W}$'s and one massless $\hat{B}$, which is eight degrees of freedom in all, together with four $\hat{\phi}$-fields. After symmetry breaking, we have three massive vector fields $\hat{W}_1$, $\hat{W}_2$ and $\hat{Z}$ with nine degrees of freedom, one massless vector field $\hat{A}$ with two, and one massive scalar $\hat{H}$. Of course, the physical application will be to identify the $\hat{W}$ and $\hat{Z}$ fields with those physical particles, and the $\hat{A}$ field with the massless photon. In the gauge (19.87), the W and Z particles have propagators of the form (19.22).

The identification of $\hat{A}^\mu$ with the photon field is made clearer if we look at the form of $D_\mu\hat{\phi}$ written in terms of $\hat{A}_\mu$ and $\hat{Z}_\mu$, discarding the $\hat{W}_1$, $\hat{W}_2$ pieces:

$$D_\mu\hat{\phi} = \left\{\partial_\mu + \mathrm{i}g\sin\theta_\mathrm{W}\left(\frac{1+\tau_3}{2}\right)\hat{A}_\mu + \frac{\mathrm{i}g}{\cos\theta_\mathrm{W}}\left[\frac{\tau_3}{2} - \sin^2\theta_\mathrm{W}\left(\frac{1+\tau_3}{2}\right)\right]\hat{Z}_\mu\right\}\hat{\phi}. \tag{19.97}$$

Now the operator $(1+\tau_3)$ acting on $\langle 0|\hat{\phi}|0\rangle$ gives zero, as observed in (19.84), and this is why $\hat{A}_\mu$ does not acquire a mass when $\langle 0|\hat{\phi}|0\rangle \neq 0$ (gauge fields coupled to *unbroken* symmetries of $\langle 0|\hat{\phi}|0\rangle$ do *not* become massive). Although certainly not unique, this choice of $\hat{\phi}$ and $\langle 0|\hat{\phi}|0\rangle$ is undoubtedly very economical and natural. We are interpreting the zero eigenvalue of $(1+\tau_3)$ as the electromagnetic charge of the vacuum, which we do not wish to be non-zero. We then make the identification

$$e = g\sin\theta_\mathrm{W} \tag{19.98}$$

in order to get the right 'electromagnetic $D_\mu$' in (19.97).

We emphasize once more that the particular form of (19.88) corresponds to a *choice of gauge*, namely the unitary one (cf the discussions in sections 19.3 and 19.5). There is always the possibility of using other gauges, as in the Abelian case, and this will, in general, be advantageous when doing loop calculations involving renormalization. We would then return to a general parametrization such as (cf (19.65) and (17.95))

$$\hat{\phi} = \begin{pmatrix} 0 \\ v/\sqrt{2} \end{pmatrix} + \frac{1}{\sqrt{2}}\begin{pmatrix} \hat{\phi}_2 - \mathrm{i}\hat{\phi}_1 \\ \hat{\sigma} - \mathrm{i}\hat{\phi}_3 \end{pmatrix} \tag{19.99}$$

and add 't Hooft gauge-fixing terms

$$-\frac{1}{2\xi}\left\{\sum_{i=1,2}(\partial_\mu\hat{W}_i^\mu + \xi M_\mathrm{W}\hat{\phi}_i)^2 + (\partial_\mu\hat{Z}^\mu + \xi M_\mathrm{Z}\hat{\phi}_3)^2 + (\partial_\mu\hat{A}^\mu)^2\right\}. \tag{19.100}$$

In this case the gauge boson propagators are all of the form (19.75) and $\xi$-dependent. In such gauges, the Feynman rules will have to involve graphs corresponding to exchange of quanta of the 'unphysical' fields $\hat{\phi}_i$, as well as those of the physical Higgs scalar $\hat{\sigma}$. There will also have to be suitable ghost interactions in the non-Abelian sector as discussed in section 13.5.3. The complete Feynman rules of the electroweak theory are given in appendix B of Cheng and Li (1984).

The model introduced here is actually the 'Higgs sector' of the Standard Model but without any couplings to fermions. We have seen how, by supposing that the potential in (19.79) has the symmetry-breaking sign of the parameter $\mu^2$, the $W^{\pm}$ and $Z^0$ gauge bosons can be given masses. This seems to be an ingenious and even elegant 'mechanism' for arriving at a renormalizable theory of massive vector bosons. One may, of course, wonder whether this 'mechanism' is after all purely phenomenological, somewhat akin to the GL theory of a superconductor. In the latter case, we know that it can be derived from 'microscopic' BCS theory and this naturally leads to the question whether there could be a similar underlying 'dynamical' theory, behind the Higgs sector. It is, in fact, quite simple to construct a theory in which the Higgs fields $\hat{\phi}$ appear as bound, or composite, states of heavy fermions.

But generating masses for the gauge bosons is not the only job that the Higgs sector does in the Standard Model: it also generates masses for all the fermions. As we will see in chapter 22, the gauge symmetry of the weak interactions is a *chiral* one which requires that there should be no explicit fermion masses in the Lagrangian. We saw in chapter 18 how it is likely that the strong QCD interactions do, in fact, break chiral symmetry for the quarks, spontaneously. But, of course, the leptons are not coupled to QCD, and even as far as the quarks are concerned we saw that some small Lagrangian mass was required (to give a finite mass to the pion, for example). Thus for both quarks and leptons a chiral-symmetry-breaking mass seems unavoidable. To preserve the weak gauge symmetry, this must—in its turn—be interpreted as arising 'spontaneously' also; that is, *not* via an explicit mass term in the Lagrangian. The dynamical generation of fermion masses would, in fact, be closely analogous to the generation of the energy gap in the BCS theory, as we saw in section 18.1. So we may ask: is it possible to find a dynamical theory which generates *both* masses for the vector bosons *and* for the fermions? Such theories are generically known as 'technicolour models' (Weinberg 1979, Susskind 1979) and they have been intensively studied (see, for example, Peskin (1997)). One problem is that such theories are already tightly constrained by the precision electroweak experiments (see chapter 22), and meeting these constraints seems to require rather elaborate kinds of models. However, technicolour theories do offer the prospect of a new strongly interacting sector, which could be probed in the next generation of colliders. But such ideas take us beyond the scope of the present volume. Within the Standard Model, one proceeds along what seems a more phenomenological route, attributing the masses of fermions to their couplings with the Higgs field, in a way quite analogous to that in which the nucleon acquired a mass in the linear $\sigma$-model of section 18.3, and which will be explained in chapter 22.

We now turn, in the last part of the book, to weak interactions and the electroweak theory.

## Problems

**19.1** Show that

$$[(-k^2 + M^2)\, g^{\nu\mu} + k^\nu k^\mu] \left( \frac{-g_{\mu\rho} + k_\mu k_\rho / M^2}{k^2 - M^2} \right) = g^\nu_\rho.$$

**19.2** Verify (19.18).

**19.3** Verify (19.19).

**19.4** Verify (19.46).

**19.5** Insert (19.55) into $\hat{\mathcal{L}}_{\rm H}$ of (19.40) and derive (19.56) for the quadratic terms.

**19.6** Insert (19.65) into $\hat{\mathcal{L}}_{\rm H}$ of (19.40) and derive the quadratic terms of (19.69).

**19.7** Derive (19.71).

**19.8** Write the left-hand side of (19.74) in momentum space (as in (19.4)), and show that the inverse of the factor multiplying $\tilde{\hat{A}}^\mu$ is (19.75) without the 'i' (cf problem 19.1).

**19.9** Verify (19.88).

# PART 8

# WEAK INTERACTIONS AND THE ELECTROWEAK THEORY

# 20

# INTRODUCTION TO THE PHENOMENOLOGY OF WEAK INTERACTIONS

Public letter to the group of the Radioactives at the district society meeting in Tübingen:

Physikalisches Institut
der Eidg. Technischen Hochschule
Gloriastr.
Zürich

Zürich, 4 December 1930

Dear Radioactive Ladies and Gentlemen,
As the bearer of these lines, to whom I graciously ask you to listen, will explain to you in more detail, how because of the 'wrong' statistics of the N and $^6$Li nuclei and the continuous $\beta$-spectrum, I have hit upon a desperate remedy to save the 'exchange theorem' of statistics and the law of conservation of energy. Namely, the possibility that there could exist in the nuclei electrically neutral particles, that I wish to call neutrons, which have the spin $\frac{1}{2}$ and obey the exclusion principle and which further differ from light quanta in that they do not travel with the velocity of light. The mass of the neutrons should be of the same order of magnitude as the electron mass and in any event not larger than 0.01 proton masses.—The continuous $\beta$-spectrum would then become understandable by the assumption that in $\beta$-decay, a neutron is emitted in addition to the electron such that the sum of the energies of the neutron and electron is constant......
I admit that on a first look my way out might seem to be quite unlikely, since one would certainly have seen the neutrons by now if they existed. But nothing ventured nothing gained, and the seriousness of the matter with the continuous $\beta$-spectrum is illustrated by a quotation of my honoured predecessor in office, Mr Debye, who recently told me in Brussels: 'Oh, it is best not to think about it, like the new taxes.' Therefore one should earnestly discuss each way of salvation.—So, dear Radioactives, examine and judge it.—Unfortunately I cannot appear in Tübingen personally, since I am indispensable here in Zürich because of a ball on the night of 6/7 December.—With my best regards to you, and also Mr Back, your humble servant,

W Pauli

Quoted from Winter (2000), pp 4–5.

At the end of the previous chapter we arrived at an important part of the Lagrangian of the Standard Model, namely the terms involving just the gauge and Higgs fields. The full electroweak Lagrangian also includes, of course, the couplings of these fields to the quarks and leptons. We could at this point simply write these couplings down, with little motivation, and proceed at once to discuss the empirical consequences. But such an approach, though economical, would assume considerable knowledge of weak interaction phenomenology on the reader's part. We prefer to keep this book as self-contained as possible and so in the present chapter we shall provide an introduction to this phenomenology, following a 'semi-historical' route (for fuller historical treatments we refer the reader to Marshak *et al* (1969) or to Winter (2000), for example).

Much of what we shall discuss is still, for many purposes, a very useful approximation to the full theory at energies well below the masses of the $W^{\pm}$ ($\sim$80 GeV) and $Z^0$ ($\sim$90 GeV), as will be explained in section 21.2. Besides, as we shall see, in the neutrino sector especially the 'historical' data need to be carefully interpreted in order to understand the focus of much ongoing research.

## 20.1 Fermi's 'current–current' theory of nuclear $\beta$-decay and its generalizations

The first quantum field theory of a weak interaction process was proposed by Fermi (1934a, b) for nuclear $\beta$-decay, building on the 'neutrino hypothesis' of Pauli. In 1930, Pauli (in his 'Dear Radioactive Ladies and Gentlemen' letter) had suggested that the continuous $e^-$ spectrum in $\beta$-decay could be understood by supposing that, in addition to the $e^-$, the decaying nucleus also emitted a light, spin-$\frac{1}{2}$, electrically neutral particle, which he called the 'neutron'. In this first version of the proposal, Pauli regarded his hypothetical particle as a constituent of the nucleus. This had the attraction of solving not only the problem with the continuous $e^-$ spectrum but a second problem as well—what he called the 'wrong' statistics of the $^{14}N$ and $^{6}Li$ nuclei. Taking $^{14}N$ for definiteness, the problem was as follows. Assuming that the nucleus was somehow composed of the only particles (other than the photon) known in 1930, namely electrons and protons, one requires 14 protons and seven electrons for the known charge of seven. This implies a half-odd integer value for the total nuclear spin. But data from molecular spectra indicated that the nitrogen nuclei obeyed Bose–Einstein, not Fermi–Dirac statistics, so that—if the usual 'spin-statistics' connection were to hold—the spin of the nitrogen nucleus should be an integer, not a half-odd integer. This second part of Pauli's hypothesis was quite soon overtaken by the discovery of the (real) neutron by Chadwick (1932), after which it was rapidly accepted that nuclei consisted of protons and (Chadwick's) neutrons.

However, the $\beta$-spectrum problem remained and, at the Solvay Conference in 1933, Pauli restated his hypothesis (Pauli 1934), using now the name 'neutrino' which had meanwhile been suggested by Fermi. Stimulated by the discussions at the Solvay meeting, Fermi then developed his theory of $\beta$-decay. In the new

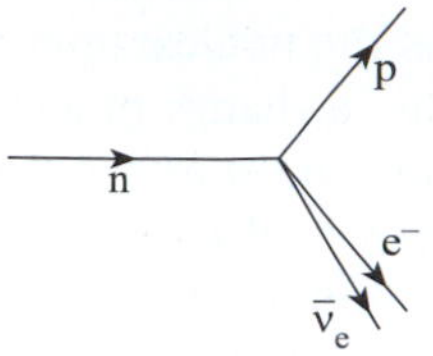


**Figure 20.1.** Four-fermion interaction for neutron $\beta$-decay.

picture of the nucleus, neither the electron nor the neutrino were to be thought of as nuclear constituents. Instead, the electron–neutrino pair had somehow to be created and emitted in the transition process of the nuclear decay, much as a photon is created and emitted in nuclear $\gamma$-decay. Indeed, Fermi relied heavily on the analogy with electromagnetism. The basic process was assumed to be the transition neutron→proton, with the emission of an $e^-\nu$ pair, as shown in figure 20.1. The n and p were then regarded as 'elementary' and without structure (point-like); the whole process took place at a single spacetime point, like the emission of a photon in QED. Further, Fermi conjectured that the nucleons participated via a weak interaction analogue of the electromagnetic transition *currents* frequently encountered in volume 1 for QED. In this case, however, rather than having the 'charge conserving' form of $\bar{u}_p\gamma^\mu u_p$ for instance, the 'weak current' had the form $\bar{u}_p\gamma^\mu u_n$, in which the charge of the nucleon changed. The lepton pair was also charged, obviously. The whole interaction then had to be Lorentz invariant, implying that the $e^-\nu$ pair had also to appear in a similar (4-vector) 'current' form. Thus a 'current–current' amplitude was proposed, of the form

$$A\bar{u}_p\gamma^\mu u_n\bar{u}_{e^-}\gamma_\mu u_\nu \tag{20.1}$$

where $A$ was constant. Correspondingly, the process was described field theoretically in terms of the local interaction density

$$A\bar{\hat{\psi}}_p(x)\gamma^\mu\hat{\psi}_n(x)\bar{\hat{\psi}}_e(x)\gamma_\mu\hat{\psi}_\nu(x). \tag{20.2}$$

The discovery of positron $\beta$-decay soon followed and then electron capture: these processes were easily accommodated by adding to (20.2) its Hermitian conjugate

$$A\bar{\hat{\psi}}_n(x)\gamma^\mu\hat{\psi}_p(x)\bar{\hat{\psi}}_\nu(x)\gamma^\mu\hat{\psi}_e(x) \tag{20.3}$$

taking $A$ to be real. The sum of (20.2) and (20.3) gave a good account of many observed characteristics of $\beta$-decay, when used to calculate transition probabilities in first-order perturbation theory.

Soon after Fermi's theory was presented, however, it became clear that the observed selection rules in some nuclear transitions could not be accounted for by the forms (20.2) and (20.3). Specifically, in 'allowed' transitions (where the orbital angular momentum carried by the leptons is zero), it was found that, while for many transitions the nuclear spin did not change ($\Delta J = 0$), for others—of comparable strength—a change of nuclear spin by one unit ($\Delta J = 1$) occurred. Now, in nuclear decays the energy release is very small ($\sim$ few MeV) compared to the mass of a nucleon, and so the non-relativistic limit is an excellent approximation for the nucleon spinors. It is then easy to see (problem 20.1) that, in this limit, the interactions (20.2) and (20.3) imply that the nucleon spins cannot 'flip'. Hence some other interaction(s) must be present. Gamow and Teller (1936) introduced the general four-fermion interaction, constructed from bilinear combinations of the nucleon pair and of the lepton pair, but not their derivatives. For example, the combination

$$\bar{\hat{\psi}}_{\mathrm{p}}(x)\hat{\psi}_{\mathrm{n}}(x)\bar{\hat{\psi}}_{\mathrm{e}}(x)\hat{\psi}_{\nu}(x) \tag{20.4}$$

could occur, and also

$$\bar{\hat{\psi}}_{\mathrm{p}}(x)\sigma_{\mu\nu}\hat{\psi}_{\mathrm{n}}(x)\bar{\hat{\psi}}_{\mathrm{e}}\sigma^{\mu\nu}\hat{\psi}_{\nu}(x) \tag{20.5}$$

where

$$\sigma_{\mu\nu} = \frac{\mathrm{i}}{2}(\gamma_\mu\gamma_\nu - \gamma_\nu\gamma_\mu). \tag{20.6}$$

The non-relativistic limit of (20.4) gives $\Delta J = 0$, but (20.5) allows $\Delta J = 1$. Other combinations are also possible, as we shall discuss shortly. Note that the interaction must always be Lorentz invariant.

Thus began a long period of difficult experimentation to establish the correct form of the $\beta$-decay interaction. With the discovery of the muon (section 1.3.1) and the pion (section 2.2), more weak decays became experimentally accessible, for example $\mu$ decay

$$\mu^- \to \mathrm{e}^- + \nu + \nu \tag{20.7}$$

and $\pi$ decay

$$\pi^- \to \mathrm{e}^- + \nu. \tag{20.8}$$

Note that we have deliberately called all the neutrinos just '$\nu$', without any particle/anti-particle indication or lepton flavour label: we shall have more to say on these matters in section 20.6. There were hopes that the couplings of the pairs (p,n), ($\nu$, $\mathrm{e}^-$) and ($\nu$, $\mu^-$) might have the same form ('universality') but the data were incomplete and, in part, apparently contradictory.

The breakthrough came in 1956, when Lee and Yang (1956) suggested that parity was not conserved in all weak decays. Hitherto, it had always been assumed that any physical interaction had to be such that parity was conserved, and this assumption had been built into the structure of the proposed $\beta$-decay interactions, such as (20.2), (20.4) or (20.5). Once it was looked for properly, following the analysis of Lee and Yang, parity violation was indeed found to be a strikingly evident feature of weak interactions.

## 20.2 Parity violation in weak interactions

In 1957, the experiment of Wu *et al* (1957) established, for the first time, that parity was violated in a weak interaction, specifically nuclear $\beta$-decay. The experiment involved a sample of $^{60}$Co ($J = 5$) cooled to 0.01 K in a solenoid. At this temperature most of the nuclear spins are aligned by the magnetic field and so there is a net polarization $\langle \boldsymbol{J} \rangle$, which is in the direction opposite to the applied magnetic field. $^{60}$Co decays to $^{60}$Ni ($J = 4$), a $\Delta J = 1$ transition. The degree of $^{60}$Co alignment was measured from observations of the angular distribution of $\gamma$-rays from $^{60}$Ni. The relative intensities of electrons emitted along and against the magnetic field direction were measured, and the results were consistent with a distribution of the form

$$I(\theta) = 1 - \langle \boldsymbol{J} \rangle \cdot \boldsymbol{p}/E \tag{20.9}$$

$$= 1 - v \cos\theta \tag{20.10}$$

where $v$, $\boldsymbol{p}$ and $E$ are, respectively, the electron speed, momentum and energy, and $\theta$ is the angle of emission of the electron with respect to $\langle \boldsymbol{J} \rangle$.

Why does this indicate parity violation? To see this, we must first recall the definition of vectors ('polar vectors') and pseudovectors ('axial vectors'). A polar vector is one which transforms in the same way as the coordinate $\boldsymbol{x}$ under the parity operator **P**

$$\mathbf{P} : \boldsymbol{x} \to -\boldsymbol{x}. \tag{20.11}$$

Thus a polar vector $\boldsymbol{V}$ is defined by the behaviour

$$\mathbf{P} : \boldsymbol{V} \to -\boldsymbol{V} \tag{20.12}$$

under parity. Examples are the velocity $\boldsymbol{v}$, momentum $\boldsymbol{p}$ and electromagnetic current $\boldsymbol{j}_{\text{em}}$. The vector product of two such vectors defines the behaviour of an axial vector

$$\mathbf{P} : \boldsymbol{U} \times \boldsymbol{V} \to (-\boldsymbol{U}) \times (-\boldsymbol{V}) = \boldsymbol{U} \times \boldsymbol{V} \tag{20.13}$$

under parity. In contrast to (20.11) and (20.12), an axial vector does not reverse sign under **P**: the most common example is angular momentum $\boldsymbol{l} = \boldsymbol{x} \times \boldsymbol{p}$. By extension, any angular momentum, such as spin, is also an axial vector. In forming scalar products therefore, we must now distinguish between a *scalar* such as $\boldsymbol{U} \cdot \boldsymbol{V}$ (the dot product of two polar vectors) and a *pseudoscalar* such as $\boldsymbol{U} \cdot (\boldsymbol{V} \times \boldsymbol{W})$ (the triple scalar product of three polar vectors):

$$\mathbf{P} : \boldsymbol{U} \cdot \boldsymbol{V} \to +\boldsymbol{U} \cdot \boldsymbol{V} \tag{20.14}$$

$$\mathbf{P} : \boldsymbol{U} \cdot (\boldsymbol{V} \times \boldsymbol{W}) \to -\boldsymbol{U} \cdot (\boldsymbol{V} \times \boldsymbol{W}). \tag{20.15}$$

The scalar remains the same under **P** but the pseudoscalar changes sign.

Consider now how the distribution (20.10) would be described in a parity-transformed coordinate system. Applying the rules just stated, $\langle \boldsymbol{J} \rangle \to \langle \boldsymbol{J} \rangle$ and

$\boldsymbol{p} \to -\boldsymbol{p}$ so that, as described by the new system, the distribution would have the form

$$I_{\mathbf{P}}(\theta) = 1 + v\cos\theta. \tag{20.16}$$

The difference between (20.16) and (20.10) implies that, by performing the measurement, we can *determine* which of the two coordinate systems we must, in fact, be using. The two are inequivalent, in contrast to all the other coordinate system equivalences which we have previously studied (e.g. under three-dimensional rotations and Lorentz transformations). This is an operational consequence of 'parity violation'. The crucial point in this example, evidently, is the appearance of the *pseudoscalar* quantity $\langle \boldsymbol{J} \rangle \cdot \boldsymbol{p}$ in (20.9), alongside the obviously scalar quantity '1'.

The Fermi theory, employing only vector currents, needs a modification to account for this result. To see how this may be done, we need to consider the behaviour of the Dirac spinors and fermion fields under **P**.

## 20.3 Parity transformation of Dirac wavefunctions and field operators

We consider the behaviour of the free-particle Dirac equation

$$\mathrm{i}\frac{\partial \psi(\boldsymbol{x}, t)}{\partial t} = -\mathrm{i}\boldsymbol{\alpha} \cdot \boldsymbol{\nabla} \psi(\boldsymbol{x}, t) + \beta m \psi(\boldsymbol{x}, t) \tag{20.17}$$

under the coordinate transformation

$$\mathbf{P} : \boldsymbol{x} \to \boldsymbol{x}' = -\boldsymbol{x}, t \to t. \tag{20.18}$$

Equation (20.17) will be *covariant* under (20.18) (see appendix D of volume 1 and also section 4.4 of volume 1) if we can find a wavefunction $\psi_{\mathbf{P}}(\boldsymbol{x}', t)$ for observers using the transformed coordinate system such that 'their' Dirac equation has exactly the same form in their system as (20.17):

$$\mathrm{i}\frac{\partial \psi_{\mathbf{P}}}{\partial t}(\boldsymbol{x}', t) = -\mathrm{i}\boldsymbol{\alpha} \cdot \boldsymbol{\nabla}' \psi_{\mathbf{P}}(\boldsymbol{x}', t) + \beta m \psi_{\mathbf{P}}(\boldsymbol{x}', t). \tag{20.19}$$

Now we know that $\boldsymbol{\nabla}' = -\boldsymbol{\nabla}$, since $\boldsymbol{x}' = -\boldsymbol{x}$. Hence, (20.19) becomes

$$\mathrm{i}\frac{\partial \psi_{\mathbf{P}}}{\partial t}(\boldsymbol{x}', t) = \mathrm{i}\boldsymbol{\alpha} \cdot \boldsymbol{\nabla} \psi_{\mathbf{P}}(\boldsymbol{x}', t) + \beta m \psi_{\mathbf{P}}(\boldsymbol{x}', t). \tag{20.20}$$

Multiplying this equation from the left by $\beta$ and using $\beta\boldsymbol{\alpha} = -\boldsymbol{\alpha}\beta$, we find that

$$\frac{\mathrm{i}\partial}{\partial t}[\beta \psi_{\mathbf{P}}(\boldsymbol{x}', t)] = -\mathrm{i}\boldsymbol{\alpha} \cdot \boldsymbol{\nabla}[\beta \psi_{\mathbf{P}}(\boldsymbol{x}', t)] + \beta m[\beta \psi(\boldsymbol{x}', t)]. \tag{20.21}$$

Comparing (20.21) and (20.17), it follows that we may consistently 'translate' between $\psi$ and $\psi_{\mathbf{P}}$ using the relation

$$\psi(\boldsymbol{x}, t) = \beta \psi_{\mathbf{P}}(-\boldsymbol{x}, t) \tag{20.22}$$

or, equivalently,

$$\psi_{\mathbf{P}}(\boldsymbol{x}, t) = \beta\psi(-\boldsymbol{x}, t). \tag{20.23}$$

In fact, we could include an arbitrary phase factor $\eta_{\mathbf{P}}$ on the right-hand side of (20.23): such a phase leaves the normalization of $\psi$ and all bilinears of the form $\bar{\psi}$ (gamma matrix) $\psi$ unaltered. The possibility of such a phase factor did not arise in the case of Lorentz transformations, since we can insist that for infinitesimal ones the transformed $\psi'$ and the original $\psi$ differ only infinitesimally (not by a finite phase factor). But the parity transformation cannot be built up out of infinitesimal steps—the coordinate system is either reflected or it is not. Parity is said to be a discrete transformation in contrast to the continuous Lorentz transformations (and rotations).

As an example of (20.23), consider the free-particle solutions in the standard form (4.40), (4.46):

$$\psi(\boldsymbol{x}, t) = N\begin{pmatrix} \phi \\ \frac{\boldsymbol{\sigma}\cdot\boldsymbol{p}}{E+m}\phi \end{pmatrix}\exp(-\mathrm{i}Et + \mathrm{i}\boldsymbol{p}\cdot\boldsymbol{x}). \tag{20.24}$$

Then

$$\psi_{\mathbf{P}}(\boldsymbol{x}, t) = \beta\psi(-\boldsymbol{x}, t) = N\begin{pmatrix} \phi \\ \frac{-\boldsymbol{\sigma}\cdot\boldsymbol{p}}{E+m}\phi \end{pmatrix}\exp(-\mathrm{i}Et - \mathrm{i}\boldsymbol{p}\cdot\boldsymbol{x}) \tag{20.25}$$

which can be conveniently summarized by the simple statement that the three-momentum $\boldsymbol{p}$ as seen in the parity transformed system is minus that in the original one, as expected. Note that $\boldsymbol{\sigma}$ does not change sign.

In the same way we can introduce the idea of a unitary quantum field operator $\hat{\mathbf{P}}$ which transforms Dirac field operators $\hat{\psi}(\boldsymbol{x}, t)$ according to

$$\hat{\psi}_{\mathbf{P}}(\boldsymbol{x}, t) \equiv \hat{\mathbf{P}}\hat{\psi}(\boldsymbol{x}, t)\hat{\mathbf{P}}^{-1} = \beta\hat{\psi}(-\boldsymbol{x}, t). \tag{20.26}$$

An explicit form for $\hat{\mathbf{P}}$ is given in section 15.11 of Bjorken and Drell (1965), for example.

Consider now the behaviour under $\hat{\mathbf{P}}$ of a 4-vector current of the form $\bar{\hat{\psi}}_1(x)\gamma^\mu\hat{\psi}_2(x)$. We have, for $\mu = 0$,

$$\begin{aligned} \bar{\hat{\psi}}_{1\mathbf{P}}(\boldsymbol{x}, t)\gamma^0\hat{\psi}_{2\mathbf{P}}(\boldsymbol{x}, t) &= \hat{\psi}^\dagger_{1\mathbf{P}}(\boldsymbol{x}, t)\hat{\psi}_{2\mathbf{P}}(\boldsymbol{x}, t) \\ &= \hat{\psi}^\dagger_1(-\boldsymbol{x}, t)\beta\cdot\beta\hat{\psi}_2(-\boldsymbol{x}, t) \\ &= \bar{\hat{\psi}}_1(-\boldsymbol{x}, t)\hat{\psi}_2(-\boldsymbol{x}, t) \end{aligned} \tag{20.27}$$

showing that the $\mu = 0$ component is a scalar under $\mathbf{P}$: this is to be expected, as the electric charge density, $\rho$, is also a scalar. For the spatial parts, we have

$$\begin{aligned} \bar{\hat{\psi}}_{1\mathbf{P}}(\boldsymbol{x}, t)\boldsymbol{\gamma}\hat{\psi}_{2\mathbf{P}}(\boldsymbol{x}, t) &= \hat{\psi}^\dagger_{1\mathbf{P}}(\boldsymbol{x}, t)\beta\boldsymbol{\gamma}\hat{\psi}_{2\mathbf{P}}(\boldsymbol{x}, t) \\ &= \hat{\psi}^\dagger_1(-\boldsymbol{x}, t)\beta\beta\boldsymbol{\gamma}\beta\hat{\psi}_2(-\boldsymbol{x}, t) \\ &= -\bar{\hat{\psi}}_1(-\boldsymbol{x}, t)\boldsymbol{\gamma}\hat{\psi}(-\boldsymbol{x}, t) \end{aligned} \tag{20.28}$$

using $\beta\boldsymbol{\gamma} = -\boldsymbol{\gamma}\beta$. Thus the spatial parts transform as a polar vector, like the current density $\boldsymbol{j}$.

To accommodate parity violation, we must, however, have both axial and polar vectors, so as to create pseudoscalars well as scalars. For Dirac particles, this is done via the $\gamma_5$ matrix already introduced in section 12.3.2. We recall the definition

$$\gamma_5 = \mathrm{i}\gamma^0\gamma^1\gamma^2\gamma^3. \tag{20.29}$$

$\gamma_5$ can easily be shown to anti-commute with the other four $\gamma$ matrices:

$$\{\gamma_5, \gamma_\mu\} = 0. \tag{20.30}$$

With the usual choice of the Dirac matrices used in chapter 4, namely

$$\beta = \begin{pmatrix} 1 & 0 \\ 0 & -1 \end{pmatrix} \qquad \boldsymbol{\alpha} = \begin{pmatrix} 0 & \boldsymbol{\sigma} \\ \boldsymbol{\sigma} & 0 \end{pmatrix} \tag{20.31}$$

and $\gamma_0 = \beta$, $\boldsymbol{\gamma} = \beta\boldsymbol{\alpha}$, we easily find that

$$\gamma_5 = \begin{pmatrix} 0 & 1 \\ 1 & 0 \end{pmatrix}. \tag{20.32}$$

Consider now the quantity $\bar{\hat{\psi}}_1(\boldsymbol{x}, t)\gamma_5\hat{\psi}_2(\boldsymbol{x}, t)$. Under the parity transformation, this becomes

$$\begin{aligned} \bar{\hat{\psi}}_{1\mathbf{P}}(\boldsymbol{x}, t)\gamma_5\hat{\psi}_{2\mathbf{P}}(\boldsymbol{x}, t) &= \hat{\psi}^\dagger_{1\mathbf{P}}(\boldsymbol{x}, t)\beta\gamma_5\hat{\psi}_{2\mathbf{P}}(\boldsymbol{x}, t) \\ &= \hat{\psi}^\dagger_1(-\boldsymbol{x}, t)\beta\beta\gamma_5\beta\hat{\psi}_2(-\boldsymbol{x}, t) \\ &= -\bar{\hat{\psi}}_1(-\boldsymbol{x}, t)\gamma_5\hat{\psi}_2(-\boldsymbol{x}, t) \end{aligned} \tag{20.33}$$

using $\beta\gamma_5 = -\gamma_5\beta$ and $\beta^2 = 1$. Thus, this combination of $\bar{\hat{\psi}}_1$ and $\hat{\psi}_2$ is a *pseudoscalar*.

Finally, and most importantly, consider the combination

$$\bar{\hat{\psi}}_1(\boldsymbol{x}, t)\gamma^\mu\gamma_5\hat{\psi}_2(\boldsymbol{x}, t).$$

The reader can easily check (problem 20.2) that the $\mu = 0$ component of this is a pseudoscalar, while the spatial part is an axial vector. We call this kind of 4-vector an *axial 4-vector*, the usual '$\bar{\psi}\gamma^\mu\psi$' one being just a *vector*, for short.

Let us write the components of an axial 4-vector $\hat{A}^\mu$ as

$$\hat{A}^\mu = (\hat{A}^0, \hat{\boldsymbol{A}}). \tag{20.34}$$

Then, under parity, $\hat{A}^0 \to -\hat{A}^0$ and $\hat{\boldsymbol{A}} \to \hat{\boldsymbol{A}}$, where we are suppressing possible spacetime arguments $\boldsymbol{x}, t$. Similarly, for an ordinary 4-vector

$$\hat{V}^\mu = (\hat{V}^0, \hat{\boldsymbol{V}}), \tag{20.35}$$

the components transform by $\hat{V}^0 \to \hat{V}^0$, $\hat{\boldsymbol{V}} \to -\hat{\boldsymbol{V}}$ under parity, as we have seen. It follows that the Lorentz invariant product

$$\hat{A}_\mu \hat{V}^\mu = \hat{A}^0 \hat{V}^0 - \hat{\boldsymbol{A}} \cdot \hat{\boldsymbol{V}} \tag{20.36}$$

transforms as a *pseudoscalar* under parity, while $\hat{A}_\mu \hat{A}^\mu$ and $\hat{V}_\mu \hat{V}^\mu$ both transform as scalars. We learn, therefore, that one way to introduce a Lorentz invariant pseudoscalar interaction is to form the 'dot product' of a $\hat{V}^\mu$ and an $\hat{A}^\mu$ type object. This proves to be the key to unlocking the structure of the weak interaction. Indeed, after many years of careful experiments, and many false trails, it was eventually established (always, of course, to within some experimental error) that the currents participating in Fermi's current–current interaction are, in fact, certain combinations of V-type and A-type currents, for both nucleons and leptons.

## 20.4 V – A theory: chirality and helicity

Quite soon after the discovery of parity violation, Sudarshan and Marshak (1958) and then Feynman and Gell-Mann (1958) and Sakurai (1958) proposed a specific form for the current–current interaction, namely the V – A ('V minus A') structure. For example, in place of the leptonic combination $\bar{u}_{\mathrm{e}^-}\gamma_\mu u_\nu$, these authors proposed the form $\bar{u}_{\mathrm{e}^-}\gamma_\mu(1-\gamma_5)u_\nu$, being the difference (with equal weight) of a V-type and an A-type current. For the part involving the nucleons, the proposal was slightly more complicated, having the form $\bar{u}_{\mathrm{p}}\gamma_\mu(1-r\gamma_5)u_{\mathrm{n}}$ where $r$ had the empirical value $\sim 1.2$. From our present perspective, of course, the hadronic transition is actually occurring at the quark level, so that rather than a transition $\mathrm{n} \to \mathrm{p}$ we now think in terms of a $\mathrm{d} \to \mathrm{u}$ one. In this case, the remarkable fact is that the appropriate current to use is, once again, essentially the simple 'V – A' one, $\bar{u}_{\mathrm{u}}\gamma_\mu(1-\gamma_5)u_{\mathrm{d}}$.[1] *This* V – A *structure for quarks and leptons is fundamental to the Standard Model.*

We must now at once draw the reader's attention to a rather remarkable feature of this V – A structure, which is that the $(1-\gamma_5)$ factor can be thought of as acting either on the $u$ spinor or on the $\bar{u}$ spinor. Consider, for example, a term $\bar{u}_{\mathrm{e}^-}\gamma_\mu(1-\gamma_5)u_\nu$. We have

$$\begin{aligned}
\bar{u}_{\mathrm{e}^-}\gamma_\mu(1-\gamma_5)u_\nu &= u^\dagger_{\mathrm{e}^-}\beta\gamma_\mu(1-\gamma_5)u_\nu \\
&= u^\dagger_{\mathrm{e}^-}(1-\gamma_5)\beta\gamma_\mu u_\nu \\
&= [(1-\gamma_5)u_{\mathrm{e}^-}]^\dagger\beta\gamma_\mu u_\nu \\
&= \overline{[(1-\gamma_5)u_{\mathrm{e}^-}]}\,\gamma_\mu u_\nu.
\end{aligned} \tag{20.37}$$

To understand the significance of this, it is advantageous to work with a different representation of the Dirac matrices. We work in a representation in which $\gamma_5$ is

[1] We shall see in section 20.10 that a slight modification is necessary.

chosen to be diagonal, namely

$$\gamma_5 = \begin{pmatrix} 1 & 0 \\ 0 & -1 \end{pmatrix} \quad \boldsymbol{\alpha} = \begin{pmatrix} \boldsymbol{\sigma} & 0 \\ 0 & -\boldsymbol{\sigma} \end{pmatrix} \quad \beta = \begin{pmatrix} 0 & 1 \\ 1 & 0 \end{pmatrix} \quad \boldsymbol{\gamma} = \begin{pmatrix} 0 & -\boldsymbol{\sigma} \\ \boldsymbol{\sigma} & 0 \end{pmatrix} \tag{20.38}$$

which is related to our 'usual' choice (20.31) by a unitary transformation. The $\boldsymbol{\alpha}$ and $\beta$ of (20.38) were introduced earlier in equation (4.97) and problem 4.15: readers who have not worked through that problem are advised to do so now. We may also suggest a backward glance at section 12.3.2 and chapter 17.

First of all, it is clear that any combination '$(1-\gamma_5)u$' is an eigenstate of $\gamma_5$ with eigenvalue $-1$:

$$\gamma_5(1-\gamma_5)u = (\gamma_5 - 1)u = -(1-\gamma_5)u \tag{20.39}$$

using $\gamma_5^2 = 1$. In the terminology of section 12.3.2, '$(1-\gamma_5)u$' has definite *chirality*, namely L ('left-handed'), meaning that it belongs to the eigenvalue $-1$ of $\gamma_5$. We may introduce the projection operators $P_{\mathrm{R}}$, $P_{\mathrm{L}}$ of section 12.3.2,

$$P_{\mathrm{L}} \equiv \left(\frac{1-\gamma_5}{2}\right) \qquad P_{\mathrm{R}} \equiv \left(\frac{1+\gamma_5}{2}\right) \tag{20.40}$$

satisfying

$$P_{\mathrm{R}}^2 = P_{\mathrm{R}} \qquad P_{\mathrm{L}}^2 = P_{\mathrm{L}} \qquad P_{\mathrm{R}}P_{\mathrm{L}} = P_{\mathrm{L}}P_{\mathrm{R}} = 0 \qquad P_{\mathrm{R}} + P_{\mathrm{L}} = 1 \tag{20.41}$$

and define

$$u_{\mathrm{L}} \equiv P_{\mathrm{L}}u \qquad u_{\mathrm{R}} \equiv P_{\mathrm{R}} \tag{20.42}$$

for any $u$. Then

$$\begin{aligned} \bar{u}_1\gamma_\mu\left(\frac{1-\gamma_5}{2}\right)u_2 &= \bar{u}_1\gamma_\mu P_{\mathrm{L}}u_2 = \bar{u}_1\gamma_\mu P_{\mathrm{L}}^2 u_2 \\ &= \bar{u}_1\gamma_\mu P_{\mathrm{L}}u_{2\mathrm{L}} = \bar{u}_1 P_{\mathrm{R}}\gamma_\mu u_{2\mathrm{L}} \\ &= u_1^\dagger P_{\mathrm{L}}\beta\gamma_\mu u_{2\mathrm{L}} = \bar{u}_{1\mathrm{L}}\gamma_\mu u_{2\mathrm{L}} \end{aligned} \tag{20.43}$$

which formalizes (20.37) and emphasizes the fact that *only the chiral L components of the u spinors enter into weak interactions*, a remarkably simple statement.

To see the physical consequences of this, we need the forms of the Dirac spinors in this new representation, which we shall now derive explicitly, for convenience. As usual, positive energy spinors are defined as solutions of $(\not{p} - m)u = 0$, so that writing

$$u = \begin{pmatrix} \phi \\ \chi \end{pmatrix} \tag{20.44}$$

we obtain

$$(E - \boldsymbol{\sigma} \cdot \boldsymbol{p})\phi = m\chi$$
$$(E + \boldsymbol{\sigma} \cdot \boldsymbol{p})\chi = m\phi. \tag{20.45}$$

A convenient choice of two-component spinors $\phi$, $\chi$ is to take them to be *helicity eigenstates* (see section 4.3). For example, the eigenstate $\phi_+$ with positive helicity $\lambda = +1$ satisfies

$$\boldsymbol{\sigma} \cdot \boldsymbol{p}\phi_+ = |\boldsymbol{p}|\phi_+ \tag{20.46}$$

while the eigenstate $\phi_-$ with $\lambda = -1$ satisfies (20.46) with a minus on the right-hand side. Thus, the spinor $u(p, \lambda = +1)$ can be written as

$$u(p, \lambda = +1) = N \begin{pmatrix} \phi_+ \\ \frac{(E-|\boldsymbol{p}|)}{m}\phi_+ \end{pmatrix}. \tag{20.47}$$

The normalization $N$ is fixed as usual by requiring $\bar{u}u = 2m$, from which it follows (problem 20.3) that $N = (E + |\boldsymbol{p}|)^{1/2}$. Thus, finally, we have

$$u(p, \lambda = +1) = \begin{pmatrix} \sqrt{E + |\boldsymbol{p}|}\phi_+ \\ \sqrt{E - |\boldsymbol{p}|}\phi_+ \end{pmatrix}. \tag{20.48}$$

Similarly,

$$u(p, \lambda = -1) = \begin{pmatrix} \sqrt{E - |\boldsymbol{p}|}\phi_- \\ \sqrt{E + |\boldsymbol{p}|}\phi_- \end{pmatrix}. \tag{20.49}$$

Now we have agreed that only the chiral 'L' components of all $u$-spinors enter into weak interactions, in the Standard Model. But from the explicit form of $\gamma_5$ given in (20.38), we see that when acting on any spinor $u$, the projector $P_{\mathrm{L}}$ 'kills' the top two components:

$$P_{\mathrm{L}} \begin{pmatrix} \phi \\ \chi \end{pmatrix} = \begin{pmatrix} 0 \\ \chi \end{pmatrix}. \tag{20.50}$$

In particular,

$$P_{\mathrm{L}}u(p, \lambda = +1) = \begin{pmatrix} 0 \\ \sqrt{E - |\boldsymbol{p}|}\phi_+ \end{pmatrix} \tag{20.51}$$

and

$$P_{\mathrm{L}}u(p, \lambda = -1) = \begin{pmatrix} 0 \\ \sqrt{E + |\boldsymbol{p}|}\phi_- \end{pmatrix}. \tag{20.52}$$

Equations (20.51) and (20.52) are very important. In particular, equation (20.51) implies that in the limit of zero mass $m$ (and, hence, $E \to |\boldsymbol{p}|$), only the *negative helicity* $u$-spinor will enter. More quantitatively, using

$$\sqrt{E - |\boldsymbol{p}|} = \frac{\sqrt{E^2 - \boldsymbol{p}^2}}{\sqrt{E + |\boldsymbol{p}|}} \approx \frac{m}{2E} \qquad \text{for } m \ll E, \tag{20.53}$$

we can say that *positive helicity components of all fermions are suppressed in matrix elements by factors of order m/E*. Bearing in mind that the helicity operator $\boldsymbol{\sigma} \cdot \boldsymbol{p}/|\boldsymbol{p}|$ is a pseudoscalar, this 'unequal' treatment for $\lambda = +1$ and $\lambda = -1$ components is, of course, precisely related to the parity violation built into the $\mathrm{V} - \mathrm{A}$ structure.

A similar analysis may be done for the $v$-spinors. They satisfy $(\not p + m)v = 0$ and the normalization $\bar{v}v = -2m$. We must, however, remember the 'small subtlety' to do with the labelling of $v$-spinors, discussed in section 4.5.3: the two-component spinors $\chi$ in $v(p, \lambda = +1)$ actually satisfy $\boldsymbol{\sigma} \cdot \boldsymbol{p}\chi_- = -|\boldsymbol{p}|\chi_-$ and, similarly, the $\chi_+$'s in $v(p, \lambda = -1)$ satisfy $\boldsymbol{\sigma} \cdot \boldsymbol{p}\chi_+ = |\boldsymbol{p}|\chi_+$. We then find (problem 20.4) the results

$$v(p, \lambda = +1) = \begin{pmatrix} -\sqrt{E - |\boldsymbol{p}|}\chi_- \\ \sqrt{E + |\boldsymbol{p}|}\chi_- \end{pmatrix} \tag{20.54}$$

and

$$v(p, \lambda = -1) = \begin{pmatrix} \sqrt{E + |\boldsymbol{p}|}\chi_+ \\ -\sqrt{E - |\boldsymbol{p}|}\chi_+ \end{pmatrix}. \tag{20.55}$$

Once again, the action of $P_{\mathrm{L}}$ removes the top two components, leaving the result that, in the massless limit, only the $\lambda = +1$ state survives. Recalling the 'hole theory' interpretation of section 4.5.3, this would mean that *the positive helicity components of all anti-fermions dominate in weak interactions*, negative helicity components being suppressed[2] by factors of order $m/E$ .

We should emphasize that although these two results, stated in italics, were derived in the convenient representation (20.38) for the Dirac matrices, they actually hold independently of any choice of representation. This can be shown by using general helicity projection operators.

In Pauli's original letter, he suggested that the mass of the neutrino might be of the same order as the electron mass. Immediately after the discovery of parity violation, it was realized that the result could be elegantly explained by the assumption that the neutrinos were strictly massless particles (Landau 1957, Lee and Yang 1957, Salam 1957). In this case, $u$ and $v$ spinors satisfy the same equation $\not p(u \text{ or } v) = 0$, which reduces via (20.45) (in the $m = 0$ limit) to the two independent two-component 'Weyl' equations.

$$E\phi_0 = \boldsymbol{\sigma} \cdot \boldsymbol{p}\,\phi_0 \qquad E\chi_0 = -\boldsymbol{\sigma} \cdot \boldsymbol{p}\,\chi_0. \tag{20.56}$$

Remembering that $E = |\boldsymbol{p}|$ for a massless particle, we see that $\phi_0$ has positive helicity and $\chi_0$ negative helicity. In this strictly massless case, helicity is Lorentz invariant, since the direction of $\boldsymbol{p}$ cannot be reversed by a velocity transformation with $v < c$. Furthermore, each of the equations in (20.56) violates parity, since $E$ is clearly a scalar while $\boldsymbol{\sigma} \cdot \boldsymbol{p}$ is a pseudoscalar (note that when $m \neq 0$ we can infer

[2] The proportionality of the negative helicity amplitude to the mass of the anti-fermion is, of course, exactly as noted for $\pi^+ \to \mu^+\nu_\mu$ decay in section 18.2.

from (20.45) that, in this representation, $\phi \leftrightarrow \chi$ under **P**, which is consistent with (20.56) and with the form of $\beta$ in (20.38)). Thus the (massless) neutrino could be 'blamed' for the parity violation. In this model, neutrinos have one definite helicity, either positive or negative. As we have seen, the massless limit of the (four-component) V − A theory leads to the same conclusion.

Which helicity is actually chosen by Nature was determined in a classic experiment by Goldhaber *et al* (1958), involving the K-capture reaction

$$e^- + {}^{152}\mathrm{Eu} \rightarrow \nu + {}^{152}\mathrm{Sm}^* \tag{20.57}$$

as described by Perkins (2000), for example. They found that the helicity of the emitted neutrino was (within errors) 100% *negative*, a result taken as confirming the 'two-component' neutrino theory and the V − A theory.

We now turn to the question of whether there is a distinction to be made between neutrinos and anti-neutrinos. As a preliminary, we first introduce another discrete symmetry operation, that of charge conjugation **C**.

## 20.5 Charge conjugation for fermion wavefunctions and field operators

We begin by following rather similar steps to those in section 20.3 for parity. Consider the Dirac equation for a particle of charge $-e$ ($e > 0$) in a field $A^\mu$:

$$(\mathrm{i}\not\partial + e\not A - m)\psi = 0. \tag{20.58}$$

The equation satisfied by a particle of the same mass and opposite charge $+e$ is

$$(\mathrm{i}\not\partial - e\not A - m)\psi_{\mathrm{C}} = 0. \tag{20.59}$$

Remarkably, there is a sort of 'covariance' involved here under the transformation $e \rightarrow -e$: we can relate $\psi_{\mathrm{C}}$ to $\psi$ in such a way that (20.59) follows from (20.58). Take the complex conjugate of (20.59), so as to produce the equation

$$(\mathrm{i}\gamma^{\mu *}\partial_\mu + e\gamma^{\mu *}A_\mu + m)\psi_{\mathrm{C}}^* = 0 \tag{20.60}$$

assuming $A_\mu$ to be real. Multiplying (20.60) from the left by a matrix $C_0$ (to be determined), which is assumed to be non-singular, we obtain

$$[C_0\gamma^{\mu *}C_0^{-1}(\mathrm{i}\partial_\mu + eA_\mu) + m]C_0\psi_{\mathrm{C}}^* = 0. \tag{20.61}$$

Thus, if we can find a $C_0$ such that

$$C_0\gamma^{\mu *}C_0^{-1} = -\gamma^\mu \tag{20.62}$$

we may identify $C_0\psi_{\mathrm{C}}^*$ with $\psi$ (up to an ever-possible phase factor). In either of our two representations (20.31) or (20.38), all $\gamma$-matrices are real except $\gamma^2$ which is pure imaginary. A possible choice for $C_0$ is then $\mathrm{i}\gamma^2$, the i being inserted for convenience so as to make $C_0$ real.

The wavefunctions $\psi$ and $\psi_{\mathrm{C}}$ are then related by

$$\psi = \mathrm{i}\gamma^2\psi_{\mathrm{C}}^* \qquad \text{or} \qquad \psi_{\mathrm{C}} = \mathrm{i}\gamma^2\psi^* \tag{20.63}$$

where we have used $(\mathrm{i}\gamma^2)^2 = I$.

Consider the application of this result, taking $\psi$ to be a negative 4-momentum solution of the free-particle Dirac equation with a helicity $\lambda = +1$ $v$-spinor:

$$\begin{aligned}\psi &= v(p, \lambda = +1)\mathrm{e}^{\mathrm{i}p\cdot x} \\ &= \begin{pmatrix} -\sqrt{E - |\boldsymbol{p}|}\chi_- \\ \sqrt{E + |\boldsymbol{p}|}\chi_- \end{pmatrix} \mathrm{e}^{\mathrm{i}p\cdot x}\end{aligned} \tag{20.64}$$

using (20.54), in the representation (20.38). Then, in this case,

$$\psi_{\mathrm{C}} = \mathrm{i}\gamma^2\psi^* = \begin{pmatrix} 0 & -\mathrm{i}\sigma_2 \\ \mathrm{i}\sigma_2 & 0 \end{pmatrix} \begin{pmatrix} -\sqrt{E - |\boldsymbol{p}|}\,\chi_-^* \\ \sqrt{E + |\boldsymbol{p}|}\,\chi_-^* \end{pmatrix} \mathrm{e}^{-\mathrm{i}p\cdot x}. \tag{20.65}$$

Now the $\chi_-$ here satisfies

$$\boldsymbol{\sigma}\cdot\boldsymbol{p}\chi_- = -|\boldsymbol{p}|\chi_- \tag{20.66}$$

as agreed in section 20.3. Taking the complex conjugate of (20.66), we find that

$$\boldsymbol{\sigma}^*\cdot\boldsymbol{p}\chi_-^* = -|\boldsymbol{p}|\chi_-^* \tag{20.67}$$

and, recalling that $\sigma_1$ and $\sigma_3$ are real while $\sigma_2$ is pure imaginary, we see that (20.67) is

$$(\sigma_1 p_1 - \sigma_2 p_2 + \sigma_3 p_3)\chi_-^* = -|\boldsymbol{p}|\chi_-^*. \tag{20.68}$$

Multiplying by $\sigma_2$ from the left and, using $\sigma_1\sigma_2 = -\sigma_2\sigma_1$, $\sigma_1\sigma_3 = -\sigma_3\sigma_1$, $\sigma_2^2 = 1$, we obtain (inserting a $-\mathrm{i}$ freely)

$$\boldsymbol{\sigma}\cdot\boldsymbol{p}(-\mathrm{i}\sigma_2\chi_-^*) = |\boldsymbol{p}|(-\mathrm{i}\sigma_2\chi_-^*) \tag{20.69}$$

showing that $(-\mathrm{i}\sigma_2\chi_-^*)$ is a spinor with positive helicity, say $\phi_+$. Thus, we find for this case that

$$\begin{aligned}\psi_{\mathrm{C}} = \mathrm{i}\gamma^2 v^*(p, \lambda = +1)\mathrm{e}^{-\mathrm{i}p\cdot x} &= \begin{pmatrix} \sqrt{E + |\boldsymbol{p}|}\,\phi_+ \\ \sqrt{E - |\boldsymbol{p}|}\,\phi_+ \end{pmatrix} \mathrm{e}^{-\mathrm{i}p\cdot x} \\ &= u(p, \lambda = +1)\mathrm{e}^{-\mathrm{i}p\cdot x}\end{aligned} \tag{20.70}$$

and so the transformed wavefunction $\psi_{\mathrm{C}}$ is precisely the wavefunction of a positive 4-momentum solution with positive helicity (cf (20.48)). In a similar way, defining $\mathrm{i}\sigma_2\chi_+^*$ as $\phi_-$, we find that $\mathrm{i}\gamma^2 v^*(p, \lambda = -1) = u(p, \lambda = -1)$. Thus, our transformation takes negative 4-momentum solutions with helicity $\lambda$ into positive 4-momentum solutions with helicity $\lambda$.

This transformation is the nearest we can get, in a wavefunction theory, to a particle $\leftrightarrow$ anti-particle transformation. For the latter, we want an operator which simply changes a particle with a certain 4-momentum and helicity into the corresponding anti-particle with the *same* 4-momentum and helicity. This can only be done in a quantum field formalism. There, the required operator is $\hat{\mathbf{C}}$, with the property

$$\hat{\psi}_{\mathrm{C}} \equiv \hat{\mathbf{C}}\hat{\psi}\hat{\mathbf{C}}^{-1} = C_0\hat{\psi}^{\dagger\mathrm{T}} = \mathrm{i}\gamma^2\hat{\psi}^{\dagger\mathrm{T}}. \tag{20.71}$$

(Note that we write explicitly ${}^{\dagger\mathrm{T}}$ for ${}^*$ as $\hat{\psi}$ will contain annihilation and creation operators for which ${}^*$ is undefined).

Let us consider the effect of the transformation (20.71) on a standard normal mode expansion of a Dirac field:

$$\hat{\psi}(x) = \int \frac{\mathrm{d}^3\boldsymbol{k}}{(2\pi)^3}\frac{1}{\sqrt{2E}}\sum_{\lambda}[\hat{c}_\lambda(k)u(k,\lambda)\mathrm{e}^{-\mathrm{i}k\cdot x} + \hat{d}_\lambda^\dagger(k)v(k,\lambda)\mathrm{e}^{\mathrm{i}k\cdot x}] \tag{20.72}$$

where $E = \sqrt{m^2+\boldsymbol{k}^2}$. Using $\mathrm{i}\gamma^2u^*(k,\lambda) = v(k,\lambda)$, which follows by inverting (20.70), and the similar relation for $\lambda = -1$, we find that

$$\hat{\psi}_{\mathrm{C}}(x) = \int \frac{\mathrm{d}^3\boldsymbol{k}}{(2\pi)^3}\frac{1}{\sqrt{2E}}\sum_{\lambda}[\hat{c}_\lambda^\dagger(k)v(k,\lambda)\mathrm{e}^{\mathrm{i}k\cdot x} + \hat{d}_\lambda(k)u(k,\lambda)\mathrm{e}^{-\mathrm{i}k\cdot x}] \tag{20.73}$$

from which it is clear that the field $\hat{\psi}_{\mathrm{C}}(x)$ is just the same as $\hat{\psi}(x)$ but with $\hat{c}_\lambda(k)$ replaced by $\hat{d}_\lambda(k)$ (and $\hat{c}_\lambda^\dagger(k)$ by $\hat{d}_\lambda^\dagger(k)$)—that is, *particle* operators replaced by *anti-particle* ones, just as the $\hat{\mathbf{C}}$-conjugate field should be. In particular, the $k$ and $\lambda$ values are not altered, as required.

We have introduced the idea of particle–anti-particle conjugation within the context of electromagnetic interaction, where it is indeed a good symmetry. But we must now ask whether it is also a good symmetry in weak interactions. The answer to this question must be an immediate negative, since we have seen that the V$-$A interaction treats a positive helicity particle very differently from a negative helicity anti-particle, while one is precisely transformed into the other under $\hat{\mathbf{C}}$. In perhaps more physical terms, we know that the $\mathrm{e}^-$ emitted in $\mu^-$-decay is predominately in the $\lambda = -1$ helicity state. Particle–anti-particle symmetry would predict that an $\mathrm{e}^+$ emitted in the **C**-conjugate process should also have helicity $\lambda = -1$, but it does not.

However, it is clear that the helicity operator itself is odd under **P**. Thus the **CP**-conjugate of an $\mathrm{e}^-$ with $\lambda = -1$ is an $\mathrm{e}^+$ with $\lambda = +1$, and so the V $-$ A interaction does preserve the combined symmetry of **CP**. It may easily be verified (problem 20.5) that the 'two-component' theory of (20.56) automatically incorporates **CP** invariance. We shall discuss **CP** further in section 22.7.1.

Returning to (20.58) and (20.59), we see that, whether we use $\hat{\psi}$ or $\hat{\psi}_{\mathrm{C}}$, four distinct kinds of 'modes' are involved: there are *particles* with either sign of helicity and *anti-particles* with either sign of helicity. This is just what we need to describe fermions which carry a conserved quantum number (such as their electromagnetic charge), by which—according to some convention—'particle' can be distinguished from 'anti-particle'. Thus far, in this book, this has been the case, since we have only considered charged fermions. But in the case of the neutral neutrinos the situation is not so clear, as we shall now discuss.

## 20.6 Lepton number

In section 1.3.1 of volume 1, we gave a brief discussion of leptonic quantum numbers ('lepton flavours'), adopting a traditional approach in which the data are interpreted in terms of conserved quantum numbers carried by neutrinos, which serve to distinguish neutrinos from anti-neutrinos. We must now examine the matter more closely, in the light of what we have learned about the helicity properties of the V − A interaction.

In 1955, Davis (1955)—following a suggestion made by Pontecorvo (1948)—argued as follows. Consider the $e^-$ capture reaction $e^- + p \to \nu + n$, which was, of course, well established. Then, in principle, the inverse reaction $\nu + n \to e^- + p$ should also exist. Of course, the cross-section is extremely small but by using a large enough target volume this might perhaps be compensated. Specifically, the reaction $\nu + {}^{37}_{17}\mathrm{Cl} \to e^- + {}^{37}_{18}\mathrm{Ar}$ was proposed, the argon being detected through its radioactive decay. Suppose, however, that the 'neutrinos' actually used are those which accompany electrons in $\beta^-$-decay. If (as was supposed in section 1.3.1) these are to be regarded as anti-neutrinos, '$\bar{\nu}$', carrying a conserved lepton number, then the reaction

$$`\bar{\nu}\text{'} + {}^{37}_{17}\mathrm{Cl} \to e^- + {}^{37}_{18}\mathrm{Ar} \tag{20.74}$$

should *not* be observed. If, on the other hand, the '$\nu$' in the capture process and the '$\bar{\nu}$' in $\beta$-decay are not distinguished by the weak interaction, the reaction (20.74) should be observed. Davis found no evidence for reaction (20.74), at the expected level of cross-section, a result which could clearly be interpreted as confirming the 'conserved electron number hypothesis'.

However, another interpretation is possible. The $e^-$ in $\beta$-decay has predominately negative helicity and its accompanying '$\bar{\nu}$' has predominately positive helicity. The fraction of the other helicity present is of the order $m/E$, where $E \sim$ few Mev, and the neutrino mass is less than 1 eV; this is, therefore, an almost undetectable 'contamination' of negative helicity component in the '$\bar{\nu}$'. Now the property of the V − A interaction is that it conserves helicity in the zero mass limit (in which chirality is the same as helicity). Hence, the positive helicity '$\bar{\nu}$' from $\beta^-$-decay will (predominately) produce a positive helicity lepton, which must be the $e^+$ not the $e^-$. Thus the property of the V − A interaction, together with the very small value of the neutrino mass, conspire effectively to forbid (20.74), independently of any considerations about 'lepton number'.

Indeed, the 'helicity-allowed' reaction

$$`\bar{\nu}\text{'} + p \to e^+ + n \tag{20.75}$$

was observed by Reines and Cowan (1956) (see also Cowan *et al* (1956)). Reaction (20.75) too, of course, can be interpreted in terms of '$\bar{\nu}$' carrying a lepton number of $-1$, equal to that of the $e^+$. It was also established that only '$\nu$' produced $e^-$ via (20.74), where '$\nu$' is the helicity $-1$ state (or, on the other interpretation, the carrier of lepton number $+1$).

The situation may therefore be summarized as follows. In the case of $e^-$ and $e^+$, all four 'modes'—$e^-(\lambda = +1), e^-(\lambda = -1), e^+(\lambda = +1), e^+(\lambda = -1)$—are experimentally accessible via electromagnetic interactions, even though only two generally dominate in weak interactions ($e^-(\lambda = -1)$ and $e^+(\lambda = +1)$). Neutrinos, in contrast, seem to interact only weakly. In their case, we may if we wish say that the participating states are (in association with $e^-$ or $e^+$) $\bar{\nu}_e$ $(\lambda = +1)$ and $\nu_e(\lambda = -1)$, to a very good approximation. But we may also regard these two states as simply two different helicity states of one particle, rather than of a particle and its anti-particle. As we have seen, the helicity rules do the job required just as well as the lepton number rules. In short, the question is: are these 'neutrinos' distinguished only by their helicity, or is there an additional distinguishing characteristic ('electron number')? In the latter case, we should expect the 'other' two states $\bar{\nu}_e(\lambda = -1)$ and $\nu_e(\lambda = +1)$ to exist as well as the ones known from weak interactions.

If, in fact, no quantum number—other than the helicity—exists which distinguishes the neutrino states, then we would have to say that the **C**-conjugate of a neutrino state is a neutrino, not an anti-neutrino—that is, 'neutrinos are their own anti-particles'. A neutrino would be a fermionic state somewhat like a photon, which is, of course, also its own anti-particle. Such '**C**-self-conjugate' fermions are called *Majorana fermions*, in contrast to the *Dirac* variety, which have all four possible modes present (two helicities, two particle/anti-particle). The field operator for a Majorana fermion, $\hat{\psi}_M(x)$, will have a mode expansion of the form (20.72) but the operator $\hat{c}_\lambda^\dagger$ will appear instead of the operator $\hat{d}_\lambda^\dagger$. Such a field will then clearly obey the relation

$$\hat{\mathbf{C}}\hat{\psi}_M\hat{\mathbf{C}}^{-1} = \hat{\psi}_M \tag{20.76}$$

which is the *Majorana condition*. The quantum theory of free Majorana fermions is described in appendix P.

The distinction between the 'Dirac' and 'Majorana' neutrino possibilities becomes an essentially 'metaphysical' one in the limit of strictly massless neutrinos, since then (as we have seen) a given helicity state cannot be flipped by going to a suitably moving Lorentz frame, nor by any weak (or electromagnetic) interaction, since they both conserve chirality which is the same as helicity in the massless limit. We would have just the two states $\nu_e(\lambda = -1)$ and $\bar{\nu}_e(\lambda = +1)$ and no way of creating $\nu_e(\lambda = +1)$ or $\bar{\nu}_e(\lambda = -1)$. The '$^-$' label then becomes superfluous. Unfortunately, the massless limit is approached smoothly and it seems highly likely that neutrino masses are, in fact, so small that the 'wrong helicity' supression factors will make it very difficult to see the presence of the possible states $\nu_e(\lambda = +1)$, $\bar{\nu}_e(\lambda = -1)$, if indeed they exist.

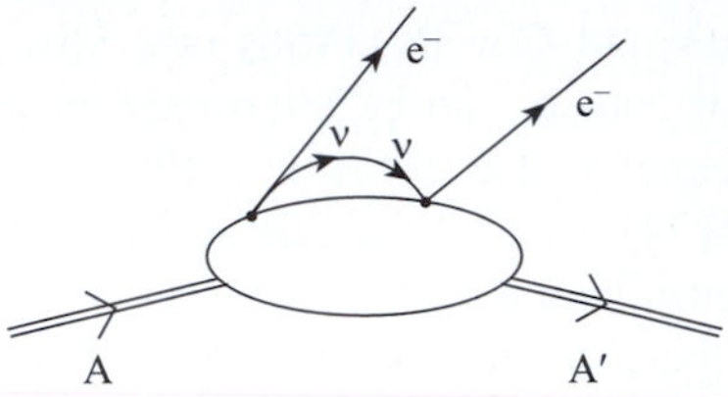


**Figure 20.2.** Double $\beta$-decay without emission of a neutrino, a test for Majorana-type neutrinos.

One much-discussed experimental test case (see, for example, the review by Kayser in Hagiwara *et al* (2002) and references therein) concerns 'neutrinoless double $\beta$-decay', which is the process $\mathrm{A} \to \mathrm{A}' + \mathrm{e}^- + \mathrm{e}^-$, where A, A′ are nuclei. If the neutrino emitted in the first $\beta$-decay carries no electron-type conserved quantum number, then in principle it can initiate a second weak interaction, exactly as in Davis' original argument, via the diagram shown in figure 20.2. The $\nu$ emitted along with the $\mathrm{e}^-$ at the first vertex will be predominately $\lambda = +1$, but in the second vertex the V − A interaction will 'want' it to have $\lambda = -1$, like the outgoing $\mathrm{e}^-$. Thus, there is bound to be one '$m/E$' suppression factor, whichever vertex we choose to make 'easy'. There is also a complicated nucleus physics overlap factor. As yet, no clear evidence for this process has been obtained.

In the same way, '$\bar{\nu}$'' particles accompanying the $\mu^-$'s in $\pi^-$ decay

$$\pi^- \to \mu^- + `\bar{\nu}' \tag{20.77}$$

are observed to produce only $\mu^+$'s when they interact with matter, not $\mu^-$'s. Again this can be interpreted either in terms of helicity conservation or in terms of conservation of a leptonic quantum number $L_\mu$. We shall assume the analogous properties are true for the $\bar{\nu}''$'s accompanying $\tau$ leptons.

On the other hand, helicity arguments alone would allow the reaction

$$`\bar{\nu}' + \mathrm{p} \to \mathrm{e}^+ + \mathrm{n} \tag{20.78}$$

to proceed, but as we saw in section 1.3.1 the experiment of Danby *et al* (1962) found no evidence for it. Thus there is evidence, in this type of reaction, for a flavour quantum number distinguishing neutrinos which interact in association with one kind of charged lepton from those which interact in association with a different charged lepton. However, a number of observations (see section 22.7.2) have combined to demonstrate convincingly that 'neutrino oscillations' do occur, in which states of one such flavour can acquire a component of another, as it propagates. It would create too big a detour to continue with the details of this interesting physics at this point: we shall return to it in section 22.7.2. For the moment we simply state that, for the simple case of $\nu_\mathrm{e} \leftrightarrow \nu_\mu$ mixing (for example), the probability that an initially pure $\nu_\mu$ state becomes a $\nu_\mathrm{e}$ state *in vacuo* is proportional to

$$\sin^2(L/L_0) \tag{20.79}$$

where $L$ is the distance from the $\nu_\mu$ production point, and the 'oscillation length' $L_0$ is given by (Perkins 2000, section 9.7)

$$L_0(\text{in km}) \simeq 0.8 \times E(\text{GeV})/\Delta m^2(\text{eV}^2) \tag{20.80}$$

where $E$ is the neutrino beam energy and $\Delta m^2$ is the squared mass difference $|m^2_{\nu_\mu} - m^2_{\nu_e}|$. If $\Delta m^2 \sim 10^{-4}\,\text{eV}^2$ and $E \sim 1$ GeV, we see that $L_0 \sim 8000$ km, so that $L$ must be of the order of the radius of the earth before (20.79) is appreciable. Just such oscillations have been observed via atmospheric or solar neutrinos, but the experiment of Danby *et al* (1962) was obviously unable to see them. It is also the case that none of the known (2003) neutrino experiments is sensitive to the difference between the Dirac and Majorana 'option'.

The upshot of all this is that, while the 'Majorana neutrino' hypothesis is interesting and still viable, and one for which some appealing theoretical arguments can be made (Gell-Mann *et al* 1979, Yanagida 1979, Mohapatra and Senjanovic 1980, 1981, see also appendix P, section P.2), it is fair to say that the Standard Model treats neutrinos as Dirac particles, and that is what we shall generally assume in the rest of this part of the book. In due course (section 22.6), we shall see why, if neutrinos are Majorana particles, the way their mass must appear would suggest an origin in 'physics beyond the Standard Model'.

## 20.7 The universal current–current theory for weak interactions of leptons

After the breakthroughs of parity violation and V − A theory, the earlier hopes (Pontecorvo 1947, Klein 1948, Puppi 1948, Lee *et al* 1949, Tiomno and Wheeler 1949) were revived of a universal weak interaction among the pairs of particles (p,n), $(\nu_e, e^-)$, $(\nu_\mu, \mu^-)$, using the V − A modification to Fermi's theory. From our modern standpoint, this list has to be changed by the replacement of (p,n) by the corresponding quarks (u,d), and by the inclusion of the third lepton pair $(\nu_\tau, \tau^-)$ as well as two other quark pairs (c,s) and (t,b). It is to these pairs that the 'V − A' structure applies, as already indicated in section 20.4, and a certain form of 'universality' does hold, as we now describe.

Because of certain complications which arise, we shall postpone the discussion of the quark currents until section 20.10, concentrating here on the leptonic currents. In this case, Fermi's original vector-like current $\hat{\bar{\psi}}_e \gamma^\mu \hat{\psi}_\nu$ becomes modified to a *total leptonic charged current*

$$\hat{j}^\mu_{\text{CC}}(\text{leptons}) = \hat{j}^\mu_{\text{wk}}(\text{e}) + \hat{j}^\mu_{\text{wk}}(\mu) + \hat{j}^\mu_{\text{wk}}(\tau) \tag{20.81}$$

where, for example,

$$\hat{j}^\mu_{\text{wk}}(\text{e}) = \hat{\bar{\nu}}_e \gamma^\mu (1 - \gamma_5)\hat{e}. \tag{20.82}$$

In (20.82) we are now adopting, for the first time, a useful shorthand whereby the field operator for the electron field, say, is denoted by $\hat{e}(x)$ rather than $\hat{\psi}_{\rm e}(x)$ and the '$x$' argument is suppressed. The 'charged' current terminology refers to the fact that these weak current operators $\hat{j}^{\mu}_{\rm wk}$ carry net charge, in contrast to an electromagnetic current operator such as $\bar{\hat{e}}\gamma^{\mu}\hat{e}$ which is electrically neutral. We shall see in section 20.9 that there are also electrically neutral weak currents.

The interaction Hamiltonian density accounting for all leptonic weak interactions is then taken to be

$$\hat{\mathcal{H}}^{\rm lep}_{\rm CC} = \frac{G_{\rm F}}{\sqrt{2}}\hat{j}^{\mu}_{\rm CC}({\rm leptons})\,\hat{j}^{\dagger}_{{\rm CC}\mu}({\rm leptons}). \tag{20.83}$$

Note that

$$(\bar{\hat{\nu}}_{\rm e}\gamma^{\mu}(1-\gamma_5)\hat{e})^{\dagger} = \bar{\hat{e}}\gamma^{\mu}(1-\gamma_5)\hat{\nu}_{\rm e} \tag{20.84}$$

and similarly for the other bilinears. The currents can also be written in terms of the chiral components of the fields (recall section 20.4) using

$$2\bar{\hat{\nu}}_{\rm eL}\gamma^{\mu}\hat{e}_{\rm L} = \bar{\hat{\nu}}_{\rm e}\gamma^{\mu}(1-\gamma_5)\hat{e} \tag{20.85}$$

for example. 'Universality' is manifest in the fact that all the lepton pairs have the same form of the V − A coupling, and the same 'strength parameter' $G_{\rm F}/\sqrt{2}$ multiplies all of the products in (20.83).

The terms in (20.83), when it is multiplied out, describe many physical processes. For example, the term

$$\frac{G_{\rm F}}{\sqrt{2}}\bar{\hat{\nu}}_{\mu}\gamma^{\mu}(1-\gamma_5)\hat{\mu}\,\bar{\hat{e}}\gamma_{\mu}(1-\gamma_5)\hat{\nu}_{\rm e} \tag{20.86}$$

describes $\mu^-$ decay:

$$\mu^- \to \nu_{\mu} + {\rm e}^- + \bar{\nu}_{\rm e} \tag{20.87}$$

as well as all the reactions related by 'crossing' particles from one side to the other, for example

$$\nu_{\mu} + {\rm e}^- \to \mu^- + \nu_{\rm e}. \tag{20.88}$$

The value of $G_{\rm F}$ can be determined from the rate for process (20.87) (see, for example, Renton 1990, section 6.1.2) and it is found to be

$$G_{\rm F} \simeq 1.166 \times 10^{-5}\ {\rm GeV}^{-2}. \tag{20.89}$$

This is a convenient moment to note that the theory is *not renormalizable* according to the criteria discussed in section 11.8 at the end of the previous volume: $G_{\rm F}$ has dimensions $({\rm mass})^{-2}$. We shall return to this aspect of Fermi-type V − A theory in section 21.4.

There are also what we might call 'diagonal' terms in which the same lepton pair is taken from $\hat{j}^{\mu}_{\rm wk}$ and $\hat{j}^{\dagger}_{{\rm wk}\mu}$, for example

$$\frac{G_{\rm F}}{\sqrt{2}}\bar{\hat{\nu}}_{\rm e}\gamma^{\mu}(1-\gamma_5)\hat{e}\,\bar{\hat{e}}\gamma_{\mu}(1-\gamma_5)\hat{\nu}_{\rm e} \tag{20.90}$$

which describes reactions such as

$$\bar{\nu}_\mathrm{e} + \mathrm{e}^- \to \bar{\nu}_\mathrm{e} + \mathrm{e}^-. \tag{20.91}$$

The cross-section for (20.91) was measured by Reines *et al* (1976) after many years of effort.

It is interesting that some seemingly rather similar processes are forbidden to occur, to first order in $\hat{\mathcal{H}}^{\mathrm{lep}}_{\mathrm{CC}}$ (see (20.83)), for example

$$\bar{\nu}_\mu + \mathrm{e}^- \to \bar{\nu}_\mu + \mathrm{e}^-. \tag{20.92}$$

For reasons which will become clearer in section 20.9, (20.92) is called a 'neutral current' process, in contrast to all the others (such as $\beta$-decay or $\mu$-decay) we have discussed so far, which are called 'charged current' processes. If the lepton pairs are arranged so as to have no net lepton number (for example $\mathrm{e}^-\bar{\nu}_\mathrm{e}$, $\mu^+\nu_\mu$, $\nu_\mu\bar{\nu}_\mu$ etc) then pairs with non-zero charge occur in charged current processes, while those with zero charge participate in neutral current processes. In the case of (20.91), the leptons can be grouped either as $(\bar{\nu}_\mathrm{e}\mathrm{e}^-)$ which is charged, or as $(\bar{\nu}_\mathrm{e}\nu_\mathrm{e})$ or $(\mathrm{e}^+\mathrm{e}^-)$ which is neutral. However, there is no way of pairing the leptons in (20.92) so as to cancel the lepton number and have non-zero charge. So (20.92) is a *purely* 'neutral current' process, while *some* 'neutral current' contribution could be present in (20.91), in principle. In 1973 such neutral current processes were discovered (Hasert *et al* 1973), generating a whole new wave of experimental activity. Their existence had, in fact, been predicted in the first version of the Standard Model, due to Glashow (1961). Today we know that charged current processes are mediated by the $\mathrm{W}^\pm$ bosons and the neutral current ones by the $\mathrm{Z}^0$. We shall discuss the neutral current couplings in section 20.9.

## 20.8 Calculation of the cross-section for $\nu_\mu + \mathrm{e}^- \to \mu^- + \nu_\mathrm{e}$

After so much qualitative discussion, it is time to calculate something. We choose the process (20.88), sometimes called inverse muon decay, which is a pure 'charged current' process. The amplitude, in the Fermi-like V − A current theory, is

$$\mathcal{M} = -\mathrm{i}(G_\mathrm{F}/\sqrt{2})\bar{u}(\mu)\gamma_\mu(1-\gamma_5)u(\nu_\mu)\bar{u}(\nu_\mathrm{e})\gamma^\mu(1-\gamma_5)u(\mathrm{e}). \tag{20.93}$$

We shall be interested in energies much greater than any of the lepton masses and so we shall work in the *massless limit*: this is mainly for ease of calculation—the full expressions for non-zero masses can be obtained with more effort.

From the general formula (6.129) for $2 \to 2$ scattering we have, neglecting all masses,

$$\frac{\mathrm{d}\sigma}{\mathrm{d}\Omega} = \frac{1}{64\pi^2 s}\overline{|\mathcal{M}|^2} \tag{20.94}$$

where $|\overline{\mathcal{M}}|^2$ is the appropriate spin-averaged matrix element squared, as in (8.184) for example. In the case of neutrino–electron scattering, we must average over initial electron states for unpolarized electrons and sum over the final muon polarization states. For the neutrinos there is no averaging over initial neutrino helicities, since only left-handed (massless) neutrinos participate in the weak interaction. Similarly, there is no sum over final neutrino helicities. However, for convenience of calculation, we can, in fact, sum over both helicity states of both neutrinos since the $(1-\gamma_5)$ factors guarantee that right-handed neutrinos contribute nothing to the cross-section. As for the e$\mu$ scattering example in section 8.7, the calculation then reduces to that of a product of traces:

$$|\overline{\mathcal{M}}|^2 = \left(\frac{G_{\mathrm{F}}^2}{2}\right) \mathrm{Tr}[\not{k}'\gamma_\mu(1-\gamma_5)\not{k}\gamma_\nu(1-\gamma_5)]\frac{1}{2}\,\mathrm{Tr}[\not{p}'\gamma^\mu(1-\gamma_5)\not{p}\gamma^\nu(1-\gamma_5)] \tag{20.95}$$

all lepton masses being neglected. We define

$$|\overline{\mathcal{M}}|^2 = \left(\frac{G_{\mathrm{F}}^2}{2}\right) N_{\mu\nu}E^{\mu\nu} \tag{20.96}$$

where the $\nu_\mu \to \mu^-$ tensor $N_{\mu\nu}$ is given by

$$N_{\mu\nu} = \mathrm{Tr}[\not{k}'\gamma_\mu(1-\gamma_5)\not{k}\gamma_\nu(1-\gamma_5)] \tag{20.97}$$

without a $1/(2s+1)$ factor, and the $\mathrm{e}^- \to \nu_{\mathrm{e}}$ tensor is

$$E^{\mu\nu} = \tfrac{1}{2}\,\mathrm{Tr}[\not{p}'\gamma^\mu(1-\gamma_5)\not{p}\gamma^\nu(1-\gamma_5)] \tag{20.98}$$

including a factor of $\frac{1}{2}$ for spin averaging.

Since this calculation involves a couple of new features, let us look at it in some detail. By commuting the $(1-\gamma_5)$ factor through two $\gamma$ matrices ($\not{p}\gamma^\nu$) and using the result that

$$(1-\gamma_5)^2 = 2(1-\gamma_5) \tag{20.99}$$

the tensor $N_{\mu\nu}$ may be written as

$$\begin{aligned} N_{\mu\nu} &= 2\,\mathrm{Tr}[\not{k}'\gamma_\mu(1-\gamma_5)\not{k}\gamma_\nu] \\ &= 2\,\mathrm{Tr}(\not{k}'\gamma_\mu\not{k}\gamma_\nu) - 2\,\mathrm{Tr}(\gamma_5\not{k}\gamma_\nu\not{k}'\gamma_\mu). \end{aligned} \tag{20.100}$$

The first trace is the same as in our calculation of e$\mu$ scattering (cf (8.185)):

$$\mathrm{Tr}(\not{k}'\gamma_\mu\not{k}\gamma_\nu) = 4[k'_\mu k_\nu + k'_\nu k_\mu + (q^2/2)g_{\mu\nu}]. \tag{20.101}$$

The second trace must be evaluated using the result

$$\mathrm{Tr}(\gamma_5\not{a}\not{b}\not{c}\not{d}) = 4\mathrm{i}\epsilon_{\alpha\beta\gamma\delta}a^\alpha b^\beta c^\gamma d^\delta \tag{20.102}$$

(see equation (J.37) in appendix J of volume 1). The totally anti-symmetric tensor $\epsilon_{\alpha\beta\gamma\delta}$ is just the generalization of $\epsilon_{ijk}$ to four dimensions and is defined by

$$\epsilon_{\alpha\beta\gamma\delta} = \begin{cases} +1 & \text{for } \epsilon_{0123} \text{ and all even permutations of } 0, 1, 2, 3 \\ -1 & \text{for } \epsilon_{1023} \text{ and all odd permutations of } 0, 1, 2, 3 \\ 0 & \text{otherwise.} \end{cases}$$

Its appearance here is a direct consequence of parity violation. Note that this definition has the consequence that

$$\epsilon_{0123} = +1 \tag{20.103}$$

but

$$\epsilon^{0123} = -1. \tag{20.104}$$

We will also need to contract two $\epsilon$ tensors. By looking at the possible combinations, it should be easy to convince yourself of the result

$$\epsilon_{ijk}\epsilon_{ilm} = \begin{vmatrix} \delta_{jl} & \delta_{jm} \\ \delta_{kl} & \delta_{km} \end{vmatrix} \tag{20.105}$$

i.e.

$$\epsilon_{ijk}\epsilon_{ilm} = \delta_{jl}\delta_{km} - \delta_{kl}\delta_{jm}. \tag{20.106}$$

For the four-dimensional $\epsilon$ tensor, one can show (see problem 20.7) that

$$\epsilon_{\mu\nu\alpha\beta}\epsilon^{\mu\nu\gamma\delta} = -2! \begin{vmatrix} \delta_\alpha^\gamma & \delta_\beta^\gamma \\ \delta_\alpha^\delta & \delta_\beta^\delta \end{vmatrix} \tag{20.107}$$

where the minus sign arises from (20.104) and the 2! from the fact that the two indices are contracted.

We can now evaluate $N_{\mu\nu}$. We obtain, after some rearrangement of indices, the result for the $\nu_\mu \to \mu^-$ tensor:

$$N_{\mu\nu} = 8[(k'_\mu k_\nu + k'_\nu k_\mu + (q^2/2)g_{\mu\nu}) - \mathrm{i}\epsilon_{\mu\nu\alpha\beta}k^\alpha k'^\beta]. \tag{20.108}$$

For the electron tensor $E^{\mu\nu}$ we have a similar result (divided by 2):

$$E^{\mu\nu} = 4[(p^{\mu\prime} p^\nu + p^{\nu\prime} p^\mu + (q^2/2)g^{\mu\nu}) - \mathrm{i}\epsilon^{\mu\nu\gamma\delta} p_\gamma p'_\delta]. \tag{20.109}$$

Now, in the approximation of neglecting all lepton masses,

$$q^\mu N_{\mu\nu} = q^\nu N_{\mu\nu} = 0 \tag{20.110}$$

as for the electromagnetic tensor $L_{\mu\nu}$ (cf (8.188)). Hence, we may replace

$$p' = p + q \tag{20.111}$$

and drop all terms involving $q$ in the contraction with $N_{\mu\nu}$. In the anti-symmetric term, however, we have

$$\epsilon^{\mu\nu\gamma\delta} p_\gamma (p_\delta + q_\delta) = \epsilon^{\mu\nu\gamma\delta} p_\gamma q_\delta \tag{20.112}$$

since the term with $p_\delta$ vanishes because of the anti-symmetry of $\epsilon_{\mu\nu\gamma\delta}$. Thus, we arrive at

$$E^{\mu\nu}_{\text{eff}} = 8p^\mu p^\nu + 2q^2 g^{\mu\nu} - 4\mathrm{i}\epsilon^{\mu\nu\gamma\delta} p_\gamma q_\delta. \tag{20.113}$$

We must now calculate the '$N \cdot E$' contraction in (20.96). Since we are neglecting all masses, it is easiest to perform the calculation in invariant form before specializing to the 'laboratory' frame. The usual Mandelstam variables are (neglecting all masses)

$$s = 2k \cdot p \tag{20.114}$$

$$u = -2k' \cdot p \tag{20.115}$$

$$t = -2k \cdot k' = q^2 \tag{20.116}$$

satisfying

$$s + t + u = 0. \tag{20.117}$$

The result of performing the contraction

$$N_{\mu\nu} E^{\mu\nu} = N_{\mu\nu} E^{\mu\nu}_{\text{eff}} \tag{20.118}$$

may be found using the result (20.107) for the contraction of two $\epsilon$ tensors (see problem 20.7): the answer for $\nu_\mu \mathrm{e}^- \to \mu^- \nu_\mathrm{e}$ is

$$N_{\mu\nu} E^{\mu\nu} = 16(s^2 + u^2) + 16(s^2 - u^2) \tag{20.119}$$

where the first term arises from the symmetric part of $N_{\mu\nu}$, similar to $L_{\mu\nu}$, and the second term from the anti-symmetric part involving $\epsilon_{\mu\nu\alpha\beta}$. We have also used

$$t = q^2 = -(s + u) \tag{20.120}$$

valid in the approximation in which we are working. Thus, for $\nu_\mu \mathrm{e}^- \to \mu^- \nu_\mathrm{e}$ we have

$$N_{\mu\nu} E^{\mu\nu} = +32s^2 \tag{20.121}$$

and with

$$\frac{\mathrm{d}\sigma}{\mathrm{d}\Omega} = \frac{1}{64\pi^2 s} \left( \frac{G_\mathrm{F}^2}{2} \right) N_{\mu\nu} E^{\mu\nu} \tag{20.122}$$

we finally obtain the result

$$\frac{\mathrm{d}\sigma}{\mathrm{d}\Omega} = \frac{G_\mathrm{F}^2 s}{4\pi^2}. \tag{20.123}$$

The total cross-section is then

$$\sigma = \frac{G_F^2 s}{\pi}. \tag{20.124}$$

Since $t = -2p^2(1 - \cos\theta)$, where $p$ is the CM momentum and $\theta$ the CM scattering angle, (20.123) can alternatively be written in invariant form as (problem 20.8)

$$\frac{d\sigma}{dt} = \frac{G_F^2}{\pi}. \tag{20.125}$$

All other purely leptonic processes may be calculated in an analogous fashion (see Bailin (1982) and Renton (1990) for further examples).

When we discuss deep inelastic neutrino scattering in section 20.11, we shall be interested in neutrino 'laboratory' cross-sections, as in the electron scattering case of chapter 9. A simple calculation gives $s \simeq 2m_e E$ (neglecting squares of lepton masses by comparison with $m_e E$), where $E$ is the 'laboratory' energy of a neutrino incident, in this example, on a stationary electron. It follows that *the total 'laboratory' cross-section in this Fermi-like current–current model rises linearly with E*. We shall return to the implications of this in section 20.11.

The process (20.88) was measured by Bergsma *et al* (CHARM collaboration) (1983) using the CERN wide-band beam ($E_\nu \sim 20$ GeV). The ratio of the observed number of events to that expected for pure V − A was quoted as $0.98 \pm 0.12$.

## 20.9 Leptonic weak neutral currents

The first observations of the weak neutral current process $\bar{\nu}_\mu e^- \to \bar{\nu}_\mu e^-$ were reported by Hasert *et al* (1973), in a pioneer experiment using the heavy-liquid bubble chamber Gargamelle at CERN, irradiated with a $\bar{\nu}_\mu$ beam. As in the case of the charged currents, much detailed experimental work was necessary to determine the precise form of the neutral current couplings. They are, of course, *predicted* by the Glashow–Salam–Weinberg (GSW) theory, as we shall explain in chapter 22. For the moment, we continue with the current–current approach, parametrizing the currents in a convenient way.

There are two types of 'neutral current' couplings: those involving neutrinos of the form $\bar{\hat{\nu}}_l \ldots \hat{\nu}_l$; and those involving the charged leptons of the form $\bar{\hat{l}} \ldots \hat{l}$. We shall assume the following form for these currents (with one eye on the GSW theory to come):

(1) neutrino neutral current

$$g_{\rm N} c^{\nu_l} \bar{\hat{\nu}}_l \gamma^\mu \left(\frac{1-\gamma_5}{2}\right) \hat{\nu}_l \qquad l = e, \mu, \tau; \tag{20.126}$$

(2) charged lepton neutral current

$$g_{\rm N} \bar{\hat{l}} \gamma^\mu \left[c_L^l \frac{(1-\gamma_5)}{2} + c_R^l \frac{(1+\gamma_5)}{2}\right] \hat{l} \qquad l = e, \mu, \tau. \tag{20.127}$$

This is, of course, by no means the most general possible parametrization. The neutrino coupling is retained as pure 'V − A', while the coupling in the charged lepton sector is now a combination of 'V − A' and 'V+A' with certain coefficients $c_{\mathrm{L}}^l$ and $c_{\mathrm{R}}^l$. We may also write the coupling in terms of 'V' and 'A' coefficients defined by $c_{\mathrm{V}}^l = c_{\mathrm{L}}^l + c_{\mathrm{R}}^l, c_{\mathrm{A}}^l = c_{\mathrm{L}}^l - c_{\mathrm{R}}^l$. An overall factor $g_{\mathrm{N}}$ determines the strength of the neutral currents as compared to the charged ones: the $c$'s determine the relative amplitudes of the various neutral current processes.

As we shall see, an essential feature of the GSW theory is its prediction of weak neutral current processes, with couplings determined in terms of one parameter of the theory called '$\theta_{\mathrm{W}}$', the 'weak mixing angle' (Glashow 1961, Weinberg 1967). The GSW predictions for the parameter $g_{\mathrm{N}}$ and the $c$'s is (see equations (22.37)–(22.40))

$$g_{\mathrm{N}} = g/\cos\theta_{\mathrm{W}} \qquad c^{\nu_l} = \tfrac{1}{2} \qquad c_{\mathrm{L}}^l = -\tfrac{1}{2} + a \qquad c_{\mathrm{R}}^l = a \tag{20.128}$$

for $l = \mathrm{e}, \mu, \tau$, where $a = \sin^2\theta_{\mathrm{W}}$ and $g$ is the SU(2) gauge coupling. Note that a strong form of 'universality' is involved here too: the coefficients are independent of the 'flavour' e, $\mu$ or $\tau$, for both neutrinos and charged leptons.

The following reactions are available for experimental measurement (in addition to the charged current process (20.88) already discussed):

$$\nu_\mu \mathrm{e}^- \to \nu_\mu \mathrm{e}^- \qquad \bar{\nu}_\mu \mathrm{e}^- \to \bar{\nu}_\mu \mathrm{e}^- \qquad (\mathrm{NC}) \tag{20.129}$$

$$\nu_{\mathrm{e}} \mathrm{e}^- \to \nu_{\mathrm{e}} \mathrm{e}^- \qquad \bar{\nu}_{\mathrm{e}} \mathrm{e}^- \to \bar{\nu}_{\mathrm{e}} \mathrm{e}^- \qquad (\mathrm{NC} + \mathrm{CC}) \tag{20.130}$$

where 'NC' means neutral current and 'CC' charged current. Formulas for these cross-sections are given in section 22.4. The experiments are discussed and reviewed in Commins and Bucksbaum (1983), Renton (1990) and, most recently, by Winter (2000). All observations are in excellent agreement with the GSW predictions, with $\theta_{\mathrm{W}}$ determined as $\sin^2\theta_{\mathrm{W}} \simeq 0.23$. The reader must note, however, that modern precision measurements are sensitive to higher-order (loop) corrections, which must now be included in comparing the full GSW theory with experiment (see section 22.8). The simultaneous fit of data from all four reactions in terms of the single parameter $\theta_{\mathrm{W}}$ provides already strong confirmation of the theory—and indeed such confirmation was already emerging in the late 1970s and early 1980s, before the actual discovery of the $\mathrm{W}^{\pm}$ and $\mathrm{Z}^0$ bosons. It is also interesting to note that the presence of vector (V) interactions in the neutral current processes may suggest the possibility of some kind of link with electromagnetic interactions which are, of course, also 'neutral' (in this sense) and vector-like. In the GSW theory, this linkage is provided essentially through the parameter $\theta_{\mathrm{W}}$, as we shall see.

## 20.10 Quark weak currents

The original version of V − A theory was framed in terms of a nucleonic current of the form $\bar{\hat{\psi}}_{\mathrm{p}}\gamma^\mu(1 - r\gamma_5)\hat{\psi}_{\mathrm{n}}$. With the acceptance of quark substructure it was natural to re-interpret such a hadronic transition by a charged current of the form $\bar{\hat{u}}\gamma^{\mu}(1-\gamma_5)\hat{d}$, very similar to the charged lepton currents: indeed, here was a further example of 'universality', this time between quarks and leptons. Detailed comparison with experiment showed, however, that such d → n transitions were very slightly weaker than the analogous leptonic ones: this could be established by comparing the rates for $\mathrm{n} \rightarrow \mathrm{pe}^{-}\bar{\nu}_{\mathrm{e}}$ and $\bar{\mu} \rightarrow \nu_{\mu}\mathrm{e}^{-}\bar{\nu}_{\mathrm{e}}$.

But for quarks (or their hadronic composites), there is a further complication, which is the very familiar phenomenon of flavour change in weak hadronic processes (recall the discussion in section 1.3.2). The first step towards the modern theory of quark currents was taken by Cabibbo (1963). He postulated that the strength of the hadronic weak interaction was *shared* between the $\Delta S = 0$ and $\Delta S = 1$ transitions (where $S$ is the strangeness quantum number), the latter being relatively suppressed as compared to the former. According to Cabibbo's hypothesis, phrased in terms of quarks, the total weak charged current for u, d and s quarks is

$$\hat{j}^{\mu}_{\mathrm{Cab}}(\mathrm{u, d, s}) = \cos\theta_{\mathrm{C}}\bar{\hat{u}}\gamma^{\mu}\frac{(1-\gamma_5)}{2}\hat{d} + \sin\theta_{\mathrm{C}}\bar{\hat{u}}\gamma^{\mu}\frac{(1-\gamma_5)}{2}\hat{s}. \tag{20.131}$$

We can now postulate a total weak charged current

$$\hat{j}^{\mu}_{\mathrm{CC}}(\text{total}) = \hat{j}^{\mu}_{\mathrm{CC}}(\text{leptons}) + \hat{j}^{\mu}_{\mathrm{Cab}}(\mathrm{u, d, s}) \tag{20.132}$$

where $\hat{j}^{\mu}_{\mathrm{CC}}(\text{leptons})$ is given by (20.81), and then generalize (20.83) to

$$\hat{\mathcal{H}}^{\mathrm{tot}}_{\mathrm{CC}} = \frac{G_{\mathrm{F}}}{\sqrt{2}}\hat{j}^{\mu}_{\mathrm{CC}}(\text{total})\hat{j}^{\dagger}_{\mathrm{CC}\mu}(\text{total}). \tag{20.133}$$

The effective interaction (20.133) describes a great many processes. The purely leptonic ones discussed previously are, of course, present in $\hat{j}^{\mu}_{\mathrm{CC}}(\text{leptons})\hat{j}_{\mathrm{CC}\mu}(\text{leptons})$. But there are also now all the *semi-leptonic* processes such as the $\Delta S = 0$ (strangeness conserving) one

$$\mathrm{d} \rightarrow \mathrm{u} + \mathrm{e}^{-} + \bar{\nu}_{\mathrm{e}} \tag{20.134}$$

and the $\Delta S = 1$ (strangeness changing) one

$$\mathrm{s} \rightarrow \mathrm{u} + \mathrm{e}^{-} + \bar{\nu}_{\mathrm{e}}. \tag{20.135}$$

The notion that the 'total current' should be the sum of a hadronic and a leptonic part is already familiar from electromagnetism—see, for example, equation (8.90).

The transition (20.135), for example, is the underlying process in semi-leptonic decays such as

$$\Sigma^{-} \rightarrow \mathrm{n} + \mathrm{e}^{-} + \bar{\nu}_{\mathrm{e}} \tag{20.136}$$

and

$$\mathrm{K}^{-} \rightarrow \pi^{0} + \mathrm{e}^{-} + \bar{\nu}_{\mathrm{e}} \tag{20.137}$$

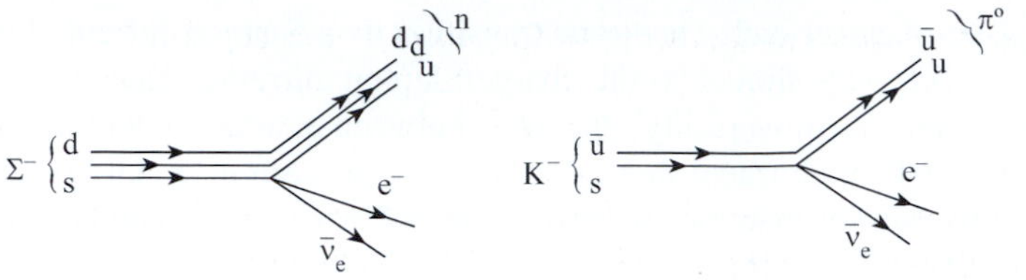


**Figure 20.3.** Strangeness-changing semi-leptonic weak decays.

as indicated in figure 20.3.

The 's' quark is assigned $S = -1$ and charge $-\frac{1}{3}e$. The s $\rightarrow$ u transition is then referred to as one with '$\Delta S = \Delta Q$', meaning that the change in the quark (or hadronic) strangeness is equal to the change in the quark (or hadronic) charge: both the strangeness and the charge increase by one unit. Prior to the advent of the quark model and the Cabibbo hypothesis, it had been established empirically that all known strangeness-changing semileptonic decays satisfied the rules $|\Delta S| = 1$ and $\Delta S = \Delta Q$. The u–s current in (20.131) satisfies these rules automatically. Note, for example, that the process apparently similar to (20.136), $\Sigma^+ \rightarrow \mathrm{n} + \mathrm{e}^+ + \nu_\mathrm{e}$, is forbidden in the lowest order (it requires a double quark transition from suu to udd). All known data on such decays can be fit with a value $\sin\theta_\mathrm{C} \simeq 0.23$ for the 'Cabibbo angle' $\theta_\mathrm{C}$ (not to be confused with $\theta_\mathrm{W}$). This relatively small angle is, therefore, a measure of the suppression of $|\Delta S| = 1$ processes relative to $\Delta S = 0$ ones.

The Cabibbo current can be written in a more compact form by introducing the 'mixed' field

$$\hat{d}' \equiv \cos\theta_\mathrm{C}\hat{d} + \sin\theta_\mathrm{C}\hat{s}. \tag{20.138}$$

Then

$$\hat{j}^\mu_{\mathrm{Cab}}(\mathrm{u, d, s}) = \bar{\hat{u}}\gamma^\mu \frac{(1-\gamma_5)}{2}\hat{d}'. \tag{20.139}$$

In 1970 Glashow, Iliopoulos and Maiani (GIM) (1970) drew attention to a theoretical problem with the interaction (20.133) if used in *second* order. Now it is, of course, the case that this interaction is not renormalizable, as noted previously for the purely leptonic one (20.83), since $G_\mathrm{F}$ has dimensions of an inverse mass squared. As we saw in section 11.7, this means that one-loop diagrams will typically diverge quadratically, so that the contribution of such a second-order process will be of order $(G_\mathrm{F}.G_\mathrm{F}\Lambda^2)$ where $\Lambda$ is a cut-off, compared to the first-order amplitude $G_\mathrm{F}$. Recalling from (20.89) that $G_\mathrm{F} \sim 10^{-5}\ \mathrm{GeV}^{-2}$, we see that for $\Lambda \sim 10$ GeV such a correction could be significant if accurate enough data existed. GIM pointed out, in particular, that some second-order processes could be found which violated the (hitherto) well-established phenomenological selection rules, such as the $|\Delta S| = 1$ and $\Delta S = \Delta Q$ rules already discussed. For example, there could be $\Delta S = 2$ amplitudes contributing to the $\mathrm{K_L} - \mathrm{K_S}$ mass difference (see Renton 1990, section 9.1.6, for example), as well as contributions to unobserved decay modes such as

$$\mathrm{K}^+ \to \pi^+ + \nu + \bar{\nu} \tag{20.140}$$

which has a *neutral* lepton pair in association with a *strangeness change* for the hadron. In fact, experiment placed very tight limits on the non-existence of (20.140)—and still does: the present limit on the branching is less than about $10^{-9}\%$. This seemed to imply a surprisingly low value for the cut-off, say $\sim 3$ GeV (Mohapatra *et al* 1968).

Partly in order to address this problem and partly as a revival of an earlier lepton-quark symmetry proposal (Bjorken and Glashow 1964), GIM introduced a fourth quark, now called c (the charm quark) with charge $\frac{2}{3}e$. Note that in 1970 the $\tau$-lepton had not been discovered, so only two lepton family pairs ($\nu_{\mathrm{e}}$, e), ($\nu_\mu$, $\mu$) were known: this fourth quark, therefore, did restore the balance, via the two-quark family pairs (u,d), (c,s). In particular, a second quark current could now be hypothesized, involving the (c,s) pair. GIM postulated that the c-quark was coupled to the 'orthogonal' d–s combination (cf (20.138))

$$\hat{s}' = -\sin\theta_{\mathrm{C}}\hat{d} + \cos\theta_{\mathrm{C}}\hat{s}. \tag{20.141}$$

The complete four-quark charged current is then

$$\hat{j}^\mu_{\mathrm{GIM}}(\mathrm{u, d, c, s}) = \bar{\hat{u}}\gamma^\mu \frac{(1-\gamma_5)}{2}\hat{d}' + \bar{\hat{c}}\gamma^\mu \frac{(1-\gamma_5)}{2}\hat{s}'. \tag{20.142}$$

The form (20.142) had already been suggested by Bjorken and Glashow (1964). The new feature of GIM was the observation that, assuming an exact $\mathrm{SU(4)_f}$ symmetry for the four quarks (in particular, equal masses), all second-order contributions which could have violated the $|\Delta S| = 1$, $\Delta S = \Delta Q$ selection rules now vanished. Further, to the extent that the (unknown) mass of the charm quark functioned as an effective cut-off $\Lambda$, due to breaking of the $\mathrm{SU(4)_f}$ symmetry, they estimated $m_{\mathrm{c}}$ to lie in the range 3–4 GeV, from the observed $\mathrm{K_L} - \mathrm{K_S}$ mass difference.

GIM went on to speculate that the non-renormalizability could be overcome if the weak interactions were described by an SU(2) Yang–Mills gauge theory, involving a triplet ($\mathrm{W}^+$, $\mathrm{W}^-$, $\mathrm{W}^0$) of gauge bosons. In this case, it is natural to introduce the idea of (weak) 'isospin', in terms of which the pairs ($\nu_{\mathrm{e}}$, e), ($\nu_\mu$, $\mu$), (u,d′), (c, s′) are all $t = \frac{1}{2}$ doublets with $t_3 = \pm\frac{1}{2}$. Charge-changing currents then involve the 'raising' matrix

$$\frac{1}{2}\tau_+ \equiv \frac{1}{2}(\tau_1 + \mathrm{i}\tau_2) = \begin{pmatrix} 0 & 1 \\ 0 & 0 \end{pmatrix} \tag{20.143}$$

and charge-lowering ones the matrix $\tau_-/2 = (\tau_1 - \mathrm{i}\tau_2)/2$. The full symmetry must also involve the matrix $\tau_3/2$, given by the commutator $[\tau_+/2, \tau_-/2] = \tau_3$. Whereas $\tau_+$ and $\tau_-$ would (in this model) be associated with transitions mediated by $W^{\pm}$, transitions involving $\tau_3$ would be mediated by $W^0$ and would correspond to 'neutral current' transitions for quarks. We now know that things are slightly more complicated than this: the correct symmetry is the SU(2) × U(1) of Glashow (1961), also invoked by GIM. Skipping, therefore, some historical steps, we parametrize the *weak quark neutral current* as (cf (20.127) for the leptonic analogue)

$$g_{\mathrm{N}} \sum_{q=\mathrm{u,c,d',s'}} \bar{\hat{q}}\gamma^{\mu}\left[c_{\mathrm{L}}^{q}\frac{(1-\gamma_5)}{2}+c_{\mathrm{R}}^{q}\frac{(1+\gamma_5)}{2}\right]\hat{q} \tag{20.144}$$

for the four flavours so far in play. In the GSW theory, the $c_{\mathrm{L}}^{q}$'s are predicted to be

$$c_{\mathrm{L}}^{\mathrm{u,c}} = \tfrac{1}{2} - \tfrac{2}{3}a \qquad c_{\mathrm{R}}^{\mathrm{u,c}} = -\tfrac{2}{3}a \tag{20.145}$$

$$c_{\mathrm{L}}^{\mathrm{d,s}} = -\tfrac{1}{2} + \tfrac{1}{3}a \qquad c_{\mathrm{R}}^{\mathrm{d,s}} = \tfrac{1}{3}a \tag{20.146}$$

where $a = \sin^2\theta_{\mathrm{W}}$ as before and $g_{\mathrm{N}} = g/\cos\theta_{\mathrm{W}}$.

One feature of (20.144) is worth nothing. Consider the terms

$$\bar{\hat{d}}'\{\ldots\}\hat{d}' + \bar{\hat{s}}'\{\ldots\}\hat{s}'. \tag{20.147}$$

It is simple to verify that, whereas either part of (20.147) alone contains a *strangeness-changing neutral* combination such as $\bar{\hat{d}}\{\ldots\}\hat{s}$ or $\bar{\hat{s}}\{\ldots\}\hat{d}$, such combinations vanish in the sum, leaving the result *diagonal in quark flavour*. Thus, there are no first-order neutral flavour-changing currents in this model, a result which will be extended to three flavours in sections 22.3 and 22.7.1.

In 1974, Gaillard and Lee (1974) performed a full one-loop calculation of the $K_{\mathrm{L}} - K_{\mathrm{S}}$ mass difference in the GSW model as extended by GIM to quarks and using the renormalization techniques recently developed by 't Hooft (1971b). They were able to predict $m_{\mathrm{c}} \sim 1.5$ GeV for the charm quark mass, a result spectacularly confirmed by the subsequent discovery of the $\mathrm{c\bar{c}}$ states in charmonium, and of charmed mesons and baryons of the appropriate mass.

## 20.11 Deep inelastic neutrino scattering

We now present another illustrative calculation within the framework of the 'current–current' model, this time involving neutrinos and quarks. We shall calculate cross-sections for deep inelastic neutrino scattering from nucleons, using the parton model introduced (for electromagnetic interactions) in chapter 9. In particular, we shall consider the processes

$$\nu_\mu + \mathrm{N} \to \mu^- + \mathrm{X} \tag{20.148}$$

$$\bar{\nu}_\mu + \mathrm{N} \to \mu^+ + \mathrm{X} \tag{20.149}$$

which, of course, involve the charged currents for both leptons and quarks. Studies of these reactions at Fermilab and CERN in the 1970s and 1980s played

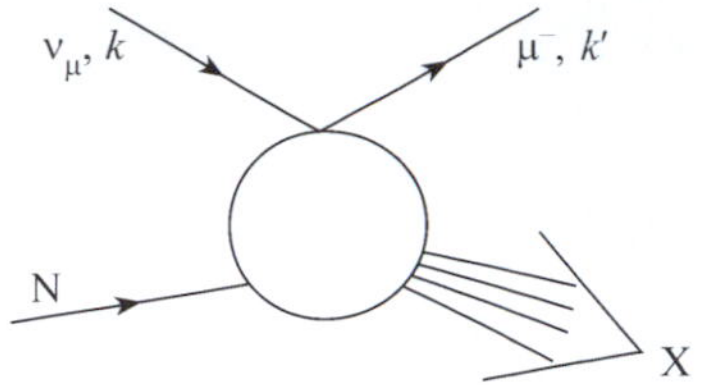


**Figure 20.4.** Inelastic neutrino scattering from a nucleon.

a crucial part in establishing the quark structure of the nucleon, in particular the quark distribution functions.

The general process is illustrated in figure 20.4. By now we are becoming accustomed to the idea that such processes are, in fact, mediated by the $W^+$, but we shall assume that the momentum transfers are such that the W-propagator is effectively constant (see the discussion in section 21.2). The effective lepton–quark interaction will then take the form

$$\hat{\mathcal{H}}_{\nu q}^{\text{eff}} = \frac{G_F}{\sqrt{2}} \bar{\hat{\mu}} \gamma_\mu (1-\gamma_5) \hat{\nu}_\mu [\bar{\hat{u}} \gamma^\mu (1-\gamma_5) \hat{d} + \bar{\hat{c}} \gamma^\mu (1-\gamma_5) \hat{s}] \tag{20.150}$$

leading to expressions for the parton-level subprocess amplitudes which are exactly similar to that in (20.93) for $\nu_\mu + e^- \to \mu^- + \nu_e$. Note that we are considering only the four flavours u, d, c, s to be 'active', and we have set $\theta_C \approx 0$.

As in (20.96), the $\nu_\mu$ cross-section will have the general form

$$d\sigma^{(\nu)} \propto N_{\mu\nu} W_{(\nu)}^{\mu\nu}(q, p) \tag{20.151}$$

where $N_{\mu\nu}$ is the neutrino tensor of (20.108). The form of the weak hadron tensor $W_{(\nu)}^{\mu\nu}$ is deduced from Lorentz invariance. In the approximation of neglecting lepton masses, we can ignore any dependence on the 4-vector $q$ since

$$q^\mu N_{\mu\nu} = q^\nu N_{\mu\nu} = 0. \tag{20.152}$$

Just as $N_{\mu\nu}$ contains the pseudotensor $\epsilon_{\mu\nu\alpha\beta}$, so too will $W_{(\nu)}^{\mu\nu}$ since parity is not conserved. In a manner similar to equation (9.10) for the case of electron scattering, we define neutrino structure functions by

$$W_{(\nu)}^{\mu\nu} = (-g^{\mu\nu}) W_1^{(\nu)} + \frac{1}{M^2} p^\mu p^\nu W_2^{(\nu)} - \frac{i}{2M^2} \epsilon^{\mu\nu\gamma\delta} p_\gamma q_\delta W_3^{(\nu)}. \tag{20.153}$$

In general, the structure functions depend on two variables, say $Q^2$ and $\nu$, where $Q^2 = -(k-k')^2$ and $\nu = p \cdot q / M$; but in the Bjorken limit approximate scaling is observed, as in the electron case:

$$\left.\begin{array}{c} Q^2 \to \infty \\ \nu \to \infty \end{array}\right\} \qquad x = Q^2/2M\nu \text{ fixed} \tag{20.154}$$

$$\nu W_2^{(\nu)}(Q^2, \nu) \to F_2^{(\nu)}(x) \tag{20.155}$$

$$M W_1^{(\nu)}(Q^2, \nu) \to F_1^{(\nu)}(x) \tag{20.156}$$

$$\nu W_3^{(\nu)}(Q^2, \nu) \to F_3^{(\nu)}(x) \tag{20.157}$$

where, as with (9.21) and (9.22), the physics lies in the assertion that the $F$'s are finite. This scaling can again be interpreted in terms of point-like scattering from partons—which we shall take to have quark quantum numbers.

In the 'laboratory' frame (in which the nucleon is at rest) the cross-section in terms of $W_1$, $W_2$ and $W_3$ may be derived in the usual way from (cf equation (9.11))

$$\mathrm{d}\sigma^{(\nu)} = \left(\frac{G_{\mathrm{F}}}{\sqrt{2}}\right)^2 \frac{1}{4k \cdot p} 4\pi M N_{\mu\nu} W_{(\nu)}^{\mu\nu} \frac{\mathrm{d}^3 \boldsymbol{k}'}{2k'(2\pi)^3}. \tag{20.158}$$

In terms of 'laboratory' variables, one obtains (problem 20.10)

$$\frac{\mathrm{d}^2\sigma^{(\nu)}}{\mathrm{d}Q^2\,\mathrm{d}\nu} = \frac{G_{\mathrm{F}}^2}{2\pi}\frac{k'}{k}\left(W_2^{(\nu)}\cos^2(\theta/2) + W_1^{(\nu)} 2\sin^2(\theta/2) + \frac{k+k'}{M}\sin^2(\theta/2) W_3^{(\nu)}\right). \tag{20.159}$$

For an incoming anti-neutrino beam, the $W_3$ term changes sign.

In neutrino scattering it is common to use the variables $x, \nu$ and the 'inelasticity' $y$ where

$$y = p \cdot q / p \cdot k. \tag{20.160}$$

In the 'laboratory' frame, $\nu = E - E'$ (the energy transfer to the nucleon) and $y = \nu/E$. The cross-section can be written in the form (see problem 20.10)

$$\frac{\mathrm{d}^2\sigma^{(\nu)}}{\mathrm{d}x\,\mathrm{d}y} = \frac{G_{\mathrm{F}}^2}{2\pi} s \left(F_2^{(\nu)} \frac{1+(1-y)^2}{2} + x F_3^{(\nu)} \frac{1-(1-y)^2}{2}\right) \tag{20.161}$$

in terms of the Bjorken scaling functions, and we have assumed the relation

$$2x F_1^{(\nu)} = F_2^{(\nu)} \tag{20.162}$$

appropriate for spin-$\frac{1}{2}$ constituents.

We now turn to the parton-level subprocesses. Their cross-sections can be straightforwardly calculated in the same way as for $\nu_\mu \mathrm{e}^-$ scattering in section 20.8. We obtain (problem 20.11)

$$\nu\mathrm{q}, \bar{\nu}\bar{\mathrm{q}}: \quad \frac{\mathrm{d}^2\sigma}{\mathrm{d}x\,\mathrm{d}y} = \frac{G_{\mathrm{F}}^2}{\pi} s x \delta\left(x - \frac{Q^2}{2M\nu}\right) \tag{20.163}$$

$$\nu\bar{\mathrm{q}}, \bar{\nu}\mathrm{q}: \quad \frac{\mathrm{d}^2\sigma}{\mathrm{d}x\,\mathrm{d}y} = \frac{G_{\mathrm{F}}^2}{\pi} s x (1-y)^2 \delta\left(x - \frac{Q^2}{2M\nu}\right). \tag{20.164}$$

**Figure 20.5.** Suppression of $\nu_\mu \bar{q} \to \mu^- \bar{q}$ for $y = 1$: (*a*) initial-state helicities; (*b*) final-state helicities at $y = 1$.

The factor $(1 - y)^2$ in the $\nu\bar{q}$, $\bar{\nu}q$ cases means that the reaction is forbidden at $y = 1$ (backwards in the CM frame). This follows from the V − A nature of the current, and angular momentum conservation, as a simple helicity argument shows. Consider, for example, the case $\nu\bar{q}$ shown in figure 20.5, with the helicities marked as shown. In our current–current interaction there are no gradient coupling terms and, therefore, no momenta in the momentum-space matrix element. This means that no orbital angular momentum is available to account for the reversal of net helicity in the initial and final states in figure 20.5. The lack of orbital angular momentum can also be inferred physically from the 'point-like' nature of the current–current coupling. For the $\nu q$ or $\bar{\nu}\bar{q}$ cases, the initial and final helicities add to zero and backward scattering is allowed.

The contributing processes are

$$\nu \mathrm{d} \to l^- \mathrm{u} \qquad \bar{\nu}\bar{\mathrm{d}} \to l^+ \bar{\mathrm{u}} \tag{20.165}$$

$$\nu \bar{\mathrm{u}} \to l^- \bar{\mathrm{d}} \qquad \bar{\nu}\mathrm{u} \to l^+ \mathrm{d} \tag{20.166}$$

the first pair having the cross-section (20.163), the second (20.164). Following the same steps as in the electron scattering case (sections 9.2 and 9.3), we obtain

$$F_2^{\nu \mathrm{p}} = F_2^{\bar{\nu}\mathrm{n}} = 2x[d(x) + \bar{u}(x)] \tag{20.167}$$

$$F_3^{\nu \mathrm{p}} = F_3^{\bar{\nu}\mathrm{n}} = 2[d(x) - \bar{u}(x)] \tag{20.168}$$

$$F_2^{\nu \mathrm{n}} = F_2^{\bar{\nu}\mathrm{p}} = 2x[u(x) + \bar{d}(x)] \tag{20.169}$$

$$F_3^{\nu \mathrm{n}} = F_3^{\bar{\nu}\mathrm{p}} = 2[u(x) - \bar{d}(x)]. \tag{20.170}$$

Inserting (20.167) and (20.168) into (20.161), for example, we find that

$$\frac{\mathrm{d}^2\sigma^{(\nu \mathrm{p})}}{\mathrm{d}x\mathrm{d}y} = 2\sigma_0 x[d(x) + (1-y)^2 \bar{u}(x)] \tag{20.171}$$

where

$$\sigma_0 = \frac{G_{\mathrm{F}}^2 s}{2\pi} = \frac{G_{\mathrm{F}}^2 ME}{\pi} \simeq 1.5 \times 10^{-42} (E/\mathrm{GeV})\ \mathrm{m}^2 \tag{20.172}$$

is the basic 'point-like' total cross-section (compare (20.124)). Similarly, one finds that

$$\frac{\mathrm{d}^2\sigma^{(\bar{\nu} \mathrm{p})}}{\mathrm{d}x\,\mathrm{d}y} = 2\sigma_0 x[(1-y)^2 u(x) + \bar{d}(x)]. \tag{20.173}$$

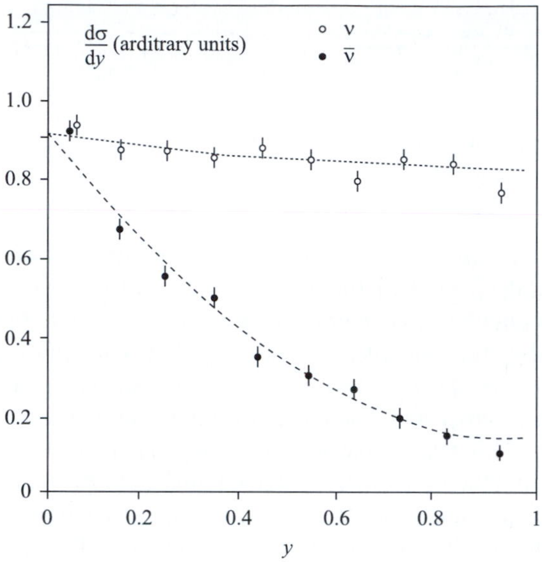


**Figure 20.6.** Charged-current inelasticity ($y$) distribution as measured by CDHS (from Winter 2000, p 443).

The corresponding results for $\nu$n and $\bar{\nu}$n are given by interchanging $u(x)$ and $d(x)$ and $\bar{u}(x)$ and $\bar{d}(x)$.

The target nuclei usually have approximately equal numbers of protons and neutrons and it is appropriate to average the 'n' and 'p' results to obtain an 'isoscalar' cross-section $\sigma^{(\nu\text{N})}$ or $\sigma^{(\bar{\nu}\text{N})}$:

$$\frac{\mathrm{d}^2\sigma^{(\nu\text{N})}}{\mathrm{d}x\,\mathrm{d}y} = \sigma_0 x[q(x) + (1-y)^2\bar{q}(x)] \tag{20.174}$$

$$\frac{\mathrm{d}^2\sigma^{(\bar{\nu}\text{N})}}{\mathrm{d}x\,\mathrm{d}y} = \sigma_0 x[(1-y)^2 q(x) + \bar{q}(x)] \tag{20.175}$$

where $q(x) = u(x) + d(x)$ and $\bar{q}(x) = \bar{u}(x) + \bar{d}(x)$.

Many simple and striking predictions now follow from these quark parton results. For example, by integrating (20.174) and (20.175) over $x$, we can write

$$\frac{\mathrm{d}\sigma^{(\nu\text{N})}}{\mathrm{d}y} = \sigma_0[Q + (1-y^2)\bar{Q}] \tag{20.176}$$

$$\frac{\mathrm{d}\sigma^{(\bar{\nu}\text{N})}}{\mathrm{d}y} = \sigma_0[(1-y)^2 Q + \bar{Q}] \tag{20.177}$$

where $Q = \int xq(x)\,\mathrm{d}x$ is the fraction of the nucleon's momentum carried by quarks and similarly for $\bar{Q}$. These two distributions in $y$ ('inelasticity

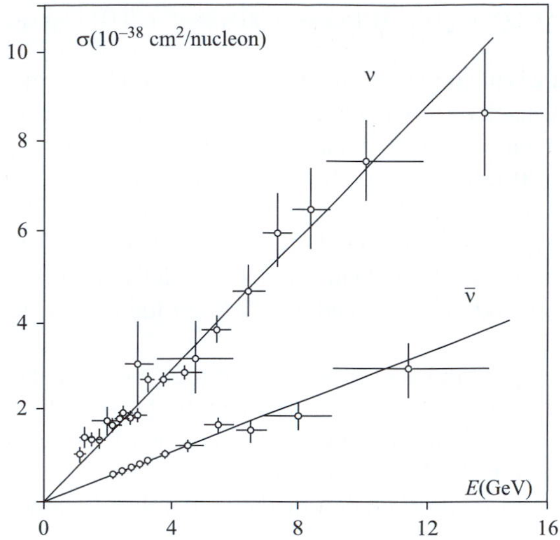


**Figure 20.7.** Low-energy $\nu$ and $\bar{\nu}$ cross-sections (from Winter 2000, p 427).

distributions'), therefore, give a direct measure of the quark and anti-quark composition of the nucleon. Figure 20.6 shows the inelasticity distributions as reported by the CDHS collaboration (de Groot *et al* 1979), from which the authors extracted the ratio

$$\bar{Q}/(Q+\bar{Q}) = 0.15 \pm 0.03 \tag{20.178}$$

after applying radiative corrections. An even more precise value can be obtained by looking at the region near $y = 1$ for $\bar{\nu}$N which is dominated by $\bar{Q}$, the small $Q$ contribution ($\propto (1-y)^2$) being subtracted out using $\nu$N data at the same $y$. This method yields

$$\bar{Q}/(Q+\bar{Q}) = 0.15 \pm 0.01. \tag{20.179}$$

Integrating (20.176) and (20.177) over $y$ gives

$$\sigma^{(\nu \mathrm{N})} = \sigma_0(Q + \tfrac{1}{3}\bar{Q}) \tag{20.180}$$

$$\sigma^{(\bar{\nu} \mathrm{N})} = \sigma_0(\tfrac{1}{3}Q + \bar{Q}) \tag{20.181}$$

and hence

$$Q + \bar{Q} = 3(\sigma^{(\nu \mathrm{N})} + \sigma^{(\bar{\nu} \mathrm{N})})/4\sigma_0 \tag{20.182}$$

while

$$\bar{Q}/(Q+\bar{Q}) = \frac{1}{2}\left(\frac{3r-1}{1+r}\right) \tag{20.183}$$

where $r = \sigma^{(\bar{\nu}\mathrm{N})}/\sigma^{(\nu\mathrm{N})}$. From total cross-section measurements and including c and s contributions, the CHARM collaboration (Allaby *et al* 1988) reported

$$Q + \bar{Q} = 0.492 \pm 0.006(\mathrm{stat}) \pm 0.019(\mathrm{syst}) \tag{20.184}$$

$$\bar{Q}/(Q+\bar{Q}) = 0.154 \pm 0.005(\text{stat}) \pm 0.011(\text{syst}). \tag{20.185}$$

The second figure is in good agreement with (20.179) and the first shows that only about 50% of the nucleon momentum is carried by charged partons, the rest being carried by the gluons, which do not have weak or electromagnetic interactions.

Equations (20.180) and (20.181), together with (20.172), predict that the total cross-sections $\sigma^{(\nu\mathrm{N})}$ and $\sigma^{(\bar{\nu}\mathrm{N})}$ rise linearly with the energy $E$. Figure 20.7 shows how this (parton model) prediction received spectacular confirmation as early as 1975 (Perkins 1975), soon after the model's success in deep inelastic scattering. In fact, both $\sigma^{(\nu\mathrm{N})}/E$ and $\sigma^{(\bar{\nu}\mathrm{N})}/E$ are found to be independent of $E$ up to $E \sim 200$ GeV.

Detailed comparison between the data at high energies and the earlier data of figure 20.7 at $E_\nu$ up to 15 GeV reveals that the $\bar{Q}$ fraction is increasing with energy. This is in accordance with the expectation of QCD corrections to the parton model (section 15.7): the $\bar{Q}$ distribution is large at small $x$ and scaling violations embodied in the evolution of the parton distributions predict a rise at small $x$ as the energy scale increases.

Returning now to (20.167)–(20.170), the two sum rules of (9.65) and (9.66) can be combined to give

$$3 = \int_0^1 \mathrm{d}x\,[u(x) + d(x) - \bar{u}(x) - \bar{d}(x)] \tag{20.186}$$

$$= \tfrac{1}{2}\int_0^1 \mathrm{d}x\,(F_3^{\nu\mathrm{p}} + F_3^{\nu\mathrm{n}}) \tag{20.187}$$

$$\equiv \int_0^1 \mathrm{d}x\, F_3^{\nu\mathrm{N}} \tag{20.188}$$

which is the Gross–Llewellyn Smith sum rule (1969), expressing the fact that the number of valence quarks per nucleon is three. The CDHS collaboration (de Groot *et al* 1979), quoted

$$I_{\mathrm{GLLS}} \equiv \int_0^1 \mathrm{d}x\, F_3^{\nu\mathrm{N}} = 3.2 \pm 0.5. \tag{20.189}$$

In perturbative QCD, there are corrections expressible as a power series in $\alpha_{\mathrm{s}}$, so that the parton model result is only reached as $Q^2 \to \infty$:

$$I_{\mathrm{GLLS}}(Q^2) = 3[1 + d_1\alpha_{\mathrm{s}}/\pi + d_2\alpha_{\mathrm{s}}^2/\pi^2 + \cdots] \tag{20.190}$$

where $d_1 = -1$ (Altarelli *et al* 1978a, b), $d_2 = -55/12 + N_{\mathrm{f}}/3$ (Gorishny and Larin 1986) where $N_{\mathrm{f}}$ is the number of active flavours. The CCFR collaboration (Shaevitz *et al* 1995) has measured $I_{\mathrm{GLLS}}$ in anti-neutrino–nucleon scattering at $\langle Q^2\rangle \sim 3$ GeV$^2$. It obtained

$$I_{\mathrm{GLLS}}(\langle Q^2\rangle = 3\ \mathrm{GeV}^2) = 2.50 \pm 0.02 \pm 0.08 \tag{20.191}$$

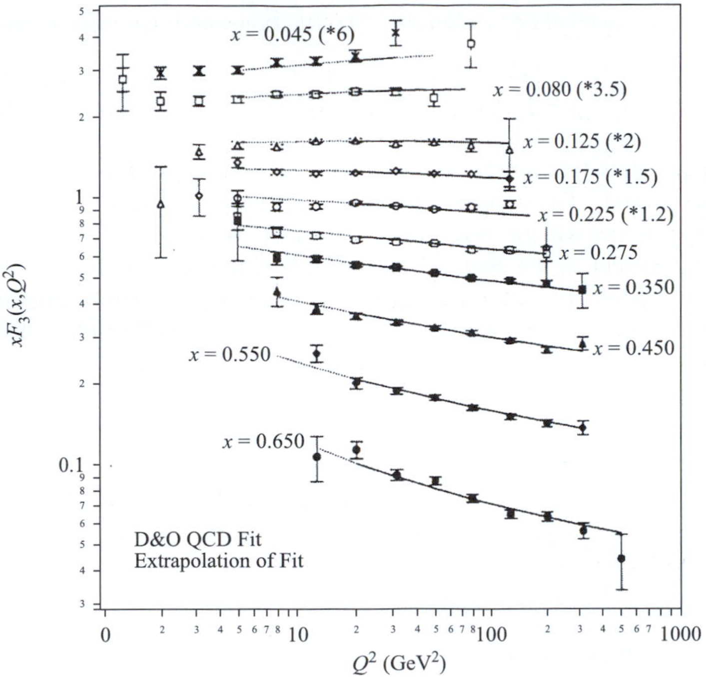


**Figure 20.8.** CCFR neutrino–iron structure functions $xF_3^{(\nu)}$ (Shaevitz *et al* 1995). The full line is the next-to-leading order (one-loop) QCD prediction and the dotted line is an extrapolation to regions outside the kinematic cuts for the fit.

in agreement with the $O(\alpha_s^3)$ calculation of Larin and Vermaseren (1991) using $\Lambda_{\overline{MS}}^{\text{QCD}} = 250 \pm 50$ MeV.

The predicted $Q^2$ evolution of $xF_3$ is particularly simple since it is not coupled to the gluon distribution. To leading order, the $xF_3$ evolution is given by (cf (15.106))

$$\frac{\mathrm{d}}{\mathrm{d}\ln Q^2}(xF_3(x,Q^2)) = \frac{\alpha_s(Q^2)}{\pi}\int_x^1 P_{qq}(z)xF_3\left(\frac{x}{z},Q^2\right)\frac{\mathrm{d}z}{z}. \tag{20.192}$$

Figure 20.8, taken from Shaevitz *et al* (1995), shows a comparison of the CCFR data with the next-to-leading order calculation of Duke and Owens (1984). This fit yields a value of $\alpha_s$ at $Q^2 = M_Z^2$ given by

$$\alpha_s(M_Z^2) = 0.111 \pm 0.002 \pm 0.003. \tag{20.193}$$

The Adler sum rule (Adler 1963) involves the functions $F_2^{\bar{\nu}p}$ and $F_2^{\nu p}$:

$$I_A = \int_0^1 \frac{\mathrm{d}x}{x}(F_2^{\bar{\nu}p} - F_2^{\nu p}). \tag{20.194}$$

In the simple model of (20.167)–(20.170), the right-hand side of $I_A$ is just

$$2\int_0^1 \mathrm{d}x\,(u(x)+\bar{d}(x)-d(x)-\bar{u}(x)) \tag{20.195}$$

which represents four times the average of $I_3$ (isospin) of the target, which is $\frac{1}{2}$ for the proton. This sum rule follows from the conservation of the charged weak current (as will be true in the Standard Model, since this is a gauge symmetry current, as we shall see in the following chapter). Its measurement, however, depends precisely on separating the non-isoscalar contribution ($I_A$ vanishes for the isoscalar average 'N'). The only published result is that of the BEBC collaboration (Allasia *et al* 1984, 1985):

$$I_A = 2.02 \pm 0.40 \tag{20.196}$$

in agreement with the expected value 2.

Relations (20.167)–(20.170) allow the $F_2$ functions for electron (muon) and neutrino scattering to be simply related. From (9.58) and (9.61), we have

$$F_2^{\mathrm{eN}} = \tfrac{1}{2}(F_2^{\mathrm{ep}} + F_2^{\mathrm{en}}) = \tfrac{5}{18}x(u+\bar{u}+d+\bar{d}) + \tfrac{1}{9}x(s+\bar{s}) + \cdots \tag{20.197}$$

while (20.167) and (20.169) give

$$F_2^{\nu\mathrm{N}} \equiv \tfrac{1}{2}(F_2^{\nu\mathrm{p}} + F_2^{\nu\mathrm{n}}) = x(u+d+\bar{u}+\bar{d}). \tag{20.198}$$

Assuming that the non-strange contributions dominate, the neutrino and charged lepton structure functions should be approximately in the ratio 18/5, which is the reciprocal of the mean squared charge of the u and d quarks in the nucleon. Figure 20.9 shows the neutrino results on $F_2$ and $xF_3$ together with those from several $\mu$N experiments scaled by the factor 18/5. The agreement is reasonably good, and this gives further confirmation of the quark parton picture.

From (20.167)–(20.170), we see that the differences $F_2^{\nu} - xF_3^{\nu}$ involve the anti-quark (sea) contribution, which from the data are concentrated at small $x$, as we already inferred in section 9.3.

We have mentioned QCD corrections to the simple parton model at several points. Clearly the full machinery introduced in chapter 15, in the context of deep inelastic charged lepton scattering, can be employed for the case of neutrino scattering also. For further access to this area we refer to Ellis *et al* (1996, chapter 4), and Winter (2000, chapter 5).

## 20.12 Non-leptonic weak interactions

The effective weak Hamiltonian of (20.133) (as modified by GIM) clearly contains the term

$$\hat{\mathcal{H}}_{\mathrm{CC}}^{q}(x) = \frac{G_{\mathrm{F}}}{\sqrt{2}}\hat{j}^{\mu}_{\mathrm{GIM}}(x)\hat{j}^{\dagger}_{\mu\mathrm{GIM}}(x) \tag{20.199}$$

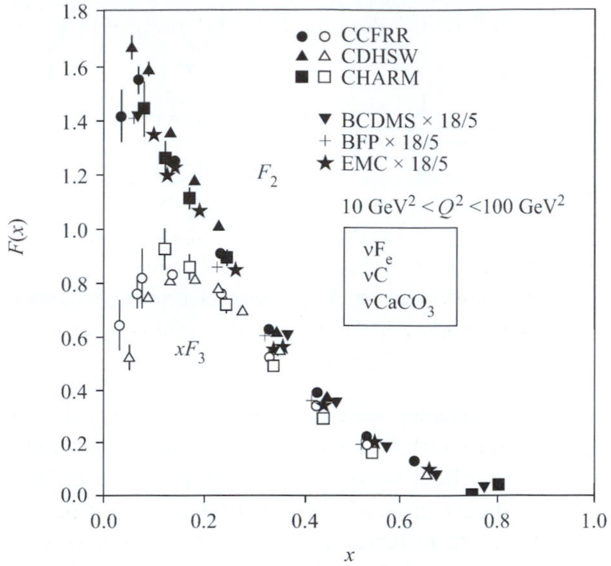


**Figure 20.9.** Comparison of neutrino results on $F_2(x)$ and $xF_3(x)$ with those from muon production properly rescaled by the factor 18/5, for a $Q^2$ ranging between 10 and 1000 $\mathrm{GeV}^2$ (from Winter 2000, p 455).

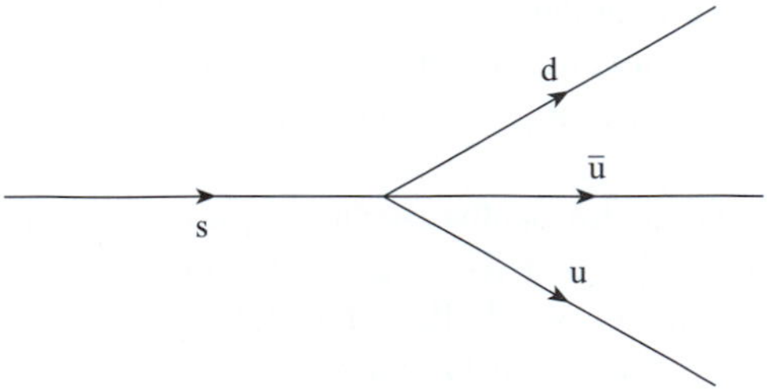


**Figure 20.10.** Effective four-fermion non-leptonic weak transition at the quark level.

in which no lepton fields are present (just as there are no quarks in (20.83)). This interaction is responsible, at the quark level, for transitions involving four-quark (or anti-quark) fields at a point. For example, the process shown in figure 20.10 can occur. By 'adding on' another two quark lines u and d, which undergo no weak interaction, we arrive at figure 20.11, which represents the non-leptonic decay $\Lambda^0 \to \mathrm{p}\pi^-$.

This figure is, of course, rather schematic since there are strong QCD interactions (not shown) which are responsible for binding the three-quark systems into baryons and the $\mathrm{q}\bar{\mathrm{q}}$ system into a meson. Unlike the case of deep inelastic lepton scattering, these QCD interactions cannot be treated

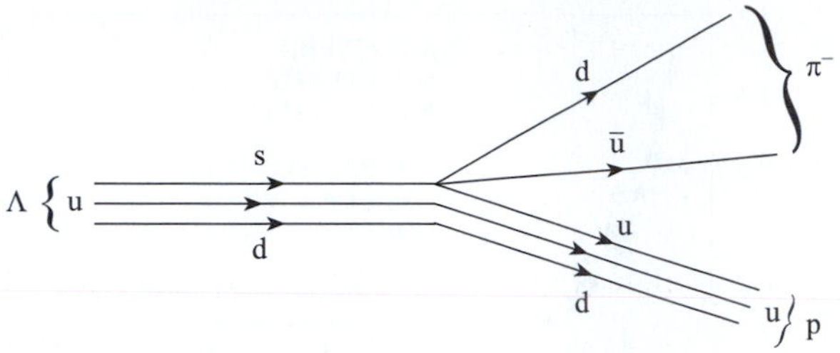


**Figure 20.11.** Non-leptonic weak decay of $\Lambda^0$ using the process of figure 20.10, with the addition of two 'spectator' quarks.

perturbatively, since the distance scales involved are typically those of the hadron sizes ($\sim$ 1 fm), where perturbation theory fails. This means that non-leptonic weak interactions among hadrons are difficult to analyse quantititively, though substantial progress is being made via lattice QCD. Similar difficulties also arise, evidently, in the case of semi-leptonic decays. In general, one has to proceed in a phenomenological way, parametrizing the decay amplitudes in terms of appropriate form factors (which are analogous to the electromagnetic form factors introduced in chapter 8). In the case of transitions involving at least one heavy quark Q, Isgur and Wise (1989, 1990) noticed that a considerable simplification occurs in the linit $m_{\mathrm{Q}} \to \infty$. For example, one universal function (the 'Isgur–Wise form factor') is sufficient to describe a large number of hadronic form factors introduced for semi-leptonic transitions between two heavy pseudoscalar ($0^-$) or vector ($1^-$) mesons. For an introduction to the Isgur–Wise theory, we refer to Donoghue *et al* (1992).

The non-leptonic sector is the scene of some very interesting physics, however, such as $\mathrm{K}^0 - \bar{\mathrm{K}}^0$ and $\mathrm{B}^0 - \bar{\mathrm{B}}^0$ oscillations, and **CP** violation in the $\mathrm{K}^0$–$\bar{\mathrm{K}}^0$ and $\mathrm{B}^0$–$\bar{\mathrm{B}}^0$ systems. We shall see how **CP** violation arises in the Standard Model in section 22.7.1. For a convenient review of this large and important area, we refer to Leader and Predazzi (1996).

## Problems

**20.1** Show that in the non-relativistic limit ($|\boldsymbol{p}| \ll M$) the matrix element $\bar{u}_{\mathrm{p}}\gamma^{\mu}u_{\mathrm{n}}$ of (20.1) vanishes if p and n have different spin states.

**20.2** Show that $\bar{\hat{\psi}}_1(\boldsymbol{x},t)\gamma^0\hat{\psi}_2(\boldsymbol{x},t)$ is a pseudoscalar under $\hat{\boldsymbol{P}}$, and that $\bar{\hat{\psi}}(\boldsymbol{x},t)\boldsymbol{\gamma}\hat{\psi}_2(\boldsymbol{x},t)$ is an axial vector.

**20.3** Verify the normalization $N = (E + |\boldsymbol{p}|)^{1/2}$ in (20.47).

**20.4** Verify (20.54) and (20.55).

**20.5** Verify that equations (20.56) are invariant under **CP**.

**20.6** The matrix $\gamma_5$ is defined by $\gamma_5 = \mathrm{i}\gamma^0\gamma^1\gamma^2\gamma^3$. Prove the following properties:

(a) $\gamma_5^2 = 1$ and hence

$$(1+\gamma_5)(1-\gamma_5) = 0;$$

(b) from the anti-commutation relations of the other $\gamma$ matrices, show that

$$\{\gamma_5, \gamma_\mu\} = 0$$

and hence that

$$(1+\gamma_5)\gamma_0 = \gamma_0(1-\gamma_5)$$

and

$$(1+\gamma_5)\gamma_0\gamma_\mu = \gamma_0\gamma_\mu(1+\gamma_5).$$

**20.7**

(a) Consider the two-dimensional anti-symmetric tensor $\epsilon_{ij}$ defined by

$$\epsilon_{12} = +1, \epsilon_{21} = -1 \qquad \epsilon_{11} = \epsilon_{22} = 0.$$

By explicitly enumerating all the possibilities (if necessary), convince yourself of the result

$$\epsilon_{ij}\epsilon_{kl} = +1(\delta_{ik}\delta_{jl} - \delta_{il}\delta_{jk}).$$

Hence prove that

$$\epsilon_{ij}\epsilon_{il} = \delta_{jl} \qquad \text{and} \qquad \epsilon_{ij}\epsilon_{ij} = 2$$

(remember, in two dimensions, $\sum_i \delta_{ii} = 2$).

(b) By similar reasoning to that in part (a) of this question, it can be shown that the product of two three-dimensional anti-symmetric tensors has the form

$$\epsilon_{ijk}\epsilon_{lmn} = \begin{vmatrix} \delta_{il} & \delta_{im} & \delta_{in} \\ \delta_{jl} & \delta_{jm} & \delta_{jn} \\ \delta_{kl} & \delta_{km} & \delta_{kn} \end{vmatrix}.$$

Prove the results

$$\epsilon_{ijk}\epsilon_{imn} = \begin{vmatrix} \delta_{jm} & \delta_{jn} \\ \delta_{km} & \delta_{kn} \end{vmatrix} \qquad \epsilon_{ijk}\epsilon_{ijn} = 2\delta_{kn} \qquad \epsilon_{ijk}\epsilon_{ijk} = 3!$$

(c) Extend these results to the case of the four-dimensional (Lorentz) tensor $\epsilon_{\mu\nu\alpha\beta}$ (remember that a minus sign will appear as a result of $\epsilon_{0123} = +1$ but $\epsilon^{0123} = -1$).

**20.8** Starting from the amplitude for the process

$$\nu_\mu + \mathrm{e}^- \to \mu^- + \nu_\mathrm{e}$$

given by the current–current theory of weak interactions,

$$\mathcal{M} = -\mathrm{i}(G_\mathrm{F}/\sqrt{2})\bar{u}(\mu)\gamma_\mu(1-\gamma_5)u(\nu_\mu)g^{\mu\nu}\bar{u}(\nu_\mathrm{e})\gamma_\nu(1-\gamma_5)u(\mathrm{e})$$

verify the intermediate results given in section 20.8 leading to the result

$$\mathrm{d}\sigma/\mathrm{d}t = G_\mathrm{F}^2/\pi$$

(neglecting all lepton masses). Hence, show that the local total cross-section for this process rises linearly with $s$:

$$\sigma = G_\mathrm{F}^2 s/\pi.$$

**20.9** The invariant amplitude for $\pi^+ \to e^+\nu$ decay may be written as

$$\mathcal{M} = (G_\mathrm{F}/\sqrt{2}) f_\pi p^\mu \bar{u}(\nu)\gamma_\mu(1-\gamma_5)v(\mathrm{e})$$

where $p^\mu$ is the 4-momentum of the pion and the neutrino is taken to be massless. Evaluate the decay rate in the rest frame of the pion using the decay rate formula

$$\Gamma = (1/2M_\pi)|\overline{\mathcal{M}}|^2\mathrm{dLips}(M_\pi^2; k_\mathrm{e}, k_\nu)$$

where the phase space factor 'dLips' is defined in (16.111). Show that the ratio of $\pi^+ \to \mathrm{e}^+\nu$ and $\pi^+ \to \mu^+\nu$ rates is given by

$$\frac{\Gamma(\pi^+ \to \mathrm{e}^+\nu)}{\Gamma(\pi^+ \to \mu^+\nu)} = \left(\frac{M_\mathrm{e}}{M_\mu}\right)^2\left(\frac{M_\pi^2 - M_\mathrm{e}^2}{M_\pi^2 - M_\mu^2}\right)^2.$$

Repeat the calculation using the amplitude

$$\mathcal{M}' = (G_\mathrm{F}/\sqrt{2}) f_\pi p^\mu \bar{u}(\nu)\gamma_\mu(g_\mathrm{V} + g_\mathrm{A}\gamma_5)v(\mathrm{e})$$

and retaining a finite neutrino mass. Discuss the $\mathrm{e}^+/\mu^+$ ratio in the light of your result.

**20.10**

(a) Verify that the inclusive inelastic neutrino–proton scattering differential cross-section has the form

$$\frac{\mathrm{d}^2\sigma^{(\nu)}}{\mathrm{d}Q^2\,\mathrm{d}\nu} = \frac{G_\mathrm{F}^2 k'}{2\pi k}\left(W_2^{(\nu)}\cos^2(\theta/2) + W_1^{(\nu)} 2\sin^2(\theta/2) + \frac{(k+k')}{M}\sin^2(\theta/2)W_3^{(\nu)}\right)$$

in the notation of section 20.11.

(b) Using the Bjorken scaling behaviour

$$\nu W_2^{(\nu)} \to F_2^{(\nu)} \qquad M W_1^{(\nu)} \to F_1^{(\nu)} \qquad \nu W_3^{(\nu)} \to F_3^{(\nu)}$$

rewrite this expression in terms of the scaling functions. In terms of the variables $x$ and $y$, neglect all massses and show that

$$\frac{\mathrm{d}^2\sigma^{(\nu)}}{\mathrm{d}x\,\mathrm{d}y} = \frac{G_\mathrm{F}^2}{2\pi} s[F_2^{(\nu)}(1-y) + F_1^{(\nu)} x y^2 + F_3^{(\nu)}(1-y/2)yx].$$

Remember that

$$\frac{k' \sin^2(\theta/2)}{M} = \frac{xy}{2}.$$

(c) Insert the Callan–Gross relation

$$2x F_1^{(\nu)} = F_2^{(\nu)}$$

to derive the result quoted in section 20.11:

$$\frac{\mathrm{d}^2\sigma^{(\nu)}}{\mathrm{d}x\,\mathrm{d}y} = \frac{G_\mathrm{F}^2}{2\pi} s F_2^{(\nu)} \left( \frac{1+(1-y)^2}{2} + \frac{x F_3^{(\nu)}}{F_2^{(\nu)}} \frac{1-(1-y)^2}{2} \right).$$

**20.11** The differential cross-section for $\nu_\mu \mathrm{q}$ scattering by charged currents has the same form (neglecting masses) as the $\nu_\mu \mathrm{e}^- \to \mu^- \nu_\mathrm{e}$ result of problem 20.8, namely

$$\frac{\mathrm{d}\sigma}{\mathrm{d}t}(\nu \mathrm{q}) = \frac{G_\mathrm{F}^2}{\pi}.$$

(a) Show that the cross-section for scattering by anti-quarks $\nu_\mu \bar{\mathrm{q}}$ has the form

$$\frac{\mathrm{d}\sigma}{\mathrm{d}t}(\nu \bar{\mathrm{q}}) = \frac{G_\mathrm{F}^2}{\pi}(1-y)^2.$$

(b) Hence prove the results quoted in section 20.11:

$$\frac{\mathrm{d}^2\sigma}{\mathrm{d}x\,\mathrm{d}y}(\nu \mathrm{q}) = \frac{G_\mathrm{F}^2}{\pi} s x \delta(x - Q^2/2M\nu)$$

and

$$\frac{\mathrm{d}^2\sigma}{\mathrm{d}x\,\mathrm{d}y}(\nu \bar{\mathrm{q}}) = \frac{G_\mathrm{F}^2}{\pi} s x (1-y^2) \delta(x - Q^2/2M\nu)$$

(where $M$ is the nucleon mass).

(c) Use the parton model prediction

$$\frac{\mathrm{d}^2}{\mathrm{d}x\,\mathrm{d}y} = \frac{G_\mathrm{F}^2}{\pi} s x [q(x) + \bar{q}(x)(1-y)^2]$$

to show that

$$F_2^{(\nu)} = 2x[q(x) + \bar{q}(x)]$$

and

$$\frac{xF_3^{(\nu)}(x)}{F_2^{(\nu)}(x)} = \frac{q(x) - \bar{q}(x)}{q(x) + \bar{q}(x)}.$$

# 21

# DIFFICULTIES WITH THE CURRENT–CURRENT AND 'NAIVE' INTERMEDIATE VECTOR BOSON MODELS

In the preceding chapter we developed the '$V - A$ current–current' phenomenology of weak interactions. We saw that this gives a remarkably accurate account of a wide range of data—so much so, in fact, that one might well wonder why it should not be regarded as a fully-fledged theory. One good reason for wanting to do this would be in order to carry out calculations beyond the lowest order, which is essentially all we have used it for so far (with the significant exception of the GIM argument). Such higher-order calculations are indeed required by the precision attained in modern high-energy experiments. But the electroweak theory of Glashow, Salam and Weinberg, now recognized as one of the pillars of the Standard Model, was formulated long before such precision measurements existed, under the impetus of quite compelling theoretical arguments. These had to do, mainly, with certain in-principle difficulties associated with the current–current model, if viewed as a 'theory'. Since we now believe that the GSW theory is the correct description of electroweak interactions up to currently tested energies, further discussions of these old issues concerning the current–current model might seem irrelevant. However, these difficulties do raise several important points of principle. An understanding of them provides valuable motivation for the GSW theory—and some idea of what is 'at stake' in regard to experiments relating to those parts of it (notably the Higgs sector) which have still not been experimentally established.

Before reviewing the difficulties, however, it is worth emphasizing once again a more positive motivation for a gauge theory of weak interactions. This is the remarkable 'universality' structure noted in the previous chapter, not only as between different types of lepton but also (within the context of Cabibbo–GIM 'mixing') between the quarks and the leptons. This recalls very strongly the 'universality' property of QED, and the generalization of this property in the non-Abelian theories of chapter 13. A gauge theory would provide a natural framework for such universal couplings.

**Figure 21.1.** Current-current amplitude for $\bar{\nu}_\mu + \mu^- \to \bar{\nu}_e + e^-$.

## 21.1 Violation of unitarity in the current–current model

We have seen several examples, in the previous chapter, in which cross-sections were predicted to rise indefinitely as a function of the invariant variable $s$, which is the square of the total energy in the CM frame. We begin by showing why this is ultimately an unacceptable behaviour.

Consider the process (figure 21.1)

$$\bar{\nu}_\mu + \mu^- \to \bar{\nu}_e + e^- \tag{21.1}$$

in the current–current model, regarding it as fundamental interaction, treated to lowest order in perturbation theory. A similar process was discussed in chapter 20. Since the troubles we shall find occur at high energies, we can simplify the expressions by neglecting the lepton masses without altering the conclusions. In this limit the invariant amplitude is (problem 21.1), up to a numerical factor,

$$\mathcal{M} = G_F E^2 (1 + \cos\theta) \tag{21.2}$$

where $E$ is the CM energy and $\theta$ is the CM scattering angle of the $e^-$ with respect to the direction of the incident $\mu^-$. This leads to the following behaviour of the cross-section:

$$\sigma \sim G_F^2 E^2. \tag{21.3}$$

Consider now a partial wave analysis of this process. For spinless particles the total cross-section may be written as a sum of partial wave cross-sections

$$\sigma = \frac{4\pi}{k^2} \sum_J (2J + 1)|f_J|^2 \tag{21.4}$$

where $f_J$ is the partial wave amplitude for angular momentum $J$ and $k$ is the CM momentum. It is a consequence of *unitarity* or flux conservation (see, for example, Merzbacher 1998, chapter 13) that the partial wave amplitude may be written in terms of a phase shift $\delta_J$:

$$f_J = e^{i\delta_J} \sin\delta_J \tag{21.5}$$

so that

$$|f_J| \leq 1. \tag{21.6}$$

Thus, the cross-section in each partial wave is bounded by

$$\sigma_J \leq 4\pi(2J + 1)/k^2 \tag{21.7}$$

which falls as the CM energy rises. By contrast, in (21.3) we have a cross-section that rises with CM energy:

$$\sigma \sim E^2. \tag{21.8}$$

Moreover, since the amplitude (equation (21.2)) only involves $(\cos\theta)^0$ and $(\cos\theta)^1$ contributions, it is clear that this rise in $\sigma$ is associated with only a few partial waves, and is not due to more and more partial waves contributing to the sum in $\sigma$. Therefore, at some energy $E$, the unitarity bound will be violated by this lowest-order (Born approximation) expression for $\sigma$.

This is the essence of the 'unitarity disease' of the current–current model. To fill in all the details, however, involves a careful treatment of the appropriate partial wave analysis for the case when all particles carry spin. We shall avoid those details and instead sketch the conclusions of such an analysis. For massless spin-$\frac{1}{2}$ particles interacting via a V − A interaction we have seen that helicity is conserved. The net effect of the spin structure is to produce the $(1 + \cos\theta)$ factor in equation (21.2). This embodies the fact that the initial state with $J_z = -1$ ($J_z$ quantized along the $\mu^-$ direction in the CM system) is forbidden by angular momentum conservation to go to a $J_z = +1$ state at $\theta = \pi$, which is the state required by the V − A interaction. Extracting this angular-momentum-conserving kinematic factor, the remaining amplitude can be regarded as that appropriate to a $J = 0$, spinless process (since there are no other factors of $\cos\theta$) so that

$$f_{\rm eff}^{J=0} \sim G_{\rm F} E^2. \tag{21.9}$$

The unitarity bound (21.6) is therefore violated for CM energies

$$E \geq G_{\rm F}^{-1/2} \sim 300\ {\rm GeV}. \tag{21.10}$$

This difficulty with the current–current theory can be directly related to the fact that the Fermi coupling constant $G_{\rm F}$ is *not dimensionless.* From calculated decay rates, $G_{\rm F}$ is found to have the value (Hagiwara *et al* 2002) :

$$G_{\rm F} \simeq 1.166\,39(1) \times 10^{-5}\ {\rm GeV}^{-2} \tag{21.11}$$

and has the dimensions of $[M]^{-2}$. Given this fact, we can arrive at the form for the cross-section for $\bar{\nu}_\mu \mu^- \to \bar{\nu}_{\rm e} {\rm e}^-$ at high energy without calculation. The cross-section has dimensions of $[L]^2 = [M]^{-2}$ but must involve $G_{\rm F}^2$ which has dimension $[M]^{-4}$. It must also be relativistically invariant. At energies well above lepton masses, the only invariant quantity available to restore the correct dimensions to $\sigma$ is $s$, the square of the CM energy $E$, so that $\sigma \sim G_{\rm F}^2 E^2$.

At this point the reader may recall a very similar-sounding argument made in section 11.8, which led to the same estimate of the 'dangerous' energy scale (21.10). In that case, the discussion referred to a hypothetical 'four-fermion' interaction without the V − A structure and it was concerned with renormalization rather than unitarity. The gamma-matrix structure is irrelevant to these issues, which ultimately have to do with the dimensionality of the coupling constant in

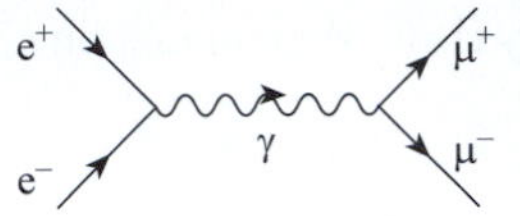


**Figure 21.2.** One-photon annihilation graph for $e^+e^- \to \mu^+\mu^-$.

both cases. In fact, as we shall see, unitarity and renormalizability are actually rather closely related.

Faced with this unitarity difficulty, we appeal to the most successful theory we have, and ask: what happens in QED? We consider an apparently quite similar process, namely $e^+e^- \to \mu^+\mu^-$ in lowest order (figure 21.2). In chapter 8 the total cross-section for this process, neglecting lepton masses, was found to be (see problem 8.19 and equation (9.88))

$$\sigma = 4\pi\alpha^2/3E^2 \tag{21.12}$$

which obediently falls with energy as required by unitarity. In this case the coupling constant $\alpha$, analogous to $G_{\rm F}$, is dimensionless, so that a factor $E^2$ is required in the denominator to give $\sigma \sim [L]^2$.

If we accept this clue from QED, we are led to search for a theory of weak interactions that involves a dimensionless coupling constant. Pressing the analogy with QED further will help us to see how one might arise. Fermi's current–current model was, as we said, motivated by the vector currents of QED. But, in Fermi's case, the currents interact directly with each other, whereas in QED they interact only indirectly via the mediation of the electromagnetic field. More formally, the Fermi current–current interaction has the 'four-point' structure

$$\text{`}G_{\rm F}(\bar{\hat{\psi}}\hat{\psi})\cdot(\bar{\hat{\psi}}\hat{\psi})\text{'} \tag{21.13}$$

while QED has the 'three-point' (Yukawa) structure

$$\text{`}e\bar{\hat{\psi}}\hat{\psi}\hat{A}\text{'}. \tag{21.14}$$

Dimensional analysis easily shows, once again, that $[G_{\rm F}] = M^{-2}$ while $[e] = M^0$. This strongly suggests that we should take Fermi's analogy further and look for a weak interaction analogue of (21.14), having the form

$$\text{`}g\bar{\hat{\psi}}\hat{\psi}\hat{W}\text{'} \tag{21.15}$$

where $\hat{W}$ is a bosonic field. Dimensional analysis shows, of course, that $[g] = M^0$.

Since the weak currents are, in fact, vector-like, we must assume that the $\hat{W}$ fields are also vectors (spin-1) so as to make (21.15) Lorentz invariant. And because the weak interactions are plainly *not* long-range, like electromagnetic

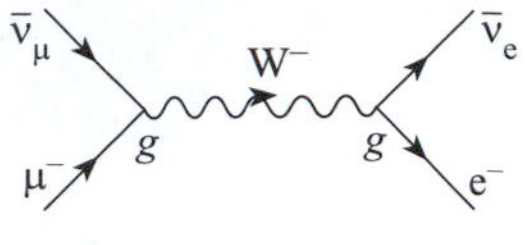


**Figure 21.3.** One-$W^-$ annihilation graph for $\bar{\nu}_\mu + \mu^- \to \bar{\nu}_e + e^-$.

ones, the mass of the W quanta cannot be zero. So we are led to postulate the existence of a massive weak analogue of the photon, the 'intermediate vector boson' (IVB), and to suppose that weak interactions are mediated by the exchange of IVB's.

There is, of course, one further difference with electromagnetism, which is that the currents in $\beta$-decay, for example, carry charge (e.g. $\hat{\bar{\psi}}_e\gamma^\mu(1-\gamma_5)\hat{\psi}_{\nu_e}$ creates negative charge or destroys positive charge). The 'companion' current carries the opposite charge (e.g. $\hat{\bar{\psi}}_p\gamma_\mu(1-r\gamma_5)\hat{\psi}_n$ destroys negative charge or creates positive charge), so as to make the total effective interaction charge-conserving, as required. It follows that the $\hat{W}$ fields must then be charged, so that expressions of the form (21.15) are neutral. Because both charge-raising and charge-lowering currents exist, we need both $W^+$ and $W^-$. The reaction (21.1), for example, is then conceived as proceeding via the Feynman diagram shown in figure 21.3, quite analogous to figure 21.2.

Because we also have weak neutral currents, we need a neutral vector boson as well, $Z^0$. In addition to all these, there is the familiar massless neutral vector boson, the photon. Despite the fact that they are *not* massless, the $W^\pm$ and $Z^0$ can be understood as gauge quanta, thanks to the symmetry-breaking mechanism explained in section 19.6. For the moment, however, we are going to follow a more scenic route and accept (as Glashow did in 1961) that we are dealing with ordinary 'unsophisticated' massive vector particles, charged and uncharged.

## 21.2 The IVB model

As discussed in section 19.1, the classical wave equation for a massive vector particle, described by the field $W^\mu$, is

$$(\Box + M_W^2)W^\mu - \partial^\mu\partial^\nu W_\nu = 0 \tag{21.16}$$

and the propagator (ignoring the i$\epsilon$) is

$$\mathrm{i}\frac{(-g^{\mu\nu} + k^\mu k^\nu/M_W^2)}{k^2 - M_W^2}. \tag{21.17}$$

Let us first see how the IVB model relates to the current–current one. Matrix elements will have the general form (up to constant factors)

$$g^2 j_1^\mu \frac{(-g_{\mu\nu} + k_\mu k_\nu/M_W^2)}{k^2 - M_W^2} j_2^\nu \tag{21.18}$$

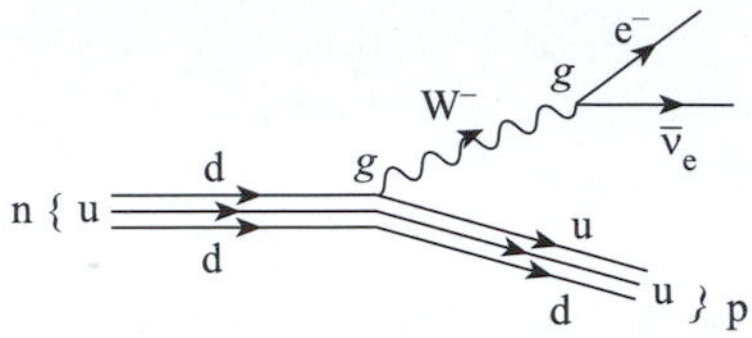


**Figure 21.4.** One-$W^-$ exchange process in $\beta$-decay.

where $j_1^\mu$ and $j_2^\nu$ are certain weak-current matrix elements, and '$g$' is playing a role analogous to '$e$'(compare (21.14) and (21.15)). Such a structure will appear not only for processes such as that shown in figure 21.3 but also for the IVB version of $\beta$-decay shown in figure 21.4.

For typical $\beta$-decay energies (and knowing that $M_W \sim 80$ GeV!), we certainly have $k^2 << M_W^2$ so that all $k$-dependence in the W-propagator (21.18) can be ignored, and we arrive back at Fermi's current–current amplitude, with the important qualitative connection

$$G_F \sim g^2/M_W^2. \tag{21.19}$$

This is a fundamental relation—the exact version of it in the GSW theory is given in equation (22.29) of chapter 22. It shows us *why* the Fermi constant has dimension $[\text{mass}]^{-2}$ and even, in a sense, why the weak interactions are weak! They are so (i.e. $G_F$ is 'small') principally because $M_W$ is so large. Indeed, out of so much apparent dissimilarity between the weak and electromagnetic interactions, perhaps some simple similarity can, after all, be rescued. Maybe the intrinsic strengths $g$ and $e$ are roughly equal:

$$g \sim e. \tag{21.20}$$

This would then lead, via (21.19), to an order-of-magnitude estimate of $M_W$:

$$M_W \sim e/G_F^{1/2} \sim 90 \text{ GeV} \tag{21.21}$$

which is indeed quite close to the true value. This simple idea is essentially correct: the precise relation between $g$ and $e$ in the GSW theory will be given in (22.46).

We now investigate whether the IVB model can do any better with unitarity than the current–current model. The analysis will bear a close similarity to the discussion of the renormalizability of the model in section 19.1, and we shall take up that issue again in section 21.4.

## 21.3 Violation of unitarity bounds in the IVB model

As the section heading indicates, matters will turn out to be fundamentally no better in the IVB model, but the demonstration is instructive. We begin by considering the process (21.1), viewed as proceeding as in figure 21.3, the amplitude for which we take to be

$$\mathrm{i}\frac{g^2}{2}\bar{u}(\mathrm{e})\gamma_\mu\frac{(1-\gamma_5)}{2}v(\nu_\mathrm{e})\left(\frac{-g^{\mu\nu}+k^\mu k^\nu/M^2}{k^2-M^2}\right)\bar{v}(\nu_\mu)\gamma_\nu\frac{(1-\gamma_5)}{2}u(\mu). \tag{21.22}$$

For convenience, the factors of two have been chosen to be those that would actually appear in the GSW theory in unitary gauge, as will be explained in section 22.1.2.

We may compare (21.22) with the amplitude for figure 21.2, which is

$$\mathrm{i}e^2\bar{v}(\mathrm{e})\gamma_\mu u(\mathrm{e})\left(\frac{-g^{\mu\nu}}{k^2}\right)\bar{u}(\mu)\gamma_\nu v(\mu) \tag{21.23}$$

where $k$ is the 4-momentum of the photon. At first sight, we might conclude that the high-energy behaviour of (21.22) is going to be considerably 'worse' than that of (21.23) in view of the presence of the $k^\mu k^\nu$ factors in the numerator of (21.22). However, they turn out to be harmless, as we can see as follows. Using 4-momentum conservation,

$$k^\mu\bar{u}(\mathrm{e})\gamma_\mu(1-\gamma_5)v(\nu_\mathrm{e}) = (p_\mathrm{e}^\mu + p_{\bar{\nu}}^\mu)\bar{u}(\mathrm{e})\gamma_\mu(1-\gamma_5)v(\nu_\mathrm{e}) \tag{21.24}$$

$$= \bar{u}(\mathrm{e})\not{p}_\mathrm{e}(1-\gamma_5)v(\nu_\mathrm{e}) + \bar{u}(\mathrm{e})\not{p}_{\bar{\nu}}(1-\gamma_5)v(\nu_\mathrm{e}). \tag{21.25}$$

Using the Dirac equation for the spinors in the forms $\bar{u}(p)(\not{p}-m) = 0$ and $(\not{p}+m)v(p) = 0$ (see problem 4.11), together with $\{\gamma^\mu, \gamma_5\} = 0$, (21.25) becomes

$$m_\mathrm{e}\bar{u}(\mathrm{e})(1-\gamma_5)v(\nu_\mathrm{e}) - m_\nu\bar{u}(\mathrm{e})(1+\gamma_5)v(\nu_\mathrm{e}). \tag{21.26}$$

A similar result holds for the $k^\nu$ factor; thus the $k^\mu k^\nu$ factors have disappeared. Indeed, neglecting the lepton masses by comparison with $M_\mathrm{W}$, the effect of the IVB is simply to replace the photon propagator $-g^{\mu\nu}/k^2$ by $-g^{\mu\nu}/(k^2 - M_\mathrm{W}^2)$. It was this photon propagator which was really responsible, in a dynamical sense, for the fall with energy of the QED cross-section (21.12) and hence we conclude that, at least for this process, the IVB modification of the four-fermion model does avoid the violation of unitarity in lowest order.

Does the IVB modification ensure that the unitarity bounds are not violated for *any* Born (i.e. tree-graph) process? The answer is no. The unitarity-violating processes turn out to be those involving *external* W particles (rather than internal ones, as in figure 21.3). Consider, for example, the process

$$\nu_\mu + \bar{\nu}_\mu \to \mathrm{W}^+ + \mathrm{W}^- \tag{21.27}$$

proceeding via the graph shown in figure 21.5. The fact that this is experimentally a somewhat esoteric reaction is irrelevant for the subsequent argument: the

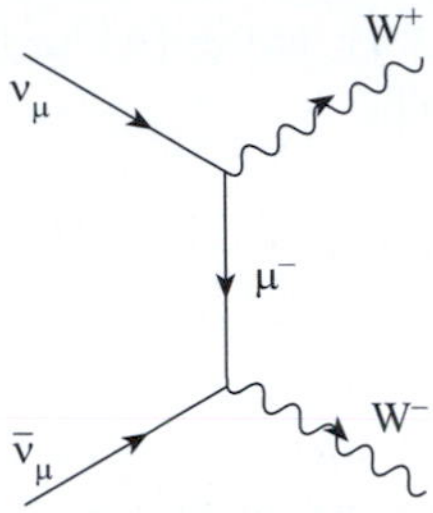


**Figure 21.5.** $\mu^-$-exchange graph for $\nu_\mu + \bar{\nu}_\mu \to W^+ + W^-$.

proposed theory, represented by the IVB modification of the four-fermion model, will necessarily generate the amplitude shown in figure 21.5, and since this amplitude violates unitarity, the theory is unacceptable. The amplitude for this process is proportional to

$$\mathcal{M}_{\lambda_1\lambda_2} = g^2 \epsilon_\mu^{-*}(k_2, \lambda_2)\epsilon_\nu^{+*}(k_1, \lambda_1)\bar{v}(p_2)\gamma^\mu(1-\gamma_5) \times \frac{(\not{p}_1 - \not{k}_1 + m_\mu)}{(p_1 - k_1)^2 - m_\mu^2}\gamma^\nu(1-\gamma_5)u(p_1) \tag{21.28}$$

where the $\epsilon^\pm$ are the polarization vectors of the W's: $\epsilon_\mu^{-*}(k_2, \lambda_2)$ is that associated with the outgoing $W^-$ with 4-momentum $k_2$ and polarization state $\lambda_2$, and similarly for $\epsilon_\nu^{+*}$.

To calculate the total cross-section, we must form $|\mathcal{M}|^2$ and sum over the three states of polarization for each of the W's. To do this, we need the result

$$\sum_{\lambda=0,\pm1} \epsilon_\mu(k, \lambda)\epsilon_\nu^*(k, \lambda) = -g_{\mu\nu} + k_\mu k_\nu / M_W \tag{21.29}$$

already given in (19.19). Our interest will, as usual, be in the high-energy behaviour of the cross-section, in which regime it is clear that the $k_\mu k_\nu / M_W^2$ term in (21.29) will dominate the $g_{\mu\nu}$ term. It is therefore worth looking a little more closely at this term. From (19.17) and (19.18) we see that in a frame in which $k^\mu = (k^0, 0, 0, |\boldsymbol{k}|)$, the transverse polarization vectors $\epsilon^\mu(k, \lambda = \pm 1)$ involve no momentum dependence, which is, in fact, carried solely in the longitudinal polarization vector $\epsilon^\mu(k, \lambda = 0)$. We may write this as

$$\epsilon(k, \lambda = 0) = \frac{k^\mu}{M_W} + \frac{M_W}{(k^0 + |\boldsymbol{k}|)}(-1, \hat{\boldsymbol{k}}) \tag{21.30}$$

which at high energy tends to $k^\mu / M_W$. Thus, it is clear that it is the longitudinal polarization states which are responsible for the $k^\mu k^\nu$ parts of the polarization sum (12.21), and which will dominate real production of W's at high energy.

Concentrating therefore on the production of longitudinal W's, we are led to examine the quantity

$$\frac{g^4}{M_W^4(p_1-k_1)^4}\operatorname{Tr}[\not{k}_2(1-\gamma_5)(\not{p}_1-\not{k}_1)\not{k}_1\not{p}_1\not{k}_1(\not{p}_1-\not{k}_1)\not{k}_2\not{p}_2] \qquad (21.31)$$

where we have neglected $m_\mu$, commuted the $(1-\gamma_5)$ factors through, and neglected neutrino masses in forming $\sum_{\text{spins}}|\mathcal{M}_{00}|^2$. Retaining only the leading powers of energy, we find (see problem 21.2) that

$$\sum_{\text{spins}}|\mathcal{M}_{00}|^2 \sim (g^4/M_W^4)(p_1\cdot k_2)(p_2\cdot k_2) = (g^4/M_W^4)E^4(1-\cos^2\theta) \quad (21.32)$$

where $E$ is the CM energy and $\theta$ the CM scattering angle. Recalling (21.19), we see that the (unsquared) amplitude must behave essentially as $G_F E^2$, precisely as in the four-fermion model (equation (21.2)). In fact, putting all the factors in, one obtains (Gastmans 1975)

$$\frac{d\sigma}{d\Omega} = G_F^2\frac{E^2\sin^2\theta}{8\pi^2} \qquad (21.33)$$

and, hence, a total cross-section which rises with energy as $E^2$, just as before. The production of longitudinal polarized W's is actually a pure $J=1$ process and the $J=1$ partial wave amplitude is

$$f_1 = G_F E^2/6\pi. \qquad (21.34)$$

The unitarity bound $|f_1| \leq 1$ is therefore violated for $E \geq (6\pi/G_F)^{1/2} \sim 10^3$ GeV (cf (21.10)).

Other unitarity-violating processes can easily be invented, and we have to conclude that the IVB model is, in this respect, no more fit to be called a theory than was the four-fermion model. In the case of the latter, we argued that the root of the disease lay in the fact that $G_F$ was not dimensionless, yet somehow this was not a good enough cure after all: perhaps (it is indeed so) 'dimensionlessness' is necessary but not sufficient (see the following section). Why is this? Returning to $\mathcal{M}_{\lambda_1,\lambda_2}$ for $\nu\bar{\nu}\to W^+W^-$ (equation (21.28)) and setting $\epsilon_\mu = k_\mu/M$ for the *longitudinal* polarization vectors, we see that we are involved with an effective amplitude

$$\frac{g^2}{M_W^2}\bar{v}(p_2)\not{k}_2(1-\gamma_5)\frac{\not{p}_1-\not{k}_1}{(p_1-k_1)^2}\not{k}_1(1-\gamma_5)u(p_1). \qquad (21.35)$$

Using the Dirac equation $\not{p}_1 u(p_1)=0$ and $p_1^2=0$, this can be reduced to

$$-\frac{g^2}{M_W^2}\bar{v}(p_2)\not{k}_2(1-\gamma_5)u(p_1). \qquad (21.36)$$

We see that the longitudinal $\epsilon$'s have brought in the factors $M_W^{-2}$, which are 'compensated' by the factor $\not{k}_2$, and it is this latter factor which causes the rise

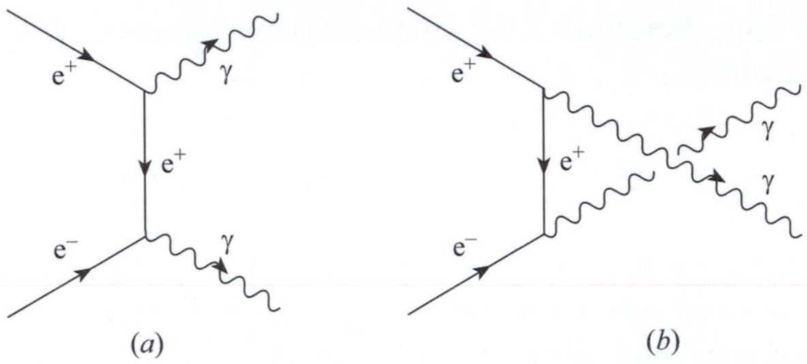


**Figure 21.6.** Lowest-order amplitudes for $e^+e^- \to \gamma\gamma$: (*a*) direct graph, (*b*) crossed graph.

with energy. The longitudinal polarization states have effectively reintroduced a dimensional coupling constant $g/M_W$.

Once more we turn in our distress to the trusty guide, QED. The analogous electromagnetic process to consider is $e^+e^- \to \gamma\gamma$. In QED there are two graphs contributing in lowest order (the 'crossed' versions of figure 8.14). These are shown in figures 21.6(*a*) and (*b*). The fact that there are two rather than one will turn out to be significant, as we shall see in section 21.4. For the moment we just concentrate on figure 21.6(*a*) which is directly analogous to the diagram for $\nu\bar{\nu} \to W^+W^-$. The amplitude for this diagram is the same as before except that $(1-\gamma_5)/2$ is replaced by 1, $g/2^{1/2}$ by $e$ and the $\epsilon$ vectors now refer to photons:

$$\mathcal{M}_{\lambda_1,\lambda_2} = e^2\epsilon^*_\mu(k_2,\lambda_2)\epsilon^*_\nu(k_1,\lambda_1)\bar{v}(p_2)\gamma^\mu \frac{\not{p}_1 - \not{k}_1 + m_e}{(p_1-k_1)^2 - m_e^2}\gamma^\nu u(p_1). \tag{21.37}$$

In the cross-section we would need to sum over the photon polarization states. For the massive spin-1 particles, we used

$$\sum_\lambda \epsilon_\mu(k,\lambda)\epsilon^*_\nu(k,\lambda) = -g_{\mu\nu} + k_\mu k_\nu/M_W^2 \tag{21.38}$$

and so we would need the analogue of this result for massless photons. This is a non-trivial point. Clearly the answer is *not* to take the $M_W \to 0$ limit of (21.38), since this diverges. However, the 'dangerous' term $k_\mu k_\nu/M_W^2$ arises entirely from the *longitudinal* polarization vectors, and we learnt in section 7.3 that, for real photons, the longitudinal state of polarization is absent altogether! We might well suspect, therefore, that since it was the longitudinal W's that caused the 'bad' high-energy behaviour of the IVB model, the 'good' high-energy behaviour of QED might have its origin in the absence of such states for photons. And this circumstance can, in its turn, be traced (cf section 7.3.1 ) to the *gauge invariance* property of QED.

Indeed, in section 8.6.3 we saw that in the analogue of (21.38) for photons (this time involving only the two transverse polarization states), the right-hand side could be taken to be just $-g_{\mu\nu}$, *provided that* the Ward identity (8.165) held, a condition directly following from gauge invariance.

We have arrived here at an important theoretical indication that what we really need is a *gauge theory* of the weak interactions, in which the W's are gauge quanta. It must, however, be a peculiar kind of gauge theory, since normally gauge invariance requires the gauge field quanta to be massless. However, we have already seen how this 'peculiarity' can indeed arise, if the local symmetry is spontaneously broken (chapter 19). But before proceeding to implement that idea in the GSW theory, we discuss one further disease (related to the unitarity one) possessed by both current–current and IVB models—that of non-renormalizability.

## 21.4 The problem of non-renormalizability in weak interactions

The preceding line of argument about unitarity violations is open to the following objection. It is an argument conducted entirely within the framework of perturbation theory. What it shows, in fact, is simply that perturbation theory must fail in theories of the type considered at some sufficiently high energy. The essential reason is that the effective expansion parameter for perturbation theory is $EG_{\rm F}^{1/2}$. Since $EG_{\rm F}^{1/2}$ becomes large at high energy, arguments based on lowest-order perturbation theory are irrelevant. The objection is perfectly valid, and we shall take account of it by linking high-energy behaviour to the problem of renormalizability, rather than unitarity. We might, however, just note in passing that yet another way of stating the results of the previous two sections is to say that, for both the current–current and IVB theories, 'weak interactions become strong at energies of order 1 TeV'.

We gave an elementary introduction to renormalization in chapters 10 and 11 of volume 1. In particular, we discussed in some detail, in section 11.8, the difficulties that arise when one tries to do higher-order calculations in the case of a four-fermion interaction with the same form (apart from the V − A structure) as the current–current model. Its coupling constant, which we called $G_{\rm F}$, also had dimension $({\rm mass})^{-2}$. The 'non-renormalizable' problem was essentially that, as one approached the 'dangerous' energy scale (21.10), one needed to supply the values of an ever-increasing number of parameters from experiment and the theory lost predictive power.

Does the IVB model fare any better? In this case, the coupling constant is dimensionless, just as in QED. 'Dimensionlessness' alone is not enough, it turns out: the IVB model is not renormalizable either. We gave an indication of why this is so in section 19.1 but we shall now be somewhat more specific, relating the discussion to the previous one about unitarity.

Consider, for example, the fourth-order processes shown in figure 21.7 and 21.8—the former for the QED process $e^+e^- \to e^+e^-$ via an intermediate $2\gamma$ state, the latter for the IVB-mediated process $\nu_\mu\bar{\nu}_\mu \to \nu_\mu\bar{\nu}_\mu$. It seems plausible

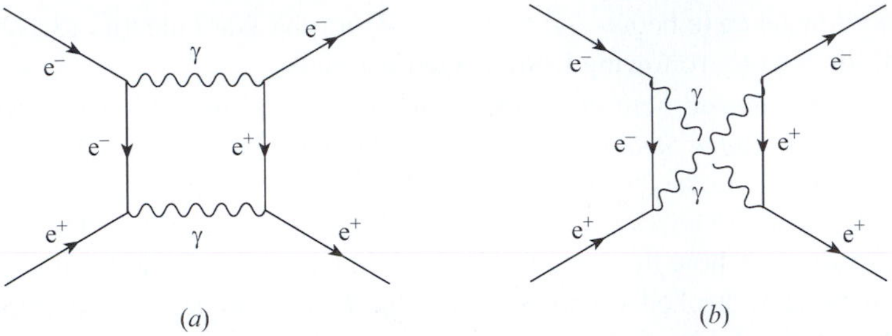


**Figure 21.7.** $O(e^4)$ contributions to $e^+e^- \to e^+e^-$.

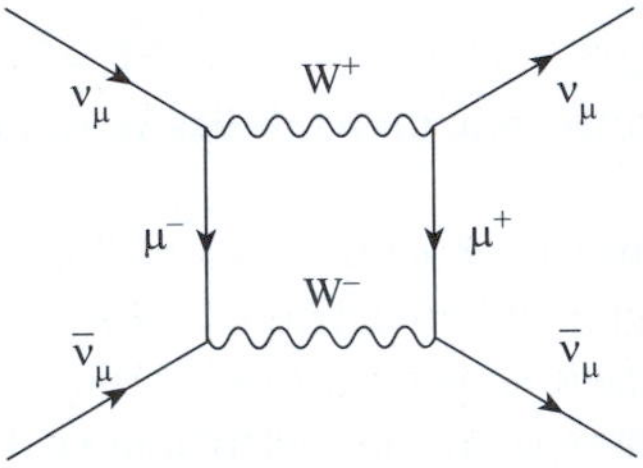


**Figure 21.8.** $O(g^4)$ contribution to $\nu_\mu\bar{\nu}_\mu \to \nu_\mu\bar{\nu}_\mu$.

from the diagrams that, in each case, the amplitudes must be formed by somehow 'sticking together' two of the lower-order graphs shown in figures 21.5 and 21.6.[1] Accepting this for the moment, and knowing that ultraviolet divergences of loop graphs originate in the high-$k$ behaviour of their integrands, we are led to compare the high-energy behaviour of the process $e^+e^- \to \gamma\gamma$, on the one hand, and $\nu_\mu\bar{\nu}_\mu \to W^+W^-$, on the other. This is exactly the comparison we were making in the previous section, but now we have arrived at it from considerations of renormalizability, rather than unitarity.

Indeed, we saw in section 20.3 that the high-energy behaviour of the amplitude $\nu\bar{\nu} \to W^+W^-$ (figure 21.5) grew as $E^2$, due to the $k$ dependence of the longitudinal polarization vectors, and this turns out to produce, via figure 21.8, a non-renormalizable divergence, for the reason indicated in section 19.1—namely, the 'bad' behaviour of the $k^\mu k^\nu/M_W^2$ factors in the W-propagators, at large $k$.

So it is plain that, once again, the blame lies with the longitudinal polarization states for the W's. Let us see how QED—a renormalizable theory—manages to avoid this problem. In this case, it seems clear from inspection of figure 21.7 that there are *two* graphs corresponding to figure 21.5, namely figures 21.6($a$)and ($b$), already discussed in the previous section. Consider, therefore, mimicking for figures 21.6($a$) and ($b$) the calculation we did for figure 21.4. We would obtain the leading high-energy behaviour by replacing the

[1] The reader may here usefully recall the discussion of unitarity for one-loop graphs in section 13.5.3.

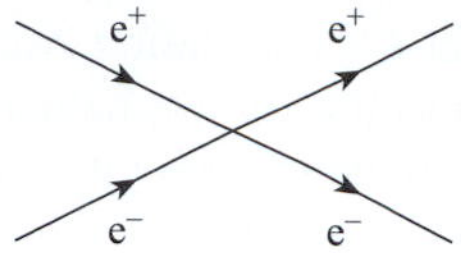


**Figure 21.9.** Four-point $e^+e^-$ vertex.

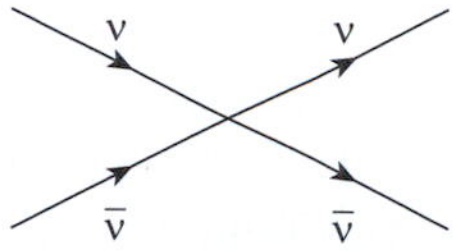


**Figure 21.10.** Four-point $\nu\bar{\nu}$ vertex.

photon polarization vectors by the corresponding momenta and it can be checked (problem 21.5) that when this replacement is made for each photon the complete amplitude for the sum of figures 21.6(*a*) and (*b*) *vanishes*.

In physical terms, of course, this result was expected, since we knew in advance that it is always possible to choose polarization vectors for *real* photons such that they are purely transverse, so that no physical process can depend on a part of $\epsilon_\mu$ proportional to $k_\mu$. Nevertheless, the calculation is highly relevant to the question of renormalizing figure 21.7. The photons in this process are not real external particles but are instead virtual, internal ones. This has the consequence that we should, in general, include their longitudinal ($\epsilon_\mu \propto k_\mu$) states as well as the transverse ones (see section 13.5.3 for something similar in the case of unitarity for one-loop diagrams). The calculation of problem 21.3 then suggests that these longitudinal states are harmless, provided that both contributions in figure 21.6 are included.

Indeed, the sum of these two contributions is *not divergent*. If it were, an infinite counter term proportional to a four-point vertex $e^+e^- \rightarrow e^+e^-$ (figure 21.9) would have to be introduced and the original QED theory, which of course lacks such a fundamental interaction, would not be renormalizable. This is exactly what *does* happen in the case of figure 21.8. The bad high-energy behaviour of $\nu\bar{\nu} \rightarrow W^+W^-$ translates into a divergence of figure 21.8—and this time there is no 'crossed' amplitude to cancel it. This divergence entails the introduction of a new vertex, figure 21.10, not present in the original IVB theory. Thus, the theory without this vertex is non-renormalizable—and if we include it, we are landed with a four-field point-like vertex which is non-renormalizable, as in the Fermi (current–current) case.

Our presentation hitherto has emphasized the fact that, in QED, the bad high-energy behaviour is rendered harmless by a cancellation between contributions from figures 21.6(*a*) and (*b*) (or figures 21.7(*a*) and (*b*)). Thus, one way to 'fix' the IVB theory might be to hypothesize a new physical process to be added to figure 21.5 in such a way that a cancellation occurred at high energies. The search for such high-energy cancellation mechanisms can, indeed, be pushed to a successful conclusion, given sufficient ingenuity and, arguably, a little hindsight. However, we are in possession of a more powerful principle. In QED, we have already seen (section 8.6.2) that the vanishing of amplitudes when an $\epsilon_\mu$ is replaced by the corresponding $k_\mu$ is due to *gauge invariance*: in other words, the potentially harmful longitudinal polarization states are, in fact, harmless in a gauge-invariant theory.

We have therefore arrived once more, after a somewhat more leisurely discussion than that of section 19.1, at the idea that we need a *gauge* theory of massive vector bosons, so that the offending $k^\mu k^\nu$ part of the propagator can be 'gauged away' as in the photon case. This is precisely what is provided by the 'spontaneously broken' gauge theory concept, as developed in chapter 19. There we saw that, taking the U(1) case for simplicity, the general expression for the gauge boson propagator in such a theory (in a 't Hooft gauge) is

$$\mathrm{i}\left[-g^{\mu\nu} + \frac{(1-\xi)k^\mu k^\nu}{k^2 - \xi M_{\mathrm{W}}^2}\right] \Big/ (k^2 - M_{\mathrm{W}}^2 + \mathrm{i}\epsilon) \tag{21.39}$$

where $\xi$ is a gauge parameter. Our IVB propagator corresponds to the $\xi \to \infty$ limit and with this choice of $\xi$, all the troubles we have been discussing appear to be present. But for any finite $\xi$ (for example $\xi = 1$) the high-energy behaviour of the propagator is actually $\sim 1/k^2$, the same as in the renormalizable QED case. This strongly suggests that such theories—in particular non-Abelian ones—are, in fact, renormalizable. 't Hooft's proof that they are ('t Hooft 1971b) triggered an explosion of theoretical work, as it became clear that, for the first time, it would be possible to make higher-order calculations for weak interaction processes using consistent renormalization procedures of the kind that had worked so well for QED.

We now have all the pieces in place, and can proceed to introduce the GSW theory based on the local gauge symmetry of SU(2) × U(1).

## Problems

### 21.1

(a) Using the representation for $\boldsymbol{\alpha}$, $\beta$ and $\gamma_5$ introduced in section 20.4 (equation (20.38)), massless particles are described by spinors of the form

$$u = E^{1/2}\begin{pmatrix}\phi_+ \\ \phi_-\end{pmatrix} \qquad \text{(normalized to } u^\dagger u = 2E\text{)}$$

where $\boldsymbol{\sigma}\cdot\hat{\boldsymbol{p}}\phi_\pm = \pm\phi_\pm$, $\hat{\boldsymbol{p}} = \boldsymbol{p}|\boldsymbol{p}|$. Find the explicit form of $u$ for the case $\hat{\boldsymbol{p}} = (\sin\theta, 0, \cos\theta)$.

(b) Consider the process $\bar{\nu}_\mu + \mu^- \to \bar{\nu}_e + e^-$, discussed in section 21.1, in the limit in which all masses are neglected. The amplitude is proportional to

$$G_F \bar{v}(\bar{\nu}_\mu, R)\gamma_\mu(1-\gamma_5)u(\mu^-, L)\bar{u}(e^-, L)\gamma^\mu(1-\gamma_5)v(\bar{\nu}_e, R)$$

where we have explicitly indicated the appropriate helicities R or L (note that, as explained in section 20.4, $(1-\gamma_5)/2$ is the projection operator for a right-handed anti-neutrino). In the CM frame, let the initial $\mu^-$ momentum be $(0, 0, E)$ and the final $e^-$ momentum be $E(\sin\theta, 0, \cos\theta)$. Verify that the amplitude is proportional to $G_F E^2(1+\cos\theta)$. (*Hint*: evaluate the 'easy' part $\bar{v}(\bar{\nu}_\mu)\gamma_\mu(1-\gamma_5)u(\mu^-)$ first. This will show that the components $\mu = 0, z$ vanish, so that only the $\mu = x, y$ components of the dot product need to be calculated.)

**21.2** Verify equation (21.32).

**21.3** Check that when the polarization vectors of each photon in figures 21.6(*a*) and (*b*) is replaced by the corresponding photon momentum, the sum of these two amplitudes vanishes.

# 22

# THE GLASHOW–SALAM–WEINBERG GAUGE THEORY OF ELECTROWEAK INTERACTIONS

Given the preceding motivations for considering a gauge theory of weak interactions, the remaining question is this: what is the relevant symmetry group of local phase transformations, i.e. the relevant *weak gauge group*? Several possibilities were suggested, but it is now very well established that the one originally proposed by Glashow (1961), subsequently treated as a spontaneously broken gauge symmetry by Weinberg (1967) and by Salam (1968), and later extended by other authors, produces a theory which is in remarkable agreement with currently known data. We shall not give a critical review of all the experimental evidence but instead proceed directly to an outline of the GSW theory, introducing elements of the data at illustrative points.

## 22.1 Weak isospin and hypercharge: the SU(2) × U(1) group of the electroweak interactions: quantum number assignments and W and Z masses

An important clue to the symmetry group involved in the weak interactions is provided by considering the transitions induced by these interactions. This is somewhat analogous to discovering the multiplet structure of atomic levels and hence the representations of the rotation group, a prominent symmetry of the Schrödinger equation, by studying electromagnetic transitions. However, there is one very important difference between the 'weak multiplets' we shall be considering and those associated with symmetries which are not spontaneously broken. We saw in chapter 12 how an unbroken non-Abelian symmetry leads to multiplets of states which are degenerate in mass, but in section 17.1 we learned that that result only holds provided the vacuum is left invariant under the symmetry transformation. When the symmetry is spontaneously broken, the vacuum is *not* invariant and we must expect that the degenerate multiplet structure will then, in general, disappear completely. This is precisely the situation in the electroweak theory.

Nevertheless, as we shall see, essential consequences of the weak symmetry group—specifically, the relations it requires between otherwise unrelated masses and couplings—are accessible to experiment. Moreover, despite the fact that members of a multiplet of a global symmetry which is spontaneously broken will, in general, no longer have even approximately the same mass, the concept of a multiplet is still useful. This is because when the symmetry is made a *local* one, we shall find (in sections 22.2 and 22.3) that the associated gauge quanta still mediate interactions between members of a given symmetry multiplet, just as in the manifest local non-Abelian symmetry example of QCD. Now, the leptonic transitions associated with the weak charged currents are, as we saw in chapter 20, $\nu_e \leftrightarrow e$, $\nu_\mu \leftrightarrow \mu$ etc. This suggests that these pairs should be regarded as *doublets* under some group. Further, we saw in section 20.10 how weak transitions involving charged quarks suggested a similar doublet structure for them also. The simplest possibility is therefore to suppose that, in both cases, a 'weak SU(2) group' called 'weak isospin', is involved. We emphasize once more that this weak isospin is distinct from the hadronic isospin of chapter 12, which is part of $SU(3)_f$. We use the symbols $t$, $t_3$ for the quantum numbers of weak isospin and make the following specific assignments for the leptonic fields:

$$t = \frac{1}{2} \quad \begin{cases} t_3 = +1/2 \\ t_3 = -1/2 \end{cases} \quad \begin{pmatrix} \hat{\nu}_e \\ \hat{e}^- \end{pmatrix}_L \quad \begin{pmatrix} \hat{\nu}_\mu \\ \hat{\mu}^- \end{pmatrix}_L \quad \begin{pmatrix} \hat{\nu}_\tau \\ \hat{\tau}^- \end{pmatrix}_L \tag{22.1}$$

where $\hat{e}_L = \frac{1}{2}(1 - \gamma_5)\hat{e}$ etc, and for the quark fields

$$t = \frac{1}{2} \quad \begin{cases} t_3 = +1/2 \\ t_3 = -1/2 \end{cases} \quad \begin{pmatrix} \hat{u} \\ \hat{d}' \end{pmatrix}_L \quad \begin{pmatrix} \hat{c} \\ \hat{s}' \end{pmatrix}_L \quad \begin{pmatrix} \hat{t} \\ \hat{b}' \end{pmatrix}_L . \tag{22.2}$$

As discussed in section 20.4, the subscript 'L' refers to the fact that only the left–handed chiral components of the fields enter as a consequence of the V − A structure. For this reason, the weak isospin group is referred to as $SU(2)_L$, to show that the weak isospin assignments and corresponding transformation properties apply only to these left–handed parts: for example, under an $SU(2)_L$ transformation

$$\begin{pmatrix} \hat{\nu}_e \\ \hat{e}^- \end{pmatrix}'_L = \exp(-\mathrm{i}\boldsymbol{\alpha} \cdot \boldsymbol{\tau}/2) \begin{pmatrix} \hat{\nu}_e \\ \hat{e}^- \end{pmatrix}_L . \tag{22.3}$$

Note that, as anticipated for a spontaneously broken symmetry, these doublets all involve pairs of particles which are not mass degenerate. In (22.2), the prime indicates that these fields are quantum–mechanical superposition of the fields $\hat{d}$, $\hat{s}$, $\hat{c}$ which are classified by their *strong* interaction quantum numbers. This is a generalization to $3\times3$ mixing of the $2\times2$ GIM mixing introduced in section 20.10, and it will be discussed further in section 22.7.1. For the moment, we ignore the corresponding mixing in the neutrino sectors but return to it in section 22.7.2.

Making this $SU(2)_L$ into a local phase invariance (following the logic of chapter 13) will entail the introduction of three gauge fields, transforming as a $t = 1$ multiplet (a triplet) under the group. Because (as with the ordinary $SU(2)_f$ of hadronic isospin) the members of a weak isodoublet differ by one unit of charge, the two gauge fields associated with transitions between doublet members will have charge $\pm1$. The quanta of these fields will, of course, be the now familiar $W^{\pm}$ bosons mediating the charged current transitions, and associated with the weak isospin raising and lowering operators $t_{\pm}$. What about the third gauge boson of the triplet? This will be electrically neutral, and a very economical and appealing idea would be to associate this neutral vector particle with the photon—thereby *unifying* the weak and electromagnetic interactions. A model of this kind was originally suggested by Schwinger (1957). Of course, the W's must somehow acquire mass, while the photon remains massless. Schwinger arranged this by introducing appropriate couplings of the vector bosons to additional scalar and pseudoscalar fields. These couplings were arbitrary and no prediction of the W masses could be made. We now believe, as emphasized in section 21.4, that the W mass must arise via the spontaneous breakdown of a non-Abelian gauge symmetry and, as we saw in section 19.6, this *does* constrain the W mass.

Apart from the question of the W mass in Schwinger's model, we now know (see chapter 20) that there exist *neutral current* weak interactions, in addition to those of the charged currents. We must also include these in our emerging gauge theory, and an obvious suggestion is to have these currents mediated by the neutral member $W^0$ of the $SU(2)_L$ gauge field triplet. Such a scheme was indeed proposed by Bludman (1958), again pre-Higgs, so that W masses were put in 'by hand'. In this model, however, the neutral currents will have the same pure left-handed $V - A$ structure as the charged currents; but, as we saw in chapter 20, the neutral currents are *not* pure $V - A$. Furthermore, the attractive feature of including the photon, and thus unifying weak and electromagnetic interactions, has been lost.

A key contribution was made by Glashow (1961); similar ideas were also advanced by Salam and Ward (1964). Glashow suggested enlarging the Schwinger–Bludman SU(2) schemes by inclusion of an additional U(1) gauge group, resulting in an '$SU(2)_L \times U(1)$' group structure. The new Abelian U(1) group is associated with a weak analogue of hypercharge—'weak hypercharge'—just as $SU(2)_L$ was associated with 'weak isospin'. Indeed, Glashow proposed that the Gell–Mann-Nishijima relation for charges should also hold for these weak analogues, giving

$$eQ = e(t_3 + y/2) \tag{22.4}$$

for the electric charge $Q$ (in units of $e$) of the $t_3$ member of a weak isomultiplet, assigned a weak hypercharge $y$. Clearly, therefore, the lepton doublets, $(\nu_e, e^-)$ etc then have $y = -1$, while the quark doublets $(u, d_C)$ etc have $y = +\frac{1}{3}$. Now, when *this* group is gauged, everything falls marvellously into place: the charged vector bosons appear as before but there are now *two* neutral vector bosons, which between them will be responsible for the weak neutral current processes and for electromagnetism. This is exactly the piece of mathematics we went through in section 19.6, which we now appropriate as an important part of the Standard Model.

For convenience, we reproduce here the main results of section 19.6. The

Higgs field $\hat{\phi}$ is an SU(2) doublet

$$\hat{\phi} = \begin{pmatrix} \hat{\phi}^+ \\ \hat{\phi}^0 \end{pmatrix} \tag{22.5}$$

with an assumed vacuum expectation value (in unitary gauge) given by

$$\langle 0|\hat{\phi}|0\rangle = \begin{pmatrix} 0 \\ v/\sqrt{2} \end{pmatrix}. \tag{22.6}$$

Fluctuations about this value are parametrized in this gauge by

$$\hat{\phi} = \begin{pmatrix} 0 \\ \frac{1}{\sqrt{2}}(v + \hat{H}) \end{pmatrix} \tag{22.7}$$

where $\hat{H}$ is the (physical) Higgs field. The Lagrangian for the sector consisting of the gauge fields and the Higgs fields is

$$\mathcal{L}_{\mathrm{G}\Phi} = (\hat{D}_\mu\hat{\phi})^\dagger(\hat{D}^\mu\hat{\phi}) + \mu^2\hat{\phi}^\dagger\hat{\phi} - \frac{\lambda}{4}(\hat{\phi}^\dagger\hat{\phi})^2 - \frac{1}{4}\hat{\boldsymbol{F}}_{\mu\nu}\cdot\hat{\boldsymbol{F}}^{\mu\nu} - \frac{1}{4}\hat{G}_{\mu\nu}\hat{G}^{\mu\nu} \tag{22.8}$$

where $\hat{\boldsymbol{F}}_{\mu\nu}$ is the SU(2) field strength tensor (19.81) for the gauge fields $\hat{\boldsymbol{W}}^\mu$, $\hat{G}_{\mu\nu}$ is the U(1) field strength tensor (19.82) for the gauge field $B^\mu$, and $\hat{D}^\mu\hat{\phi}$ is given by (19.80). After symmetry breaking (i.e. the insertion of (22.7) in (22.8)) the quadratic parts of (22.8) can be written in unitary gauge as (see problem 19.9)

$$\hat{\mathcal{L}}_{\mathrm{G}\Phi}^{\mathrm{free}} = \tfrac{1}{2}\partial_\mu\hat{H}\partial^\mu\hat{H} - \mu^2\hat{H}^2 \tag{22.9}$$

$$- \tfrac{1}{4}(\partial_\mu\hat{W}_{1\nu} - \partial_\nu\hat{W}_{1\mu})(\partial^\mu\hat{W}_1^\nu - \partial^\nu\hat{W}_1^\mu) + \tfrac{1}{8}g^2v^2\hat{W}_{1\mu}\hat{W}_1^\mu \tag{22.10}$$

$$- \tfrac{1}{4}(\partial_\mu\hat{W}_{2\nu} - \partial_\nu\hat{W}_{2\mu})(\partial^\mu\hat{W}_2^\nu - \partial^\nu\hat{W}_2^\mu) + \tfrac{1}{8}g^2v^2\hat{W}_{2\mu}\hat{W}_2^\mu \tag{22.11}$$

$$- \tfrac{1}{4}(\partial_\mu\hat{Z}_\nu - \partial_\nu\hat{Z}_\mu)(\partial^\mu\hat{Z}^\nu - \partial^\nu\hat{Z}^\mu) + \frac{v^2}{8}(g^2 + g^{2\prime})\hat{Z}_\mu\hat{Z}^\mu \tag{22.12}$$

$$- \tfrac{1}{4}\hat{F}_{\mu\nu}\hat{F}^{\mu\nu} \tag{22.13}$$

where

$$\hat{Z}^\mu = \cos\theta_{\mathrm{W}}\hat{W}_3^\mu - \sin\theta_{\mathrm{W}}\hat{B}^\mu \tag{22.14}$$

$$\hat{A}^\mu = \sin\theta_{\mathrm{W}}\hat{W}_3^\mu + \cos\theta_{\mathrm{W}}\hat{B}^\mu \tag{22.15}$$

and

$$\hat{F}^{\mu\nu} = \partial^\mu A^\nu - \partial^\nu A^\mu \tag{22.16}$$

with

$$\cos\theta_{\mathrm{W}} = g/(g^2 + g'^2)^{1/2} \qquad \sin\theta_{\mathrm{W}} = g'/(g^2 + g'^2)^{1/2}. \tag{22.17}$$

**Table 22.1.** Weak isospin and hypercharge assignments.

| | $t$ | $t_3$ | $y$ | $Q$ |
|---|---|---|---|---|
| $\nu_{eL}$, $\nu_{\mu L}$, $\nu_{\tau L}$ | 1/2 | 1/2 | $-1$ | 0 |
| $\nu_{eR}$, $\nu_{\mu R}$, $\nu_{\tau R}$ | 0 | 0 | 0 | 0 |
| $e_L$, $\mu_L$, $\tau_L$ | 1/2 | $-1/2$ | $-1$ | $-1$ |
| $e_R$, $\mu_R$, $\tau_R$ | 0 | 0 | $-2$ | $-1$ |
| $u_L$, $c_L$, $t_L$ | 1/2 | 1/2 | 1/3 | 2/3 |
| $u_R$, $c_R$, $t_R$ | 0 | 0 | 4/3 | 2/3 |
| $d'_L$, $s'_L$, $b'_L$ | 1/2 | $-1/2$ | 1/3 | $-1/3$ |
| $d'_R$, $s'_R$, $b'_R$ | 0 | 0 | $-2/3$ | $-1/3$ |
| $\phi^+$ | 1/2 | 1/2 | 1 | 1 |
| $\phi^0$ | 1/2 | $-1/2$ | 1 | 0 |

Feynman rules for the vector boson propagators (in unitary gauge) and couplings, and for the Higgs couplings, can be read off from (22.8), and are given in appendix Q.

Equations (22.9)–(22.13) give the tree-level masses of the Higgs boson and the gauge bosons: (22.9) tells us that the mass of the Higgs boson is

$$m_H = \sqrt{2}\mu = \sqrt{2\lambda}v \tag{22.18}$$

where $v/\sqrt{2}$ is the (tree-level) Higgs vacuum value; (22.10) and (22.11) show that the charged W's have a mass

$$M_W = gv/2 \tag{22.19}$$

where $g$ is the $SU(2)_L$ gauge coupling constant; (22.12) gives the mass of the $Z^0$ as

$$M_Z = M_W / \cos\theta_W \tag{22.20}$$

and (22.13) shows that the $A^\mu$ field describes a massless particle (to be identified with the photon).

Still unaccounted for are the *right–handed* chiral components of the fermion fields. There is, at present, no evidence for any weak interactions coupling to the right-handed field components and it is therefore natural—and a basic assumption of the electroweak theory—that all 'R' components are singlets under the weak isospin group. Crucially, however, the 'R' components do interact via the U(1) field $\hat{B}^\mu$: it is this that allows electromagnetism to emerge free of parity-violating $\gamma_5$ terms, as we shall see. With the help of the weak charge formula (equation (22.4)), we arrive at the assignments shown in table 22.1.

We have included 'R' components for the neutrinos in the table. It is, however, fair to say that in the original Standard Model the neutrinos were taken to be massless, with no neutrino mixing. We have seen in chapter 20 that it is for many purposes an excellent approximation to treat the neutrinos as massless, except when discussing experimental situations specifically sensitive to neutrino oscillations. We shall mention their masses again in section 22.7.2, but for the moment we proceed in the 'massless neutrinos' approximation. In this case, there are *no* 'R' components for neutrinos.

## 22.2 The leptonic currents (massless neutrinos): relation to current–current model

We write the $\mathrm{SU(2)_L \times U(1)}$ covariant derivative in terms of the fields $\hat{\boldsymbol{W}}^{\mu}$ and $\hat{B}^{\mu}$ of section 19.6 as

$$\hat{D}^{\mu} = \partial^{\mu} + \mathrm{i}g\boldsymbol{\tau}\cdot\hat{\boldsymbol{W}}^{\mu}/2 + \mathrm{i}g'y\hat{B}^{\mu}/2 \qquad \text{on 'L' SU(2) doublets} \tag{22.21}$$

and as

$$\hat{D}^{\mu} = \partial^{\mu} + \mathrm{i}g'yB^{\mu}/2 \qquad \text{on 'R' SU(2) singlets.} \tag{22.22}$$

The leptonic couplings to the gauge fields therefore arise from the 'gauge-covariantized' free leptonic Lagrangian:

$$\hat{\mathcal{L}}_{\mathrm{lept}} = \sum_{f=\mathrm{e},\mu,\tau} \bar{\hat{l}}_{f\mathrm{L}}\mathrm{i}\hat{\not{D}}\hat{l}_{f\mathrm{L}} + \sum_{f=\mathrm{e},\mu,\tau} \bar{\hat{l}}_{f\mathrm{R}}\mathrm{i}\hat{\not{D}}\hat{l}_{f\mathrm{R}} \tag{22.23}$$

where the $\hat{l}_{f\mathrm{L}}$ are the left-handed doublets

$$\hat{l}_{f\mathrm{L}} = \begin{pmatrix} \hat{\nu}_f \\ \hat{f}^- \end{pmatrix}_{\mathrm{L}} \tag{22.24}$$

and $\hat{l}_{f\mathrm{R}}$ are the singlets $\hat{l}_{\mathrm{eR}} = \hat{e}_{\mathrm{R}}$ etc.

Consider, first, the *charged leptonic currents*. The correct normalization for the charged fields is that $\hat{W}^{\mu} \equiv (\hat{W}_1^{\mu} - \mathrm{i}\hat{W}_2^{\mu})/\sqrt{2}$ destroys the $\mathrm{W}^+$ or creates the $\mathrm{W}^-$ (cf (7.15)). The '$\boldsymbol{\tau}\cdot\hat{\boldsymbol{W}}/2$' terms can be written as

$$\boldsymbol{\tau}\cdot\hat{\boldsymbol{W}}^{\mu}/2 = \frac{1}{\sqrt{2}}\left\{\tau_+\frac{(\hat{W}_1^{\mu} - \mathrm{i}\hat{W}_2^{\mu})}{\sqrt{2}} + \tau_-\frac{(\hat{W}_1^{\mu} + \mathrm{i}\hat{W}_2^{\mu})}{\sqrt{2}}\right\} + \frac{\tau_3}{2}W_3^{\mu} \tag{22.25}$$

where $\tau_{\pm} = (\tau_1 \pm \mathrm{i}\tau_2)/2$ are the usual raising and lowering operators for the doublets. Thus, the '$f = \mathrm{e}$' contribution to the first term in (22.23) picks out the process $\mathrm{e}^- \to \nu_{\mathrm{e}} + \mathrm{W}^-$ for example, with the result that the corresponding vertex is given by

$$-\frac{\mathrm{i}g}{\sqrt{2}}\gamma^{\mu}\frac{(1-\gamma_5)}{2}. \tag{22.26}$$

The 'universality' of the single coupling constant '$g$' ensures that (22.26) is also the amplitude for the $\mu - \nu_{\mu} - \mathrm{W}$ and $\tau - \nu_{\tau} - \mathrm{W}$ vertices. Thus the amplitude

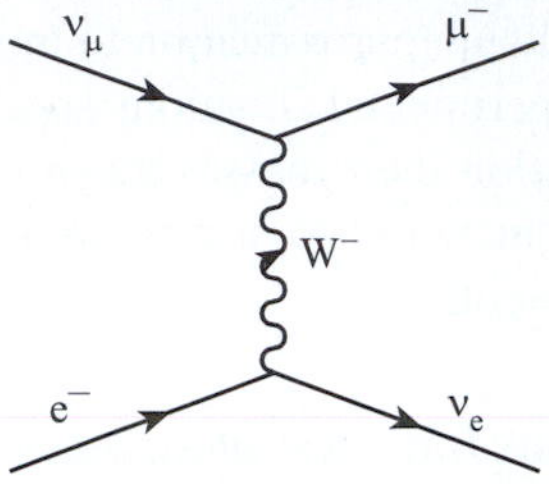


**Figure 22.1.** W-exchange process in $\nu_\mu + e^- \to \mu^- + \nu_e$.

for the $\nu_\mu + e^- \to \mu^- + \nu_e$ process considered in section 20.8 is

$$-\frac{ig}{\sqrt{2}}\bar{u}(\mu)\gamma_\mu\frac{(1-\gamma_5)}{2}u(\nu_\mu)\frac{i[-g^{\mu\nu}+k^\mu k^\nu/M_W^2]}{k^2-M_W^2} - i\frac{g}{\sqrt{2}}\bar{u}(\nu_e)\gamma_\nu\frac{(1-\gamma_5)}{2}u(e) \tag{22.27}$$

corresponding to the Feynman graph of figure 22.1.

For $k^2 \ll M_W^2$ we can replace the W-propagator by the constant value $g^{\mu\nu}/M_W^2$, leading to the amplitude

$$-\frac{ig^2}{8M_W^2}\bar{u}(\mu)\gamma_\mu(1-\gamma_5)u(\nu_\mu)\bar{u}(\nu_e)\gamma^\mu(1-\gamma_5)u(e) \tag{22.28}$$

which may be compared with the form we used in the current–current theory, equation (20.93). This comparison gives

$$\frac{G_F}{\sqrt{2}} = \frac{g^2}{8M_W^2}. \tag{22.29}$$

This is an important equation giving the precise version, in the GSW theory, of the qualitative relation (21.19) introduced earlier.

Putting together (22.19) and (22.29), we can deduce

$$G_F/\sqrt{2} = 1/(2v^2) \tag{22.30}$$

so that from the known value (21.11) of $G_F$, there follows the value of $v$:

$$v \simeq 246 \text{ GeV}. \tag{22.31}$$

Alternatively, we may quote $v/\sqrt{2}$ (the vacuum value of the Higgs field):

$$v/\sqrt{2} \simeq 174 \text{ GeV}. \tag{22.32}$$

This parameter sets the scale of electroweak symmetry breaking, but as yet no theory is able to predict its value. It is related to the parameters $\lambda$, $\mu$ of (22.8) by $v/\sqrt{2} = \sqrt{2}\mu/\lambda^{1/2}$ (cf (17.98)).

In general, the charge-changing part of (22.23) can be written as

$$-\frac{g}{\sqrt{2}}\left\{\bar{\hat{\nu}}_{e}\gamma^{\mu}\frac{(1-\gamma_5)}{2}\hat{e}+\bar{\hat{\nu}}_{\mu}\gamma^{\mu}\frac{(1-\gamma_5)}{2}\hat{\mu}+\bar{\hat{\nu}}_{\tau}\gamma^{\mu}\frac{(1-\gamma_5)}{2}\hat{\tau}\right\}\hat{W}_{\mu}$$
$$+\text{ Hermitian conjugate} \qquad (22.33)$$

where $\hat{W}^{\mu}=(\hat{W}_1^{\mu}-\mathrm{i}\hat{W}_2^{\mu})/\sqrt{2}$. (22.33) has the form

$$-\hat{j}^{\mu}_{\mathrm{CC}}(\text{leptons})\hat{W}_{\mu}-\hat{j}^{\mu\dagger}_{\mathrm{CC}}(\text{leptons})\hat{W}^{\dagger}_{\mu} \qquad (22.34)$$

where the *leptonic weak charged current* $\hat{j}^{\mu}_{\mathrm{CC}}$(leptons) is precisely thatused in the current–current model (equation (20.81)), up to the usual factors of $g$'s and $\sqrt{2}$'s. Thus the dynamical symmetry currents of the $\mathrm{SU(2)_L}$ gauge theory are exactly the 'phenomenological' currents of the earlier current–current model. The Feynman rules for the lepton–W couplings (appendix Q) can be read off from (22.33).

Turning now to the *leptonic weak neutral current*, this will appear via the couplings to the $Z^0$, written as

$$-\hat{j}^{\mu}_{\mathrm{NC}}(\text{leptons})\hat{Z}^{\mu}. \qquad (22.35)$$

Referring to (22.14) for the linear combination of $\hat{W}_3^{\mu}$ and $\hat{B}^{\mu}$ which represents $\hat{Z}^{\mu}$, we find (problem 22.1) that

$$\hat{j}^{\mu}_{\mathrm{NC}}(\text{leptons})=\frac{g}{\cos\theta_{\mathrm{W}}}\sum_{l}\bar{\hat{\psi}}_{l}\gamma^{\mu}\left[t_3^{l}\left(\frac{1-\gamma_5}{2}\right)-\sin^2\theta_{\mathrm{W}}Q_l\right]\hat{\psi}_{l} \qquad (22.36)$$

where the sum is over the six lepton fields $\nu_{\mathrm{e}}, \mathrm{e}^-, \nu_{\mu}, \ldots, \tau^-$. For the $Q=0$ neutrinos with $t_3=+\frac{1}{2}$,

$$\hat{j}^{\mu}_{\mathrm{NC}}(\text{neutrinos})=\frac{g}{2\cos\theta_{\mathrm{W}}}\sum_{l}\bar{\hat{\nu}}_{l}\gamma^{\mu}\frac{(1-\gamma_5)}{2}\hat{\nu}_{l} \qquad (22.37)$$

where now $l=\mathrm{e},\mu,\tau$. For the other (negatively charged) leptons, we shall have both L and R couplings from (22.36), and we can write

$$\hat{j}^{\mu}_{\mathrm{NC}}(\text{charged leptons})=\frac{g}{\cos\theta_{\mathrm{W}}}\sum_{l=\mathrm{e},\mu,\tau}\bar{\hat{l}}\gamma^{\mu}\left[c_{\mathrm{L}}^{l}\left(\frac{1-\gamma_5}{2}\right)+c_{\mathrm{R}}^{l}\left(\frac{1+\gamma_5}{2}\right)\right]\hat{l} \qquad (22.38)$$

where

$$c_{\mathrm{L}}^{l}=t_3^{l}-\sin^2\theta_{\mathrm{W}}Q_l=-\tfrac{1}{2}+\sin^2\theta_{\mathrm{W}} \qquad (22.39)$$

$$c_{\mathrm{R}}^{l}=-\sin^2\theta_{\mathrm{W}}Q_l=\sin^2\theta_{\mathrm{W}}. \qquad (22.40)$$

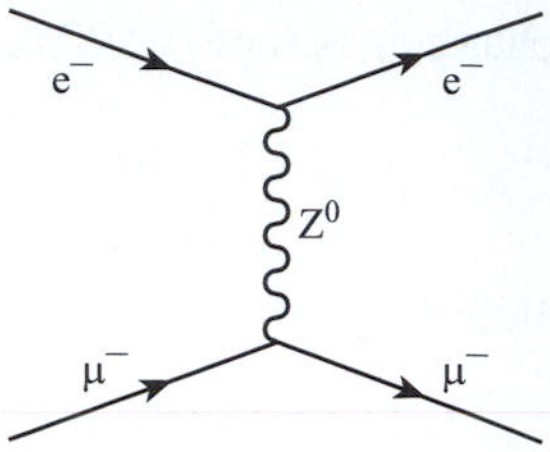


**Figure 22.2.** $Z^0$-exchange process in $e^-\mu^- \to e^-\mu^-$.

As noted earlier, the $Z^0$ coupling is not pure 'V − A'. These relations (22.37)–(22.40) are exactly the ones given earlier, in (20.126)–(20.128); in particular, the couplings are independent of '$l$' and hence exhibit lepton universality. The alternative notation

$$\hat{j}^{\mu}_{\mathrm{NC}}(\text{charged leptons}) = \frac{g}{2\cos\theta_{\mathrm{W}}} \sum_{l} \bar{\hat{l}}\gamma^{\mu}(g^{l}_{\mathrm{V}} - g^{l}_{\mathrm{A}}\gamma_5)\hat{l} \tag{22.41}$$

is often used, where

$$g^{l}_{\mathrm{V}} = -\tfrac{1}{2} + 2\sin^2\theta_{\mathrm{W}} \qquad g^{l}_{\mathrm{A}} = -\tfrac{1}{2}. \tag{22.42}$$

Note that $g^{l}_{\mathrm{V}}$ vanishes for $\sin^2\theta_{\mathrm{W}} = 0.25$. Again, the Feynman rules for lepton–Z couplings (appendix Q) are contained in (22.37) and (22.38).

As in the case of W-mediated charge-changing processes, $Z^0$-mediated processes reduce to the current–current form at low $k^2$. For example, the amplitude for $e^-\mu^- \to e^-\mu^-$ via $Z^0$ exchange (figure 22.2) reduces to

$$\begin{aligned} -\frac{\mathrm{i}g^2}{4\cos^2\theta_{\mathrm{W}} M_{\mathrm{Z}}^2} &\bar{u}(\mathrm{e})\gamma_{\mu}[c^{l}_{\mathrm{L}}(1-\gamma_5) + c^{l}_{\mathrm{R}}(1+\gamma_5)]u(\mathrm{e}) \\ &\times \bar{u}(\mu)\gamma^{\mu}[c^{l}_{\mathrm{L}}(1-\gamma_5) + c^{l}_{\mathrm{R}}(1+\gamma_5)]u(\mu). \end{aligned} \tag{22.43}$$

It is customary to define the parameter

$$\rho = M_{\mathrm{W}}^2/(M_{\mathrm{Z}}^2\cos^2\theta_{\mathrm{W}}) \tag{22.44}$$

which is unity at tree level (see (22.20)) in the absence of loop corrections. The ratio of factors in front of the $\bar{u}\ldots u$ expressions in (22.43) and (22.28) (i.e. 'neutral current process'/'charged current process') is then $2\rho$.

We may also check the electromagnetic current in the theory, by looking for the piece that couples to $\hat{A}^{\mu}$. We find that

$$\hat{j}^{\mu}_{\mathrm{emag}} = -g\sin\theta_{\mathrm{W}} \sum_{l=\mathrm{e},\mu,\tau} \bar{\hat{l}}\gamma^{\mu}\hat{l} \tag{22.45}$$

which allows us to identify the electromagnetic charge $e$ as

$$e = g \sin\theta_{\mathrm{W}} \tag{22.46}$$

as already suggested in (19.98) of chapter 19. Note that all the $\gamma_5$'s cancel from (22.45), as is of course required.

## 22.3 The quark currents

The charge-changing quark currents which are coupled to the $\mathrm{W}^{\pm}$ fields have a form very similar to that of the charged leptonic currents, except that the $t_3 = -\frac{1}{2}$ components of the L-doublets have to be understood as the flavour-mixed (weakly interacting) states

$$\begin{pmatrix} \hat{d}' \\ \hat{s}' \\ \hat{b}' \end{pmatrix}_{\mathrm{L}} = \begin{pmatrix} V_{\mathrm{ud}} & V_{\mathrm{us}} & V_{\mathrm{ub}} \\ V_{\mathrm{cd}} & V_{\mathrm{cs}} & V_{\mathrm{cb}} \\ V_{\mathrm{td}} & V_{\mathrm{ts}} & V_{\mathrm{tb}} \end{pmatrix} \begin{pmatrix} \hat{d} \\ \hat{s} \\ \hat{b} \end{pmatrix}_{\mathrm{L}}, \tag{22.47}$$

where $\hat{d}, \hat{s}$ and $\hat{b}$ are the strongly interacting fields with masses $m_{\mathrm{d}}, m_{\mathrm{s}}$ and $m_{\mathrm{b}}$ and the $V$-matrix is the Cabibbo–Kobayashi–Maskawa matrix (Cabibbo 1963, Kobayashi and Maskawa 1973) which generalizes the $2 \times 2$ GIM mixing introduced in section 20.10. We shall discuss this matrix further in section 22.7.1. Thus, the *charge-changing weak quark current* is

$$\hat{j}^{\mu}_{\mathrm{CC}}(\text{quarks}) = \frac{g}{\sqrt{2}} \left\{ \bar{\hat{u}}\gamma^{\mu}\frac{(1-\gamma_5)}{2}\hat{d}' + \bar{\hat{c}}\gamma^{\mu}\frac{(1-\gamma_5)}{2}\hat{s}' + \bar{\hat{t}}\gamma^{\mu}\frac{(1-\gamma_5)}{2}\hat{b}' \right\} \tag{22.48}$$

which generalizes (20.131) to three generations and supplies the factor $g/\sqrt{2}$, as for the leptons.

The neutral currents are diagonal in flavour if the matrix $V$ is unitary (see also section 22.7.1). Thus $\hat{j}^{\mu}_{\mathrm{NC}}(\text{quarks})$ will be given by the same expression as (20.144), except that now the sum will be over all six quark flavours. The *neutral weak quark current* is thus

$$\hat{j}^{\mu}_{\mathrm{NC}}(\text{quarks}) = \frac{g}{\cos\theta_{\mathrm{W}}} \sum_q \bar{\hat{q}}\gamma^{\mu}\left[c^q_{\mathrm{L}}\frac{(1-\gamma_5)}{2} + c^q_{\mathrm{R}}\frac{(1+\gamma_5)}{2}\right]\hat{q} \tag{22.49}$$

where

$$c^q_{\mathrm{L}} = t^q_3 - \sin^2\theta_{\mathrm{W}} Q_q \tag{22.50}$$

$$c^q_{\mathrm{R}} = -\sin^2\theta_{\mathrm{W}} Q_q. \tag{22.51}$$

These expressions are exactly as given in (20.144)–(20.146). As for the charged leptons, we can alternatively write (22.49) as

$$\hat{j}^{\mu}_{\mathrm{NC}}(\text{quarks}) = \frac{g}{2\cos\theta_{\mathrm{W}}} \sum_q \bar{\hat{q}}\gamma^{\mu}(g^q_{\mathrm{V}} - g^q_{\mathrm{A}}\gamma_5)\hat{q} \tag{22.52}$$

where

$$g_{\mathrm{V}}^{q} = t_3^q - 2\sin^2\theta_{\mathrm{W}} Q_q \tag{22.53}$$

$$g_{\mathrm{A}}^{q} = t_3^q. \tag{22.54}$$

Before proceeding to discuss some simple phenomenological consequences, we note one important feature of the Standard Model currents in general. Reading (22.1) and (22.2) together 'vertically', the leptons and quarks are grouped in three *families*, each with two leptons and two quarks. The theoretical motivation for such family grouping is that *anomalies* are cancelled within each complete family. Anomalies were discussed in section 18.4. While they can be tolerated in global (non-gauged) currents—and indeed we saw that an anomaly in the strong isospin current is responsible for $\pi^0 \to \gamma\gamma$ decay—they must cancel in the symmetry currents of a gauge theory or else renormalizability is destroyed. The condition that anomalies cancel in the gauged currents of the Standard Model is a remarkably simple one (Ryder 1996, p 384):

$$N_{\mathrm{c}}(Q_{\mathrm{u}} + Q_{\mathrm{d}}) + Q_{\mathrm{e}} = 0 \tag{22.55}$$

where $N_{\mathrm{c}}$ is the number of colours and $Q_{\mathrm{u}}$, $Q_{\mathrm{d}}$ and $Q_{\mathrm{e}}$ are the charges (in units of $e$) of the 'u', 'd' and 'e' type fields in each family. Clearly (22.55) is true for the families in (22.1) and (22.2) and indicates a remarkable connection, at some deep level, between the facts that quarks occur in three colours and have charges which are $1/3$ fractions. The Standard Model provides no explanation of this connection.

## 22.4 Simple (tree-level) predictions

We noted in section 20.9 that, before the discovery of the W and Z particles, the then known data were consistent with a single value of $\theta_{\mathrm{W}}$ given (using a modern value) by $\sin^2\theta_{\mathrm{W}} \simeq 0.23$. Using (22.29) and (22.46), we may then predict the value of $M_{\mathrm{W}}$:

$$M_{\mathrm{W}} = \left(\frac{\pi\alpha}{\sqrt{2}G_{\mathrm{F}}}\right)^{1/2} \frac{1}{\sin\theta_{\mathrm{W}}} \simeq \frac{37.28}{\sin\theta_{\mathrm{W}}}\,\mathrm{GeV} \simeq 77.73\,\mathrm{GeV}. \tag{22.56}$$

Similarly, using (22.20) we predict

$$M_{\mathrm{Z}} = M_{\mathrm{W}}/\cos\theta_{\mathrm{W}} \simeq 88.58\,\mathrm{GeV}. \tag{22.57}$$

These predictions of the theory (at lowest order) indicate the power of the underlying symmetry to tie together many apparently unrelated quantitities, which are all determined in terms of only a few basic parameters.

The width for $\mathrm{W}^- \to \mathrm{e}^- + \bar{\nu}_{\mathrm{e}}$ can be calculated using the vertex (22.26), with the result (problem 22.2)

$$\Gamma(\mathrm{W}^- \to \mathrm{e}^-\bar{\nu}_{\mathrm{e}}) = \frac{1}{12}\frac{g^2}{4\pi}M_{\mathrm{W}} = \frac{G_{\mathrm{F}}}{2^{1/2}}\frac{M_{\mathrm{W}}^3}{6\pi} \simeq 205\,\mathrm{MeV} \tag{22.58}$$

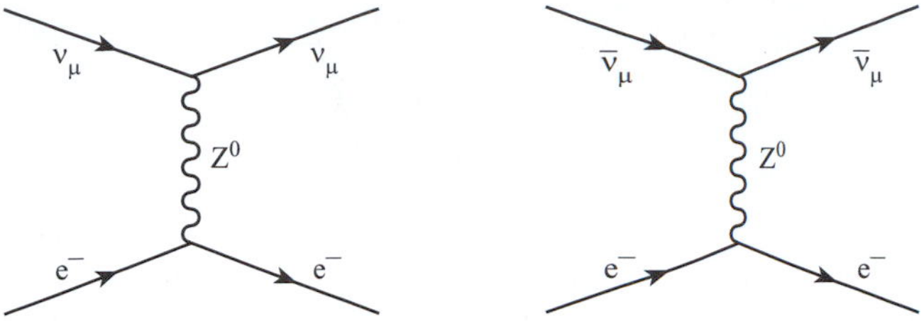


**Figure 22.3.** Neutrino–electron graphs involving $Z^0$ exchange.

using (22.56). The widths to $\mu^-\bar{\nu}_\mu$, $\tau^-\bar{\nu}_\tau$ are the same. Neglecting GIM-type flavour mixing among the two energetically allowed quark channels $\bar{\mathrm{u}}\mathrm{d}$ and $\bar{\mathrm{c}}\mathrm{s}$, their widths would also be the same, apart from a factor of three for the different colour channels. The total W width for all these channels will therefore be about 1.85 GeV, while the branching ratio for $\mathrm{W} \to \mathrm{e}\nu$ is

$$B(\mathrm{e}\nu) = \Gamma(\mathrm{W} \to \mathrm{e}\nu)/\Gamma(\text{total}) \simeq 11\%. \tag{22.59}$$

In making these estimates we have neglected all fermion masses.

The width for $\mathrm{Z}^0 \to \nu\bar{\nu}$ can be found from (22.58) by replacing $g/2^{1/2}$ by $g/2\cos\theta_{\mathrm{W}}$, and $M_{\mathrm{W}}$ by $M_{\mathrm{Z}}$, giving

$$\Gamma(\mathrm{Z}^0 \to \nu\bar{\nu}) = \frac{1}{24}\frac{g^2}{4\pi}\frac{M_{\mathrm{Z}}}{\cos^2\theta_{\mathrm{W}}} = \frac{G_{\mathrm{F}}}{2^{1/2}}\frac{M_{\mathrm{Z}}^3}{12\pi} \simeq 152\ \mathrm{MeV} \tag{22.60}$$

using (22.57). Charged lepton pairs couple with both $c_{\mathrm{L}}^l$ and $c_{\mathrm{R}}^l$ terms, leading (with neglect of lepton masses) to

$$\Gamma(\mathrm{Z}^0 \to l\bar{l}) = \left(\frac{|c_{\mathrm{L}}^l|^2 + |c_{\mathrm{R}}^l|^2}{6}\right)\frac{g^2}{4\pi}\frac{M_{\mathrm{Z}}}{\cos^2\theta_{\mathrm{W}}}. \tag{22.61}$$

The values $c_{\mathrm{L}}^\nu = \frac{1}{2}$, $c_{\mathrm{R}}^\nu = 0$ in (22.61) reproduce (22.60). With $\sin^2\theta_{\mathrm{W}} \simeq 0.23$, we find that

$$\Gamma(\mathrm{Z}^0 \to l\bar{l}) \simeq 76.5\ \mathrm{MeV}. \tag{22.62}$$

Quark pairs couple as in (22.49), the GIM mechanism ensuring that all flavour-changing terms cancel. The total width to $\mathrm{u}\bar{\mathrm{u}}$, $\mathrm{d}\bar{\mathrm{d}}$, $\mathrm{c}\bar{\mathrm{c}}$, $\mathrm{s}\bar{\mathrm{s}}$ and $\mathrm{b}\bar{\mathrm{b}}$ channels (allowing three for colour and neglecting masses) is then 1538 MeV, producing an estimated total width of approximately 2.22 GeV. (QCD corrections will increase these estimates by a factor of order 1.1.) The branching ratio to charged leptons is approximately 3.4%, to the three (invisible) neutrino channels 20.5% and to hadrons (via hadronization of the $\mathrm{q}\bar{\mathrm{q}}$ channels) about 69.3%.

Cross-sections for lepton–lepton scattering proceeding via $\mathrm{Z}^0$ exchange can be calculated (for $k^2 \ll M_{\mathrm{Z}}^2$) using the currents (22.37) and (22.38), and the

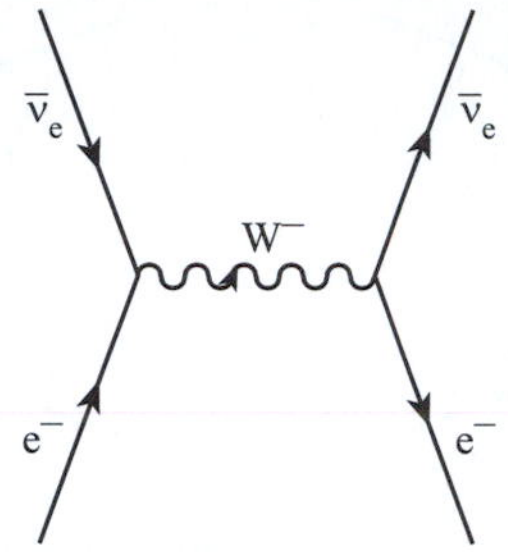


**Figure 22.4.** One-W annihilation graph in $\bar{\nu}_e e^- \rightarrow \bar{\nu}_e e^-$.

method of section 20.8. Examples are

$$\nu_\mu e^- \rightarrow \nu_\mu e^- \tag{22.63}$$

and

$$\bar{\nu}_\mu e^- \rightarrow \bar{\nu}_\mu e^- \tag{22.64}$$

as shown in figure 22.3. Since the neutral current for the electron is not pure V−A, as was the charged current, we expect to see terms involving both $|c_L^l|^2$ and $|c_R^l|^2$, and possibly an interference term. The cross-section for (22.63) is found to be ('t Hooft 1971c)

$$\mathrm{d}\sigma/\mathrm{d}y = (2G_F^2 E m_e/\pi)[|c_L^l|^2 + |c_R^l|^2(1-y)^2 - \tfrac{1}{2}(c_R^{l\,*}c_L^l + c_L^{l\,*}c_R^l)ym_e/E] \tag{22.65}$$

where $E$ is the energy of the incident neutrino in the 'laboratory' system and $y = (E - E')/E$ as before, where $E'$ is the energy of the outgoing neutrino in the 'laboratory' system.[1] Equation (22.65) may be compared with the $\nu_\mu e^- \rightarrow \mu^- \nu_e$ (charged current) cross-section of (20.125) by noting that $t = -2m_e E y$: the $|c_L^l|^2$ term agrees with the pure V − A result (20.125), while the $|c_R^l|^2$ term involves the same $(1-y)^2$ factor discussed for $\nu\bar{q}$ scattering in section 20.11. The interference term is negligible for $E \gg m_e$ . The cross-section for the anti-neutrino process (22.64) is found from (22.65) by interchanging $c_L^l$ and $c_R^l$.

A third lepton–lepton process is experimentally available,

$$\bar{\nu}_e e^- \rightarrow \bar{\nu}_e e^-. \tag{22.66}$$

In this case there is a single W intermediate state graph to consider as well as the $Z^0$ one, as shown in figure 22.4. The cross-section for (22.66) turns out to be given by an expression of the form (22.65) but with the replacements

$$c_L^l \rightarrow \tfrac{1}{2} + \sin^2\theta_W, c_R^l \rightarrow \sin^2\theta_W. \tag{22.67}$$

[1] In the kinematics, lepton masses have been neglected wherever possible.

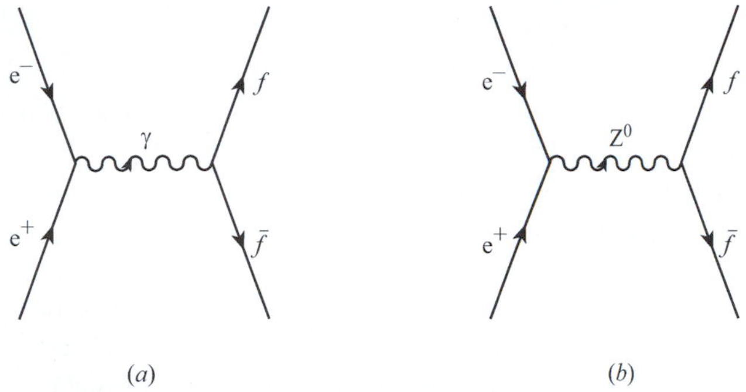


**Figure 22.5.** (*a*) One-$\gamma$ and (*b*) one-$Z^0$ annihilation graphs in $e^+e^- \to f\bar{f}$.

We emphasize once more that all these cross-sections are determined in terms of the Fermi constant $G_F$ and only one further parameter, $\sin^2\theta_W$. As mentioned in section 20.9, experimental fits to these predictions are reviewed by Commins and Bucksbaum (1983), Renton (1990) and Winter (2000).

Particularly precise determinations of the Standard Model parameters may be made at the $e^+e^-$ colliders, LEP and SLC. Consider the reaction $e^+e^- \to f\bar{f}$ where $f$ is $\mu$ or $\tau$, at energies where the lepton masses may be neglected in the final answers. In lowest order, the process is mediated by both $\gamma$- and $Z^0$-annihilation as shown in figure 22.5. Calculations of the cross-section were made some 30 years ago (for example Budny (1975)). In modern notation, the differential cross-section for the scattering of unpolarized $e^-$ and $e^+$ is given by

$$\frac{d\sigma}{d\cos\theta} = \frac{\pi\alpha^2}{2s}[(1+\cos^2\theta)A + \cos\theta B] \tag{22.68}$$

where $\theta$ is the CM scattering angle of the final-state lepton, $s = (p_{e^-} + p_{e^+})^2$ and

$$A = 1 + 2g_V^e g_V^f \operatorname{Re}\chi(s) + [(g_A^e)^2 + (g_V^e)^2][(g_A^f)^2 + (g_V^f)^2]|\chi(s)|^2 \tag{22.69}$$

$$B = 4g_A^e g_A^f \operatorname{Re}\chi(s) + 8g_A^e g_V^e g_A^f g_V^f |\chi(s)|^2 \tag{22.70}$$

$$\chi(s) = s/[4\sin^2\theta_W\cos^2\theta_W(s - M_Z^2 + i\Gamma_Z M_Z)]. \tag{22.71}$$

Note that the term surviving when all the $g$'s are set to zero, which is therefore the pure single photon contribution, is exactly as calculated in problem 8.19. The presence of the $\cos\theta$ term leads to the forward–backward asymmetry noted in that problem.

The forward–backward asymmetry $A_{FB}$ may be defined as

$$A_{FB} \equiv (N_F - N_B)/(N_F + N_B) \tag{22.72}$$

where $N_F$ is the number scattered into the forward hemisphere $0 \le \cos\theta \le 1$ and $N_B$ that into the backward hemisphere $-1 \le \cos\theta \le 0$. Integrating (22.68) one easily finds that

$$A_{\rm FB} = 3B/8A. \tag{22.73}$$

For $\sin^2\theta_{\rm W} = 0.25$ we noted after (22.42) that the $g_{\rm V}^l$'s vanish, so they are very small for $\sin^2\theta_{\rm W} \simeq 0.23$. The effect is therefore controlled essentially by the first term in (22.70). At $\sqrt{s} = 29$ GeV, for example, the asymmetry is $A_{\rm FB} \simeq -0.063$.

However, QED alone produces a small positive $A_{\rm FB}$, through interference between $1\gamma$ and $2\gamma$ annihilation processes (which have different charge conjugation parity), as well as between initial- and final-state bremsstrahlung corrections to figure 22.5($a$). Indeed, *all* one-loop radiative effects must clearly be considered, in any comparison with modern high precision data.

Many such measurements have been made 'on the Z peak', i.e. at $s = M_{\rm Z}^2$ in the parametrization (22.71). In that case, (22.73) becomes (neglecting the photon contribution)

$$A_{\rm FB}({\rm Z}^0\ {\rm peak}) = \frac{3g_{\rm A}^{\rm e}g_{\rm V}^{\rm e}g_{\rm A}^{f}g_{\rm V}^{f}}{\{[(g_{\rm A}^{\rm e})^2 + g_{\rm V}^{e})^2][(g_{\rm A}^{f})^2 + (g_{\rm V}^{f})^2]\}}. \tag{22.74}$$

Another important asymmetry observable is that involving the difference of the cross-sections for left- and right-handed incident electrons:

$$A_{\rm LR} \equiv (\sigma_{\rm L} - \sigma_{\rm R})/(\sigma_{\rm L} + \sigma_{\rm R}) \tag{22.75}$$

for which the tree-level prediction is

$$A_{\rm LR} = 2g_{\rm V}^{\rm e}g_{\rm A}^{\rm e}/[(g_{\rm V}^{\rm e})^2 + (g_{\rm A}^{\rm e})^2]. \tag{22.76}$$

A similar combination of the $g$'s for the final-state leptons can be measured by forming the 'L–R F–B' asymmetry

$$A_{\rm LR}^{\rm FB} = [(\sigma_{\rm LF} - \sigma_{\rm LB}) - (\sigma_{\rm RF} - \sigma_{\rm RB})]/(\sigma_{\rm R} + \sigma_{\rm L}) \tag{22.77}$$

for which the tree-level prediction is

$$A_{\rm LR}^{\rm FB} = 2g_{\rm V}^{f}g_{\rm A}^{f}/[(g_{\rm V}^{f})^2 + (g_{\rm A}^{2})^2]. \tag{22.78}$$

The quantity on the right-hand side of (22.78) is usually denoted by $A_f$:

$$A_f = 2g_{\rm V}^{f}g_{\rm A}^{f}/[(g_{\rm V}^{f})^2 + (g_{\rm A}^{f})^2]. \tag{22.79}$$

The asymmetry $A_{\rm FB}$ is not, in fact, direct evidence for parity violation in $e^+e^- \to \mu^+\mu^-$, since we see from (22.69) and (22.70) that it is even under $g_{\rm A}^l \to -g_{\rm A}^l$, whereas a true parity-violating effect would involve terms odd (linear) in $g_{\rm A}^l$. However, electroweak-induced parity violation effects in an apparently electromagnetic process were observed in a remarkable experiment by Prescott *et al* (1978). Longitudinally polarized electrons were inelastically scattered from deuterium and the flux of scattered electrons was measured for incident electrons of definite helicity. An asymmetry between the results, depending on the helicities, was observed—a clear signal for parity violation. This was the first demonstration of parity-violating effects in an 'electromagnetic' process: the corresponding value of $\sin^2\theta_W$ is in agreement with that determined from $\nu$ data.

We now turn to some of the main experimental evidence, beginning with the discoveries of the $W^\pm$ and $Z^0$ 1983.

## 22.5 The discovery of the $W^\pm$ and $Z^0$ at the CERN $p\bar{p}$ collider

### 22.5.1 Production cross-sections for W and Z in $p\bar{p}$ colliders

The possibility of producing the predicted $W^\pm$ and $Z^0$ particles was the principal motivation for transforming the CERN SPS into a $p\bar{p}$ collider using the stochastic cooling technique (Rubbia *et al* 1977, Staff of the CERN $\bar{p}p$ project 1981). Estimates of W and $Z^0$ production in $\bar{p}p$ collisions may be obtained (see, for example, Quigg 1977) from the parton model, in a way analogous to that used for the Drell–Yan process in section 9.4 with $\gamma$ replaced by W or $Z^0$, as shown in figure 22.6 (cf figure 9.12) and for two-jet cross-sections in section 14.3.1. As in (14.44), we denote by $\hat{s}$ the subprocess invariant

$$\hat{s} = (x_1 p_1 + x_2 p_2)^2 = x_1 x_2 s \tag{22.80}$$

for massless partons. With $\hat{s}^{1/2} = M_W \sim 80$ GeV and $s^{1/2} = 630$ GeV for the $\bar{p}p$ collider energy, we see that the $x$'s are typically $\sim$0.13, so that the valence q's in the proton and $\bar{q}$'s in the anti-proton will dominate (at $\sqrt{s} = 1.8$ TeV, appropriate to the Fermilab Tevatron, $x \simeq 0.04$ and the sea quarks will be expected to contribute). The parton model cross-section $p\bar{p} \to W^\pm +$ anything is then (setting $V_{ud} = 1$ and all other $V_{ij} = 0$)

$$\sigma(p\bar{p} \to W^\pm + X) = \frac{1}{3}\int_0^1 dx_1 \int_0^1 dx_2\, \hat{\sigma}(x_1, x_2) \left\{ \begin{array}{c} u(x_1)\bar{d}(x_2) + \bar{d}(x_1)u(x_2) \\ \bar{u}(x_1)d(x_2) + d(x_1)\bar{u}(x_2) \end{array} \right\} \tag{22.81}$$

where the $\frac{1}{3}$ is the same colour factor as in the Drell–Yan process, and the subprocess cross-section $\hat{\sigma}$ for $q\bar{q} \to W^\pm + X$ is

$$\hat{\sigma} = 4\pi^2\alpha(1/4\sin^2\theta_W)\delta(\hat{s} - M_W^2) \tag{22.82}$$

$$= \pi 2^{1/2} G_F M_W^2 \delta(x_1 x_2 s - M_W^2). \tag{22.83}$$

QCD corrections to (22.81) must, as usual, be included. Leading logarithms will make the distributions $Q^2$-dependent, and they should be evaluated at $Q^2 = M_W^2$. There will be further ($O(\alpha_s^2)$) corrections, which are often accounted for by a multiplicative factor '$K$', which is of order 1.5–2 at these energies. $O(\alpha_s^2)$ calculations are presented in Hamberg *et al* (1991) and by van der Neerven and Zijlstra (1992); see also Ellis *et al* (1996) section 9.4. The total cross-section

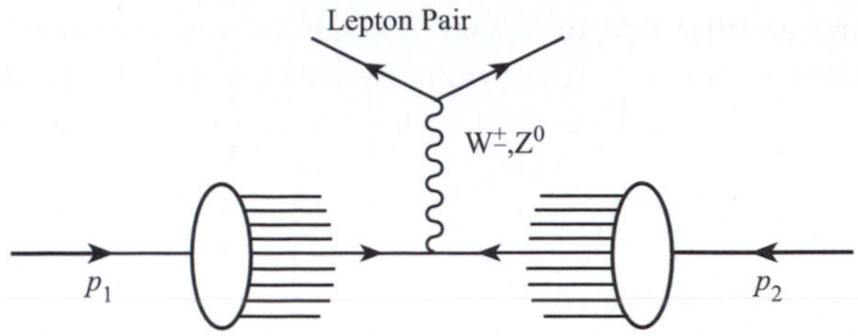


**Figure 22.6.** Parton model amplitude for $W^{\pm}$ or $Z^0$ production in $\bar{p}p$ collisions.

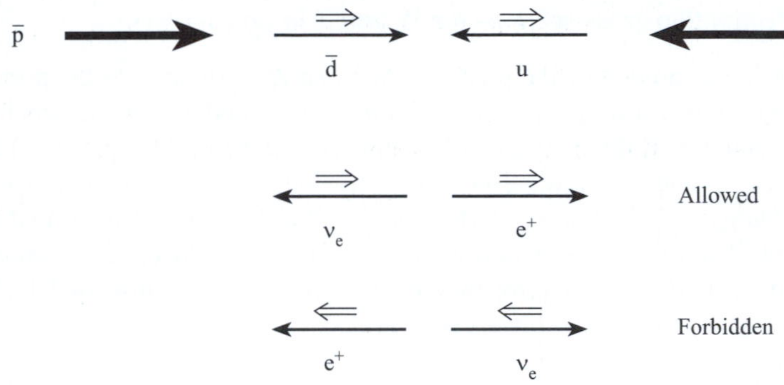


**Figure 22.7.** Preferred direction of leptons in $W^+$ decay.

for production of $W^+$ and $W^-$ at $\sqrt{s} = 630$ MeV is then of order 6.5 nb, while a similar calculation for the $Z^0$ gives about 2 nb. Multiplying these by the branching ratios gives

$$\sigma(p\bar{p} \to W + X \to e\nu X) \simeq 0.7 \text{ nb} \tag{22.84}$$

$$\sigma(p\bar{p} \to Z^0 + X \to e^+e^-X) \simeq 0.07 \text{ nb} \tag{22.85}$$

at $\sqrt{s} = 630$ MeV.

The total cross-section for $p\bar{p}$ is about 70 mb at these energies; hence (22.84) represents $\sim 10^{-8}$ of the total cross-section, and (22.85) is 10 times smaller. The rates could, of course, be increased by using the $q\bar{q}$ modes of W and $Z^0$, which have bigger branching ratios. But the detection of these is very difficult, being very hard to distinguish from conventional two-jet events produced via the mechanism discussed in section 14.3.1, which has a cross-section some $10^3$ higher than (22.84). W and $Z^0$ would appear as slight shoulders on the edge of a very steeply falling invariant mass distribution, similar to that shown in figure 9.13, and the calorimetric jet energy resolution capable of resolving such an effect is hard to achieve. Thus, despite the unfavourable branching ratios, the leptonic modes provide the better signatures, as discussed further in section 22.5.3.

### 22.5.2 Charge asymmetry in $W^{\pm}$ decay

At energies such that the simple valence quark picture of (22.81) is valid, the $W^+$ is created in the annihilation of a left-handed u quark from the proton and a right-handed $\bar{d}$ quark from the $\bar{p}$ (neglecting fermion masses). In the $W^+ \to e^+\nu_e$ decay, a right-handed $e^+$ and a left-handed $\nu_e$ are emitted. Referring to figure 22.7, we see that angular momentum conservation allows $e^+$ production parallel to the direction of the anti-proton but forbids it parallel to the direction of the proton. Similarly, in $W^- \to e^-\bar{\nu}_e$, the $e^-$ is emitted preferentially parallel to the proton (these considerations are exactly similar to those mentioned in section 20.11 with reference to $\nu q$ and $\bar{\nu} q$ scattering). The actual distribution has the form $\sim (1+\cos\theta_e^*)^2$, where $\theta_e^*$ is the angle between the $e^-$ and the p (for $W^- \to e^-\bar{\nu}_e$) or the $e^+$ and the $\bar{p}$ (for $W^+ \to e^+\nu_e$).

### 22.5.3 Discovery of the $W^{\pm}$ and $Z^0$ at the $p\bar{p}$ collider and their properties

As already indicated in section 22.5.1, the best signatures for W and Z production in $p\bar{p}$ collisions are provided by the leptonic modes

$$p\bar{p} \to W^{\pm}X \to e^{\pm}\nu X \tag{22.86}$$

$$p\bar{p} \to Z^0X \to e^+e^-X. \tag{22.87}$$

Reaction (22.86) has the larger cross-section, by a factor of 10 (cf (22.84) and (22.85)) and was observed first (UA1, Arnison *et al* 1983a; UA2, Banner *et al* 1983). However, the kinematics of (22.87) is simpler and so the $Z^0$ discovery (UA1, Arnison *et al* 1983b; UA2, Bagnaia *et al* 1983) will be discussed first.

The signature for (22.87) is, of course, an isolated and approximately back-to-back, $e^+e^-$ pair with invariant mass peaked around 90 GeV (cf (22.57)). Very clean events can be isolated by imposing a modest transverse energy cut—the $e^+e^-$ pairs required come from the decay of a massive relatively slowly moving $Z^0$. Figure 22.8 shows the transverse energy distribution of a candidate $Z^0$ event from the first UA2 sample. Figure 22.9 shows (Geer 1986) the invariant mass distribution for a later sample of 14 UA1 events in which both electrons have well-measured energies, together with the Breit–Wigner resonance curve appropriate to $M_Z = 93\ \mathrm{GeV}/c^2$, with experimental mass resolution folded in. The UA1 result for the $Z^0$ mass was

$$M_Z = 93.0 \pm 1.4(\mathrm{stat}) \pm 3.2(\mathrm{syst.})\ \mathrm{GeV}. \tag{22.88}$$

The corresponding UA2 result (DiLella 1986), based on 13 well-measured pairs, was

$$M_Z = 92.5 \pm 1.3(\mathrm{stat.}) \pm 1.5(\mathrm{syst.})\ \mathrm{GeV}. \tag{22.89}$$

In both cases, the systematic error reflects the uncertainty in the absolute calibration of the calorimeter energy scale. Clearly the agreement with (22.57) is good, but there is a suggestion that the tree-level prediction is on the low side.

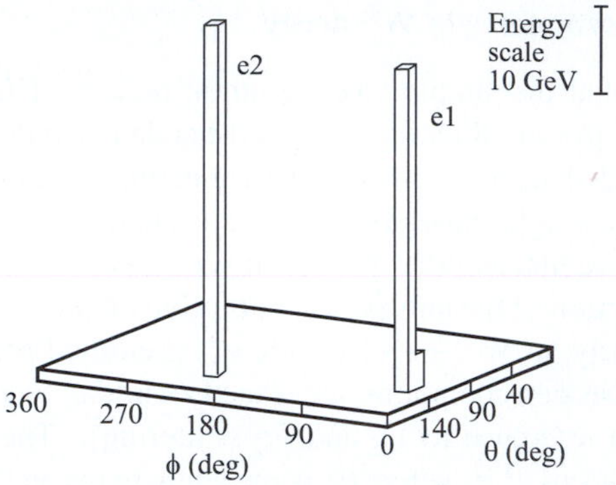


**Figure 22.8.** The cell transverse energy distribution for a $Z^0 \to e^+e^-$ event (UA2, Bagnaia *et al* 1983) in the $\theta$ and $\phi$ plane, where $\theta$ and $\phi$ are the polar and azimuth angles relative to the beam axis.

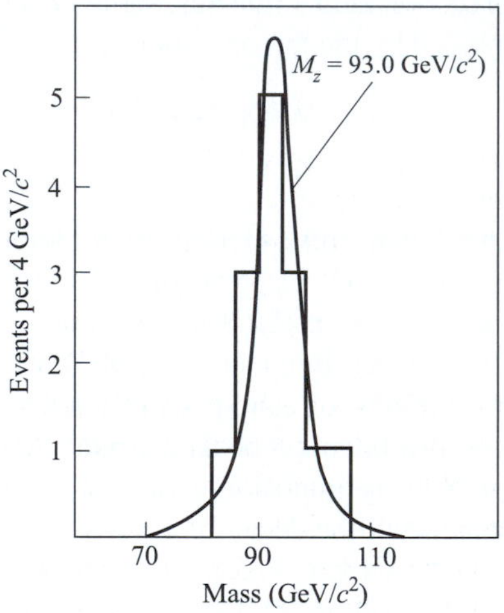


**Figure 22.9.** Invariant mass distribution for 14 well-measured $Z^0 \to e^+e^-$ decay (UA1, Geer 1986).

Indeed, loop corrections adjust (22.57) to a value $M_Z^{\text{th}} \simeq 91.19$ GeV, in excellent agreement with the current experimental value (Hagiwara *et al* 2002).

The total $Z^0$ width $\Gamma_Z$ is an interesting quantity. If we assume that, for any fermion family additional to the three known ones, only the neutrinos are significantly less massive than $M_Z/2$, we have

$$\Gamma_Z \simeq (2.5 + 0.16\Delta N_\nu) \text{ GeV} \tag{22.90}$$

from section 22.4, where $\Delta N_\nu$ is the number of additional light neutrinos (i.e. beyond $\nu_e$, $\nu_\mu$ and $\nu_\tau$) which contribute to the width through the process $Z^0 \to$

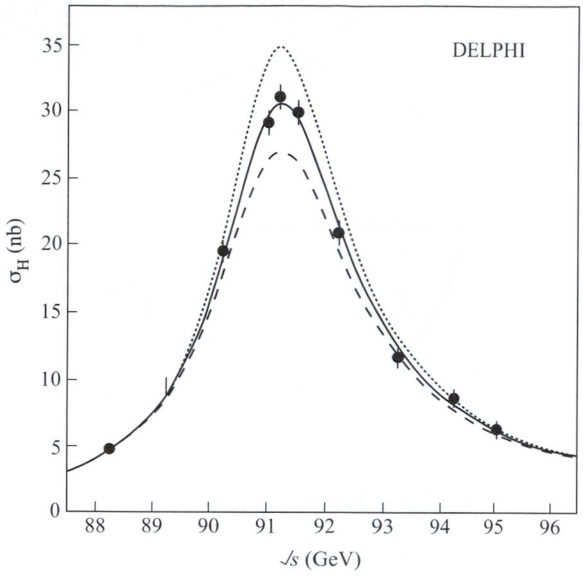


**Figure 22.10.** The cross-section for $e^+e^- \rightarrow$ hadrons around the $Z^0$ mass (DELPHI collaboration, Abreu *et al* 1990). The dotted, full and dashed lines are the predictions of the Standard Model assuming two, three and four massless neutrino species respectively. (From Abe, 1991.)

$\nu\bar{\nu}$. Thus (22.90) can be used as an important measure of such neutrinos (i.e. generations) if $\Gamma_Z$ can be determined accurately enough. The mass resolution of the $\bar{p}p$ experiments was of the same order as the total expected $Z^0$ width, so that (22.90) could not be used directly. The advent of LEP provided precision checks on (22.90): at the cost of departing from the historical development, we show data from DELPHI (Abreu *et al* 1990, Abe 1991) in figure 22.10, which established $N_\nu = 3$.

We turn now to the $W^{\pm}$. In this case an invariant mass plot is impossible, since we are looking for the $e\nu$ ($\mu\nu$) mode, and cannot measure the $\nu$'s. However, it is clear that—as in the case of $Z^0 \rightarrow e^+e^-$ decay—slow moving massive W's will emit isolated electrons with high transverse energy. Further, such electrons should be produced in association with large *missing* transverse energy (corresponding to the $\nu$'s), which can be measured by calorimetry and which should balance the transverse energy of the electrons. Thus, electrons of high $E_T$ accompanied by balancing high missing $E_T$ (i.e. similar in magnitude to that of the $e^-$ but opposite in azimuth) were the signatures used for the early event samples (UA1, Arnison *et al* 1983a; UA2, Banner *et al* 1983).

The determination of the mass of the W is not quite so straightforward as that of the Z, since we cannot construct directly an invariant mass plot for the

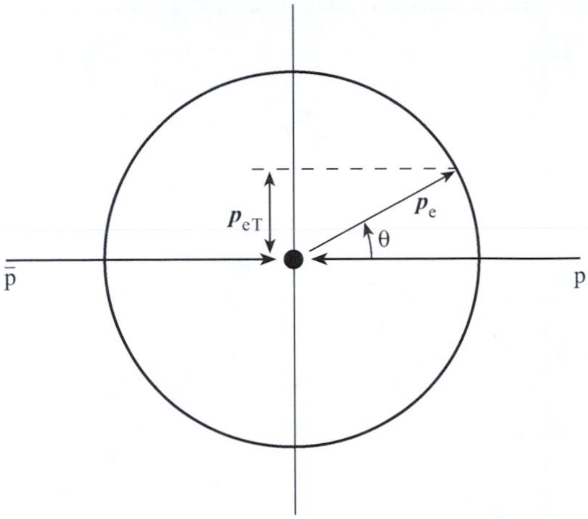


**Figure 22.11.** Kinematics of W → eν decay.

eν pair: only the missing transverse momentum (or energy) can be attributed to the ν, since some unidentified longitudinal momentum will always be lost down the beam pipe. In fact, the distribution of events in $p_{eT}$, the magnitude of the transverse momentum of the $e^-$, should show a pronounced peaking towards the maximum kinematically allowed value, which is $p_{eT} \approx \frac{1}{2}M_W$, as may be seen from the following argument. Consider the decay of a W at rest (figure 22.11). We have $|\boldsymbol{p}_e| = \frac{1}{2}M_W$ and $|\boldsymbol{p}_{eT}| = \frac{1}{2}M_W \sin\theta \equiv p_{eT}$. Thus, the transverse momentum distribution is given by

$$\frac{d\sigma}{dp_{eT}} = \frac{d\sigma}{d\cos\theta}\frac{d\cos\theta}{dp_{eT}} = \frac{d\sigma}{d\cos\theta}\left(\frac{2p_{eT}}{M_W}\right)\left(\frac{1}{4}M_W^2 - p_{eT}^2\right)^{-1/2} \tag{22.91}$$

and the last (Jacobian) factor in (22.91) produces a strong peaking towards $p_{eT} = \frac{1}{2}M_W$. This peaking will be smeared by the width and transverse motion of the W. Early determinations of $M_W$ used (22.91), but sensitivity to the transverse momentum of the W can be much reduced (Barger *et al* 1983) by considering instead the distribution in 'transverse mass', defined by

$$M_T^2 = (E_{eT} + E_{\nu T})^2 - (\boldsymbol{p}_{eT} + \boldsymbol{p}_{\nu T})^2 \simeq 2p_{eT}p_{\nu T}(1 - \cos\phi) \tag{22.92}$$

where $\phi$ is the azimuthal separation between $p_{eT}$ and $p_{\nu T}$. A Monte Carlo simulation was used to generate $M_T$ distributions for different values of $M_W$, and the most probable value was found by a maximum likelihood fit. The quoted results were

$$\text{UA1 (Geer 1986)}: \quad M_W = 83.5 \pm^{1.1}_{1.0} \text{ (stat.)} \pm 2.8 \text{ (syst.) GeV} \tag{22.93}$$

$$\text{UA2 (DiLella 1986)}: \quad M_W = 81.2 \pm 1.1 \text{ (stat.)} \pm 1.3 \text{ (syst.) GeV} \tag{22.94}$$

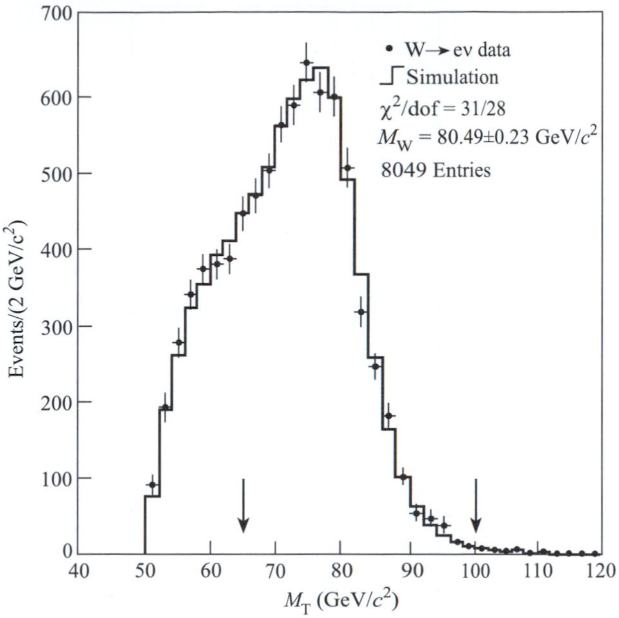


**Figure 22.12.** W $\to$ e$\nu$ transverse mass distribution measured by the CDF collaboration (Abe *et al* 1995a).

the systematic errors again reflecting uncertainty in the absolute energy scale of the calorimeters. The two experiments also quoted (Geer 1986, DiLella 1986)

$$\left.\begin{array}{ll} \text{UA1}: & \Gamma_W < 6.5 \text{ GeV} \\ \text{UA2}: & \Gamma_W < 7.0 \text{ GeV} \end{array}\right\} 90\% \text{ c.l.} \tag{22.95}$$

Once again, the agreement between the experiments and of both with (22.56) is good, the predictions again being on the low side. Loop corrections adjust (22.56) to $M_W \simeq 80.39$ GeV (Hagiwara *et al* 2002). We show in figure 22.12 a more modern determination of $M_W$ by the CDF collaboration (Abe *et al* 1995a).

The W and Z mass values may be used together with (22.20) to obtain $\sin^2\theta_W$ via

$$\sin^2\theta_W = 1 - M_W^2/M_Z^2. \tag{22.96}$$

The weighted average of UA1 and UA2 yielded

$$\sin^2\theta_W = 0.212 \pm 0.022 \text{ (stat.)}. \tag{22.97}$$

Radiative corrections have, in general, to be applied but one renormalization scheme (see section 22.8) promotes (22.96) to a definition of the renormalized $\sin^2\theta_W$ to all orders in perturbation theory. Using this scheme and quoted values of $M_W$ and $M_Z$ (Hagiwara *et al* 2002), one finds that $\sin^2\theta_W \simeq 0.222$.

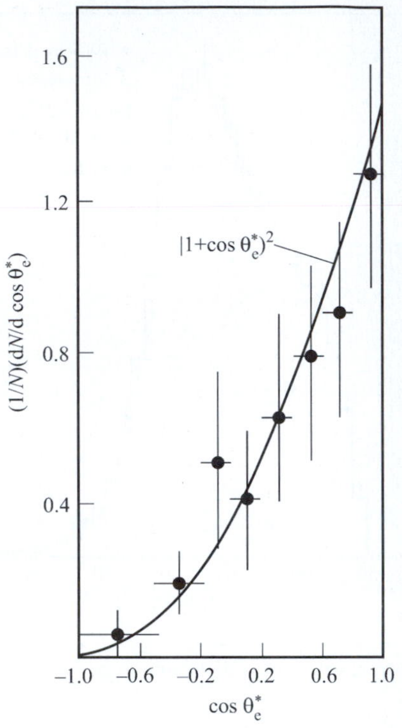


**Figure 22.13.** The W decay angular distribution of the emission angle $\theta_e^*$ of the positron (electron) with respect to the anti-proton (proton) beam direction, in the rest frame of the W, for a total of 75 events; background subtracted and acceptance corrected.

Finally, figure 22.13 shows (Arnison *et al* 1986) the angular distribution of the charged lepton in $W \to e\nu$ decay (see section 22.5.2); $\theta_e^*$ is the $e^+(e^-)$ angle in the W rest frame, measured with respect to a direction parallel (anti-parallel) to the $\bar{p}(p)$ beam. The expected form $(1 + \cos\theta_e^*)^2$ is followed very closely.

In summary, we may say that the early discovery experiments provided remarkably convincing confirmation of the principal expectations of the GSW theory, as outlined in the preceding sections.

We now consider some further aspects of the theory.

## 22.6 The fermion mass problem

The fact that the $SU(2)_L$ gauge group acts only on the L components of the fermion fields, immediately appears to create a fundamental problem as far as the *masses* of these particles are concerned: we mentioned this briefly at the end of section 19.6. Let us recall first that the standard way to introduce the interactions of gauge fields with matter fields (e.g. fermions) is via the covariant derivative replacement

$$\partial^{\mu} \to D^{\mu} \equiv \partial^{\mu} + ig\boldsymbol{\tau} \cdot \boldsymbol{W}^{\mu}/2 \tag{22.98}$$

for SU(2) fields $\boldsymbol{W}^{\mu}$ acting on $t = 1/2$ doublets. Now it is a simple exercise (compare problem 18.3) to check that the ordinary 'kinetic' part of a free Dirac fermion does not mix the L and R components of the field:

$$\bar{\hat{\psi}} \not\partial \hat{\psi} = \bar{\hat{\psi}}_{\mathrm{R}} \not\partial \psi_{\mathrm{R}} + \bar{\hat{\psi}}_{\mathrm{L}} \not\partial \psi_{\mathrm{L}}. \tag{22.99}$$

Thus we can, in principle, contemplate 'gauging' the L and the R components differently. Of course, in the case of QCD (cf (18.39)) the replacement $\not\partial \to \not D$ was made equally in each term on the right-hand side of (22.99) but this was because QCD conserves parity and must, therefore, treat L and R components the same. Weak interactions are parity-violating and the $\mathrm{SU(2)_L}$ covariant derivative acts only in the *second* term of (22.99). However, a Dirac mass term has the form

$$-m(\bar{\hat{\psi}}_{\mathrm{L}} \hat{\psi}_{\mathrm{R}} + \bar{\hat{\psi}}_{\mathrm{R}} \hat{\psi}_{\mathrm{L}}) \tag{22.100}$$

(see equation (18.41) for example) and it precisely *couples* the L and R components. It is easy to see that if only $\hat{\psi}_{\mathrm{L}}$ is subject to a transformation of the form (22.3), then (22.100) is not invariant. Thus, mass terms for Dirac fermions will *explicitly* break $\mathrm{SU(2)_L}$. The same is also true for Majorana fermions (see appendix P) which might describe the neutrinos.

This kind of explicit breaking of the gauge symmetry cannot be tolerated, in the sense that it will lead, once again, to violations of unitarity and then of renormalizability. Consider, for example, a fermion–anti-fermion annihilation process of the form

$$f\bar{f} \to \mathrm{W}_0^+ \mathrm{W}_0^-, \tag{22.101}$$

where the subscript indicates the $\lambda = 0$ (longitudinal) polarization state of the $\mathrm{W}^{\pm}$. We studied such a reaction in section 21.3 in the context of unitarity violations (in lowest-order perturbation theory) for the IVB model. Appelquist and Chanowitz (1987) considered first the case in which '$f$' is a lepton with $t = \frac{1}{2}$, $t_3 = -\frac{1}{2}$ coupling to W's, $\mathrm{Z}^0$ and $\gamma$ with the usual $\mathrm{SU(2)_L} \times \mathrm{U(1)}$ couplings, but having an explicit (Dirac) mass $m_f$. They found that in the 'right' helicity channels for the leptons ($\lambda = +1$ for $\bar{f}$, $\lambda = -1$ for $f$) the bad high-energy behaviour associated with a fermion-exchange diagram of the form of figure 21.5 was *cancelled* by that of the diagram shown in figure 22.14. The sum of the amplitudes tends to a constant as $s$ (or $E^2$) $\to \infty$. Such cancellations are a feature of gauge theories, as we indicated at the end of section 21.4, and represent one aspect of the renormalizability of the theory. But suppose, following Appelquist and Chanowitz (1987), we examine channels involving the 'wrong' helicity component, for example $\lambda = +1$ for the fermion $f$. Then it is found that the cancellation no longer occurs and we shall ultimately have a 'non-renormalizable' problem on our hands, all over again.

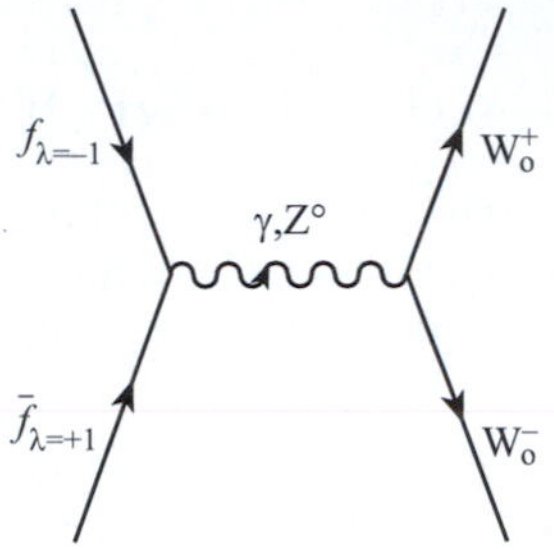


**Figure 22.14.** One-$Z^0$ and one-$\gamma$ annihilation contribution to $f_{\lambda=-1}\bar{f}_{\lambda=1} \to W_0^+W_0^-$.

An estimate of the energy at which this will happen can be made by recalling that the 'wrong' helicity state participates only by virtue of a factor ($m_f$/energy) (recall section 20.4), which here we can take to be $m_f/\sqrt{s}$. The typical bad high-energy behaviour for an amplitude $\mathcal{M}$ was $\mathcal{M} \sim G_F s$, which we expect to be modified here to

$$\mathcal{M} \sim G_F s m_f/\sqrt{s} \sim G_F m_f \sqrt{s}. \tag{22.102}$$

The estimate obtained by Appelquist and Chanowitz differs only by a factor of $\sqrt{2}$. Attending to all the factors in the partial wave expansion gives the result that the unitarity bound will be saturated at $E = E_f$ (TeV) $\sim \pi/m_f$ (TeV). Thus, for $m_t \sim 175$ GeV, $E_t \sim 18$ TeV. This would constitute a serious flaw in the theory, even though the breakdown occurs at energies beyond those currently reachable.

However, in a theory with spontaneous symmetry breaking, there is a way of giving fermion masses without introducing an explicit mass term in the Lagrangian. The linear $\sigma$-model of (18.72) shows how, if a fermion has a 'Yukawa'-type coupling to a scalar field which acquires a vev, then this will generate a fermion mass. Consider the electron, for example, and let us hypothesize such a coupling between the electron-type SU(2) doublet

$$\hat{l}_{eL} = \begin{pmatrix} \hat{\nu}_e \\ \hat{e}^- \end{pmatrix}_L \tag{22.103}$$

the Higgs doublet $\hat{\phi}$ and the R-component of the electron field:

$$\hat{\mathcal{L}}^e_{Yuk} = -g_e(\bar{\hat{l}}_{eL}\hat{\phi}\hat{e}_R + \bar{\hat{e}}_R\hat{\phi}^\dagger\hat{l}_{eL}). \tag{22.104}$$

In each term of (22.104), the two $SU(2)_L$ doublets are 'dotted together' so as to form an $SU(2)_L$ scalar, which multiplies the $SU(2)_L$ scalar R-component. Thus, (22.104) is $SU(2)_L$-invariant, and the symmetry is preserved, at the Lagrangian level, by such a term. But now insert just the vacuum value (22.6) of $\hat{\phi}$ into (22.104): we find the result

$$\hat{\mathcal{L}}^e_{Yuk}(\text{vac}) = -g_e\frac{v}{\sqrt{2}}(\bar{\hat{e}}_L\hat{e}_R + \bar{\hat{e}}_R\hat{e}_L) \tag{22.105}$$

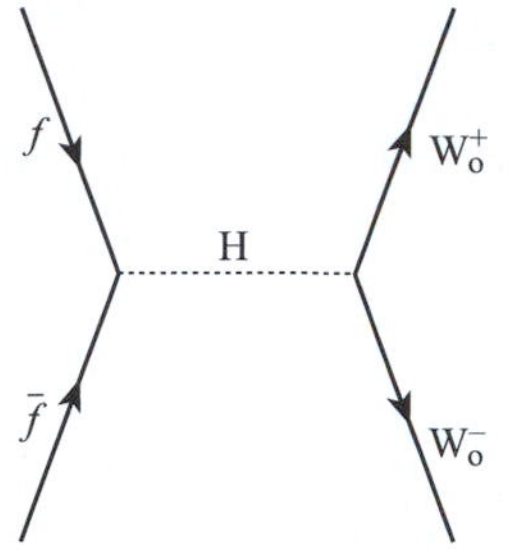


**Figure 22.15.** One-H annihilation graph.

which is exactly a (Dirac) mass of the form (22.100), allowing us to make the identification

$$m_{\mathrm{e}} = g_{\mathrm{e}} v/\sqrt{2}. \tag{22.106}$$

When oscillations about the vacuum value are considered via the replacement (22.7), the term (22.104) will generate a coupling between the electron and the Higgs fields of the form

$$-g_{\mathrm{e}}\bar{\hat{e}}\hat{e}\hat{H}/\sqrt{2} = -(m_{\mathrm{e}}/v)\bar{\hat{e}}\hat{e}\hat{H} \tag{22.107}$$

$$= -(gm_{\mathrm{e}}/2M_{\mathrm{W}})\bar{\hat{e}}\hat{e}\hat{H}. \tag{22.108}$$

Such a coupling, if present for the process $f\bar{f} \to \mathrm{W}_0^+\mathrm{W}_0^-$ considered earlier, will mean that, in addition to the $f$-exchange graph analogous to figure 21.5 and the annihilation graph of figure 22.14, a further graph shown in figure 22.15, must be included. The presence of the fermion mass in the coupling to H suggests that this graph might be just what is required to cancel the 'bad' high-energy behaviour found in (22.102)—and by this time the reader will not be surprised to be told that this is indeed the case.

At first sight it might seem that this stratagem will only work for the $t_3 = -\frac{1}{2}$ components of doublets, because of the form of $\langle 0|\hat{\phi}|0\rangle$. But we learned in section 12.1.3 that if a pair of states $\begin{pmatrix} u \\ d \end{pmatrix}$ forming an SU(2) doublet transform by

$$\begin{pmatrix} u \\ d \end{pmatrix}' = \mathrm{e}^{-\mathrm{i}\boldsymbol{\alpha}\cdot\boldsymbol{\tau}/2}\begin{pmatrix} u \\ d \end{pmatrix} \tag{22.109}$$

then the charge conjugate states $\mathrm{i}\tau_2\begin{pmatrix} u^* \\ d^* \end{pmatrix}$ transform in exactly the same way. Thus if, in our case, $\hat{\phi}$ is the SU(2) doublet

$$\hat{\phi} = \begin{pmatrix} \frac{1}{\sqrt{2}}(\hat{\phi}_1 - \mathrm{i}\hat{\phi}_2) \equiv \hat{\phi}^+ \\ \frac{1}{\sqrt{2}}(\hat{\phi}_3 - \mathrm{i}\hat{\phi}_4) \equiv \hat{\phi}^0 \end{pmatrix} \tag{22.110}$$

then the charge conjugate field

$$\hat{\phi}_{\mathrm{C}} \equiv \mathrm{i}\tau_2\hat{\phi}^* = \begin{pmatrix} \frac{1}{\sqrt{2}}(\phi_3 + \mathrm{i}\phi_4) \\ -\frac{1}{\sqrt{2}}(\phi_1 + \mathrm{i}\phi_2) \end{pmatrix} \equiv \begin{pmatrix} \bar{\hat{\phi}}^0 \\ -\hat{\phi}^- \end{pmatrix} \tag{22.111}$$

is also an SU(2) doublet, transforming in just the same way as $\hat{\phi}$. ((22.110) and (22.111) may be thought of as analogous to the ($\mathrm{K}^+$, $\mathrm{K}^0$) and ($\bar{\mathrm{K}}$, $\mathrm{K}^-$) isospin doublets in $\mathrm{SU(3)_f}$). Note that the vacuum value (22.6) will now appear in the upper component of (22.111). With the help of $\hat{\phi}_{\mathrm{C}}$ we can write down another SU(2)-invariant coupling in the $\nu_{\mathrm{e}}$–e sector, namely

$$-g_{\nu_{\mathrm{e}}}(\bar{\hat{l}}_{\mathrm{eL}}\hat{\phi}_{\mathrm{C}}\hat{\nu}_{\mathrm{eR}} + \bar{\hat{\nu}}_{\mathrm{eR}}\hat{\phi}^{\dagger}_{\mathrm{C}}\hat{l}_{\mathrm{eL}}) \tag{22.112}$$

assuming now the existence of the field $\hat{\nu}_{\mathrm{eR}}$. In the Higgs vacuum (22.6), (22.112) then yields

$$-(g_{\nu_{\mathrm{e}}}v/\sqrt{2})(\bar{\hat{\nu}}_{\mathrm{eL}}\hat{\nu}_{\mathrm{eR}} + \bar{\hat{\nu}}_{\mathrm{eR}}\hat{\nu}_{\mathrm{eL}}) \tag{22.113}$$

which is precisely a (Dirac) mass for the neutrino, if we set $g_{\nu_{\mathrm{e}}}v/\sqrt{2} = m_{\nu_{\mathrm{e}}}$.

It is clearly possible to go on like this and arrange for all the fermions, quarks as well as leptons, to acquire a mass by the same 'mechanism'. We will look more closely at the quarks in the next section. But one must admit to a certain uneasiness concerning the enormous difference in magnitudes represented by the couplings $g_{\nu_{\mathrm{e}}}, \ldots g_{\mathrm{e}}, \ldots g_{\mathrm{t}}$. If $m_{\nu_{\mathrm{e}}} < 1$ eV, then $g_{\nu_{\mathrm{e}}} < 10^{-11}$, while $g_{\mathrm{t}} \sim 1$! Besides, whereas the use of the Higgs field 'mechanism' in the W–Z sector is quite economical, in the present case it seems rather unsatisfactory simply to postulate a different '$g$' for each fermion–Higgs interaction. This does appear to indicate that we are dealing here with a 'phenomenological model', once more, rather than a 'theory'.

As far as the neutrinos are concerned, however, there is another possibility, as indicated in section 20.6, which is that they could be Majorana (not Dirac) fermions. In this case, rather than the four degrees of freedom ($\nu_{\mathrm{eL}}$, $\nu_{\mathrm{eR}}$, and their anti-particles) which exist for massive Dirac particles, only two possibilities exist for neutrinos, which we may take to be $\nu_{\mathrm{eL}}$ and $\nu_{\mathrm{eR}}$. With these, it is certainly possible to construct a Dirac-type mass term of the form (22.113). But since, after all, the $\nu_{\mathrm{eR}}$ component has zero quantum members both for $\mathrm{SU(2)_L}$ W-interactions and for U(1) B-interactions (see table 22.1), we could consider economically dropping it altogether, making do with just the $\nu_{\mathrm{eL}}$ component.

Suppose, then, that we keep only the field $\hat{\nu}_{\mathrm{eL}}$. Its charge-conjugate is defined by (see (20.71))

$$(\hat{\nu}_{\mathrm{eL}})_{\mathrm{C}} = \mathrm{i}\gamma_2\gamma_0\bar{\hat{\nu}}^{\mathrm{T}}_{\mathrm{eL}} = \mathrm{i}\gamma_2\hat{\nu}^{\dagger\mathrm{T}}_{\mathrm{L}}. \tag{22.114}$$

Now we know that the charge-conjugate field transforms under Lorentz transformations in the same way as the original field (see appendix P) and so we can use $(\hat{\nu}_{\mathrm{eL}})_{\mathrm{C}}$ to form a Lorentz invariant

$$\overline{(\hat{\nu}_{\mathrm{eL}})_{\mathrm{C}}}\ \nu_{\mathrm{eL}} \tag{22.115}$$

which has mass dimension $M^3$. Hence, we may write a 'Majorana mass term' in the form

$$-\tfrac{1}{2}m_{\mathrm{M}}[\overline{(\hat{\nu}_{\mathrm{eL}})_{\mathrm{C}}}\,\hat{\nu}_{\mathrm{eL}} + \bar{\hat{\nu}}_{\mathrm{eL}}(\hat{\nu}_{\mathrm{eL}})_{\mathrm{C}}] \tag{22.116}$$

where the $\frac{1}{2}$ is conventional. Written out in more detail, we have

$$\overline{(\hat{\nu}_{\mathrm{eL}})_{\mathrm{C}}}\,\hat{\nu}_{\mathrm{eL}} = \hat{\nu}_{\mathrm{eL}}^{\mathrm{T}}\cdot - \mathrm{i}\gamma_2^{\dagger}\gamma_0\hat{\nu}_{\mathrm{eL}} = \hat{\nu}_{\mathrm{eL}}^{\mathrm{T}}\mathrm{i}\gamma_2\gamma_0\hat{\nu}_{\mathrm{eL}} \tag{22.117}$$

in our representation (20.38). Now

$$\mathrm{i}\gamma_2\gamma_0 = \begin{pmatrix} -\mathrm{i}\sigma_2 & 0 \\ 0 & \mathrm{i}\sigma_2 \end{pmatrix}. \tag{22.118}$$

But since $\hat{\nu}_{\mathrm{eL}}$ is an L-chiral field, only its two lower components are present (cf (20.50)) and (22.117) is effectively

$$\overline{(\hat{\nu}_{\mathrm{eL}})_{\mathrm{C}}}\,\hat{\nu}_{\mathrm{eL}} = \hat{\nu}_{\mathrm{eL}}^{\mathrm{T}}(\mathrm{i}\sigma_2)\hat{\nu}_{\mathrm{eL}}. \tag{22.119}$$

Note that $\mathrm{i}\sigma_2$ is an anti-Hermitian $2 \times 2$ matrix, so that (22.119) would vanish for classical (commuting) fields.

It is at once apparent that the mass term (22.116) is *not* invariant under a global U(1) phase transformation

$$\hat{\nu}_{\mathrm{eL}} \rightarrow \mathrm{e}^{-\mathrm{i}\alpha}\hat{\nu}_{\mathrm{eL}} \tag{22.120}$$

which would correspond to lepton number (if accompanied by a similar transformation for the electron fields). Thus—as, in fact, we already knew—Majorana neutrinos do not carry a lepton number.

There is a further interesting aspect to (22.119) which is that, since two $\hat{\nu}_{\mathrm{eL}}$ operators appear rather than a $\hat{\nu}_{\mathrm{e}}$ and a $\hat{\nu}_{\mathrm{e}}^{\dagger}$ (which would lead to $L_{\mathrm{e}}$ conservation), the $(t, t_3)$ quantum numbers of the term are (1,1). This means that we cannot form an $\mathrm{SU(2)_L}$ invariant with it, using only the Standard Model Higgs $\hat{\phi}$, since the latter has $t = \frac{1}{2}$ and cannot combine with the (1,1) operator to form a singlet. Thus, we cannot make a 'tree-level' Majorana mass by the mechanism of Yukawa coupling to the Higgs field, followed by symmetry breaking.

However, we could generate suitable 'effective' operators via loop corrections, perhaps, much as we generated an effective operator representing an anomalous magnetic moment interaction in QED (cf section 11.7). But whatever it is, the operator would have to violate lepton number conservation, which is actually conserved by all the Standard Model interactions. Thus, such an effective operator could not be generated in perturbation theory. It could arise, however, as a low-energy limit of a theory defined at a higher mass scale, as the current–current model is the low energy limit of the GSW one. The typical form of such operator we need, in order to generate a term $\hat{\nu}_{\mathrm{eL}}^{\mathrm{T}}\mathrm{i}\sigma_2\hat{\nu}_{\mathrm{eL}}$, is

$$-\frac{g_{\mathrm{eM}}}{M}(\bar{\hat{l}}_{\mathrm{eL}}\hat{\phi}_{\mathrm{C}})^{\mathrm{T}}\mathrm{i}\sigma_2(\hat{\phi}_{\mathrm{C}}^{\dagger}\hat{l}_{\mathrm{eL}}). \tag{22.121}$$

Note, most importantly, that the operator '$(l\phi)(\phi l)$' in (22.121) has mass dimension *five*, which is why we introduced the factor $M^{-1}$ in the coupling: it is indeed a non-renormalizable effective interaction, just like the current–current one. We may interpret $M$ as the mass scale at which 'new physics' enters, in the spirit of the discussion in section 11.7. Suppose, for the sake of argument, this was $M \sim 10^{16}$ GeV (a scale typical of Grand Unified Theories). After symmetry breaking, then, (22.121) will generate the required Majorana mass term, with

$$m_{\mathrm{M}} \sim g_{\mathrm{eM}} \frac{v^2}{M} \sim g_{\mathrm{eM}} 10^{-2} \text{ eV}. \tag{22.122}$$

Thus, an effective coupling of 'natural' size $g_{\mathrm{eM}} \sim 1$ emerges from this argument, if indeed the mass of the $\nu_{\mathrm{e}}$ is of order $10^{-2}$ eV. Further discussion of neutrino masses is contained in appendix P, section P.2.

These considerations are tending to take us 'beyond the Standard Model', so we shall not pursue them at any greater length. Instead, we must now generalize the discussion to the three-family case.

## 22.7 Three-family mixing

### 22.7.1 Quark flavour mixing

We introduce three doublets of left-handed fields:

$$\hat{q}_{\mathrm{L}1} = \begin{pmatrix} \hat{u}_{\mathrm{L}1} \\ \hat{d}_{\mathrm{L}1} \end{pmatrix} \qquad \hat{q}_{\mathrm{L}2} = \begin{pmatrix} \hat{u}_{\mathrm{L}2} \\ \hat{d}_{\mathrm{L}2} \end{pmatrix} \qquad \hat{q}_{\mathrm{L}3} = \begin{pmatrix} \hat{u}_{\mathrm{L}3} \\ \hat{d}_{\mathrm{L}3} \end{pmatrix} \tag{22.123}$$

and the corresponding six singlets

$$\hat{u}_{\mathrm{R}1} \qquad \hat{d}_{\mathrm{R}1} \qquad \hat{u}_{\mathrm{R}2} \qquad \hat{d}_{\mathrm{R}2} \qquad \hat{u}_{\mathrm{R}3} \qquad \hat{d}_{\mathrm{R}3} \tag{22.124}$$

which transform in the now familiar way under $\mathrm{SU}(2)_L \times \mathrm{U}(1)$. The $\hat{u}$-fields correspond to the $t_3 = +\frac{1}{2}$ components of $\mathrm{SU}(2)_{\mathrm{L}}$, the $\hat{d}$ ones to the $t_3 = -\frac{1}{2}$ components, and to their 'R' partners. The labels 1, 2 and 3 refer to the family number; for example, with no mixing at all, $\hat{u}_{\mathrm{L}1} = \hat{u}_{\mathrm{L}}, \hat{d}_{\mathrm{L}1} = \hat{d}_{\mathrm{L}}$, etc. We have to consider what is the most general $\mathrm{SU}(2)_{\mathrm{L}}\times$ U(1)-invariant interaction between the Higgs field (assuming we can still get by with only one) and these various fields. Apart from the symmetry, the only other theoretical requirement is renormalizability—for, after all, if we drop this we might as well abandon the whole motivation for the 'gauge' concept. This implies (as in the discussion of the Higgs potential $\hat{V}$) that we cannot have terms like $(\bar{\hat{\psi}}\hat{\psi}\hat{\phi})^2$ appearing—which would have a coupling with dimensions $(\text{mass})^{-4}$ and would be non-renormalizable. In fact, the only renormalizable Yukawa coupling is of the form '$\bar{\hat{\psi}}\hat{\psi}\hat{\phi}$', which has a dimensionless coupling (as in the $g_{\mathrm{e}}$ and $g_{\nu_{\mathrm{e}}}$ of (22.104) and (22.112)). However, there is no *a priori* requirement for it to be 'diagonal' in the weak interaction family index $i$. The allowed generalization of (22.104) and (22.112) is, therefore, an interaction of the form (summing on repeated indices)

$$\hat{\mathcal{L}}_{\psi\phi} = a_{ij}\bar{\hat{q}}_{\mathrm{L}i}\hat{\phi}_{\mathrm{C}}\hat{u}_{\mathrm{R}j} + b_{ij}\bar{\hat{q}}_{\mathrm{L}i}\hat{\phi}\hat{d}_{\mathrm{R}j} + \text{h.c.} \tag{22.125}$$

where 'h.c.' stands for 'Hermitian conjugate',

$$\hat{q}_{\mathrm{L}i} = \begin{pmatrix} \hat{u}_{\mathrm{L}i} \\ \hat{d}_{\mathrm{L}i} \end{pmatrix} \tag{22.126}$$

and a sum on the family indices $i$ and $j$ (from 1 to 3) in (22.125) is assumed. After symmetry breaking, using the gauge (22.7), we find (problem 22.3) that

$$\hat{\mathcal{L}}_{f\phi} = -\left(1 + \frac{\hat{H}}{v}\right)[\bar{\hat{u}}_{\mathrm{L}i}m^{\mathrm{u}}_{ij}\hat{u}_{\mathrm{R}j} + \bar{\hat{d}}_{\mathrm{L}i}m^{\mathrm{d}}_{ij}\hat{d}_{\mathrm{R}j} + \text{h.c.}] \tag{22.127}$$

where the 'mass matrices' are

$$m^{\mathrm{u}}_{ij} = -\frac{v}{\sqrt{2}}a_{ij} \qquad m^{\mathrm{d}}_{ij} = -\frac{v}{\sqrt{2}}b_{ij}. \tag{22.128}$$

Although we have not indicated it, the $m^{\mathrm{u}}$ and $m^{\mathrm{d}}$ matrices could involve a '$\gamma_5$' part as well as a '1' part in Dirac space. It can be shown (Weinberg 1973, Feinberg *et al* 1959) that $m^{\mathrm{u}}$ and $m^{\mathrm{d}}$ can both be made Hermitean, $\gamma_5$-free and diagonal by making four separate unitary transformations on the 'family triplets':

$$\hat{u}_{\mathrm{L}} = \begin{pmatrix} \hat{u}_{\mathrm{L}1} \\ \hat{u}_{\mathrm{L}2} \\ \hat{u}_{\mathrm{L}3} \end{pmatrix} \qquad \hat{d}_{\mathrm{L}} = \begin{pmatrix} \hat{d}_{\mathrm{L}1} \\ \hat{d}_{\mathrm{L}2} \\ \hat{d}_{\mathrm{L}3} \end{pmatrix} \qquad \text{etc} \tag{22.129}$$

via

$$\hat{u}_{\mathrm{L}\alpha} = (U^{(\mathrm{u})}_{\mathrm{L}})_{\alpha i}\hat{u}_{\mathrm{L}i} \qquad \hat{u}_{\mathrm{R}\alpha} = (U^{(\mathrm{u})}_{\mathrm{R}})_{\alpha i}\hat{u}_{\mathrm{R}i} \tag{22.130}$$

$$\hat{d}_{\mathrm{L}\alpha} = (U^{(\mathrm{d})}_{\mathrm{L}})_{\alpha i}\hat{d}_{\mathrm{L}i} \qquad \hat{d}_{\mathrm{R}\alpha} = (U^{(\mathrm{d})}_{\mathrm{R}})_{\alpha i}\hat{d}_{\mathrm{R}i}. \tag{22.131}$$

In this notation, '$\alpha$' is the index of the 'mass diagonal' basis and '$i$' is that of the 'weak interaction' basis.[2] Then (22.127) becomes

$$\hat{\mathcal{L}}_{qH} = -\left(1 + \frac{\hat{H}}{v}\right)[m_{\mathrm{u}}\bar{\hat{u}}\hat{u} + \cdots + m_{\mathrm{b}}\bar{\hat{b}}\hat{b}]. \tag{22.132}$$

Rather remarkably, we can still manage with only the one Higgs field. It couples to each fermion with a strength proportional to the mass of that fermion, divided by $M_{\mathrm{W}}$.

[2] So, for example, $\hat{u}_{\mathrm{L}\alpha=\mathrm{t}} \equiv \hat{t}_{\mathrm{L}}$, $\hat{d}_{\mathrm{L}\alpha=\mathrm{s}} \equiv \hat{s}_{\mathrm{L}}$, etc.

Now consider the $SU(2)_L \times U(1)$ gauge-invariant interaction part of the Lagrangian. Written out in terms of the 'weak interaction' fields $\hat{u}_{L,Ri}$ and $\hat{d}_{L,Ri}$ (cf (22.21) and (22.22)), it is

$$\hat{\mathcal{L}}_{f,W,B} = i(\bar{\hat{u}}_{Lj}, \bar{\hat{d}}_{Lj})\gamma^\mu(\partial_\mu + ig\boldsymbol{\tau}\cdot\hat{\boldsymbol{W}}_\mu/2 + ig'y\hat{B}_\mu/2)\begin{pmatrix}\hat{u}_{Lj}\\ \hat{d}_{Lj}\end{pmatrix}$$
$$+ i\bar{\hat{u}}_{Rj}\gamma^\mu(\partial_\mu + ig'y\hat{B}_\mu/2)\hat{u}_{Rj} + i\bar{\hat{d}}_{Rj}\gamma^\mu(\partial_\mu + ig'y\hat{B}_\mu/2)\hat{d}_{Rj} \tag{22.133}$$

where a sum on $j$ is understood. This now has to be rewritten in terms of the mass-eigenstate fields $\hat{u}_{L,R\alpha}$ and $\hat{d}_{L,R\alpha}$.

Problem 22.4 shows that the neutral current part of (22.133) is diagonal in the mass basis, provided the $U$ matrices of (22.130) and (22.131) are unitary; that is, the neutral current interactions do not change the flavour of the physical (mass eigenstate) quarks. The charged current processes, however, involve the *non*-diagonal matrices $\tau_1$ and $\tau_2$ in (22.133) and this spoils the argument used in problem 22.4. Indeed, using (22.25) we find that the charged current piece is

$$\begin{aligned}\hat{\mathcal{L}}_{CC} &= -\frac{g}{\sqrt{2}}(\bar{\hat{u}}_{Lj}, \bar{\hat{d}}_{Lj})\gamma_\mu\tau_+\hat{W}_\mu\begin{pmatrix}\hat{u}_{Lj}\\ \hat{d}_{Lj}\end{pmatrix} + \text{h.c.}\\ &= -\frac{g}{\sqrt{2}}\bar{\hat{u}}_{Lj}\gamma^\mu\hat{d}_{Lj}\hat{W}_\mu + \text{h.c.}\\ &= -\frac{g}{\sqrt{2}}\bar{\hat{u}}_{L\alpha}[(U_L^{(u)})_{\alpha j}(U_L^{(d)\dagger})_{j\beta}]\gamma^\mu\hat{d}_{L\beta}\hat{W}_\mu + \text{h.c.}\end{aligned} \tag{22.134}$$

where the matrix

$$V_{\alpha\beta} \equiv [U_L^{(u)}U_L^{(d)\dagger}]_{\alpha\beta} \tag{22.135}$$

is not diagonal, though it is unitary. This is the well known CKM matrix (Cabibbo 1963, Kobayashi and Maskawa 1973). The interaction (22.134) then has the form

$$-\frac{g}{\sqrt{2}}\hat{W}_\mu[\bar{\hat{u}}_L\gamma^\mu\hat{d}'_L + \bar{\hat{c}}_L\gamma^\mu\hat{s}'_L + \bar{\hat{t}}_L\gamma^\mu\hat{b}'_L] + \text{h.c.} \tag{22.136}$$

where

$$\begin{pmatrix}\hat{d}'_L\\ \hat{s}'_L\\ \hat{b}'_L\end{pmatrix} = \begin{pmatrix}V_{ud} & V_{us} & V_{ub}\\ V_{cd} & V_{cs} & V_{cb}\\ V_{td} & V_{ts} & V_{tb}\end{pmatrix}\begin{pmatrix}\hat{d}_L\\ \hat{s}_L\\ \hat{b}_L\end{pmatrix}. \tag{22.137}$$

It is important to know how many independent parameters the CKM matrix contains. A general $3\times3$ complex matrix has $2\times3^2=18$ real parameters. The unitarity relation $V^\dagger V = I$ provides nine conditions, namely

$$|V_{ud}|^2 + |V_{cd}|^2 + |V_{td}|^2 = 1 \tag{22.138}$$

together with two other similar diagonal equations, and

$$V_{ub}^*V_{ud} + V_{cb}^*V_{cd} + V_{tb}^*V_{td} = 0, \tag{22.139}$$

and five other similar 'off–diagonal' equations. If all the elements of the matrix were real, it would be orthogonal ($V^{\mathrm{T}}V = I$) and could be thought of as a 'rotation', parametrized by three real numbers. The remaining six parameters in the unitary $V$ are, therefore, phases. The phase of each of the quark fields is arbitrary and may be freely changed (though $\hat{q}_{\mathrm{L}}$ and $\hat{q}_{\mathrm{R}}$ must change in the same way to keep the mass terms in (22.132) real). This would seem to give enough freedom to redefine away six phases from $V$ as defined in (22.135). However, we can see that $V$ is left invariant if we change all the quarks by the same phase. Thus, there are really only five removable phases, and $V$ may therefore be parametrized in terms of three real 'rotation angles' and one phase. The standard parametrization of $V$ (Hagiwara *et al* 2002) is

$$V = \begin{pmatrix} c_{12}c_{13} & s_{12}c_{13} & s_{13}e^{-\mathrm{i}\delta_{13}} \\ -s_{12}c_{23} - c_{12}s_{23}s_{13}\mathrm{e}^{\mathrm{i}\delta_{13}} & c_{12}c_{23} - s_{12}s_{23}s_{13}\mathrm{e}^{\mathrm{i}\delta_{13}} & s_{23}c_{13} \\ s_{12}s_{23} - c_{12}c_{23}s_{13}\mathrm{e}^{\mathrm{i}\delta_{13}} & -c_{12}s_{23} - s_{12}c_{23}s_{13}\mathrm{e}^{\mathrm{i}\delta_{13}} & c_{23}c_{13} \end{pmatrix} \tag{22.140}$$

where $c_{ij} = \cos\theta_{ij}$ and $s_{ij} = \sin\theta_{ij}$.

Before proceeding further, it is helpful to note the 90% confidence limits on the elements of the matrix $V$ as quoted by Hagiwara *et al* (2002), taking into account the unitarity constraints:

$$\begin{pmatrix} (0.9741-0.9756) & (0.219-0.226) & (0.0025-0.0048) \\ (0.219-0.226) & (0.9732-0.9748) & (0.038-0.044) \\ (0.004-0.014) & (0.037-0.044) & (0.9990-0.9993) \end{pmatrix}. \tag{22.141}$$

From this it follows that the mixing angles are small and, moreover, satisfy a definite hierarchy

$$1 \gg \theta_{12} \gg \theta_{23} \gg \theta_{13}. \tag{22.142}$$

Thus a small-angle approximation to the angles $\theta_{ij}$ in (22.140) is often satisfactory. This leads to a parametrization due to Wolfenstein (1983). Identifying $s_{12} = \lambda$, we write $V_{\mathrm{cb}} \simeq s_{23} = A\lambda^2$ and $V_{\mathrm{ub}} = s_{13}\mathrm{e}^{-\mathrm{i}\delta_{13}} = A\lambda^3(\rho - \mathrm{i}\eta)$ with $A \simeq 1$ and $|\rho - \mathrm{i}\eta| < 1$. This gives

$$V \simeq \begin{pmatrix} 1-\lambda^2/2 & \lambda & A\lambda^3(\rho-\mathrm{i}\eta) \\ -\lambda & 1-\lambda^2/2 & A\lambda^2 \\ A\lambda^3(1-\rho-\mathrm{i}\eta) & -A\lambda^2 & 1 \end{pmatrix} \tag{22.143}$$

retaining terms up to $O(\lambda^3)$.

From (22.141) we see that the elements which are least well determined are $V_{\mathrm{td}}$ and $V_{\mathrm{ub}}$. We may use the unitarity relation (22.139), with $V_{\mathrm{ud}} \approx 1 \approx V_{\mathrm{tb}}$, to relate these two elements by

$$V_{\mathrm{ub}}^* + V_{\mathrm{td}} \simeq s_{12}V_{\mathrm{cb}}^*. \tag{22.144}$$

This relation may be represented as a triangle in the complex plane as shown in figure 22.16 where, without loss of generality, $V_{\mathrm{cd}}V_{\mathrm{cb}}^*$ has been chosen to lie along

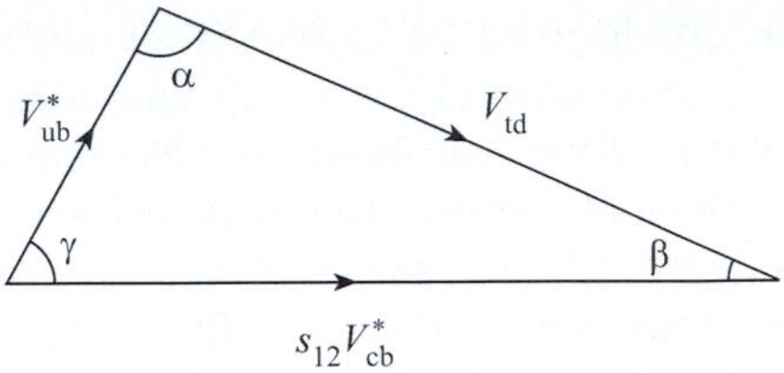


**Figure 22.16.** The 'unitarity triangle' representing equation (22.144).

the horizontal. This is known as the 'unitarity triangle'. The angles $\beta$ and $\gamma$ are defined by

$$V_{td} = |V_{td}|e^{-i\beta} \qquad V_{ub} = |V_{ub}|e^{-i\gamma} \tag{22.145}$$

and from (22.140) $\gamma = \delta_{13}$. The triangle can also be rescaled so that its base is of unit length: the apex is then at $\bar{\rho} = \rho(1 - \lambda^2/2)$, $\bar{\eta} = \eta(1 - \lambda^2/2)$ in terms of the parameters in (22.143).

The existence of the phase $\delta_{13}$ represents a fundamental difference from the simple $2 \times 2$ mixing of the original Cabibbo–GIM type considered in section 20.10. In that case, the $2 \times 2$ mixing matrix (following the same counting of parameters as before) can have only one real parameter, the Cabibbo (GIM) angle $\theta$. This corresponds to setting all angles $\theta_{i3}$ to zero in (22.140). The reason that the appearance of the phase $\delta_{13}$ in (22.140) is so significant is that it implies (if $\delta_{13} \neq 0$) that **CP** is violated. The action of $\hat{\mathbf{P}}$ and $\hat{\mathbf{C}}$ on fermionic fields was discussed in sections 20.3 and 20.5:

$$\hat{\mathbf{P}}\hat{\psi}(\boldsymbol{x}, t)\hat{\mathbf{P}}^{-1} = \gamma_0\hat{\psi}(-\boldsymbol{x}, t) \tag{22.146}$$

$$\hat{\mathbf{C}}\hat{\psi}(\boldsymbol{x}, t)\hat{\mathbf{P}}^{-1} = i\gamma^2\hat{\psi}^{\dagger T}(\boldsymbol{x}, t). \tag{22.147}$$

Hence,

$$\hat{\mathbf{C}}\hat{\mathbf{P}}\hat{\psi}(\boldsymbol{x}, t)(\hat{\mathbf{C}}\hat{\mathbf{P}})^{-1} = i\gamma^2\gamma^0\hat{\psi}^{\dagger T}(-\boldsymbol{x}, t). \tag{22.148}$$

We also have

$$\hat{\mathbf{C}}\hat{\mathbf{P}}\hat{\boldsymbol{W}}_\mu(\boldsymbol{x}, t)(\hat{\mathbf{C}}\hat{\mathbf{P}})^{-1} = \hat{\boldsymbol{W}}^\dagger_\mu(-\boldsymbol{x}, t). \tag{22.149}$$

Careful application of these relations then shows (problem 22.5) that (for example)

$$(\hat{\mathbf{C}}\hat{\mathbf{P}})\bar{\hat{u}}_L V_{ud}\gamma^\mu\hat{d}_L\hat{\boldsymbol{W}}_\mu(\hat{\mathbf{C}}\hat{\mathbf{P}})^{-1} = \bar{\hat{d}}_L V_{ud}\gamma^\mu\hat{u}_L\hat{\boldsymbol{W}}^\dagger_\mu \tag{22.150}$$

where the fact that the fields are evaluated at $\boldsymbol{x}$ will not matter since we integrate over $\boldsymbol{x}$ in the action. But the corresponding term from the 'h.c.' part in (22.136) is

$$\bar{\hat{d}}_L V^*_{ud}\gamma^\mu\hat{u}_L\hat{\boldsymbol{W}}^\dagger_\mu. \tag{22.151}$$

It follows that, for **CP** to be conserved, the entries in the $V$-matrix must be real. With three families (though not with only two) the possibility exists of having a

**CP**-violating phase parameter $\delta_{13}$, as was stressed by Kobayashi and Maskawa (1973).

In fact, **CP** violation in the decays of neutral K mesons is an intensively studied phenomenon, since its discovery in 1964 (Christenson *et al* 1964). In the Standard Model, it is interpreted as arising from the phases in the $V_{\mathrm{ub}}$ and $V_{\mathrm{td}}$ elements (see (22.143)). It is, however, a very small effect, as seen from the magnitudes of $V_{\mathrm{ub}}$ and $V_{\mathrm{td}}$ in (22.143) (they involve third-generation mixing and are $O(\lambda^3)$). It can be shown (Jarlskog 1985a, b) that **CP**-violating amplitudes, or differences of rates, are all proportional to the quantity

$$s_{12}s_{13}s_{23}c_{12}c_{13}^2c_{23}\sin\delta_{13} \tag{22.152}$$

which is just twice the area of the unitarity triangle. For a full introduction to the phenomenology of **CP** violation we refer the reader to Leader and Predazzi (1996, chapters 18 and 19), and to Branco *et al* (1999) and Bigi and Sanda (2000).

Larger effects are expected in the $\mathrm{B}^0$ system, which is a major motivation for the construction of B factories. As an indication, present experimental results from Babar (Aubert *et al* 2002) and Belle (Abe *et al* 2002) yield

$$\sin 2\beta = 0.78 \pm 0.08 \tag{22.153}$$

when averaged.

According to the summary in the current *Review of Particle Properties* (Hagiwara *et al* 2002) all processes can be quantitatively understood by one value of the CKM phase $\delta_{13}(=\gamma) = 59° \pm 13°$. The value of $\beta = 24° \pm 4°$ as determined from the overall fit is consistent with (22.153).

### 22.7.2 Neutrino flavour mixing

An analysis similar to the previous one can be carried out in the leptonic sector. We would then have leptonic flavour mixing in charged current processes, via interactions of the form

$$\bar{\hat{\nu}}_{\mathrm{L}\alpha}V_{\alpha\beta}^{(l)}\gamma^\mu\hat{e}_{\mathrm{L}\beta}\hat{W}_\mu + \mathrm{h.c.} \tag{22.154}$$

where $V^{(l)}$ is the leptonic analogue of the CKM matrix, namely the MNS matrix (Maki *et al* 1962; see also Pontecorvo 1967). We would also have lepton mass terms $(\bar{\hat{\nu}}_{\mathrm{L}i}m_{ij}^{\nu}\hat{\nu}_{\mathrm{R}j} + \mathrm{h.c.})$ and $(\bar{\hat{e}}_{\mathrm{L}i}m_{ij}^{\mathrm{e}}\hat{e}_{\mathrm{R}j} + \mathrm{h.c.})$ in analogy with (22.127). If the neutrino mass matrix $m^\nu$ was identically zero, or if—improbably—its eigenvalues were all equal, we would be free to redefine $\hat{\nu}_{\mathrm{L}}$ and $\hat{\nu}_{\mathrm{R}}$ by

$$\hat{\hat{\nu}}_{\mathrm{L}} = V^{(l)\dagger}\hat{\nu}_{\mathrm{L}} \qquad \hat{\hat{\nu}}_{\mathrm{R}} = V^{(l)\dagger}\hat{\nu}_{\mathrm{R}} \tag{22.155}$$

so as to reduce the charged current term to family-diagonal form. It is these ˜states that we would identify with the physical neutrino states, in that case. But it is now clear, experimentally, that neutrino flavour mixing does take place, indicating that neutrinos do have (different) masses.

The reader will recall that it is an open question whether neutrinos are Dirac or Majorana particles (sections 20.6 and 22.6). If they are of Majorana type, this has an interesting consequence for the parametrization of the matrix $V^{(l)}$. We saw following equation (22.119) that global phase transformations (ordinarily corresponding to a number conservation law) cannot be freely made on Majorana fields, as they carry no lepton number. Thus, two of the phases which could be removed from the $3 \times 3$ quark mixing matrix $V$, cannot be removed from the leptonic analogue matrix $V^{(l)}$ if the neutrinos are Majorana particles. In the Majorana case, the matrix $V^{(l)}$ can be parametrized as

$$V^{(l)}(\text{Majorana}) = V(\text{CKM type}) \times \text{diag}(e^{i\alpha_1/2}, e^{i\alpha_2/2}, 1) \tag{22.156}$$

and we have *three* **CP**-violating phases.

Because the neutrino mass differences are (apparently) so small, quantum-mechanical oscillations between neutrinos of different (leptonic) flavour can be observed to occur over macroscopic distances. The subject is extensively covered in a number of books and reviews, for example Mohapatra and Pal (1991), Boehm and Vogel (1987), Kayser *et al* (1989), Bahcall (1989), Bilenky (2000) and Kayser in Hagiwara *et al* (2002). We note that the extra **CP**-violating phases in (22.156) do not affect neutrino oscillations but do affect the rate for neutrinoless double $\beta$-decay (see section 20.6).

## 22.8 Higher-order corrections

The $Z^0$ mass is presently (2002) determined to be

$$M_Z = 91.1876 \pm 0.0021 \text{ GeV} \tag{22.157}$$

from the Z lineshape at LEP1 (Tournefier 2001). The W mass (CDF: Affolder *et al* (2001); D0: Abbott *et al* (2000); UA2: Alitti *et al* (1992)) is

$$M_W = 80.451 \pm 0.061 \text{ GeV}. \tag{22.158}$$

The asymmetry parameter $A_e$ (see (22.79)) is (Abe *et al* 2000)

$$A_e = 0.15138 \pm 0.00216 \tag{22.159}$$

from measurements at SLD. These are just three examples from the table of 35 observables listed in the review of the electroweak model by Erler and Langacker in Hagiwara *et al* (2002). Such remarkable precision is a triumph of machine design and experimental art—and it is the reason why we need a renormalizable electroweak theory. The overall fit to the data, including higher-order corrections, is quoted by Erler and Langacker as $\chi^2$/degree of freedom

$= 47.3/38$. The probability of a larger $\chi^2$ is 14%: one of the major discrepancies is a $3.2\sigma$ deviation in the hadronic charge asymmetry $(3/4)A_e A_b$; another is a $2.5\sigma$ deviation in the muon anomalous magnetic moment, $g_\mu - 2$. This reasonably strong numerical consistency lends impressive support to the belief that we are indeed dealing with a renormalizable spontaneously broken gauge theory, *because no extra parameters, not in the original Lagrangian, have had to be introduced.*

In fact, one can turn this around, in more than one way. First, the one remaining unobserved element in the theory—the Higgs boson—has a mass $M_H$ which is largely unconstrained by theory (see section 22.10.2) and it is therefore a parameter in the fits. Some information about $M_H$ can therefore be gained by seeing how the fits vary with $M_H$. Actually, we shall see in equation (22.174) that the dependence on $M_H$ is only logarithmic—it acts rather like a cut-off, so the fits are not very sensitive to $M_H$. By contrast, some loop corrections are proportional to the square of the top mass (see (22.173)), and consequently very tight bounds could be placed on $m_t$ via its *virtual* presence (in loops) before its *real* presence was confirmed, as we shall discuss shortly and in section 22.9. Second, very careful analysis of small discrepancies between precision data and electroweak predictions may indicate the presence of 'new physics'.

After all this (and earlier) emphasis on the renormalizability of the electroweak theory, and the introduction to one-loop calculations in QED at the end of volume 1, the reader perhaps now has a right to expect an exposition of loop corrections in the electroweak theory. But the fact is that this is a very complicated and technical story, requiring quite a bit more formal machinery which would be outside the intended scope of this book (suitable references include Altarelli *et al* (1989), especially the pedagogical account by Consoli *et al* (1989); and the equally approachable lectures by Hollik (1991)). Instead, we want to touch on just a few of the simpler and more important aspects of one-loop corrections, especially insofar as they have phenomenological implications.

As we have seen, we obtain cut-off independent results from loop corrections in a renormalizable theory by taking the values of certain parameters—those appearing in the original Lagrangian—from experiment, according to a well-defined procedure ('renormalization scheme'). In the electroweak case, the parameters in the Lagrangian are

$$\text{gauge couplings } g, g' \tag{22.160}$$
$$\text{Higgs potential parameters } \lambda, \mu^2 \tag{22.161}$$
$$\text{Higgs–fermion Yukawa couplings } g_f \tag{22.162}$$
$$\text{CKM angles } \theta_{12}, \theta_{13}, \theta_{23} \qquad \text{phase } \delta_{13} \tag{22.163}$$
$$\text{MNS angles } \theta^\nu_{12}, \theta^\nu_{13}, \theta^\nu_{23} \qquad \text{phase } \delta^\nu_{13}(+\alpha_1, \alpha_2?). \tag{22.164}$$

The fermion masses and mixings, and the Higgs mass, can be separated off, leaving $g$, $g'$ and one combination of $\lambda$ and $\mu^2$ (for instance, the tree-level vacuum value $v$). These three parameters are usually replaced by the equivalent and more convenient set

$$\alpha \quad \text{(Mohr and Taylor 2000)} \tag{22.165}$$

$$G_{\text{F}} \quad \text{(Marciano and Sirlin 1988, van Ritbergen and Stuart 1999)} \tag{22.166}$$

$$M_{\text{Z}} \text{ (Tournefier 2001).} \tag{22.167}$$

These are, of course, related to $g$, $g'$ and $v$; for example, at tree level

$$\alpha = g^2 g'^2/(g^2 + g'^2)4\pi \qquad M_{\text{Z}} = \frac{1}{2}v\sqrt{g^2 + g'^2} \qquad G_{\text{F}} = \frac{1}{\sqrt{2}v^2} \tag{22.168}$$

but these relations become modified in higher order. The renormalized parameters will 'run' in the way described in chapters 15 and 16: the running of $\alpha$, for example, has been observed directly (TOPAZ: Levine *et al* 1997; VENUS: Okada *et al* 1998; OPAL: Abbiendi *et al* 2000; L3: Acciari *et al* 2000).

After renormalization, one can derive radiatively-corrected values for physical quantities in terms of the set (22.165)–(22.167) (together with $M_{\text{H}}$ and the fermion masses and mixings). But a renormalization scheme has to be specified, at any finite order (though, in practice, the differences are very small). One conceptually simple scheme is the 'on-shell' one (Sirlin 1980, 1984, Kennedy *et al* 1989, Kennedy and Lynn 1989, Bardin *et al* 1989, Hollik 1990; for reviews see Langacker 1995). In this scheme, the tree-level formula

$$\sin^2\theta_{\text{W}} = 1 - M_{\text{W}}^2/M_{\text{Z}}^2 \tag{22.169}$$

is promoted into a *definition* of the renormalized $\sin^2\theta_{\text{W}}$ to all orders in perturbation theory, it being then denoted by $s_{\text{W}}^2$:

$$s_{\text{W}}^2 = 1 - M_{\text{W}}^2/M_{\text{Z}}^2. \tag{22.170}$$

The radiatively-corrected value for $M_{\text{W}}$ is then

$$M_{\text{W}}^2 = \frac{(\pi\alpha/\sqrt{2}G_{\text{F}})}{s_{\text{W}}^2(1-\Delta r)} \tag{22.171}$$

where $\Delta r$ includes the radiative corrections relating $\alpha$, $\alpha(M_{\text{Z}})$, $G_{\text{F}}$, $M_{\text{W}}$ and $M_{\text{Z}}$. Another scheme is the modified minimal subtraction ($\overline{\text{MS}}$) scheme (section 15.5) which introduces the quantity $\sin^2\hat{\theta}_{\text{W}}(\mu) \equiv \hat{g}'^2(\mu)/[\hat{g}'^2(\mu) + \hat{g}^2(\mu)]$ where the couplings $\hat{g}$ and $\hat{g}'$ are defined in the $\overline{\text{MS}}$ scheme and $\mu$ is chosen to be $M_{\text{Z}}$ for most electroweak processes. Attention is then focused on $\hat{s}_{\text{Z}}^2 \equiv \sin^2\hat{\theta}_{\text{W}}(M_{\text{Z}})$. This is the scheme used by Erler and Langacker in Hagiwara *et al* (2002).

We shall continue here with the scheme defined by (22.170). We cannot go into detail about all the contributions to $\Delta r$ but we do want to highlight two features of the result—which are surprising, important phenomenologically, and related to an interesting symmetry. It turns out (Consoli *et al* 1989, Hollik 1991) that the leading terms in $\Delta r$ have the form

$$\Delta r = \Delta r_0 - \frac{(1 - s_{\mathrm{W}}^2)}{s_{\mathrm{W}}^2} \Delta\rho + (\Delta r)_{\mathrm{rem}}. \tag{22.172}$$

In (22.172), $\Delta r_0 = 1 - \alpha/\alpha(M_Z)$ is due to the running of $\alpha$, and has the value $\Delta r_0 = 0.0664(2)$ (see section 11.5.3). $\Delta\rho$ is given by (Veltman 1977)

$$\Delta\rho = \frac{3G_{\mathrm{F}}(m_{\mathrm{t}}^2 - m_{\mathrm{b}}^2)}{8\pi^2\sqrt{2}} \tag{22.173}$$

while the 'remainder' $(\Delta r)_{\mathrm{rem}}$ contains a significant term proportional to $\ln(m_{\mathrm{t}}/m_Z)$, and a contribution from the Higgs boson which is (for $m_{\mathrm{H}} \gg M_{\mathrm{W}}$)

$$(\Delta r)_{\mathrm{rem,H}} \approx \frac{\sqrt{2}G_{\mathrm{F}}M_{\mathrm{W}}^2}{16\pi^2} \frac{11}{3} \left[ \ln\left(\frac{m_{\mathrm{H}}^2}{M_{\mathrm{W}}^2}\right) - \frac{5}{6} \right]. \tag{22.174}$$

As the notation suggests, $\Delta\rho$ is a leading contribution to the parameter $\rho$ introduced in (22.44). As explained there, it measures the strength of neutral current processes relative to charged current ones. $\Delta\rho$ is then a radiative correction to $\rho$. It turns out that, to good approximation, electroweak radiative corrections in $e^+e^- \to Z^0 \to f\bar{f}$ can be included by replacing the fermionic couplings $g_{\mathrm{V}}^f$ and $g_{\mathrm{A}}^f$(see (22.42), (22.53) and (22.54)) by

$$\bar{g}_{\mathrm{V}}^f = \sqrt{\rho_f}(t_3^{(f)} - 2Q_f\kappa_f s_{\mathrm{W}}^2) \tag{22.175}$$

and

$$\bar{g}_{\mathrm{A}}^f = \sqrt{\rho_f} t_3^{(f)} \tag{22.176}$$

together with corrections to the $Z^0$ propagator. The corrections have the form (in the on-shell scheme) $\rho_f \approx 1 + \Delta\rho$ (of equation (22.173)) and $\kappa_f \approx 1 + [s_{\mathrm{W}}^2/(1-s_{\mathrm{W}}^2)]\Delta\rho$, for $f \neq \mathrm{b}, \mathrm{t}$. For the b-quark, there is an additional contribution coming from the presence of the virtual top quark in vertex corrections to $Z \to \mathrm{b}\bar{\mathrm{b}}$ (Akhundov *et al* 1986, Beenakker and Hollik 1988).

The running of $\alpha$ in $\Delta r_0$ is expected, but (22.173) and (22.174) contain surprising features. As regards (22.173), it is associated with top–bottom quark loops in vacuum polarization amplitudes, of the kind discussed for $\bar{\Pi}_\gamma^{[2]}$ in section 11.5 but this time in weak boson propagators. In the QED case, referring to equation (11.38) for example, we saw that the contribution of heavy fermions '$(|q^2| \ll m_f^2)$' was suppressed, appearing as $O(|q^2|/m_f^2)$. In such a situation (which is the usual one), the heavy particles are said to 'decouple'. But the correction (22.173) is quite different, the fermion masses being in the *numerator*. Clearly, with a large value of $m_{\mathrm{t}}$, this can make a relatively big difference. This is why some precision measurements are surprisingly sensitive to the value of

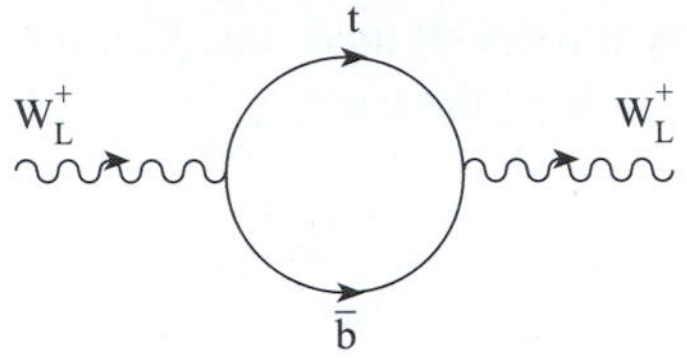


**Figure 22.17.** t–$\bar{\text{b}}$ vacuum polarization contribution.

$m_t$ in the range near (as we now know) the physical value. Second, as regards the dependence on $m_H$, we might well have expected it to involve $m_H^2$ in the numerator if we considered the typical divergence of a scalar particle in a loop (we shall return to this after discussing (22.173)). $\Delta r$ would then have been very sensitive to $m_H$ but, in fact, the sensitivity is only logarithmic.

We can understand the appearance of the fermion masses (squared) in the numerator of (22.173) as follows. The shift $\Delta\rho$ is associated with vector boson vacuum polarization contributions, for example the one shown in figure 22.17. Consider, in particular, the contribution from the longitudinal polarization components of the W's. As we have seen, these components are nothing but three of the four Higgs components which the $W^\pm$ and $Z^0$ 'swallowed' to become massive. But the couplings of these 'swallowed' Higgs fields to fermions are determined by just the same Higgs–fermion Yukawa couplings as we introduced to generate the fermion masses via spontaneous symmetry breaking. Hence we expect the fermion loops to contribute (to these longitudinal W states) something of order $g_f^2/4\pi$, where $g_f$ is the Yukawa coupling. Since $g_f \sim m_f/v$ (see (22.106)) we arrive at an estimate $\sim m_f^2/4\pi v^2 \sim G_F m_f^2/4\pi$ as in (22.173). An important message is that *particles which acquire their mass spontaneously do not 'decouple'*.

But we now have to explain why $\Delta\rho$ in (22.173) would vanish if $m_t^2 = m_b^2$—and why only $\ln m_H^2$ appears in (22.174). Both these facts are related to a symmetry of the assumed minimal Higgs sector which we have not yet discussed. Let us first consider the situation at tree level, where $\rho = 1$. It may be shown (Ross and Veltman 1975) that $\rho = 1$ is a natural consequence of having the symmetry broken by an $SU(2)_L$ doublet Higgs field (rather than a triplet, say)—or indeed by any number of doublets. The nearness of the measured $\rho$ parameter to 1 is, in fact, good support for the hypothesis that there are only doublet Higgs fields. Problem 22.6 explores a simple model with a Higgs field in the triplet representation.

At tree level, it is simplest to think of $\rho$ in connection with the mass ratio (22.44). To see the significance of this, let us go back to the Higgs gauge field Lagrangian $\hat{\mathcal{L}}_{G\Phi}$ of (22.8) which produced the masses. With the doublet Higgs of the form (22.110), it is a striking fact that the Higgs potential only involves the highly symmetrical combination of fields

$$\hat{\phi}_1^2 + \hat{\phi}_2^2 + \hat{\phi}_3^2 + \hat{\phi}_4^2 \tag{22.177}$$

as does the vacuum condition (17.102). This suggests that there may be some extra symmetry in (22.8) which is special to the doublet structure. But of course, to be of any interest, this symmetry has to be present in the $(D_\mu\hat{\phi})^\dagger(D^\mu\hat{\phi})$ term as well.

The nature of this symmetry is best brought out by introducing a change of notation for Higgs doublet $\hat{\phi}^+$ and $\hat{\phi}^0$: instead of (22.110), we now write

$$\hat{\phi} = \begin{pmatrix} (\hat{\pi}_2 + \mathrm{i}\hat{\pi}_1)/\sqrt{2} \\ (\hat{\sigma} - \mathrm{i}\hat{\pi}_3)/\sqrt{2} \end{pmatrix} \tag{22.178}$$

while the $\hat{\phi}_{\mathrm{C}}$ field of (22.111) becomes

$$\hat{\phi}_{\mathrm{C}} = \begin{pmatrix} (\hat{\sigma} + \mathrm{i}\hat{\pi}_3)/\sqrt{2} \\ -(\hat{\pi}_2 - \mathrm{i}\hat{\pi}_1)/\sqrt{2} \end{pmatrix}. \tag{22.179}$$

We then find that these can be written as

$$\hat{\phi} = \frac{1}{\sqrt{2}}(\hat{\sigma} + \mathrm{i}\boldsymbol{\tau}\cdot\hat{\boldsymbol{\pi}})\begin{pmatrix} 0 \\ 1 \end{pmatrix} \qquad \hat{\phi}_{\mathrm{C}} = \frac{1}{\sqrt{2}}(\hat{\sigma} + \mathrm{i}\boldsymbol{\tau}\cdot\hat{\boldsymbol{\pi}})\begin{pmatrix} 1 \\ 0 \end{pmatrix}. \tag{22.180}$$

Consider now the covariant $\mathrm{SU(2)_L}\times\mathrm{U(1)}$ derivative acting on $\hat{\phi}$, as in (22.8), and suppose to begin with that $g' = 0$. Then

$$\begin{aligned} D_\mu\hat{\phi} &= \frac{1}{\sqrt{2}}(\partial_\mu + \mathrm{i}g\boldsymbol{\tau}\cdot\hat{\boldsymbol{W}}_\mu/2)(\hat{\sigma} + \mathrm{i}\boldsymbol{\tau}\cdot\hat{\boldsymbol{\pi}})\begin{pmatrix} 0 \\ 1 \end{pmatrix} \\ &= \frac{1}{\sqrt{2}}\left\{\partial_\mu\hat{\sigma} + \mathrm{i}\boldsymbol{\tau}\cdot\partial_\mu\hat{\boldsymbol{\pi}} + \mathrm{i}\frac{g}{2}\hat{\sigma}\boldsymbol{\tau}\cdot\hat{\boldsymbol{W}}_\mu \right. \\ &\quad \left. - \frac{g}{2}[\hat{\boldsymbol{\pi}}\cdot\hat{\boldsymbol{W}}_\mu + \mathrm{i}\boldsymbol{\tau}\cdot\hat{\boldsymbol{W}}_\mu \times \hat{\boldsymbol{\pi}}]\right\}\begin{pmatrix} 0 \\ 1 \end{pmatrix} \end{aligned} \tag{22.181}$$

using $\tau_i\tau_j = \delta_{ij} + \mathrm{i}\epsilon_{ijk}\tau_k$. Now the vacuum choice (22.6) corresponds to $\hat{\sigma} = v, \hat{\boldsymbol{\pi}} = 0$, so that when we form $(D_\mu\hat{\phi})^\dagger(D^\mu\hat{\phi})$ from (22.181), we will get just

$$\frac{1}{2}(0, 1)\left\{\frac{g^2}{4}v^2(\boldsymbol{\tau}\cdot\hat{\boldsymbol{W}}_\mu)(\boldsymbol{\tau}\cdot\hat{\boldsymbol{W}}^\mu)\right\}\begin{pmatrix} 0 \\ 1 \end{pmatrix} = \frac{1}{2}M_{\mathrm{W}}^2\hat{\boldsymbol{W}}_\mu\cdot\hat{\boldsymbol{W}}^\mu \tag{22.182}$$

with $M_{\mathrm{W}} = gv/2$ as usual. The condition $g' = 0$ corresponds (cf (22.17)) to $\theta_{\mathrm{W}} = 0$, and thus to $\hat{W}_{3\mu} = \hat{Z}_\mu$, and so (22.182) states that in the limit $g' \to 0$, $M_{\mathrm{W}} = M_{\mathrm{Z}}$, as expected if $\cos\theta_{\mathrm{W}} = 1$. It is clear from (22.181) that the three components $\hat{\boldsymbol{W}}_\mu$ are treated on a precisely equal footing by the Higgs field (22.178), and indeed the notation suggests that $\hat{\boldsymbol{W}}_\mu$ and $\hat{\boldsymbol{\pi}}$ should perhaps be regarded as some kind of *new* triplets.

It is straightforward to calculate $(D_\mu\hat{\phi})^\dagger(D^\mu\hat{\phi})$ from (22.181): one finds (problem 22.7)

$$\begin{aligned}(D_\mu\hat{\phi})^\dagger D^\mu\hat{\phi} = {} & \frac{1}{2}(\partial_\mu\hat{\sigma})^2 + \frac{1}{2}(\partial_\mu\hat{\boldsymbol{\pi}})^2 - \frac{g}{2}\partial_\mu\hat{\sigma}\hat{\boldsymbol{\pi}}\cdot\hat{\boldsymbol{W}}^\mu \\ & + \frac{g}{2}\hat{\sigma}\,\partial_\mu\hat{\boldsymbol{\pi}}\cdot\hat{\boldsymbol{W}}^\mu + \frac{g}{2}\partial_\mu\hat{\boldsymbol{\pi}}\cdot(\hat{\boldsymbol{\pi}}\times\hat{\boldsymbol{W}}^\mu) \\ & + \frac{g^2}{4}\hat{\boldsymbol{W}}_\mu^2(\hat{\sigma}^2+\hat{\boldsymbol{\pi}}^2) + \frac{g^2}{4}\hat{\boldsymbol{\pi}}^2\hat{\boldsymbol{W}}_\mu^2. \end{aligned} \tag{22.183}$$

This expression now reveals what the symmetry is: (22.183) is invariant under global SU(2) transformations under which $\hat{\boldsymbol{W}}_\mu$ and $\hat{\boldsymbol{\pi}}$ are vectors—that is (cf (3.9))

$$\left.\begin{array}{c}\hat{\boldsymbol{W}}_\mu \to \hat{\boldsymbol{W}}_\mu + \boldsymbol{\epsilon}\times\hat{\boldsymbol{W}}_\mu \\ \hat{\boldsymbol{\pi}} \to \hat{\boldsymbol{\pi}} + \boldsymbol{\epsilon}\times\hat{\boldsymbol{\pi}} \\ \hat{\sigma}\to\hat{\sigma}\end{array}\right\}. \tag{22.184}$$

This is why, from the term $\hat{\boldsymbol{W}}_\mu^2\hat{\sigma}^2$, all three W fields have the same mass in this $g' \to 0$ limit.

If we now reinstate $g'$, and use (22.14) and (22.15) to write $\hat{W}_{3\mu}$ and $\hat{B}_\mu$ in terms of the physical fields $\hat{Z}_\mu$ and $\hat{A}_\mu$ as in (19.97), (22.181) becomes

$$\begin{aligned}\frac{1}{\sqrt{2}}\bigg\{ & \partial_\mu + \mathrm{i}g\frac{\tau_1}{2}\hat{W}_{1\mu} + \mathrm{i}g\frac{\tau_2}{2}\hat{W}_{2\mu} + \mathrm{i}g\frac{\tau_3}{2}\frac{\hat{Z}_\mu}{\cos\theta_{\mathrm{W}}} + \mathrm{i}g\sin\theta_{\mathrm{W}}\left(\frac{1+\tau_3}{2}\right)\hat{A}_\mu \\ & - \frac{\mathrm{i}g}{\cos\theta_{\mathrm{W}}}\sin^2\theta_{\mathrm{W}}\left(\frac{1+\tau_3}{2}\right)\hat{Z}_\mu\bigg\}(\hat{\sigma}+\mathrm{i}\boldsymbol{\tau}\cdot\hat{\boldsymbol{\pi}})\begin{pmatrix}0\\1\end{pmatrix}. \end{aligned} \tag{22.185}$$

We see from (22.185) that $g' \neq 0$ has two effects. First, there is a '$\boldsymbol{\tau}\cdot\hat{\boldsymbol{W}}$'-like term, as in (22.181), except that the '$\hat{W}_3$' part of it is now $\hat{Z}/\cos\theta_{\mathrm{W}}$. In the vacuum $\hat{\sigma} = v, \hat{\boldsymbol{\pi}} = 0$ which simply means that the mass of the Z is $M_{\mathrm{Z}} = M_{\mathrm{W}}/\cos\theta_{\mathrm{W}}$, i.e. $\rho = 1$; and this relation is preserved under 'rotations' of the form (22.184), since they do not mix $\hat{\boldsymbol{\pi}}$ and $\hat{\sigma}$. Hence this mass relation (and $\rho = 1$) is a consequence of the global SU(2) symmetry of the interactions and the vacuum under (22.184), and of the relations (22.14) and (22.15) which embody the requirement of a massless photon.

However, there are additional terms in (22.185) which single out the '$\tau_3$' component, and therefore break this global SU(2). These terms vanish as $g' \to 0$ and do not contribute at tree level, but we expect that they will cause $O(g'^2)$ corrections to $\rho = 1$ at the one-loop level.

None of this, however, yet involves the quark masses and the question of why $m_{\mathrm{t}}^2 - b_{\mathrm{b}}^2$ appears in the numerator in (22.173). We can now answer this question. Consider a typical mass term, of the form discussed in section 22.7.1, for a quark doublet of the $i$th family

$$\hat{\mathcal{L}}_m = -g_+(\bar{\hat{u}}_{\mathrm{L}i}\bar{\hat{d}}_{\mathrm{L}i})\hat{\phi}_{\mathrm{C}}\hat{u}_{\mathrm{R}i} - g_-(\bar{\hat{u}}_{\mathrm{L}i}\bar{\hat{d}}_{\mathrm{L}i})\hat{\phi}\hat{d}_{\mathrm{R}i}. \tag{22.186}$$

Using (22.178) and (22.179), this can be written as

$$\begin{aligned}\hat{\mathcal{L}}_m &= \frac{-g_+}{\sqrt{2}}(\bar{\hat{u}}_{\mathrm{L}i}\bar{\hat{d}}_{\mathrm{L}i})(\hat{\sigma}+\mathrm{i}\boldsymbol{\tau}\cdot\hat{\boldsymbol{\pi}})\begin{pmatrix}\hat{u}_{\mathrm{R}i}\\0\end{pmatrix} - \frac{g_-}{\sqrt{2}}(\bar{\hat{u}}_{\mathrm{L}i}\bar{\hat{d}}_{\mathrm{L}i})(\hat{\sigma}+\mathrm{i}\boldsymbol{\tau}\cdot\hat{\boldsymbol{\pi}})\begin{pmatrix}0\\\hat{d}_{\mathrm{R}i}\end{pmatrix}\\ &= -\frac{(g_++g_-)}{2\sqrt{2}}(\bar{\hat{u}}_{\mathrm{L}i}\bar{\hat{d}}_{\mathrm{L}i})(\hat{\sigma}+\mathrm{i}\boldsymbol{\tau}\cdot\hat{\boldsymbol{\pi}})\begin{pmatrix}\hat{u}_{\mathrm{R}i}\\\hat{d}_{\mathrm{R}i}\end{pmatrix}\\ &\quad -\frac{(g_+-g_-)}{2\sqrt{2}}(\bar{\hat{u}}_{\mathrm{L}i}\bar{\hat{d}}_{\mathrm{L}i})(\hat{\sigma}+\mathrm{i}\boldsymbol{\tau}\cdot\hat{\boldsymbol{\pi}})\tau_3\begin{pmatrix}\hat{u}_{\mathrm{R}i}\\\hat{d}_{\mathrm{R}i}\end{pmatrix}.\end{aligned} \tag{22.187}$$

Consider now a simultaneous (infinitesimal) global SU(2) transformation on the two doublets $(\hat{u}_{\mathrm{L}i}, \hat{d}_{\mathrm{L}i})^{\mathrm{T}}$ and $(\hat{u}_{\mathrm{R}i}, \hat{d}_{\mathrm{R}i})^{\mathrm{T}}$:

$$\begin{pmatrix}\hat{u}_{\mathrm{L}i}\\\hat{d}_{\mathrm{L}i}\end{pmatrix}\to(1-\mathrm{i}\boldsymbol{\epsilon}\cdot\boldsymbol{\tau}/2)\begin{pmatrix}\hat{u}_{\mathrm{L}i}\\\hat{d}_{\mathrm{L}i}\end{pmatrix}\qquad\begin{pmatrix}\hat{u}_{\mathrm{R}i}\\\hat{d}_{\mathrm{R}i}\end{pmatrix}\to(1-\mathrm{i}\boldsymbol{\epsilon}\cdot\boldsymbol{\tau}/2)\begin{pmatrix}\hat{u}_{\mathrm{R}i}\\\hat{d}_{\mathrm{R}i}\end{pmatrix}. \tag{22.188}$$

Under (22.188), the first term of (22.187) becomes (to first order in $\boldsymbol{\epsilon}$)

$$-\frac{(g_++g_-)}{2\sqrt{2}}(\bar{\hat{u}}_{\mathrm{L}i}\bar{\hat{d}}_{\mathrm{L}i})[\hat{\sigma}+\mathrm{i}\boldsymbol{\tau}\cdot(\hat{\boldsymbol{\pi}}+\hat{\boldsymbol{\pi}}\times\boldsymbol{\epsilon})]\begin{pmatrix}\hat{u}_{\mathrm{R}i}\\\hat{d}_{\mathrm{R}i}\end{pmatrix}. \tag{22.189}$$

From (22.189) we see that if, at the same time as (22.188), we *also* make the transformation of $\boldsymbol{\pi}$ given in (22.184), then this first term in $\hat{\mathcal{L}}_m$ will be invariant under these combined transformations. The second term in (22.187), however, will not be invariant under (22.188) but only under transformations with $\epsilon_1 = \epsilon_2 = 0, \epsilon_3 \neq 0$. We conclude that the global SU(2) symmetry of (22.184), which was responsible for $\rho = 1$ at the tree level, can be extended also to the quark sector; but—because the $g_\pm$ in (22.186) are proportional to the masses of the quark doublet—this symmetry is explicitly broken by the quark mass difference. This is why a t–$\bar{\mathrm{b}}$ loop in a W vacuum polarization correction can produce the 'non-decoupled' contribution (22.173) to $\rho$, which grows as $m_{\mathrm{t}}^2 - m_{\mathrm{b}}^2$ and produces quite detectable shifts from the tree-level predictions, given the accuracy of the data.

Returning to (22.188), the transformation on the L components is just the same as a standard $\mathrm{SU(2)_L}$ transformation, except that it is global; so the gauge interactions of the quarks obey this symmetry also. As far as the R components are concerned, they are totally decoupled in the gauge dynamics, and we are free to make the transformation (22.188) if we wish. The resulting complete transformation, which does the same to both the L and R components, is a non-chiral one—in fact, it is precisely an ordinary 'isospin' transformation of the type

$$\begin{pmatrix}\hat{u}_i\\\hat{d}_i\end{pmatrix}\to(1-\mathrm{i}\boldsymbol{\epsilon}\cdot\boldsymbol{\tau}/2)\begin{pmatrix}\hat{u}_i\\\hat{d}_i\end{pmatrix}. \tag{22.190}$$

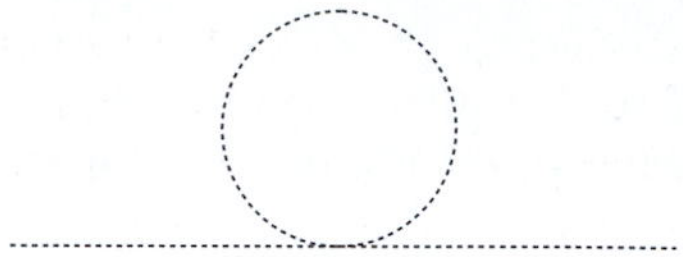

**Figure 22.18.** One-boson self-energy graph in $(\hat{\phi}^\dagger\hat{\phi})^2$.

The reader will recognize that the mathematics here is exactly the same as that in section 18.3 involving the SU(2) of isospin in the $\sigma$-model. This analysis of the symmetry of the Higgs (or a more general symmetry breaking sector) was first given by Sikivie *et al* (1980). The isospin SU(2) is frequently called 'custodial SU(2)' since it 'protects' $\rho = 1$.

What about the *absence* of $m_{\rm H}^2$ corrections? Here the position is rather more subtle. Without the Higgs particle H the theory is non-renormalizable, and hence one might expect to see some radiative correction becoming very large ($O(m_{\rm H}^2)$) as one tried to 'banish' H from theory by sending $m_{\rm H} \to \infty$ ($m_{\rm H}$ would be acting like a cut-off). The reason is that in such a $(\hat{\phi}^\dagger\hat{\phi})^2$ theory, the simplest loop we meet is that shown in figure 22.18 and it is easy to see by counting powers, as usual, that it diverges as the square of the cut-off. This loop contributes to the Higgs self-energy, and will be renormalized by taking the value of the coefficient of $\hat{\phi}^\dagger\hat{\phi}$ in (22.8) from experiment. We will return to this particular detail in section 22.10.1.

Even without a Higgs contribution however, it turns out that the electroweak theory is renormalizable at the one-loop level if the fermion masses are zero (Veltman 1968, 1970). Thus, one suspects that the large $m_{\rm H}^2$ effects will not be so dramatic after all. In fact, calculation shows (Veltman 1977, Chanowitz *et al* 1978, 1979) that one-loop radiative corrections to electroweak observables grow at most like $\ln m_{\rm H}^2$ for large $m_{\rm H}$. While there are finite corrections which are approximately $O(m_{\rm H}^2)$ for $m_{\rm H}^2 \ll M_{\rm W,Z}^2$, for $m_{\rm H}^2 \gg M_{\rm W,Z}^2$ the $O(m_{\rm H}^2)$ pieces cancel out from all observable quantities,[3] leaving only $\ln m_{\rm H}^2$ terms. This is just what we have in (22.174) and it means, unfortunately, that the sensitivity of the data to this important parameter of the Standard Model is only logarithmic. Fits to data typically give $m_{\rm H}$ in the region of 100 GeV at the minimum of the $\chi^2$ curve but the error (which is not simple to interpret) is of the order of 50 GeV. Direct searches now rule out a Higgs mass less than about 110 GeV , while the $\sim$ 2 s.d. effect seen at the close of play at LEP gave $m_{\rm H} \sim 115$ GeV (LEP Higgs (2001)). For further details, see the review of searches for Higgs bosons by Igo-Kemenes in Hagiwara *et al* (2002).

At the two-loop level, the expected $O(m_{\rm H}^4)$ behaviour becomes $O(m_{\rm H}^2)$ instead (van der Bij and Veltman 1984, van der Bij 1984)—and, of course, appears (relative to the one-loop contributions) with an additional factor of

[3] Apart from the $\hat{\phi}^\dagger\hat{\phi}$ coefficient! See section 22.10.

$O(\alpha)$. This relative insensitivity of the radiative corrections to $m_{\mathrm{H}}$, in the limit of large $m_{\mathrm{H}}$, was discovered by Veltman (1977) and called a 'screening' phenomenon by him: for large $m_{\mathrm{H}}$ (which also means, as we have seen, large $\lambda$), we have an effectively strongly interacting theory whose principal effects are screened off from observables at lower energy. It was shown by Einhorn and Wudka (1989) that this screening is also a consequence of the (approximate) isospin SU(2) symmetry we have just discussed in connection with (22.173). Phenomenologically, the upshot is that it is unfortunately very difficult to get a good handle on the value of $m_{\mathrm{H}}$ from fits to the precision data. With the top quark, the situation was very different.

## 22.9 The top quark

Having drawn attention to the relative sensitivity of radiative connections to loops containing virtual top quarks, it is worth devoting a little space to a 'backward glance' at the year immediately prior to the discovery of the t-quark (Abe *et al* 1994a, b, 1995b, Abachi *et al* 1995b) at the CDF and D0 detectors at FNAL's Tevatron, in p–$\bar{\mathrm{p}}$ collisions at $E_{\mathrm{CM}} = 1.8$ TeV.

The W and Z particles were, as we have seen, discovered in 1983 and at that time, and for some years subsequently, the data were not precise enough to be sensitive to virtual t effects. In the late 1980s and early 1990s, LEP at CERN and SLC at Stanford began to produce new and highly accurate data which did allow increasingly precise predictions to be made for the top quark mass, $m_{\mathrm{t}}$. Thus, a kind of race began, between experimentalists searching for the real top and theorists fitting ever more precise data to get tighter and tighter limits on $m_{\mathrm{t}}$ from its virtual effects.

In fact, by the time of the actual experimental discovery of the top quark, the experimental error in $m_{\mathrm{t}}$ was just about the same as the theoretical one (and—of course—the central values were consistent). Thus, in their May 1994 review of the electroweak theory (contained in Moutanet *et al* 1994, p 1304ff), Langacker and Erler gave the result of a fit to all electroweak data as

$$m_{\mathrm{t}} = 169 \pm^{16}_{18} \pm^{17}_{20} \ \mathrm{GeV}, \tag{22.191}$$

the central figure and first error being based on $m_{\mathrm{H}} = 300$ GeV, the second (+) error assuming $m_{\mathrm{H}} = 1000$ GeV and the second (−) error assuming $m_{\mathrm{H}} = 60$ GeV. At about the same time, Ellis *et al* (1994) gave the extraordinarily precise value

$$m_{\mathrm{t}} = 162 \pm 9 \ \mathrm{GeV} \tag{22.192}$$

without any assumption for $m_{\mathrm{H}}$.

A month or so earlier, the CDF collaboration (Abe *et al* 1994a, b) announced 12 events consistent with the hypothesis of production of a $\mathrm{t}\bar{\mathrm{t}}$ pair and, on this hypothesis, the mass was found to be

$$m_{\mathrm{t}} = 174 \pm 10 \pm^{13}_{12} \ \mathrm{GeV} \tag{22.193}$$

and this was followed by nine similar events from D0 (Abachi *et al* 1995a). By February 1995 both groups had amassed more data and the discovery was announced (Abe *et al* 1995b, Abachi *et al* 1995b). The 2002 experimental value for $m_t$ is 174.3 ± 5.1 GeV (Hagiwara *et al* 2002) as compared to the value predicted by fits to the electroweak data of 175.3 ± 44 GeV. This represents an extraordinary triumph for both theory and experiment. It is surely remarkable how the quantum fluctuations of a yet-to-be detected new particle could pin down its mass so precisely. It seems hard to deny that Nature has, indeed, made use of the subtle intricacies of a renormalizable, spontaneously broken, non-Abelian chiral gauge theory.

One feature of the 'real' top events is particularly noteworthy. Unlike the mass of the other quarks, $m_t$ is greater than $M_W$ and this means that it can decay to b + W via *real* W emission:

$$t \to W^+ + b. \tag{22.194}$$

In contrast, the b quark itself decays by the usual *virtual* W processes. Now we have seen that the virtual process is supressed by $\sim 1/M_W^2$ if the energy release (as in the case of b-decay) is well below $M_W$. But the real process (22.194) suffers no such suppression and proceeds very much faster. In fact (problem 22.8) the top quark lifetime from (22.194) is estimated to be $\sim 4 \times 10^{-25}$ s! This is quite similar to the lifetime of the $W^+$ itself, via $W^+ \to e^+\nu_e$ for example. Consider now the production of a $t\bar{t}$ pair in the collision between two partons. As the t and $\bar{t}$ separate, the strong interactions which should eventually 'hadronize' them will not play a role until they are $\sim$ 1 fm apart. But if they are travelling close to the speed of light, they can only travel some $10^{-16}$ m before decaying. Thus t's tend to decay before they experience the confining QCD interactions. Instead, the hadronization is associated with the b quark, which has a more typical weak lifetime ($\sim 1.5 \times 10^{-12}$ s). By the same token, this fast decay of the t quark means that there will be no detectable $t\bar{t}$ 'toponium', bound by QCD.

With the t quark now safely real, the one remaining missing particle in the Standard Model is the Higgs boson, and its discovery is of the utmost importance. It is fitting that we should end this final chapter with a brief review of Higgs physics.

## 22.10 The Higgs sector

### 22.10.1 Introduction

The Lagrangian for an *unbroken* $SU(2)_L \times U(1)$ gauge theory of vector bosons and fermions is rather simple and elegant, all the interactions being determined by just two Lagrangian parameters $g$ and $g'$ in a 'universal' way. All the particles in this hypothetical world are, however, massless. In the real world, while the electroweak interactions are undoubtedly well described by the $SU(2)_L \times U(1)$ theory, neither the mediating gauge quanta (apart from the photon) nor the fermions are massless. They must acquire mass in some way that does not break the gauge symmetry of the Lagrangian, or else the renormalizability of the theory is destroyed and its remarkable empirical success (at a level which includes loop corrections) would be some kind of freak accident. In chapter 19 we discussed how such a breaking of a gauge symmetry does happen, dynamically, in a superconductor. In that case 'electron pairing' was a crucial ingredient. In particle physics, while a lot of effort has gone into examining various analogous 'dynamical symmetry breaking' theories, none has yet emerged as both theoretically compelling and phenomenologically viable. However, a simple count of the number of degrees of freedom in a massive vector field, as opposed to a massless one, indicates that *some* additional fields must be present in order to give mass to the originally massless gauge bosons. And so, in the Standard Model, it is simply *assumed*, following the original ideas of Higgs and others (Higgs 1964, Englert and Brout 1964, Guralnik *et al* 1964; Higgs 1966) that a suitable scalar ('Higgs') field exists, with a potential which breaks the symmetry spontaneously. Furthermore, rather than (as in BCS theory) obtaining the fermion mass gaps dynamically, they too are put in 'by hand' via Yukawa-like couplings to the Higgs field.

It has to be admitted that this part of the Standard Model appears to be the least satisfactory. While the coupling of the Higgs field to the gauge fields is determined by the gauge symmetry, the Higgs self-coupling is not a gauge interaction and is unrelated to anything else in the theory. Likewise, the Yukawa-like fermion couplings are not gauge interactions either, and they are both unconstrained and uncomfortably different in orders of magnitude. True, all these are renormalizable couplings—but this basically means that their values are not calculable and have all to be taken from experiment.

Such considerations may indicate that the 'Higgs Sector' of the Standard Model is on a somewhat different footing from the rest of it—a commonly held view, indeed. Perhaps it should be regarded as more a 'phenomenology' than a 'theory', much as the current–current model was. In this connection, we may mention a point which has long worried many theorists. In section 22.8 we noted that figure 22.18 gives a quadratically divergent ($O(\Lambda^2)$) and positive contribution to the $\hat{\phi}^\dagger\hat{\phi}$ term in the Lagrangian, at one-loop order. This term would ordinarily, of course, be just the mass term of the scalar field. But in the Higgs case, the matter is much more delicate. The whole phenomenology depends on the renormalized coefficient having a *negative* value, triggering the spontaneous breaking of the symmetry. This means that the $O(\Lambda^2)$ one-loop correction must be cancelled by the 'bare' mass term $\frac{1}{2}m_{\mathrm{H},0}^2\hat{\phi}^\dagger\hat{\phi}$ so as to achieve a negative coefficient of order $-v^2$. This cancellation between $m_{\mathrm{H},0}^2$ and $\Lambda^2$ will have to be very precise indeed if $\Lambda$—the scale of 'new physics'—is very high, as is commonly assumed (say $10^{16}$ GeV).

The reader may wonder why attention should *now* be drawn to this particular piece of renormalization: aren't all divergences handled this way? In a sense they are, but the fact is that this is the first case we have had in which we have to cancel a *quadratic* divergence. The other mass corrections have all been logarithmic, for which there is nothing like such a dramatic 'fine-tuning' problem. There is a good reason for this in the case of the electron mass, upon which we remarked in section 11.2. Chiral symmetry forces self-energy corrections for fermions to be proportional to their mass and, hence, to contain only logarithms of the cut-off. Similarly, gauge invariance for the vector bosons prohibits any $O(\Lambda^2)$ connections in perturbation theory. But there is no symmetry, within the Standard Model, which 'protects' the coefficient of $\hat{\phi}^\dagger\hat{\phi}$ in this way. It is hard to understand what can be stopping it from being of order $\Lambda^2$, if we take the apparently reasonable point of view that the Standard Model will ultimately fail at some scale $\Lambda$ where new physics enters. Thus the difficulty is: why is the empirical parameter $v$ 'shielded' from the presumed high scale of new physics? This 'problem' is often referred to as the 'hierarchy problem'. We stress again that we are dealing here with an absolutely crucial symmetry-breaking term, which one would really like to understand far better.

Of course, the problem would go away if the scale $\Lambda$ were as low as, say, a few TeV. As we shall see in the next section this happens to be, not accidentally, the same scale at which the Standard Model ceases to be a perturbatively calculable theory. Various possibilities have been suggested for the kind of physics that might enter at energies of a few TeV. For example, 'technicolour' models (Peskin 1997) regard the Higgs field as a composite of some new heavy fermions, rather like the BCS-pairing idea referred to earlier. A second possibility is supersymmetry (Peskin 1997), in which there is a 'protective' symmetry operating, since scalar fields can be put alongside fermions in supermultiplets, and benefit from the protection enjoyed by the fermions. A third possibility is that of large extra dimensions (Antoniadis 2002).

These undoubtedly fascinating ideas obviously take us well beyond our proper subject, to which we must now return. Whatever may lie 'beyond' it, the Lagrangian of the Higgs sector of the Standard Model leads to many perfectly definite predictions which may be confronted with experiment, as we shall briefly discuss in section 22.10.3 (for a full account see Dawson *et al* (1990) and for more compact ones see Ellis *et al* (1996, chapter 11) and the review by Igo-Kemenes in Hagiwara *et al* (2002)). The elucidation of the mechanism of gauge symmetry breaking is undoubtedly of the greatest importance to particle physics: quite apart from the $\mathrm{SU(2)_L} \times \mathrm{U(1)}$ theory, very many of the proposed theories which go 'beyond the Standard Model' face a similar 'mass problem' and generally appeal to some variant of the 'Higgs mechanism' to deal with it.

The most significant prediction of the Higgs mechanism in $\mathrm{SU(2)_L} \times \mathrm{U(1)}$—and one originally pointed out by Peter Higgs himself (1964)—is that even after the gauge bosons have swallowed three of the scalar fields to acquire mass, *one* physical scalar field necessarily remains, with mass $m_\mathrm{H} = \sqrt{2}\mu = \sqrt{\lambda}v/\sqrt{2}$. The discovery of this *Higgs boson* has, therefore, always been a vital goal in particle physics. Before turning to experiment, however, we want to mention some theoretical considerations concerning $m_{\mathrm{H}}$ by way of orientation.

### 22.10.2 Theoretical considerations concerning $m_{\mathrm{H}}$

The coupling constant $\lambda$, which determines $m_{\mathrm{H}}$ given the known value of $v$, is unfortunately undetermined in the Standard Model. However, some quite strong theoretical arguments suggest that $m_{\mathrm{H}}$ cannot be arbitrarily large.

Like all coupling constants in a renormalizable theory, $\lambda$ must 'run'. For the $(\hat{\phi}^\dagger\hat{\phi})^2$ interaction of (22.8), a one-loop calculation of the $\beta$-function leads to

$$\lambda(E) = \frac{\lambda(v)}{1 - \frac{3\lambda(v)}{8\pi^2}\ln(E/v)}. \tag{22.195}$$

Like QED, this theory is not asymptotically free: the coupling increases with the scale $E$. In fact, the theory becomes non-perturbative at the scale $E^*$ such that

$$E^* \sim v \exp\left(\frac{8\pi^2}{3\lambda(v)}\right). \tag{22.196}$$

Note that this is exponentially sensitive to the 'low-energy' coupling constant $\lambda(v)$—and that $E^*$ decreases rapidly as $\lambda(v)$ increases. But (see (22.18)) $m_{\mathrm{H}}$ is essentially proportional to $\lambda^{1/2}(v)$. Hence, as $m_{\mathrm{H}}$ increases, non-perturbative behaviour sets in increasingly early. Suppose we say that we should like perturbative behaviour to be maintained up to an energy scale $\Lambda$. Then we require

$$m_{\mathrm{H}} < v\left[\frac{4\pi^2}{3\ln(\Lambda/v)}\right]^{1/2}. \tag{22.197}$$

For $\Lambda \sim 10^{16}$ GeV, this gives $m_{\mathrm{H}} < 160$ GeV. However, if the non-perturbative regime sets in at 1 TeV, then the bound on $m_{\mathrm{H}}$ is weaker, $m_{\mathrm{H}} < 750$ GeV.

This is an oversimplified argument for various reasons, though the essential point is correct. An important omission is the contribution of the top quark to the running of $\lambda(E)$. More refined versions place both approximate upper, and lower, bounds on $m_{\mathrm{H}}$ (Cabibbo *et al* 1979, Isidori *et al* 2001, Hambye and Riesselmann 1997). The conclusion is that for 130 GeV $< m_{\mathrm{H}} <$ 190 GeV the perturbative regime could extend to $\sim 10^{16}$ GeV, but that for $m_{\mathrm{H}} <$ 130 GeV the theory would be non-perturbative at a much lower scale. The precise critical values are sensitive to the value of $m_{\mathrm{t}}$.

There is another, independent, argument which suggests that $m_{\mathrm{H}}$ cannot be too large. We have previously considered violations of unitarity by the lowest-order diagrams for certain processes (see chapter 21 and section 22.6). As we saw, in a non-gauge theory with massive vector bosons, such violations are associated with the longitudinal polarization states of the bosons, which carry factors proportional to the 4–momentum $k^\mu$ (see (21.30)). In a gauge theory, strong cancellations in the high-energy behaviour occur between different lowest-order diagrams. This behaviour is characteristic of gauge theories (Llewellyn Smith 1973, Cornwall *et al* 1974) and is related to their renormalizability. One process of this sort which we have yet to consider, however, is that in which two longitudinally polarized W's scatter from each other. A considerable number of diagrams (seven in all) contribute to this process, in leading order: exchange of $\gamma$, Z and Higgs particles, together with the W–W self-interaction. When all these are added up the high-energy behaviour of the total amplitude turns out to be proportional to $\lambda$, the Higgs coupling constant (see, for example, Ellis *et al* (1996, chapter 8)). This, at first sight, unexpected result can be understood as follows. The longitudinal components of the W's arise from the '$\partial^\mu \hat{\phi}$' parts in (22.8) (compare equation (19.48) in the U(1) case), which produce $k^\mu$ factors. Thus, the scattering of longitudinal W's is effectively the scattering of the three Goldstone bosons in the complex Higgs doublet. These bosons have self-interactions arising from the $\lambda(\hat{\phi}^\dagger\hat{\phi})^2$ Higgs potential, for which the Feynman amplitude is just proportional to $\lambda$. Now, although such a constant term obviously cannot violate unitarity as the energy increases (as has happened in the other cases), it can do so if $\lambda$ itself is too big—and since $\lambda \propto m_{\mathrm{H}}^2$, this puts a bound on $m_{\mathrm{H}}$. A constant amplitude is pure $J = 0$ (compare (21.9)) and so, in order of magnitude, we expect unitarity to imply $\lambda < 1$. In terms of standard quantities,

$$\lambda = m_{\mathrm{H}}^2 G_{\mathrm{F}}/\sqrt{2} \tag{22.198}$$

and so we expect

$$m_{\mathrm{H}} < G_{\mathrm{F}}^{-1/2}. \tag{22.199}$$

A more refined analysis (Lee *et al* 1977a, b) gives

$$m_{\mathrm{H}} < \left(\frac{8\sqrt{2}\pi}{3G_{\mathrm{F}}}\right)^{1/2} \approx 1 \text{ TeV}. \tag{22.200}$$

Like the preceding argument, this one does not say that $m_{\mathrm{H}}$ *must* be less than some fixed number. Rather, it states that if $m_{\mathrm{H}}$ gets bigger than a certain value, perturbation theory will fail or 'new physics' will enter. It is, in fact, curiously reminiscent of the original situation with the four-fermion current–current interaction itself (compare (21.10) with (22.199)). Perhaps this is a clue that we need to replace the Higgs phenomenology. At all events, this line of reasoning seems to imply that the Higgs boson will either be found at a mass well below 1 TeV, or else some electroweak interactions will become effectively strong with new physical consequences. This 'no lose' situation provided powerful motivation for the construction of the LHC.

We now consider some simple aspects of Higgs production and decay processes at collider energies, as predicted by the Standard Model.

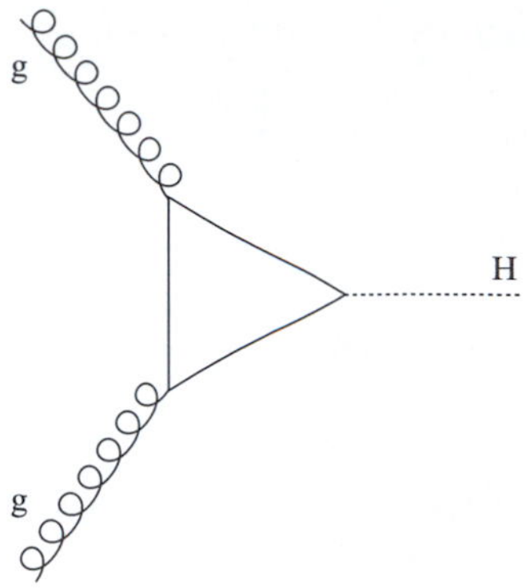


**Figure 22.19.** Higgs boson production process by 'gluon fusion'.

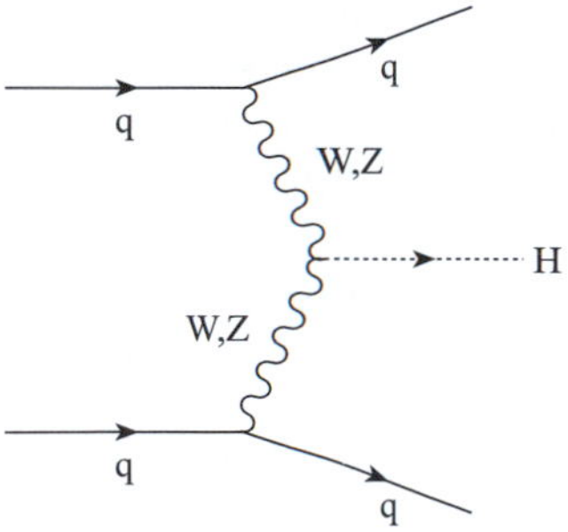


**Figure 22.20.** Higgs boson production process by 'vector boson fusion'.

### 22.10.3 Higgs phenomenology

Our discussion is based on the existing lower bound on $m_{\mathrm{H}}$ established at LEP (LEP 2003):

$$m_{\mathrm{H}} \geq 114.4 \text{ GeV (95\% c.l.).} \tag{22.201}$$

This already excludes many possibilities in both production and decay. At both the Tevatron and the LHC, the dominant production mechanism is expected to be 'gluon fusion' via an intermediate top quark loop as shown in figure 22.19 (Georgi *et al* 1978, Glashow *et al* 1978, Stange *et al* 1994a, b). Since the gluon probability distribution rises rapidly at small $x$ values, which are probed at larger collider energy $\sqrt{s}$, the cross-section for this process will rise with energy. At the Tevatron with $\sqrt{s} = 2$ TeV, the production cross-section ranges from about 1 pb for $m_{\mathrm{H}} \simeq 100$ GeV to 0.2 pb for $m_{\mathrm{H}} \simeq 200$ GeV. At an LHC energy of $\sqrt{s} = 14$ TeV, the cross-section is about 50 pb for $m_{\mathrm{H}} \simeq 100$ GeV and 1 pb for $m_{\mathrm{H}} \simeq 700$ GeV. The cross-section is the same for pp and for $\mathrm{p\bar{p}}$ colliders.

The next largest cross-section, roughly ten times smaller, is for 'vector boson fusion' via the diagram of figure 22.20. However, detection of H depends

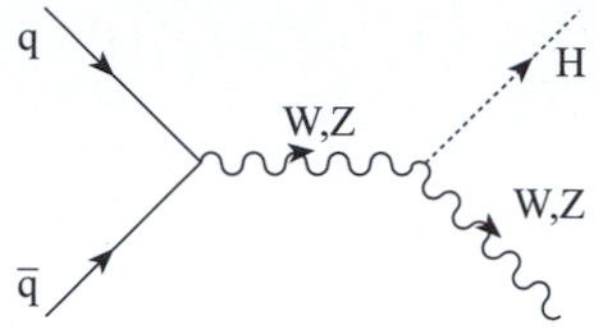


**Figure 22.21.** Higgs boson production in association with W or Z.

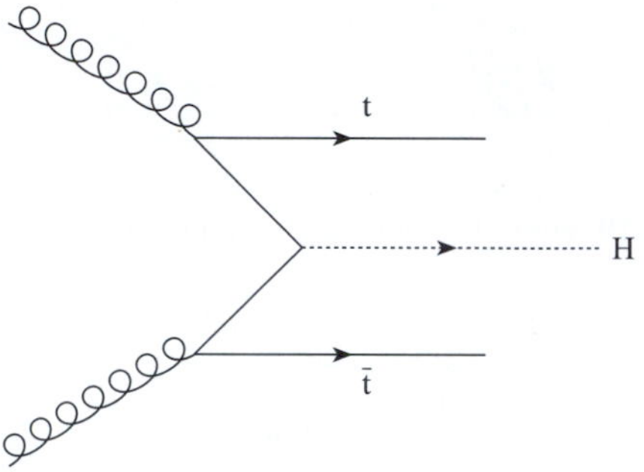


**Figure 22.22.** Higgs boson production in association with a $t\bar{t}$ pair.

crucially on being able to separate the signal from the large backgrounds expected in many of the Higgs decay channels (to be discussed shortly). For this reason, the mechanism of associated production of a Higgs boson with a vector boson, shown in figure 22.21, could also be important since the leptonic decays of the W or Z can be exploited for triggering. A $p\bar{p}$ collider gives a somewhat larger cross-section for this process than a pp collider. A fourth possibility is 'associated production with top quarks' as shown in figure 22.22, for example. Figure 22.23 (taken from Ellis *et al* 1996) shows the cross-sections for the various production processes as a function of $m_{\mathrm{H}}$.

The Higgs boson will, of course, have to be detected via its decays. For $m_{\mathrm{H}} < 140$ GeV, decays to fermion–anti-fermion pairs dominate, of which $b\bar{b}$ has the largest branching ratio. The width of $\mathrm{H} \to f\bar{f}$ is easily calculated to lowest order and is (problem 22.9)

$$\Gamma(\mathrm{H} \to f\bar{f}) = \frac{C G_{\mathrm{F}} m_f^2 m_{\mathrm{H}}}{4\pi\sqrt{2}} \left(1 - \frac{4m_f^2}{m_{\mathrm{H}}^2}\right)^{3/2} \tag{22.202}$$

where the colour factor $C$ is three for quarks and one for leptons. For such $m_{\mathrm{H}}$ values, $\Gamma(\mathrm{H} \to f\bar{f})$ is less than 10 MeV. The final state $\mathrm{ZH} \to \mu^+\mu^- b\bar{b}$ provides a clean signature through tagging b-jets using the high-$p_{\mathrm{T}}$ leptons from the decay $b \to cl^-\bar{\nu}$.

The situation changes significantly when $m_{\mathrm{H}}$ becomes greater than twice the

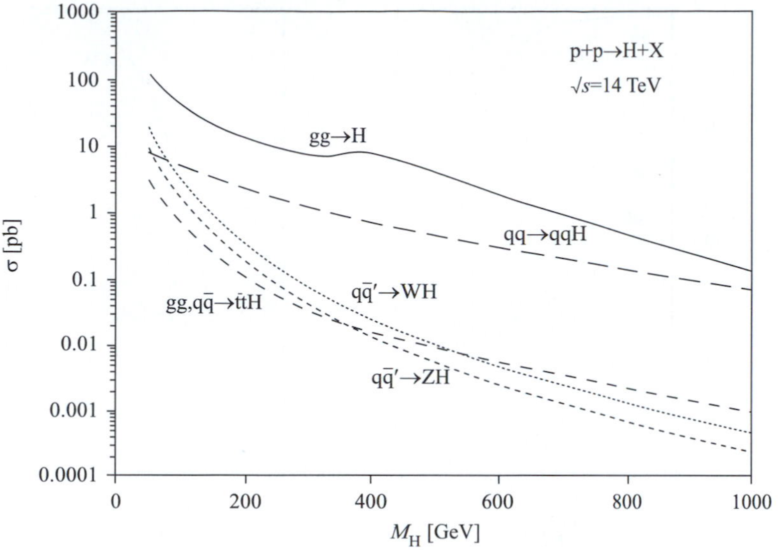


**Figure 22.23.** Higgs boson production cross-sections in pp collisions at the LHC (figure from Ellis *et al* 1996, p 399).

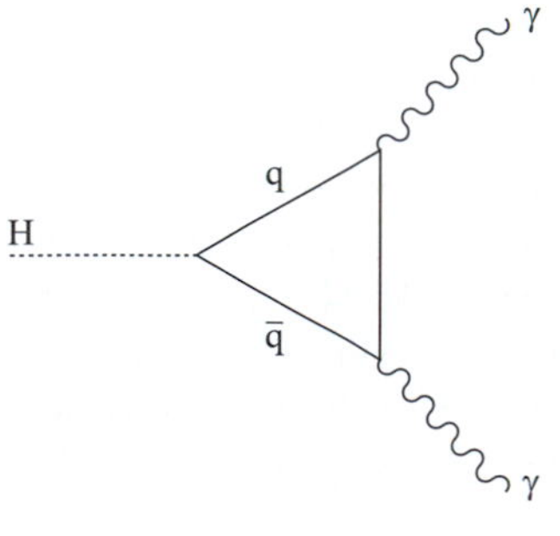


**Figure 22.24.** Higgs boson decay via quark triangle.

vector boson masses. The tree-level width for $\mathrm{H} \to \mathrm{W}^+\mathrm{W}^-$ is (problem 22.9)

$$\Gamma(\mathrm{H} \to \mathrm{W}^+\mathrm{W}^-) = \frac{G_\mathrm{F} m_\mathrm{H}^3}{8\pi\sqrt{2}} \left(1 - \frac{4M_\mathrm{W}^2}{m_\mathrm{H}^2}\right)\left(1 - \frac{4M_\mathrm{W}^2}{m_\mathrm{H}^2} + 12\frac{M_\mathrm{W}^4}{m_\mathrm{H}^4}\right) \tag{22.203}$$

and the width for $\mathrm{H} \to \mathrm{ZZ}$ is the same with $M_\mathrm{W} \to M_\mathrm{Z}$ and a factor of $\frac{1}{2}$ to allow for the two identical bosons in the final state. These widths rise rapidly with $m_\mathrm{H}$, reaching $\Gamma \sim 1$ GeV when $m_\mathrm{H} \sim 200$ GeV and $\Gamma \sim 100$ GeV for $m_\mathrm{H} \sim 500$ GeV. It is apparent that for $m_\mathrm{H}$ any larger than this (say $m_\mathrm{H} \sim 1$ TeV) the width of the state will become comparable to its mass, which is just another facet of the 'strong interaction' regime discussed earlier.

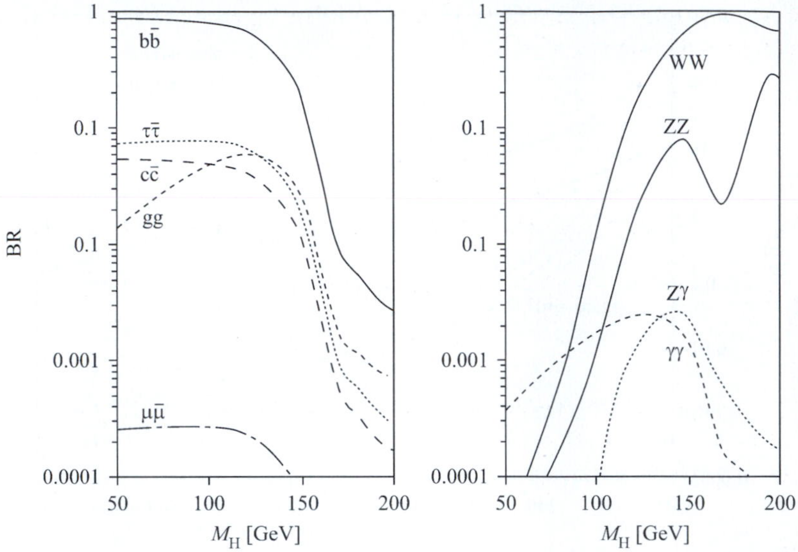


**Figure 22.25.** Branching ratios of the Higgs boson (from Ellis *et al* 1996, p 393).

The vector bosons themselves decay to leptons and quark jets, with the latter having the larger branching ratios. But final states containing hadronic jets will have to contend with large backgrounds at hadron collides. The rarer purely leptonic final states such as $l^+l^-l^+l^-$ are likely to prove the best hope for discovery. Another rare but characteristic decay is $\mathrm{H} \to \gamma\gamma$ via intermediate W and quark triangle loops (figure 22.24).

Figure 22.25, taken from Ellis *et al* (1996), shows the complete set of phenomenologically relevant Higgs branching ratios for a 'light' Higgs boson.

The LHC is scheduled to start physics runs in the year 2007. The ATLAS and CMS detectors have been optimized for Higgs boson searches and should be well able to discover a Higgs boson with $m_{\mathrm{H}}$ in the range 100 GeV to 1 TeV. Future machines, such as a high-energy $\mathrm{e}^+\mathrm{e}^-$ or $\mu^+\mu^-$ collider, will allow this crucial energy regime to be explored with high precision.

## Problems

**22.1** By identifying the part of (22.23) which has the form (22.35), derive (22.36).

**22.2** Using the vertex (22.26), verify (22.56).

**22.3** Insert (22.7) into (22.125) to derive (22.127).

**22.4** Verify that the neutral current part of (22.133) is diagonal in the 'mass' basis.

**22.5** Verify (22.150).

**22.6** Suppose that the Higgs field is a triplet of $SU(2)_L$ rather than a doublet; and suppose that its vacuum value is

$$\langle 0|\hat{\phi}|0\rangle = \begin{pmatrix} 1 & 0 & 0 \\ 0 & 0 & 0 \\ 0 & 0 & -1 \end{pmatrix}$$

in the gauge in which it is real. The non-vanishing component has $t_3 = -1$, using

$$t_3 = \begin{pmatrix} 1 & 0 & 0 \\ 0 & 0 & 0 \\ 0 & 0 & -1 \end{pmatrix}$$

in the 'angular-momentum-like' basis. Since we want the charge of the vacuum to be zero and we have $Q = t_3 + y/2$, we must assign $y(\hat{\phi}) = 2$. So the covariant derivative on $\hat{\phi}$ is

$$(\partial_\mu + \mathrm{i}g\boldsymbol{t} \cdot \hat{\boldsymbol{W}}_\mu + \mathrm{i}g'\hat{B}_\mu)\hat{\phi}$$

where

$$t_1 = \begin{pmatrix} 0 & \frac{1}{\sqrt{2}} & 0 \\ \frac{1}{\sqrt{2}} & 0 & \frac{1}{\sqrt{2}} \\ 0 & \frac{1}{\sqrt{2}} & 0 \end{pmatrix} \qquad t_2 = \begin{pmatrix} 0 & \frac{-\mathrm{i}}{\sqrt{2}} & 0 \\ \frac{\mathrm{i}}{\sqrt{2}} & 0 & \frac{-\mathrm{i}}{\sqrt{2}} \\ 0 & \frac{\mathrm{i}}{\sqrt{2}} & 0 \end{pmatrix}$$

and $t_3$ is as before (it is easy to check that these three matrices do satisfy the required SU(2) commutation relations $[t_1, t_2] = \mathrm{i}t_3$). Show that the photon and Z fields are still given by (22.14) and (22.15), with the same $\sin\theta_W$ as in (22.17), but that now

$$M_Z = \sqrt{2}M_W/\cos\theta_W.$$

What is the value of the parameter $\rho$ in this model?

**22.7** Use (22.181) to verify (22.183).

**22.8** Calculate the lifetime of the top quark to decay via $t \to W^+ + b$.

**22.9** Using the Higgs couplings given in appendix Q, verify (22.202) and (22.203).

# APPENDIX M

# GROUP THEORY

## M.1 Definition and simple examples

A group $\mathcal{G}$ is a set of elements $(a, b, c, \ldots)$ with a law for combining any two elements $a$, $b$ so as to form their ordered 'product' $ab$, such that the following four conditions hold:

(i) For every $a, b \in \mathcal{G}$, the product $ab \in \mathcal{G}$ (the symbol '$\in$' means 'belongs to' or 'is a member of').

(ii) The law of combination is associative, i.e.

$$(ab)c = a(bc). \tag{M.1}$$

iii). $\mathcal{G}$ contains a unique identity element, $e$, such that for all $a \in \mathcal{G}$,

$$ae = ea = a. \tag{M.2}$$

(iv) For all $a \in \mathcal{G}$, there is a unique inverse element, $a^{-1}$, such that

$$aa^{-1} = a^{-1}a = e. \tag{M.3}$$

Note that, in general, the law of combination is not commutative, i.e. $ab \neq ba$: if it is commutative ($ab = ba$), the group is *Abelian*; if not, it is *non-Abelian*. Any finite set of elements satisfying the conditions (i)–(iv) forms a finite group, the *order* of the group being equal to the number of elements in the set. If the set does not have a finite number of elements it is an infinite group.

As a simple example, the set of four numbers (1, i, $-1$, $-i$) form a finite Abelian group of order 4, with the law of combination being ordinary multiplication. The reader may check that each of (i)–(iv) is satisfied, with $e$ taken to be the number 1 and the inverse being the algebraic reciprocal. A second group of order 4 is provided by the matrices

$$\begin{pmatrix} 1 & 0 \\ 0 & 1 \end{pmatrix} \quad \begin{pmatrix} 0 & 1 \\ -1 & 0 \end{pmatrix} \quad \begin{pmatrix} -1 & 0 \\ 0 & -1 \end{pmatrix} \quad \begin{pmatrix} 0 & -1 \\ 1 & 0 \end{pmatrix} \tag{M.4}$$

with the combination law being matrix multiplication, '$e$' being the first (unit) matrix and the inverse being the usual matrix inverse. Although matrix multiplication is not commutative in general, it happens to be so for these particular matrices. In fact, the way these four matrices multiply together is (as the reader can verify) exactly the same as the way the four numbers (1, i, −1, −i) (in that order) do. Further, the correspondence between the elements of the two groups is 'one to one': that is, if we label the two sets of group elements by $(e, a, b, c)$ and $(e', a', b', c')$, we have the correspondences $e \leftrightarrow e'$, $a \leftrightarrow a'$, $b \leftrightarrow b'$, $c \leftrightarrow c'$. Two groups with the same multiplication structure and with a one-to-one correspondence between their elements are said to be *isomorphic*. If they have the same multiplication structure but the correspondence is not one-to-one, they are *homomorphic*.

## M.2 Lie groups

We are interested in *continuous groups*—that is, groups whose elements are labelled by a number of continuously variable real parameters $\alpha_1, \alpha_2, \ldots, \alpha_r$ : $g(\alpha_1, \alpha_2, \ldots, \alpha_r) \equiv g(\boldsymbol{\alpha})$. In particular, we are concerned with various kinds of 'coordinate transformations' (not necessarily spacetime ones but including also 'internal' transformations such as those of SU(3)). For example, rotations in three dimensions form a group whose elements are specified by three real parameters (e.g. two for defining the axis of the rotation and one for the angle of rotation about that axis). Lorentz transformations also form a group, this time with six real parameters (three for 3D rotations, three for pure velocity transformations). The matrices of SU(3) are specified by the values of eight real parameters. By convention, parametrizations are arranged in such a way that $g(\mathbf{0})$ is the identity element of the group. For a continuous group, condition (i) takes the form

$$g(\boldsymbol{\alpha})g(\boldsymbol{\beta}) = g(\boldsymbol{\gamma}(\boldsymbol{\alpha}, \boldsymbol{\beta})) \tag{M.5}$$

where the parameters $\boldsymbol{\gamma}$ are continuous functions of the parameters $\boldsymbol{\alpha}$ and $\boldsymbol{\beta}$. A more restrictive condition is that $\boldsymbol{\gamma}$ should be an *analytic* function of $\boldsymbol{\alpha}$ and $\boldsymbol{\beta}$; if this is the case, the group is a *Lie group*.

The analyticity condition implies that if we are given the form of the group elements in the neighbourhood of any one element, we can 'move out' from that neighbourhood to other nearby elements, using the mathematical procedure known as 'analytic continuation' (essentially, using a power series expansion); by repeating the process, we should be able to reach all group elements which are 'continuously connected' to the original element. The simplest group element to consider is the identity, which we shall now denote by $I$. Lie proved that the properties of the elements of a Lie group which can be reached continuously from the identity $I$ are determined from elements lying in the neighbourhood of $I$.

## M.3 Generators of Lie groups

Consider (following Lichtenberg 1970, chapter 5) a group of transformations defined by

$$x_i' = f_i(x_1, x_2, \ldots, x_N; \alpha_1, \alpha_2, \ldots, \alpha_r) \tag{M.6}$$

where the $x_i$'s ($i = 1, 2, \ldots, N$) are the 'coordinates' on which the transformations act, and the $\alpha$'s are the (real) parameters of the transformations. By convention, $\boldsymbol{\alpha} = \mathbf{0}$ is the identity transformation, so

$$x_i = f_i(\boldsymbol{x}, \mathbf{0}). \tag{M.7}$$

A transformation in the neighbourhood of the identity is then given by

$$\mathrm{d}x_i = \sum_{\nu=1}^{r} \frac{\partial f_i}{\partial \alpha_\nu} \mathrm{d}\alpha_\nu \tag{M.8}$$

where the $\{\mathrm{d}\alpha_\nu\}$ are infinitesimal parameters and the partial derivative is understood to be evaluated at the point $(\boldsymbol{x}, \mathbf{0})$.

Consider now the change in a function $F(\boldsymbol{x})$ under the infinitesimal transformation (M.8). We have

$$\begin{aligned} F \to F + \mathrm{d}F &= F + \sum_{i=1}^{N} \frac{\partial F}{\partial x_i} \mathrm{d}x_i \\ &= F + \sum_{i=1}^{N} \left[ \sum_{\nu=1}^{r} \frac{\partial f_i}{\partial \alpha_\nu} \mathrm{d}\alpha_\nu \right] \frac{\partial F}{\partial x_i} \\ &\equiv \left\{ 1 - \sum_{\nu=1}^{r} \mathrm{d}\alpha_\nu \mathrm{i}\hat{X}_\nu \right\} F \end{aligned} \tag{M.9}$$

where

$$\hat{X}_\nu \equiv \mathrm{i} \sum_{i=1}^{N} \frac{\partial f_i}{\partial \alpha_\nu} \frac{\partial}{\partial x_i} \tag{M.10}$$

is a *generator of infinitesimal transformations*.[1] Note that in (M.10) $\nu$ runs from 1 to $r$, so there are as many generators as there are parameters labelling the group elements. Finite transformations are obtained by 'exponentiating' the quantity in braces in (M.9) (compare (12.30)):

$$\hat{U}(\boldsymbol{\alpha}) = \exp\{-\mathrm{i}\boldsymbol{\alpha} \cdot \hat{\boldsymbol{X}}\} \tag{M.11}$$

where we have written $\sum_{\nu=1}^{r} \alpha_\nu \hat{X}_\nu = \boldsymbol{\alpha} \cdot \hat{\boldsymbol{X}}$.

An important theorem states that the commutator of any two generators of a Lie group is a linear combination of the generators:

$$[\hat{X}_\lambda, \hat{X}_\mu] = c_{\lambda\mu}^{\nu} \hat{X}_\nu \tag{M.12}$$

where the constants $c_{\lambda\mu}^{\nu}$ are complex numbers called the *structure constants* of the group; a sum over $\nu$ from 1 to $r$ is understood on the right-hand side. The commutation relations (M.12) are called the *algebra* of the group.

[1] Clearly there is lot of 'convention' (the sign, the i) in the definition of $\hat{X}_\nu$. It is chosen for convenient consistency with familiar generators, for example those of SO(3) (see section M.4.1).

## M.4 Examples

### M.4.1 SO(3) and three-dimensional rotations

Rotations in three dimensions are defined by

$$\boldsymbol{x}' = R\boldsymbol{x} \tag{M.13}$$

where $R$ is a real $3 \times 3$ matrix such that the length of $\boldsymbol{x}$ is preserved, i.e. $\boldsymbol{x}'^{\mathrm{T}}\boldsymbol{x}' = \boldsymbol{x}^{\mathrm{T}}\boldsymbol{x}$. This implies that $R^{\mathrm{T}}R = I$, so that $R$ is an orthogonal matrix. It follows that

$$1 = \det(R^{\mathrm{T}}R) = \det R^{\mathrm{T}} \det R = (\det R)^2 \tag{M.14}$$

and so $\det R = \pm 1$. Those $R$'s with $\det R = -1$ include a parity transformation ($\boldsymbol{x}' = -\boldsymbol{x}$), which is not continuously connected to the identity. Those with $\det R = 1$ are 'proper rotations' and they form the elements of the group SO(3): the *S* pecial *O* rthogonal group in *3* dimensions.

An $R$ close to the identity matrix $I$ can be written as $R = I + \delta R$ where

$$(I + \delta R)^{\mathrm{T}}(I + \delta R) = I. \tag{M.15}$$

Expanding this out to first order in $\delta R$ gives

$$\delta R^{\mathrm{T}} = -\delta R \tag{M.16}$$

so that $\delta R$ is an anti-symmetric $3 \times 3$ matrix (compare (12.19)). We may parametrize $\delta R$ as

$$\delta R = \begin{pmatrix} 0 & \epsilon_3 & -\epsilon_2 \\ -\epsilon_3 & 0 & \epsilon_1 \\ \epsilon_2 & -\epsilon_1 & 0 \end{pmatrix} \tag{M.17}$$

and an infinitesimal rotation is then given by

$$\boldsymbol{x}' = \boldsymbol{x} - \boldsymbol{\epsilon} \times \boldsymbol{x} \tag{M.18}$$

(compare (12.64)) or

$$\mathrm{d}x_1 = -\epsilon_2 x_3 + \epsilon_3 x_2 \qquad \mathrm{d}x_2 = -\epsilon_3 x_1 + \epsilon_1 x_3 \qquad \mathrm{d}x_3 = -\epsilon_1 x_2 + \epsilon_2 x_1. \tag{M.19}$$

Thus in (M.8), identifying $\mathrm{d}\alpha_1 \equiv \epsilon_1,\ \mathrm{d}\alpha_2 \equiv \epsilon_2,\ \mathrm{d}\alpha_3 \equiv \epsilon_3$, we have

$$\frac{\partial f_1}{\partial \alpha_1} = 0 \qquad \frac{\partial f_1}{\partial \alpha_2} = -x_3 \qquad \frac{\partial f_1}{\partial \alpha_3} = x_2 \qquad \text{etc.} \tag{M.20}$$

The generators (M.10) are then

$$\left.\begin{aligned} \hat{X}_1 &= \mathrm{i}x_3 \frac{\partial}{\partial x_2} - \mathrm{i}x_2 \frac{\partial}{\partial x_3} \\ \hat{X}_2 &= \mathrm{i}x_1 \frac{\partial}{\partial x_3} - \mathrm{i}x_3 \frac{\partial}{\partial x_1} \\ \hat{X}_3 &= \mathrm{i}x_2 \frac{\partial}{\partial x_1} - \mathrm{i}x_1 \frac{\partial}{\partial x_2} \end{aligned}\right\} \tag{M.21}$$

which are easily recognized as the quantum-mechanical angular momentum operators

$$\hat{X} = \boldsymbol{x} \times -\mathrm{i}\nabla \tag{M.22}$$

which satisfy the *SO(3) algebra*

$$[\hat{X}_i, \hat{X}_j] = \mathrm{i}\epsilon_{ijk}\hat{X}_k. \tag{M.23}$$

The action of finite rotations, parametrized by $\boldsymbol{\alpha} = (\alpha_1, \alpha_2, \alpha_3)$, on functions $F$ is given by

$$\hat{U}(\boldsymbol{\alpha}) = \exp\{-\mathrm{i}\boldsymbol{\alpha} \cdot \hat{\boldsymbol{X}}\}. \tag{M.24}$$

The operators $\hat{U}(\boldsymbol{\alpha})$ form a group which is isomorphic to SO(3). The structure constants of SO(3) are $\mathrm{i}\epsilon_{ijk}$, from (M.23).

### M.4.2 SU(2)

We write the infinitesimal SU(2) transformation (acting on a general complex two-component column vector) as (cf (12.27))

$$\begin{pmatrix} q_1' \\ q_2' \end{pmatrix} = (1 + \mathrm{i}\boldsymbol{\epsilon} \cdot \boldsymbol{\tau}/2) \begin{pmatrix} q_1 \\ q_2 \end{pmatrix} \tag{M.25}$$

so that

$$\begin{aligned} \mathrm{d}q_1 &= \frac{\mathrm{i}\epsilon_3}{2} q_1 + \left(\frac{\mathrm{i}\epsilon_1}{2} + \frac{\epsilon_2}{2}\right) q_2 \\ \mathrm{d}q_2 &= \frac{-\mathrm{i}\epsilon_3}{2} q_2 + \left(\frac{\mathrm{i}\epsilon_1}{2} - \frac{\epsilon_2}{2}\right) q_1. \end{aligned} \tag{M.26}$$

Then (with $\mathrm{d}\alpha_1 \equiv \epsilon_1$ etc)

$$\frac{\partial f_1}{\partial \alpha_1} = \frac{\mathrm{i}q_2}{2}, \quad \frac{\partial f_1}{\partial \alpha_2} = \frac{q_2}{2}, \quad \frac{\partial f_1}{\partial \alpha_3} = \frac{\mathrm{i}q_1}{2} \tag{M.27}$$

$$\frac{\partial f_2}{\partial \alpha_1} = \frac{\mathrm{i}q_1}{2}, \quad \frac{\partial f_2}{\partial \alpha_2} = -\frac{q_1}{2}, \quad \frac{\partial f_2}{\partial \alpha_3} = -\frac{\mathrm{i}q_2}{2} \tag{M.28}$$

and (from (M.10))

$$\hat{X}_1' = -\frac{1}{2}\left\{q_2 \frac{\partial}{\partial q_1} + q_1 \frac{\partial}{\partial q_2}\right\} \tag{M.29}$$

$$\hat{X}_2' = \frac{\mathrm{i}}{2}\left\{q_2 \frac{\partial}{\partial q_1} - q_1 \frac{\partial}{\partial q_2}\right\} \tag{M.30}$$

$$\hat{X}_3' = \frac{1}{2}\left\{-q_1 \frac{\partial}{\partial q_1} + q_2 \frac{\partial}{\partial q_2}\right\}. \tag{M.31}$$

It is an interesting exercise to check that the commutation relations of the $\hat{X}_i'$'s are exactly the same as those of the $\hat{X}_i$'s in (M.23). The two groups are therefore said to have the same algebra with the same structure constants, and they are in fact isomorphic in the vicinity of their respective identity elements. They are not the same for 'large' transformations, however, as we discuss in section M.7.

### M.4.3 SO(4): The special orthogonal group in four dimensions

This is the group whose elements are $4 \times 4$ matrices $S$ such that $S^{\mathrm{T}} S = I$, where $I$ is the $4 \times 4$ unit matrix, with the condition $\det S = +1$. The Euclidean (length)$^2$ $x_1^2 + x_2^2 + x_3^2 + x_4^2$ is left invariant under SO(4) transformations. Infinitesimal SO(4) transformations are characterized by the 4D analogue of those for SO(3), namely by $4 \times 4$ real anti-symmetric matrices $\delta S$, which have six real parameters. We choose to parametrize $\delta S$ in such a way that the Euclidean 4-vector $(\boldsymbol{x}, x_4)$ is transformed to (cf (18.74), (18.75), (18.79) and (18.80))

$$\begin{aligned} \boldsymbol{x}' &= \boldsymbol{x} - \boldsymbol{\epsilon} \times \boldsymbol{x} - \boldsymbol{\eta} x_4 \\ x_4' &= x_4 + \boldsymbol{\eta} \cdot \boldsymbol{x} \end{aligned} \tag{M.32}$$

where $\boldsymbol{x} = (x_1, x_2, x_3)$ and $\boldsymbol{\eta} = (\eta_1, \eta_2, \eta_3)$. Note that the first three components transform by (M.18) when $\boldsymbol{\eta} = 0$, so that SO(3) is a *subgroup* of SO(4). The six generators are (with $\mathrm{d}\alpha_1 \equiv \epsilon_1$ etc)

$$\hat{X}_1 = \mathrm{i} x_3 \frac{\partial}{\partial x_2} - \mathrm{i} x_2 \frac{\partial}{\partial x_3} \tag{M.33}$$

and similarly for $\hat{X}_2$ and $\hat{X}_3$ as in (M.21), together with (defining $\mathrm{d}\alpha_4 = \eta_1$ etc)

$$\hat{X}_4 = \mathrm{i}\left(-x_4 \frac{\partial}{\partial x_1} + x_1 \frac{\partial}{\partial x_4}\right) \tag{M.34}$$

$$\hat{X}_5 = \mathrm{i}\left(-x_4 \frac{\partial}{\partial x_2} + x_2 \frac{\partial}{\partial x_4}\right) \tag{M.35}$$

$$\hat{X}_6 = \mathrm{i}\left(-x_4 \frac{\partial}{\partial x_3} + x_3 \frac{\partial}{\partial x_4}\right). \tag{M.36}$$

Relabelling these last three generators as $\hat{Y}_1 \equiv \hat{X}_4,\ \hat{Y}_2 \equiv \hat{X}_5,\ \hat{Y}_3 \equiv \hat{X}_6$, we find the following algebra:

$$[\hat{X}_i, \hat{X}_j] = \mathrm{i}\epsilon_{ijk} \hat{X}_k \tag{M.37}$$

$$[\hat{X}_i, \hat{Y}_j] = \mathrm{i}\epsilon_{ijk} \hat{Y}_k \tag{M.38}$$

$$[\hat{Y}_i, \hat{Y}_j] = \mathrm{i}\epsilon_{ijk} \hat{X}_k \tag{M.39}$$

together with

$$[\hat{X}_1, \hat{Y}_1] = [\hat{X}_2, \hat{Y}_2] = [\hat{X}_3, \hat{Y}_3] = 0. \tag{M.40}$$

(M.37) confirms that the three generators controlling infinitesimal transformations among the first three components $\boldsymbol{x}$ obey the angular momentum commutation relations. (M.37)–(M.40) constitute the algebra of SO(4).

This algebra may be simplified by introducing the linear combinations

$$\hat{M}_i = \tfrac{1}{2}(\hat{X}_i + \hat{Y}_i) \tag{M.41}$$

$$\hat{N}_i = \tfrac{1}{2}(\hat{X}_i - \hat{Y}_i) \tag{M.42}$$

which satisfy

$$[\hat{M}_i, \hat{M}_j] = \mathrm{i}\epsilon_{ijk}\hat{M}_k \tag{M.43}$$

$$[\hat{N}_i, \hat{N}_j] = \mathrm{i}\epsilon_{ijk}\hat{N}_k \tag{M.44}$$

$$[\hat{M}_i, \hat{N}_j] = 0. \tag{M.45}$$

From (M.43)–(M.45) we see that, in this form, the six generators have separated into two sets of three, each set obeying the algebra of SO(3) (or of SU(2)) and commuting with the other set. They therefore behave like two *independent* angular momentum operators. The algebra (M.43)–(M.45) is referred to as SU(2)×SU(2).

### M.4.4 The Lorentz group

In this case the quadratic form left invariant by the transformation is the Minkowskian one $(x^0)^2 - \boldsymbol{x}^2$ (see appendix D of volume 1). We may think of infinitesimal Lorentz transformations as corresponding physically to ordinary infinitesimal 3D rotations, together with infinitesimal pure velocity transformations ('boosts'). The basic 4-vector then transforms by

$$\left.\begin{array}{rcl} x^{0\prime} & = & x^0 - \boldsymbol{\eta}\cdot\boldsymbol{x} \\ \boldsymbol{x}' & = & \boldsymbol{x} - \boldsymbol{\epsilon}\times\boldsymbol{x} - \boldsymbol{\eta}x^0 \end{array}\right\} \tag{M.46}$$

where $\boldsymbol{\eta}$ is now the infinitesimal velocity parameter (the reader may check that $(x^0)^2 - \boldsymbol{x}^2$ is indeed left invariant by (M.46), to first order in $\boldsymbol{\epsilon}$ and $\boldsymbol{\eta}$). The six generators are then $\hat{X}_1$, $\hat{X}_2$, $\hat{X}_3$ as in (M.21), together with

$$\hat{K}_1 = -\mathrm{i}\left(x^1\frac{\partial}{\partial x^0} + x^0\frac{\partial}{\partial x^1}\right) \tag{M.47}$$

$$\hat{K}_2 = -\mathrm{i}\left(x^2\frac{\partial}{\partial x^0} + x^0\frac{\partial}{\partial x^2}\right) \tag{M.48}$$

$$\hat{K}_3 = -\mathrm{i}\left(x^3\frac{\partial}{\partial x^0} + x^0\frac{\partial}{\partial x^3}\right). \tag{M.49}$$

The corresponding algebra is

$$[\hat{X}_i, \hat{X}_j] = \mathrm{i}\epsilon_{ijk}\hat{X}_k \tag{M.50}$$

$$[\hat{X}_i, \hat{K}_j] = \mathrm{i}\epsilon_{ijk}\hat{K}_k \tag{M.51}$$

$$[\hat{K}_i, \hat{K}_j] = -\mathrm{i}\epsilon_{ijk}\hat{X}_k. \tag{M.52}$$

Note the minus sign on the right-hand side of (M.52) as compared with (M.39).

### M.4.5 SU(3)

A general infinitesimal SU(3) transformation may be written as (cf (12.71) and (12.72))

$$\begin{pmatrix} q_1 \\ q_2 \\ q_3 \end{pmatrix}' = \left(1 + \mathrm{i}\frac{1}{2}\boldsymbol{\eta}\cdot\boldsymbol{\lambda}\right)\begin{pmatrix} q_1 \\ q_2 \\ q_3 \end{pmatrix} \tag{M.53}$$

where there are now eight of these $\boldsymbol{\eta}$'s, $\boldsymbol{\eta} = (\eta_1, \eta_2, \ldots, \eta_8)$ and the $\lambda$-matrices are the Gell-Mann matrices

$$\lambda_1 = \begin{pmatrix} 0 & 1 & 0 \\ 1 & 0 & 0 \\ 0 & 0 & 0 \end{pmatrix} \quad \lambda_2 = \begin{pmatrix} 0 & -\mathrm{i} & 0 \\ \mathrm{i} & 0 & 0 \\ 0 & 0 & 0 \end{pmatrix} \quad \lambda_3 = \begin{pmatrix} 1 & 0 & 0 \\ 0 & -1 & 0 \\ 0 & 0 & 0 \end{pmatrix} \tag{M.54}$$

$$\lambda_4 = \begin{pmatrix} 0 & 0 & 1 \\ 0 & 0 & 0 \\ 1 & 0 & 0 \end{pmatrix} \quad \lambda_5 = \begin{pmatrix} 0 & 0 & -\mathrm{i} \\ 0 & 0 & 0 \\ \mathrm{i} & 0 & 0 \end{pmatrix} \quad \lambda_6 = \begin{pmatrix} 0 & 0 & 0 \\ 0 & 0 & 1 \\ 0 & 1 & 0 \end{pmatrix} \tag{M.55}$$

$$\lambda_7 = \begin{pmatrix} 0 & 0 & 0 \\ 0 & 0 & -\mathrm{i} \\ 0 & \mathrm{i} & 0 \end{pmatrix} \quad \lambda_8 = \begin{pmatrix} \frac{1}{\sqrt{3}} & 0 & 0 \\ 0 & \frac{1}{\sqrt{3}} & 0 \\ 0 & 0 & -\frac{2}{\sqrt{3}} \end{pmatrix}. \tag{M.56}$$

In this parametrization the first three of the eight generators $\hat{G}_r$ $(r = 1, 2, \ldots, 8)$ are the same as $\hat{X}_1'$, $\hat{X}_2'$, $\hat{X}_3'$ of (M.29)–(M.30). The others may be constructed as usual from (M.10); for example,

$$\hat{G}_5 = \frac{\mathrm{i}}{2}\left(q_3\frac{\partial}{\partial q_1} - q_1\frac{\partial}{\partial q_3}\right) \qquad \hat{G}_7 = \frac{\mathrm{i}}{2}\left(q_3\frac{\partial}{\partial q_2} - q_2\frac{\partial}{\partial q_3}\right). \tag{M.57}$$

The SU(3) algebra is found to be

$$[\hat{G}_a, \hat{G}_b] = \mathrm{i} f_{abc}\hat{G}_c \tag{M.58}$$

where $a$, $b$ and $c$ each run from 1 to 8. The structure constants are $\mathrm{i} f_{abc}$, and the non-vanishing $f$'s are as follows:

$$f_{123} = 1 \quad f_{147} = 1/2 \quad f_{156} = -1/2 \quad f_{246} = 1/2 \quad f_{257} = 1/2 \tag{M.59}$$

$$f_{345} = 1/2 \quad f_{367} = -1/2 \quad f_{458} = \sqrt{3}/2 \quad f_{678} = \sqrt{3}/2. \tag{M.60}$$

Note that the $f$'s are anti-symmetric in all pairs of indices (Carruthers 1966, chapter 2).

## M.5 Matrix representations of generators and of Lie groups

We have shown how the generators $\hat{X}_1, \hat{X}_2, \ldots, \hat{X}_r$ of a Lie group can be constructed as differential operators, understood to be acting on functions of the 'coordinates' to which the transformations of the group refer. These generators satisfy certain commutation relations, the Lie algebra of the group. For any given Lie algebra, it is also possible to find sets of *matrices* $X_1, X_2, \ldots, X_r$ (without hats) which satisfy the same commutation relations as the $\hat{X}_\nu$'s—that is, they have the same algebra. Such matrices are said to form a (matrix) representation of the Lie algebra or, equivalently, of the generators. The idea is familiar from the study of angular momentum in quantum mechanics (Schiff 1968, section 27), where the entire theory may be developed from the commutation relations (with $\hbar = 1$)

$$[\hat{J}_i, \hat{J}_j] = \mathrm{i}\epsilon_{ijk}\hat{J}_k \tag{M.61}$$

for the angular momentum operators $\hat{J}_i$, together with the physical requirement that the $\hat{J}_i$'s (and the matrices representing them) must be Hermitian. In this case the matrices are of the form (in quantum-mechanical notation)

$$\left(J_i^{(J)}\right)_{M'_J M_J} \equiv \langle JM'_J|\hat{J}_i|JM_J\rangle \tag{M.62}$$

where $|JM_J\rangle$ is an eigenstate of $\hat{\boldsymbol{J}}^2$ and of $\hat{J}_3$ with eigenvalues $J(J+1)$ and $M_J$ respectively. Since $M_J$ and $M'_J$ each run over the $2J+1$ values defined by $-J \leq M_J,\ M'_J \leq J$, the matrices $J_i^{(J)}$ are of dimension $(2J+1) \times (2J+1)$. Clearly, since the generators of SU(2) have the same algebra as (M.61), an identical matrix representation may be obtained for them: these matrices were denoted by $T_i^{(T)}$ in section 12.1.2. It is important to note that $J$ (or $T$) can take an infinite sequence of values $J = 0, 1/2, 1, 3/2, \ldots$, corresponding physically to various 'spin' magnitudes. Thus there are infinitely many sets of three matrices $(J_1^{(J)}, J_2^{(J)}, J_3^{(J)})$ all with the same commutation relations as (M.61).

A similar method for obtaining matrix representations of Lie algebras may be followed in other cases. In physical terms, the problem amounts to finding a correct labelling of the base states, analogous to $|JM\rangle$. In the latter case, the quantum number $J$ specifies each different representation. The reason it does so is because (as should be familiar) the corresponding operator $\hat{\boldsymbol{J}}^2$ commutes with every generator:

$$[\hat{\boldsymbol{J}}^2, \hat{J}_i] = 0. \tag{M.63}$$

Such an operator is called a *Casimir operator* and by a lemma due to Schur (Hammermesh 1962, pp 100–1) it must be a multiple of the unit operator. The numerical value it has differs for each different representation and may, therefore, be used to characterize a representation (namely as '$J = 0$', '$J = 1/2$', etc).

In general, more than one such operator is needed to characterize a representation completely. For example, in SO(4), the two operators $\hat{\boldsymbol{M}}^2$ and

$\hat{N}^2$ commute with all the generators and take values $M(M+1)$ and $N(N+1)$ respectively, where $M, N = 0, 1/2, 1, \ldots$. Thus, the labelling of the matrix elements of the generators is the same as it would be for two independent particles, one of spin $M$ and the other of spin $N$. For given $M, N$ the matrices are of dimension $[(2M+1)\times(2N+1)]\times[(2M+1)+(2N+1)]$. The number of Casimir operators required to characterize a representation is called the *rank* of the group (or the algebra). This is also equal to the number of independent mutually commuting generators (though this is by no means obvious). Thus, SO(4) is a rank two group, with two commuting generators $\hat{M}_3$ and $\hat{N}_3$; so is SU(3), since $\hat{G}_3$ and $\hat{G}_8$ commute. Two Casimir operators are therefore required to characterize the representations of SU(3), which may be taken to be the 'quadratic' one

$$\hat{C}_2 \equiv \hat{G}_1^2 + \hat{G}_2^2 + \cdots + \hat{G}_8^2 \tag{M.64}$$

together with a 'cubic' one

$$\hat{C}_3 \equiv d_{abc}\hat{G}_a\hat{G}_b\hat{G}_c \tag{M.65}$$

where the coefficients $d_{abc}$ are defined by the relation

$$\{\lambda_a, \lambda_b\} = \tfrac{4}{3}\delta_{ab}I + 2d_{abc}\lambda_c \tag{M.66}$$

and are symmetric in all pairs of indices (they are tabulated in Carruthers 1966, table 2.1). In practice, for the few SU(3) representations that are actually required, it is more common to denote them (as we have in the text) by their dimensionality, which for the cases **1** (singlet), **3** (triplet), **3*** (anti-triplet), **8** (octet) and **10** (decuplet) is, in fact, a unique labelling. The values of $\hat{C}_2$ in these representations are

$$\hat{C}_2(\mathbf{1}) = 0 \qquad \hat{C}_2(\mathbf{3}, \mathbf{3}^*) = 4/3 \qquad \hat{C}_2(\mathbf{8}) = 3 \qquad \hat{C}_2(\mathbf{10}) = 6. \tag{M.67}$$

Having characterized a given representation by the eigenvalues of the Casimir operator(s), a further labelling is then required to characterize the states within a given representation (the analogue of the eigenvalue of $\hat{J}_3$ for angular momentum). For SO(4) these further labels may be taken to be the eigenvalues of $\hat{M}_3$ and $\hat{N}_3$: for SU(3) they are the eigenvalues of $\hat{G}_3$ and $\hat{G}_8$—i.e. those corresponding to the third component of isospin and hypercharge, in the flavour case (see figures 12.3 and 12.4).

In the case of groups whose elements are themselves matrices, such as SO(3), SO(4), SU(2), SU(3) and the Lorentz group, one particular representation of the generators may always be obtained by considering the general form of a matrix in the group which is infinitesimally close to the unit element. In a suitable parametrization, we may write such a matrix as

$$1 + \mathrm{i}\sum_{\nu=1}^{r} \epsilon_\nu X_\nu^{(\mathcal{G})} \tag{M.68}$$

where $(\epsilon_1, \epsilon_2, \ldots, \epsilon_r)$ are infinitesimal parameters, and $(X_1^{(\mathcal{G})}, X_2^{(\mathcal{G})}, \ldots, X_r^{(\mathcal{G})})$ are matrices representing the generators of the (matrix) group $\mathcal{G}$. This is exactly the same procedure we followed for SU(2) in section 12.1.1, where we found from (12.26) that the three $X_\nu^{(\mathrm{SU(2)})}$'s were just $\boldsymbol{\tau}/2$, satisfying the SU(2) algebra. Similarly, in section 12.2 we saw that the eight SU(3) $X_\nu^{(\mathrm{SU(3)})}$'s were just $\boldsymbol{\lambda}/2$, satisfying the SU(3) algebra. These particular two representations are called the *fundamental* representations of the SU(2) and SU(3) algebras, respectively; they are the representations of lowest dimensionality. For SO(3), the three $X_\nu^{(\mathrm{SO(3)})}$'s are (from (M.17))

$$X_1^{(\mathrm{SO(3)})} = \begin{pmatrix} 0 & 0 & 0 \\ 0 & 0 & -\mathrm{i} \\ 0 & \mathrm{i} & 0 \end{pmatrix}$$
$$X_2^{(\mathrm{SO(3)})} = \begin{pmatrix} 0 & 0 & \mathrm{i} \\ 0 & 0 & 0 \\ -\mathrm{i} & 0 & 0 \end{pmatrix}$$
$$X_3^{(\mathrm{SO(3)})} = \begin{pmatrix} 0 & -\mathrm{i} & 0 \\ \mathrm{i} & 0 & 0 \\ 0 & 0 & 0 \end{pmatrix} \tag{M.69}$$

which are the same as the $3 \times 3$ matrices $T_i^{(1)}$ of (12.48):

$$\left(T_i^{(1)}\right)_{jk} = -\mathrm{i}\epsilon_{ijk}. \tag{M.70}$$

The matrices $\tau_i/2$ and $T_i^{(1)}$ correspond to the values $J = 1/2$, $J = 1$, respectively, in angular momentum terms.

It is not a coincidence that the coefficients on the right-hand side of (M.70) are (minus) the SO(3) structure constants. One representation of a Lie algebra is always provided by a set of matrices $\{X_\nu^{(\mathrm{R})}\}$ whose elements are defined by

$$\left(X_\lambda^{(\mathrm{R})}\right)_{\mu\nu} = -c_{\lambda\mu}^{\nu} \tag{M.71}$$

where the $c$'s are the structure constants of (M.12), and each of $\mu$, $\nu$, $\lambda$ runs from 1 to $r$. Thus, these matrices are of dimensionality $r \times r$, where $r$ is the number of generators. That this prescription works is due to the fact that the generators satisfy the *Jacobi identity*

$$[\hat{X}_\lambda, [\hat{X}_\mu, \hat{X}_\nu]] + [\hat{X}_\mu, [\hat{X}_\nu, \hat{X}_\lambda]] + [\hat{X}_\nu, [\hat{X}_\lambda, \hat{X}_\mu]] = 0. \tag{M.72}$$

Using (M.12) to evaluate the commutators, and the fact that the generators are independent, we obtain

$$c_{\mu\nu}^{\alpha} c_{\lambda\alpha}^{\beta} + c_{\nu\lambda}^{\alpha} c_{\mu\alpha}^{\beta} + c_{\lambda\mu}^{\alpha} c_{\nu\alpha}^{\beta} = 0. \tag{M.73}$$

The reader may fill in the steps leading from here to the desired result:

$$\left(X_\lambda^{(R)}\right)_{\nu\alpha}\left(X_\mu^{(R)}\right)_{\alpha\beta} - \left(X_\mu^{(R)}\right)_{\nu\alpha}\left(X_\lambda^{(R)}\right)_{\alpha\beta} = c_{\lambda\mu}^{\alpha}\left(X_\alpha^{(R)}\right)_{\nu\beta}. \qquad \text{(M.74)}$$

(M.74) is, of course, precisely the $(\nu\beta)$ matrix element of

$$[X_\lambda^{(R)}, X_\mu^{(R)}] = c_{\lambda\mu}^{\alpha} X_\alpha^{(R)} \qquad \text{(M.75)}$$

showing that the $X_\mu^{(R)}$'s satisfy the group algebra (M.12), as required. The representation in which the generators are represented by (minus) the structure constants, in the sense of (M.71), is called the *regular* or *adjoint* representation.

Having obtained any particular matrix representation $\boldsymbol{X}^{(P)}$ of the generators of a group $\mathcal{G}$, a corresponding *matrix representation of the group elements* can be obtained by exponentiation, via

$$D^{(P)}(\boldsymbol{\alpha}) = \exp\{i\boldsymbol{\alpha}\cdot\boldsymbol{X}^{(P)}\} \qquad \text{(M.76)}$$

where $\boldsymbol{\alpha} = (\alpha_1, \alpha_2, \ldots, \alpha_r)$ (see (12.31) and (12.49) for SU(2), and (12.74) and (12.81) for SU(3)). In the case of the groups whose elements are matrices, exponentiating the generators $\boldsymbol{X}^{(\mathcal{G})}$ just recreates the general matrices of the group, so we may call this the 'self-representation': the one in which the group elements are represented by themselves. In the more general case (M.76), the crucial property of the matrices $D^{(P)}(\boldsymbol{\alpha})$ is that they obey the same group combination law as the elements of the group $\mathcal{G}$ they are representing; that is, if the group elements obey

$$g(\boldsymbol{\alpha})g(\boldsymbol{\beta}) = g(\boldsymbol{\gamma}(\boldsymbol{\alpha},\boldsymbol{\beta})) \qquad \text{(M.77)}$$

then

$$D^{(P)}(\boldsymbol{\alpha})D^{(P)}(\boldsymbol{\beta}) = D^{(P)}(\boldsymbol{\gamma}(\boldsymbol{\alpha},\boldsymbol{\beta})). \qquad \text{(M.78)}$$

It is a rather remarkable fact that there are certain, say, $10\times 10$ matrices which multiply together in exactly the same way as the rotation matrices of SO(3).

## M.6 The Lorentz group

Consideration of matrix representations of the Lorentz group provides insight into the equations of relativistic quantum mechanics, for example the Dirac equation. Consider the infinitesimal Lorentz transformation (M.46). The $4\times 4$ matrix corresponding to this may be written in the form

$$1 + i\boldsymbol{\epsilon}\cdot\boldsymbol{X}^{(LG)} - i\boldsymbol{\eta}\cdot\boldsymbol{K}^{(LG)} \qquad \text{(M.79)}$$

where

$$X_1^{(LG)} = \begin{pmatrix} 0 & 0 & 0 & 0 \\ 0 & 0 & 0 & 0 \\ 0 & 0 & 0 & -i \\ 0 & 0 & i & 0 \end{pmatrix} \text{ etc} \qquad \text{(M.80)}$$

(as in (M.69) but with an extra border of 0's) and

$$K_1^{(\mathrm{LG})} = \begin{pmatrix} 0 & -\mathrm{i} & 0 & 0 \\ -\mathrm{i} & 0 & 0 & 0 \\ 0 & 0 & 0 & 0 \\ 0 & 0 & 0 & 0 \end{pmatrix}$$

$$K_2^{(\mathrm{LG})} = \begin{pmatrix} 0 & 0 & -\mathrm{i} & 0 \\ 0 & 0 & 0 & 0 \\ -\mathrm{i} & 0 & 0 & 0 \\ 0 & 0 & 0 & 0 \end{pmatrix}$$

$$K_3^{(\mathrm{LG})} = \begin{pmatrix} 0 & 0 & 0 & -\mathrm{i} \\ 0 & 0 & 0 & 0 \\ 0 & 0 & 0 & 0 \\ -\mathrm{i} & 0 & 0 & 0 \end{pmatrix}. \tag{M.81}$$

In (M.80) and (M.81) the matrices are understood to be acting on the four-component vector

$$\begin{pmatrix} x^0 \\ x^1 \\ x^2 \\ x^3 \end{pmatrix}. \tag{M.82}$$

It is straightforward to check that the matrices $X_i^{(\mathrm{LG})}$ and $K_i^{(\mathrm{LG})}$ satisfy the algebra (M.50)–(M.52) as expected.

An important point to note is that the matrices $K_i^{(\mathrm{LG})}$, in contrast to $X_i^{(\mathrm{LG})}$ or $X_i^{(\mathrm{SO(3)})}$, and to the corresponding matrices of SU(2) and SU(3), are *not* Hermitian. A theorem states that only the generators of *compact* Lie groups can be represented by finite-dimensional Hermitian matrices. Here 'compact' means that the domain of variation of all the parameters is bounded (none exceeds a given positive number $p$ in absolute magnitude) and closed (the limit of every convergent sequence of points in the set also lies in the set). For the Lorentz group, the limiting velocity $c$ is not included (the $\gamma$-factor goes to infinity), and so the group is non-compact.

In a general representation of the Lorentz group, the generators $X_i$, $K_i$ will obey the algebra (M.50)–(M.52). Let us introduce the combinations

$$\boldsymbol{P} \equiv \tfrac{1}{2}(\boldsymbol{X} + \mathrm{i}\boldsymbol{K}) \tag{M.83}$$

$$\boldsymbol{Q} \equiv \tfrac{1}{2}(\boldsymbol{X} - \mathrm{i}\boldsymbol{K}). \tag{M.84}$$

Then the algebra becomes

$$[P_i, P_j] = \mathrm{i}\epsilon_{ijk} P_k \tag{M.85}$$

$$[Q_i, Q_j] = \mathrm{i}\epsilon_{ijk} Q_k \tag{M.86}$$

$$[P_i, Q_j] = 0 \tag{M.87}$$

which are apparently the same as (M.43)–(M.45). We can see from (M.81) that the matrices $\mathrm{i}\boldsymbol{K}^{(\mathrm{LG})}$ *are* Hermitian, and the same is, in fact, true in a general finite-dimensional representation. So we can appropriate standard angular momentum theory to set up the representations of the algebra of the $\boldsymbol{P}$'s and $\boldsymbol{Q}$'s—namely, they behave just like two independent (mutually commuting) angular momenta. The eigenvalues of $\boldsymbol{P}^2$ are of the form $P(P+1)$, for $P = 0, 1/2, \ldots$, and similarly for $\boldsymbol{Q}^2$: the eigenvalues of $P_3$ are $M_P$ where $-P \leq M_P \leq P$, and similarly for $Q_3$.

Consider the particular case where the eigenvalue of $\boldsymbol{Q}^2$ is zero ($Q = 0$) and the value of $P$ is 1/2. The first condition implies that the $\boldsymbol{Q}$'s are identically zero, so that

$$\boldsymbol{X} = \mathrm{i}\boldsymbol{K} \tag{M.88}$$

in this representation, while the second condition tells us that

$$\boldsymbol{P} = \tfrac{1}{2}(\boldsymbol{X} + \mathrm{i}\boldsymbol{K}) = \tfrac{1}{2}\boldsymbol{\sigma} \tag{M.89}$$

the familiar matrices for spin-$\frac{1}{2}$. We label this representation by the values of $P$ ($\frac{1}{2}$) and $Q$ (0) (these are the eigenvalues of the two Casimir operators). Then using (M.88) and (M.89) we find that

$$\boldsymbol{X}^{(\frac{1}{2},0)} = \tfrac{1}{2}\boldsymbol{\sigma} \tag{M.90}$$

and

$$\boldsymbol{K}^{(\frac{1}{2},0)} = -\frac{\mathrm{i}}{2}\boldsymbol{\sigma}. \tag{M.91}$$

Now recall that the general infinitesimal Lorentz transformation has the form

$$1 + \mathrm{i}\boldsymbol{\epsilon} \cdot \boldsymbol{X} - \mathrm{i}\boldsymbol{\eta} \cdot \boldsymbol{K}. \tag{M.92}$$

In the present case, this becomes

$$1 + \mathrm{i}\boldsymbol{\epsilon} \cdot \boldsymbol{\sigma}/2 - \boldsymbol{\eta} \cdot \boldsymbol{\sigma}/2. \tag{M.93}$$

These matrices are of dimension $2 \times 2$, and act on two-component spinors, which therefore transform under an infinitesimal Lorentz transformation by

$$\phi' = (1 + \mathrm{i}\boldsymbol{\epsilon} \cdot \boldsymbol{\sigma}/2 - \boldsymbol{\eta} \cdot \boldsymbol{\sigma}/2)\phi. \tag{M.94}$$

We say that $\phi$ 'transforms as the (1/2, 0) representation of the Lorentz group'. The '$1 + \mathrm{i}\boldsymbol{\epsilon} \cdot \boldsymbol{\sigma}/2$' part is the familiar (infinitesimal) rotation matrix for spinors, first met in section 4.4: it exponentiates to give $\exp(\mathrm{i}\boldsymbol{\alpha} \cdot \boldsymbol{\sigma}/2)$ for finite rotations. The '$-\boldsymbol{\eta} \cdot \boldsymbol{\sigma}/2$' part shows how such a spinor transforms under a pure (infinitesimal) velocity transformation. Exponentiating this part gives a transformation law

$$\phi' = \exp(-\tfrac{1}{2}v\hat{\boldsymbol{n}} \cdot \boldsymbol{\sigma})\phi \tag{M.95}$$

for a finite boost characterized by a speed $v$ and a direction $\hat{\boldsymbol{n}}$.

There is, however, a second two-dimensional representation, which is characterized by the labelling $P = 0$, $Q = 1/2$, which we denote by (0, 1/2). In this case, the previous steps yield

$$\boldsymbol{X}^{(0,\frac{1}{2})} = \tfrac{1}{2}\boldsymbol{\sigma} \tag{M.96}$$

as before, but

$$\boldsymbol{K}^{(0,\frac{1}{2})} = \frac{\mathrm{i}}{2}\boldsymbol{\sigma}. \tag{M.97}$$

So the corresponding two-component spinor $\chi$ transforms by

$$\chi' = (1 + \mathrm{i}\boldsymbol{\epsilon}\cdot\boldsymbol{\sigma}/2 + \boldsymbol{\eta}\cdot\sigma/2)\chi. \tag{M.98}$$

We see that $\phi$ and $\chi$ behave the same under rotations but 'oppositely' under boosts.

How does all this relate to the spin-$\frac{1}{2}$ wave equations we have introduced in the text—that is, the Dirac equation of chapter 4 and the Weyl equation (4.150) and its $u_{\mathrm{L}}$ analogue (see also (20.56))? Consider the Weyl equation

$$(E - \boldsymbol{\sigma}\cdot\boldsymbol{p})\phi = 0 \tag{M.99}$$

for a massless spin-$\frac{1}{2}$ particle with energy $E$ and momentum $\boldsymbol{p}$ such that $E = |\boldsymbol{p}|$. The work of section 4.4 guarantees that, under a three-dimensional rotation, $\phi$ will transform by

$$\phi' = \exp(\mathrm{i}\boldsymbol{\alpha}\cdot\boldsymbol{\sigma}/2)\phi. \tag{M.100}$$

So let us consider boosts. An infinitesimal velocity transformation (see M.46) takes $(E, \boldsymbol{p})$ to

$$E' = E - \boldsymbol{\eta}\cdot\boldsymbol{p} \tag{M.101}$$

$$\boldsymbol{p}' = \boldsymbol{p} - \boldsymbol{\eta}E. \tag{M.102}$$

In this primed frame, (M.99) becomes

$$(E' - \boldsymbol{\sigma}\cdot\boldsymbol{p}')\phi' = 0. \tag{M.103}$$

We shall verify that for (M.99), (M.101), (M.102) and (M.103) to be consistent, $\phi$ must transform as

$$\phi' = (1 - \boldsymbol{\eta}\cdot\boldsymbol{\sigma}/2)\phi \tag{M.104}$$

exactly as in (M.94), demonstrating that this $\phi$ is (as the notation has assumed) a (1/2, 0) object.

Let

$$V_\eta = (1 - \boldsymbol{\eta}\cdot\boldsymbol{\sigma}/2). \tag{M.105}$$

Then applying $V_\eta^{-1}$ to (M.99) and inserting $V_\eta^{-1}V_\eta$ gives

$$[V_\eta^{-1}(E - \boldsymbol{\sigma}\cdot\boldsymbol{p})V_\eta^{-1}]V_\eta\phi = 0. \tag{M.106}$$

The part in square brackets is, to first order in $\boldsymbol{\eta}$,

$$\begin{aligned}(1+\boldsymbol{\eta}\cdot\boldsymbol{\sigma}/2)(E-\boldsymbol{\sigma}\cdot\boldsymbol{p})(1+\boldsymbol{\eta}\cdot\boldsymbol{\sigma}/2) &= (E-\boldsymbol{\eta}\cdot\boldsymbol{p})-\boldsymbol{\sigma}\cdot(\boldsymbol{p}-E\boldsymbol{\eta}) \\ &= E'-\boldsymbol{\sigma}\cdot\boldsymbol{p}'. \end{aligned} \tag{M.107}$$

Hence (M.103) follows with the identification (M.104), showing that also under boosts $\phi$ transforms as a (1/2, 0) object. The reader who has worked through problem 4.15 will recognize these manipulations, with a sign change for the infinitesimal velocity parameters in (M.101) and (M.102) as compared with (4.151) and (4.152). In a similar way, one can verify that the Weyl spinor $\chi$ satisfying

$$(E+\boldsymbol{\sigma}\cdot\boldsymbol{p})\chi = 0 \tag{M.108}$$

transforms as a (0, 1/2) object.

The four-component Dirac wavefunction $\psi$ is put together from one $\phi$ and one $\chi$, via

$$\psi = \begin{pmatrix}\phi \\ \chi\end{pmatrix} \tag{M.109}$$

and describes a *massive* spin-$\frac{1}{2}$ particle according to the equations

$$\begin{aligned} E\phi &= \boldsymbol{\sigma}\cdot\boldsymbol{p}\phi + m\chi \\ E\chi &= -\boldsymbol{\sigma}\cdot\boldsymbol{p}\chi + m\phi \end{aligned} \tag{M.110}$$

as discussed in problem 4.15 and section 20.4. The $\phi$ and $\chi$ spinors are projected out of $\psi$ by the chirality operators

$$P_{\mathrm{R,L}} = \tfrac{1}{2}(1\pm\gamma_5) \tag{M.111}$$

as discussed in section 20.4, in the representation such that

$$\gamma_5 = \begin{pmatrix}1 & 0 \\ 0 & -1\end{pmatrix}. \tag{M.112}$$

## M.7 The relation between SU(2) and SO(3)

We have seen (sections M.4.1 and M.4.2) that the algebras of these two groups are identical. So the groups are isomorphic in the vicinity of their respective identity elements. Furthermore, matrix representations of one algebra automatically provide representations of the other. Since exponentiating these infinitesimal matrix transformations produces matrices representing group elements corresponding to finite transformations in both cases, it might appear that the groups are fully isomorphic. But actually they are not, as we shall now discuss.

We begin by re-considering the parameters used to characterize elements of SO(3) and SU(2). A general 3D rotation is described by the SO(3) matrix $R(\hat{\boldsymbol{n}}, \theta)$, where $\hat{\boldsymbol{n}}$ is the axis of the rotation and $\theta$ is the angle of rotation. For example,

$$R(\hat{z}, \theta) = \begin{pmatrix} \cos\theta & \sin\theta & 0 \\ -\sin\theta & \cos\theta & 0 \\ 0 & 0 & 1 \end{pmatrix}. \qquad \text{(M.113)}$$

On the other hand, we can write the general SU(2) matrix $\boldsymbol{V}$ in the form

$$\boldsymbol{V} = \begin{pmatrix} a & b \\ -b^* & a^* \end{pmatrix} \qquad \text{(M.114)}$$

where $|a|^2 + |b|^2 = 1$ from the unit determinant condition. It therefore depends on three real parameters, the choice of which we are now going to examine in more detail than previously. In (12.32) we wrote $\boldsymbol{V}$ as $\exp(\mathrm{i}\boldsymbol{\alpha} \cdot \boldsymbol{\tau}/2)$, which certainly involves three real parameters $\alpha_1, \alpha_2, \alpha_3$; and below (12.35) we proposed, further, to write $\boldsymbol{\alpha} = \hat{\boldsymbol{n}}\theta$, where $\theta$ is an angle and $\hat{\boldsymbol{n}}$ is a unit vector. Then, since (as the reader may verify)

$$\exp(\mathrm{i}\theta\boldsymbol{\tau} \cdot \hat{\boldsymbol{n}}/2) = \cos\theta/2 + \mathrm{i}\boldsymbol{\tau} \cdot \hat{\boldsymbol{n}} \sin\theta/2 \qquad \text{(M.115)}$$

it follows that this latter parametrization corresponds to writing, in (M.114),

$$a = \cos\theta/2 + \mathrm{i}n_z \sin\theta/2 \qquad b = (n_y + \mathrm{i}n_x)\sin\theta/2 \qquad \text{(M.116)}$$

with $n_x^2 + n_y^2 + n_z^2 = 1$. Clearly the condition $|a|^2 + |b|^2 = 1$ is satisfied, and one can convince oneself that the full range of $a$ and $b$ is covered if $\theta/2$ lies between 0 and $\pi$ (in particular, it is not necessary to extend the range of $\theta/2$ so as to include the interval $\pi$ to $2\pi$, since the corresponding region of $a, b$ can be covered by changing the orientation of $\hat{\boldsymbol{n}}$, which has not been constrained in any way). It follows that the parameters $\boldsymbol{\alpha}$ satisfy $\boldsymbol{\alpha}^2 \leq 4\pi^2$; that is, the space of the $\alpha$'s is the interior, and surface, of a sphere of radius $2\pi$, as shown in figure M.1.

What about the parameter space of SO(3)? In this case, the same parameters $\hat{\boldsymbol{n}}$ and $\theta$ specify a rotation, but now $\theta$ (rather than $\theta/2$) runs from 0 to $\pi$. However, we may allow the range of $\theta$ to extend to $2\pi$, by taking advantage of the fact that

$$R(\hat{\boldsymbol{n}}, \pi + \theta) = R(-\hat{\boldsymbol{n}}, \theta). \qquad \text{(M.117)}$$

Thus if we agree to limit $\hat{\boldsymbol{n}}$ to directions in the upper hemisphere of figure M.1, for 3D rotations, we can say that the whole sphere represents the parameter space of SU(2), but that of SO(3) is provided by the upper half only.

Now let us consider the correspondence—or *mapping*—between the matrices of SO(3) and SU(2): we want to see if it is one-to-one. The notation strongly suggests that the matrix $\boldsymbol{V}(\hat{\boldsymbol{n}}, \theta) \equiv \exp(\mathrm{i}\theta\hat{\boldsymbol{n}} \cdot \tau/2)$ of SU(2) corresponds to the matrix $R(\hat{\boldsymbol{n}}, \theta)$ of SO(3) but the way it actually works has a subtlety.

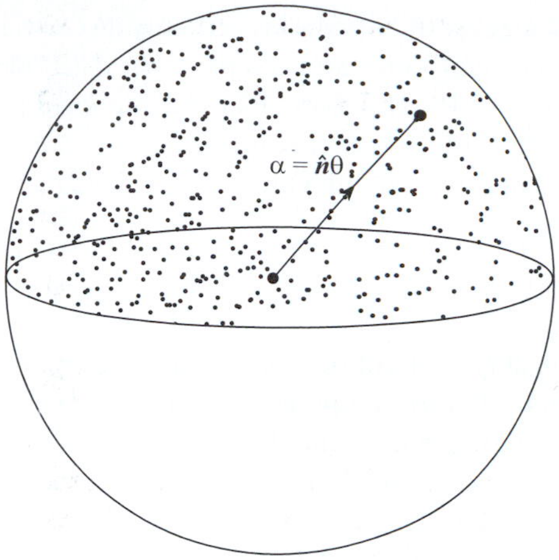


**Figure M.1.** The parameter spaces of SO(3) and SU(2): the whole sphere is the parameter space of SU(2), the upper (stippled) hemisphere that of SO(3).

We form the quantity $\boldsymbol{x} \cdot \boldsymbol{\tau}$, and assert that

$$\boldsymbol{x}' \cdot \boldsymbol{\tau} = \boldsymbol{V}(\hat{\boldsymbol{n}}, \theta)\, \boldsymbol{x} \cdot \boldsymbol{\tau}\, \boldsymbol{V}^{\dagger}(\hat{\boldsymbol{n}}, \theta) \tag{M.118}$$

where $\boldsymbol{x}' = R(\hat{\boldsymbol{n}}, \theta)\boldsymbol{x}$. We can easily verify (M.118) for the special case $R(\hat{z}, \theta)$, using (M.113): the general case follows with more labour (but the general infinitesimal case should, by now, be a familiar manipulation). (M.118) establishes a precise mapping between the elements of SU(2) and those of SO(3) but it is not one-to-one (i.e. not an isomorphism), since plainly $\boldsymbol{V}$ can always be replaced by $-\boldsymbol{V}$ and $\boldsymbol{x}'$ will be unchanged, and hence so will the associated SO(3) matrix $R(\hat{\boldsymbol{n}}, \theta)$. It is, therefore, a homomorphism.

Next, we prove a little theorem to the effect that the identity element $e$ of a group $\mathcal{G}$ must be represented by the unit matrix of the representation: $D(e) = I$. Let $D(a)$, $D(e)$ represent the elements $a, e$ of $\mathcal{G}$. Then $D(ae) = D(a)D(e)$ by the fundamental property (M.78) of representation matrices. However, $ae = a$ by the property of $e$. So we have $D(a) = D(a)D(e)$, and hence $D(e) = I$.

Now let us return to the correspondence between SU(2) and SO(3). $\boldsymbol{V}(\hat{\boldsymbol{n}}, \theta)$ corresponds to $R(\hat{\boldsymbol{n}}, \theta)$, but can an SU(2) matrix be said to provide a valid representation of SO(3)? Consider the case $\boldsymbol{V}(\hat{n} = \hat{z}, \theta = 2\pi)$. From (M.115) this is equal to

$$\begin{pmatrix} -1 & 0 \\ 0 & -1 \end{pmatrix} \tag{M.119}$$

but the corresponding rotation matrix, from (M.113), is the identity matrix. Hence, our theorem is violated, since (M.119) is plainly not the identity matrix of SU(2). Thus, the SU(2) matrices cannot be said to represent rotations, in the strict sense. Nevertheless, spin-$\frac{1}{2}$ particles certainly do exist, so Nature appears to make use of these 'not quite' representations! The SU(2) identity element is, of course, $\boldsymbol{V}(\hat{n} = \hat{z}, \theta = 4\pi)$, confirming that the rotational properties of a spinor are quite other than those of a classical object.

In fact, two and only two distinct elements of SU(2), namely

$$\begin{pmatrix} 1 & 0 \\ 0 & 1 \end{pmatrix} \quad \text{and} \quad \begin{pmatrix} -1 & 0 \\ 0 & -1 \end{pmatrix} \tag{M.120}$$

correspond to the identity element of SO(3) in the correspondence (M.118)—just as, in general, $\boldsymbol{V}$ and $-\boldsymbol{V}$ correspond to the same SO(3) element $R(\hat{\boldsymbol{n}}, \theta)$, as we saw. The failure to be a true representation is localized simply to a sign: we may indeed say that, up to a sign, SU(2) matrices provide a representation of SO(3). If we 'factor out' this sign, the groups are isomorphic. A more mathematically precise way of saying this is given in Jones (1990, chapter 8).

# APPENDIX N

## DIMENSIONAL REGULARIZATION

After combining propagator denominators of the form $(p^2 - m^2 + \mathrm{i}\epsilon)^{-1}$ by Feynman parameters (cf (10.40) and (11.16)) and shifting the origin of the loop momentum to complete the square (cf (10.42) and (11.16)), all one-loop Feynman integrals may be reduced to evaluating an integral of the form

$$I_d(\Delta, n) \equiv \int \frac{\mathrm{d}^d k}{(2\pi)^d} \frac{1}{[k^2 - \Delta + \mathrm{i}\epsilon]^n} \tag{N.1}$$

or to a similar integral with factors of $k$ (such as $k_\mu k_\nu$) in the numerator. We consider (N.1) first.

For our purposes, the case of physical interest is $d = 4$, and $n$ is commonly 2 (e.g. in one-loop self-energies). Power-counting shows that (N.1) diverges for $d \geq 2n$. The idea behind *dimensional regularization* ('t Hooft and Veltman 1972) is to treat $d$ as a variable parameter, taking values smaller than $2n$, so that (N.1) converges and can be evaluated explicitly as a function of $d$ (and, of course, the other variables, including $n$). Then the nature of the divergence as $d \to 4$ can be exposed (much as we did with the cut-off procedure in section 10.3) and dealt with by a suitable renormalization scheme. The crucial advantage of dimensional regularization is that it preserves gauge invariance, unlike the simple cut-off regularization we used in chapters 10 and 11.

We write

$$I_d = \frac{1}{(n-1)!} \left(\frac{\partial}{\partial \Delta}\right)^{n-1} \int \frac{\mathrm{d}^d k}{(2\pi)^d} \frac{1}{[k^2 - \Delta + \mathrm{i}\epsilon]}. \tag{N.2}$$

The $d$ dimensions are understood as one time-like dimension $k^0$ and $d - 1$ space-like dimensions. We begin (as discussed in connection with (15.75)) by 'Euclideanizing' the integral by setting $k^0 = \mathrm{i}k^e$ with $k^e$ real. Then the Minkowskian square $k^2$ becomes $-(k^e)^2 - \boldsymbol{k}^2 \equiv -k_{\mathrm{E}}^2$, and $\mathrm{d}^d k$ becomes $\mathrm{i}\mathrm{d}^d k_{\mathrm{E}}$, so that now

$$I_d = \frac{-\mathrm{i}}{(n-1)!} \left(\frac{\partial}{\partial \Delta}\right)^{n-1} \int \frac{\mathrm{d}^d k_{\mathrm{E}}}{(2\pi)^d} \frac{1}{(k_{\mathrm{E}}^2 + \Delta)} \tag{N.3}$$

the '$\mathrm{i}\epsilon$' may be understood as included in $\Delta$. The integral is evaluated by introducing the following way of writing $(k_{\mathrm{E}}^2 + \Delta)^{-1}$:

$$(k_{\mathrm{E}}^2 + \Delta)^{-1} = \int_0^\infty \mathrm{d}\beta \mathrm{e}^{-\beta(k_{\mathrm{E}}^2 + \Delta)} \tag{N.4}$$

which leads to

$$I_d = \frac{-\mathrm{i}}{(n-1)!}\left(\frac{\partial}{\partial \Delta}\right)^{n-1} \int_0^\infty \mathrm{d}\beta \int \frac{\mathrm{d}^d k_{\mathrm{E}}}{(2\pi)^d} \mathrm{e}^{-\beta(k_{\mathrm{E}}^2+\Delta)}. \tag{N.5}$$

The interchange of the orders of the $\beta$ and $k_{\mathrm{E}}$ integrations is permissible since $I_d$ is convergent. The $k_{\mathrm{E}}$ integrals are, in fact, a series of Gaussians:

$$\begin{aligned}\int \frac{\mathrm{d}^d k_{\mathrm{E}}}{(2\pi)^d} \mathrm{e}^{-\beta(k_{\mathrm{E}}^2+\Delta)} &= \mathrm{e}^{-\beta\Delta} \left\{ \prod_{j=1}^{d} \int \frac{\mathrm{d}k_j}{(2\pi)} \mathrm{e}^{-\beta k_j^2} \right\} \\ &= \frac{\mathrm{e}^{-\beta\Delta}}{(2\pi)^d} \left(\frac{\pi}{\beta}\right)^{d/2}. \end{aligned} \tag{N.6}$$

Hence,

$$\begin{aligned} I_d &= \frac{-\mathrm{i}}{(n-1)!} \frac{1}{(4\pi)^{d/2}} \left(\frac{\partial}{\partial \Delta}\right)^{n-1} \int \mathrm{d}\beta \mathrm{e}^{-\beta\Delta} \beta^{-d/2} \\ &= \frac{-\mathrm{i}}{(n-1)!} \frac{(-1)^{n-1}}{(4\pi)^{d/2}} \int \mathrm{d}\beta \mathrm{e}^{-\beta\Delta} \beta^{n-(d/2)-1}. \end{aligned} \tag{N.7}$$

The last integral can be written in terms of Euler's integral for the *gamma function* $\Gamma(z)$ defined by

$$\Gamma(z) = \int_0^\infty x^{z-1} \mathrm{e}^{-x}\, \mathrm{d}x. \tag{N.8}$$

Since $\Gamma(n) = (n-1)!$, it is convenient to write (N.8) entirely in terms of $\Gamma$ functions as

$$I_d = \mathrm{i} \frac{(-1)^n}{(4\pi)^{d/2}} \frac{\Gamma(n - d/2)}{\Gamma(n)} \Delta^{(d/2)-n}. \tag{N.9}$$

This formula agrees with (15.76) for the case $n = 2$ (remembering that $I_d$ was defined in Minkowski space).

Equation (N.9) gives an explicit definition of $I_d$ which can be used for any value of $d$, not necessarily an integer. As a function of $z$, $\Gamma(z)$ has isolated poles (see apppendix F of volume 1) at $z = 0, -1, -2, \ldots$. The behaviour near $z = 0$ is given by

$$\Gamma(z) = \frac{1}{z} - \gamma + O(z) \tag{N.10}$$

where $\gamma$ is the Euler–Mascheroni constant having the value $\gamma \approx 0.5772$. Using

$$z\Gamma(z) = \Gamma(z+1) \tag{N.11}$$

we find the behaviour near $z = -1$:

$$\begin{aligned} \Gamma(-1+t) &= \frac{-1}{1-t}\Gamma(t) \\ &= -\left[\frac{1}{t} + 1 - \gamma + O(t)\right] \end{aligned} \tag{N.12}$$

similarly near $z = -2$:

$$\Gamma(-2+t) = \frac{1}{2}\left[\frac{1}{z} + \frac{3}{2} - \gamma + O(t)\right]. \tag{N.13}$$

Consider now the case $n = 2$, for which $\Gamma(n - d/2)$ in (N.9) will have a pole at $d = 4$. Setting $d = 4 - \epsilon$, the divergent behaviour is given by

$$\Gamma(2 - d/2) = \frac{2}{\epsilon} - \gamma + O(\epsilon) \tag{N.14}$$

from (N.10), as stated in (15.78). $I_d(\Delta, 2)$ is then given by

$$I_d(\Delta, 2) = \frac{\mathrm{i}}{(4\pi)^{2-\epsilon/2}}\Delta^{-\epsilon/2}\left[\frac{2}{\epsilon} - \gamma + O(\epsilon)\right]. \tag{N.15}$$

When $\Delta^{-\epsilon/2}$ and $(4\pi)^{-2+\epsilon/2}$ are expanded in powers of $\epsilon$, for small $\epsilon$, the terms linear in $\epsilon$ will produce terms independent of $\epsilon$ when multiplied by the $\epsilon^{-1}$ in the bracket of (N.15). Using $x^\epsilon \approx 1 + \epsilon \ln x + O(\epsilon^2)$, we find (see (15.79)) that

$$I_d(\Delta, 2) = \frac{\mathrm{i}}{(4\pi)^2}\left[\frac{2}{\epsilon} - \gamma + \ln 4\pi - \ln \Delta + O(\epsilon)\right]. \tag{N.16}$$

Another source of $\epsilon$-dependence arises from the fact (see problem 15.7) that a gauge coupling which is dimensionless in $d = 4$ dimensions will acquire mass dimension $\mu^{\epsilon/2}$ in $d = 4 - \epsilon$ dimensions. A vacuum polarization loop with two powers of the coupling will then contain a factor $\mu^\epsilon$. When expanded in powers of $\epsilon$, this will convert the $\ln \Delta$ in (N.16) to $\ln(\Delta/\mu^2)$.

Renormalization schemes will subtract the explicit pole pieces (which diverge as $\epsilon \to 0$), but may also include in the subtraction certain finite terms as well. For example, in the $\overline{\text{MS}}$ scheme one subtracts the pole and the '$-\gamma + \ln 4\pi$' piece.

Finally, consider the integral

$$I_d^{\mu\nu}(\Delta, n) \equiv \int \frac{\mathrm{d}^d k}{(2\pi)^d} \frac{k^\mu k^\nu}{[k^2 - \Delta + \mathrm{i}\epsilon]^n}. \tag{N.17}$$

From Lorentz covariance this must be proportional to the only second-rank tensor available, namely $g^{\mu\nu}$:

$$I_d^{\mu\nu} = A g^{\mu\nu}. \tag{N.18}$$

The constant '$A$' can be determined by contracting both sides of (N.17) with $g_{\mu\nu}$, using $g^{\mu\nu}g_{\mu\nu} = d$ in $d$ dimensions. So

$$\begin{aligned} A &= \frac{1}{d}\int \frac{\mathrm{d}^d k}{(2\pi)^d} \frac{k^2}{(k^2 - \Delta + \mathrm{i}\epsilon)^n} \\ &= \frac{1}{d}\left\{\int \frac{\mathrm{d}^d k}{(2\pi)^d} \frac{1}{(k^2 - \Delta + \mathrm{i}\epsilon)^{n-1}} + \Delta \int \frac{\mathrm{d}^d k}{(2\pi)^d} \frac{1}{(k^2 - \Delta + \mathrm{i}\epsilon)^n}\right\} \end{aligned}$$

$$
\begin{aligned}
&= \frac{\mathrm{i}(-1)^n}{(4\pi)^{d/2}} \frac{\Delta^{(d/2)-n+1}}{d} \left\{ \frac{-\Gamma(n-1-d/2)}{\Gamma(n-1)} + \frac{\Gamma(n-d/2)}{\Gamma(n)} \right\} \\
&= \frac{\mathrm{i}(-1)^n}{(4\pi)^{d/2}} \frac{\Delta^{(d/2)-n+1}}{d} \frac{\Gamma(n-1-d/2)}{\Gamma(n)} \{-n + (n - d/2)\} \\
&= \frac{\mathrm{i}(-1)^{n-1} \Delta^{(d/2)-n+1}}{(4\pi)^{d/2}} \frac{1}{2} \frac{\Gamma(n-1-d/2)}{\Gamma(n)}.
\end{aligned}
\tag{N.19}
$$

Using these results, one can show straightforwardly (problem 15.5) that the gauge-non-invariant part of (11.18)—i.e. the piece in braces—vanishes. With the technique of dimensional regularization, starting from a gauge-invariant formulation of the theory, the renormalization programme can be carried out while retaining manifest gauge invariance.

# APPENDIX O

## GRASSMANN VARIABLES

In the path integral representation of quantum amplitudes (chapter 16) the fields are regarded as classical functions. Matrix elements of time-ordered products of bosonic operators could be satisfactorily represented (see the discussion following (16.71)). But something new is needed to represent, for example, the time-ordered product of two fermionic operators: there must be a sign difference between the two orderings, since the fermionic operators *anti-commute*. Thus, it seems that to represent amplitudes involving fermionic operators by path integrals we must think in terms of 'classical' anti-commuting variables.

Fortunately, the necessary mathematics was developed by Grassmann in 1855 and applied to quantum amplitudes by Berezin (1966). Any two *Grassmann numbers* $\theta_1, \theta_2$ satisfy the fundamental relation

$$\theta_1\theta_2 + \theta_2\theta_1 = 0 \tag{O.1}$$

and, of course,

$$\theta_1^2 = \theta_2^2 = 0. \tag{O.2}$$

Grassmann numbers can be added and subtracted in the ordinary way and muliplied by ordinary numbers. For our application, the essential thing we need to be able to do with Grassmann numbers is to integrate over them. It is natural to think that, as with ordinary numbers and functions, integration would be some kind of inverse of differentiation. So let us begin with differentiation.

We define

$$\frac{\partial(a\theta)}{\partial\theta} = a \tag{O.3}$$

where $a$ is any ordinary number, and

$$\frac{\partial}{\partial\theta_1}(\theta_1\theta_2) = \theta_2; \tag{O.4}$$

then necessarily

$$\frac{\partial}{\partial\theta_2}(\theta_1\theta_2) = -\theta_1. \tag{O.5}$$

Consider now a function of one such variable, $f(\theta)$. An expansion of $f$ in powers of $\theta$ terminates after only two terms because of the property (O.2):

$$f(\theta) = a + b\theta. \tag{O.6}$$

So

$$\frac{\partial f(\theta)}{\partial \theta} = b \tag{O.7}$$

but also

$$\frac{\partial^2 f}{\partial \theta^2} = 0 \tag{O.8}$$

for any such $f$. Hence, the operator $\partial/\partial\theta$ has no inverse (think of the matrix analogue $A^2 = 0$: if $A^{-1}$ existed, we could deduce $0 = A^{-1}(A^2) = (A^{-1}A)A = A$ for all $A$). Thus, we must approach Grassmann integration other than via an inverse of differentiation.

We only need to consider integrals over the complete range of $\theta$, of the form

$$\int \mathrm{d}\theta f(\theta) = \int \mathrm{d}\theta (a + b\theta). \tag{O.9}$$

Such an integral should be linear in $f$; thus, it must be a linear function of $a$ and $b$. One further property fixes its value: we require the result to be *invariant under translations of* $\theta$ *by* $\theta \to \theta + \eta$, where $\eta$ is a Grassmann number. This property is crucial to manipulations made in the path integral formalism, for instance in 'completing the square' manipulations similar to those in section 16.4, but with Grassmann numbers. So we require

$$\int \mathrm{d}\theta (a + b\theta) = \int \mathrm{d}\theta ([a + b\eta] + b\theta). \tag{O.10}$$

This has changed the constant (independent of $\theta$) term but left the linear term unchanged. The only linear function of $a$ and $b$ which behaves like this is a multiple of $b$, which is conventionally taken to be simply $b$. Thus, we define

$$\int \mathrm{d}\theta (a + b\theta) = b \tag{O.11}$$

which means that integration is, in some sense, the same as differentiation!

When we integrate over products of different $\theta$'s, we need to specify a convention about the order in which the integrals are to be performed. We adopt the convention

$$\int \mathrm{d}\theta_1 \int \mathrm{d}\theta_2\, \theta_2\theta_1 = 1 \tag{O.12}$$

that is, the innermost integral is done first, then the next, and so on.

Since our application will be to Dirac fields, which are complex-valued, we need to introduce complex Grassmann numbers, which are built out of real and imaginary parts in the usual way (this would not be necessary for Majorana fermions). Thus we may define

$$\psi = \frac{1}{\sqrt{2}}(\theta_1 + \mathrm{i}\theta_2) \qquad \psi^* = \frac{1}{\sqrt{2}}(\theta_1 - \mathrm{i}\theta_2) \tag{O.13}$$

and then

$$-\mathrm{i}\,\mathrm{d}\psi\,\mathrm{d}\psi^* = \mathrm{d}\theta_1\,\mathrm{d}\theta_2. \tag{O.14}$$

It is convenient to define complex conjugation to include reversing the order of quantities:

$$(\psi\chi)^* = \chi^*\psi^*. \tag{O.15}$$

Then (O.14) is consistent under complex conjugation.

We are now ready to evaluate some Gaussian integrals over Grassmann variables, which is essentially all we need in the path integral formalism. We begin with

$$\begin{aligned}\int\!\!\int \mathrm{d}\psi^*\,\mathrm{d}\psi\,\mathrm{e}^{-b\psi^*\psi} &= \int\!\!\int \mathrm{d}\psi^*\,\mathrm{d}\psi(1 - b\psi^*\psi)\\ &= \int\!\!\int \mathrm{d}\psi^*\,\mathrm{d}\psi(1 + b\psi\psi^*) = b.\end{aligned} \tag{O.16}$$

Note that the analogous integral with ordinary variables is

$$\int\!\!\int \mathrm{d}x\,\mathrm{d}y\,\mathrm{e}^{-b(x^2+y^2)/2} = 2\pi/b. \tag{O.17}$$

The important point here is that, in the Grassman case, $b$ appears with a positive, rather than a negative, power on the right-hand side. However, if we insert a factor $\psi\psi^*$ into the integrand in (O.16), we find that it becomes

$$\int\!\!\int \mathrm{d}\psi^*\,\mathrm{d}\psi\;\psi\psi^*(1 + b\psi\psi^*) = \int\!\!\int \mathrm{d}\psi^*\,\mathrm{d}\psi\;\psi\psi^* = 1 \tag{O.18}$$

and the insertion has effectively produced a factor $b^{-1}$. This effect of an insertion is the same in the 'ordinary variables' case:

$$\int\!\!\int \mathrm{d}x\,\mathrm{d}y(x^2 + y^2)/2\,\mathrm{e}^{-b(x^2+y^2)/2} = 2\pi/b^2. \tag{O.19}$$

Now consider a Gaussian integral involving two different Grassmann variables:

$$\int \mathrm{d}\psi_1^*\,\mathrm{d}\psi_1\,\mathrm{d}\psi_2^*\,\mathrm{d}\psi_2\,\mathrm{e}^{-\psi^{*\mathrm{T}}M\psi} \tag{O.20}$$

where

$$\psi = \begin{pmatrix}\psi_1\\ \psi_2\end{pmatrix} \tag{O.21}$$

and $M$ is a $2\times 2$ matrix, whose entries are ordinary numbers. The only terms which survive the integration are those which, in the expansion of the exponential, contain each of $\psi_1^*$, $\psi_1$, $\psi_2^*$ and $\psi_2$ exactly once. These are the terms

$$\begin{aligned}&\tfrac{1}{2}[M_{11}M_{22}(\psi_1^*\psi_1\psi_2^*\psi_2 + \psi_2^*\psi_2\psi_1^*\psi_1)\\ &\quad + M_{12}M_{21}(\psi_1^*\psi_2\psi_2^*\psi_1 + \psi_2^*\psi_1\psi_1^*\psi_2)].\end{aligned} \tag{O.22}$$

To integrate (O.22) conveniently, according to the convention (O.12), we need to re-order the terms into the form $\psi_2\psi_2^*\psi_1\psi_1^*$; this produces

$$(M_{11}M_{22} - M_{12}M_{21})(\psi_2\psi_2^*\psi_1\psi_1^*), \tag{O.23}$$

and the integral (O.20) is therefore just

$$\int\int \mathrm{d}\psi_1^* \,\mathrm{d}\psi \,\mathrm{d}\psi_2^* \,\mathrm{d}\psi_2 \,\mathrm{e}^{-\psi^{*\mathrm{T}}M\psi} = \det M. \tag{O.24}$$

The reader may show, or take on trust, the obvious generalization to $N$ independent complex Grassmann variables $\psi_1, \psi_2, \psi_3, \ldots, \psi_N$. This result is sufficient to establish the assertion made in section 16.4 concerning the integral (16.82), when written in 'discretized' form.

We may contrast (O.24) with an analogous result for two ordinary complex numbers $z_1, z_2$. In this case we consider the integral

$$\int\int \mathrm{d}z_1^* \,\mathrm{d}z_1 \,\mathrm{d}z_2^* \,\mathrm{d}z_2 \,\mathrm{e}^{-z^*Hz} \tag{O.25}$$

where $z$ is a two-component column matrix with elements $z_1$ and $z_2$. We take the matrix $H$ to be Hermitian, with positive eigenvalues $b_1$ and $b_2$. Let $H$ be diagonalized by the unitary transformation

$$\begin{pmatrix} z_1' \\ z_2' \end{pmatrix} = U \begin{pmatrix} z_1 \\ z_2 \end{pmatrix} \tag{O.26}$$

with $UU^\dagger = I$. Then

$$\mathrm{d}z_1' \,\mathrm{d}z_2' = \det U \,\mathrm{d}z_1 \,\mathrm{d}z_2 \tag{O.27}$$

and so

$$\mathrm{d}z_1' \,\mathrm{d}z_1'^* \,\mathrm{d}z_2' \,\mathrm{d}z_2'^* = \mathrm{d}z_1 \,\mathrm{d}z_1^* \,\mathrm{d}z_2 \,\mathrm{d}z_2^* \tag{O.28}$$

since $|\det U|^2 = 1$. The integral (O.25) then becomes

$$\int \mathrm{d}z_1' \,\mathrm{d}z_1'^* \mathrm{e}^{-b_1 z_1'^* z_1'} \int \mathrm{d}z_2' \,\mathrm{d}z_2'^* \mathrm{e}^{-b_2 z_2'^* z_2'} \tag{O.29}$$

the integrals converging provided $b_1, b_2 > 0$. Next, setting $z_1 = (x_1 + \mathrm{i}y_1)/\sqrt{2}$, $z_2 = (x_2 + \mathrm{i}y_2)/\sqrt{2}$, (O.29) can be evaulated using (O.17), and the result is proportional to $(b_1 b_2)^{-1}$, which is the *inverse* of the determinant of the matrix $H$, when diagonalized. Thus—compare (O.16) and (O.17)—Gaussian integrals over complex Grassmann variables are proportional to the determinant of the matrix in the exponent, while those over ordinary complex variables are proportional to the inverse of the determinant.

Returning to integrals of the form (O.20), consider now a two-variable (both complex) analogue of (O.18):

$$\int \mathrm{d}\psi_1^* \,\mathrm{d}\psi_1 \,\mathrm{d}\psi_2^* \,\mathrm{d}\psi_2 \,\psi_1\psi_2^* \,\mathrm{e}^{-\psi^{*\mathrm{T}}M\psi}. \tag{O.30}$$

This time, only the term $\psi_1^* \psi_2$ in the expansion of the exponential will survive the integration and the result is just $-M_{12}$. By exploring a similar integral (still with the term $\psi_1 \psi_2^*$) in the case of three complex Grassmann variables, the reader should be convinced that the general result is

$$\prod_i \int \mathrm{d}\psi_i^* \, \mathrm{d}\psi_i \; \psi_k \psi_l^* \mathrm{e}^{-\psi^{*\mathrm{T}} M \psi} = (M^{-1})_{kl} \det M. \tag{O.31}$$

With this result we can make plausible the fermionic analogue of (16.79), namely

$$\langle \Omega | T \left\{ \psi(x_1) \bar{\psi}(x_2) \right\} | \Omega \rangle = \frac{\int \mathcal{D}\bar{\psi} \mathcal{D}\psi \; \psi(x_1) \bar{\psi}(x_2) \exp[-\int \mathrm{d}^4 x_\mathrm{E} \, \bar{\psi}(\mathrm{i}\partial\!\!\!/ - m)\psi]}{\int \mathcal{D}\bar{\psi} \mathcal{D}\psi \exp[-\int \mathrm{d}^4 x_\mathrm{E} \, \bar{\psi}(\mathrm{i}\partial\!\!\!/ - m)\psi]} \tag{O.32}$$

note that $\bar{\psi}$ and $\psi^*$ are unitarily equivalent. The denominator of this expression is [1] $\det(\mathrm{i}\partial\!\!\!/ - m)$, while the numerator is this same determinant multiplied by the inverse of the operator $(\mathrm{i}\partial\!\!\!/ - m)$; but this is just $(p\!\!\!/ - m)^{-1}$ in momentum space, the familiar Dirac propagator.

[1] The reader may interpret this as a finite-dimensional determinant, after discretization.

# APPENDIX P

# MAJORANA FERMIONS

In this appendix we aim to give an elementary introduction to 'Majorana' fermions—that is, spin-$\frac{1}{2}$ particles with no conserved quantum number allowing an observable distinction to be made between the 'particle' and the 'anti-particle'. Such particles (which clearly cannot be electrically charged) would then be fermionic analogues of the $\pi^0$, for example, which is its own anti-particle. As we saw in section 7.1, a charged scalar field (or one carrying some other conserved quantum number) has two field degrees of freedom, whereas a neutral scalar field has only one. The two degrees of freedom correspond to the physically distinct states of particle and anti-particle, in that case. For the Dirac field, introduced in section 7.2, there is an additional doubling of the number of degrees of freedom to four in all, corresponding to particles and anti-particles with spin up or down (or helicity $+1$ or $-1$, etc). But, for neutral fermions such as neutrinos, the possibility exists that they might be their own anti-particles, and so have only two degrees of freedom corresponding just to the two possible spin states.

It is not so clear, at first sight, where there is room for such a possibility in the conventional presentation of the Dirac equation (see section 4.2), which appears to lead inevitably to the 'particle/anti-particle, up/down' description with four degrees of freedom. We therefore begin by re-considering relativistic wave equations for spin-$\frac{1}{2}$ particles.

## P.1 Spin-$\frac{1}{2}$ wave equations

Within the framework of quantum-mechanical wave equations (rather than that of quantum fields), one way of approaching the 'number of degrees of freedom' issue is via the discussion of the Dirac equation in appendix M, section M.6—that is, in terms of the way two-component spinors transform under Lorentz transformations. Specifically, we saw there that the 4-component Dirac spinor

$$\psi = \begin{pmatrix} \phi \\ \chi \end{pmatrix} \tag{P.1}$$

could be understood as being built from two different kinds of spinor. One kind ($\phi$) transforms according to the (1/2, 0) representation of the Lorentz group, via

$$\phi' = (1 + \mathrm{i}\boldsymbol{\epsilon}\cdot\boldsymbol{\sigma}/2 - \boldsymbol{\eta}\cdot\boldsymbol{\sigma}/2)\phi \tag{P.2}$$

for infinitesimal rotations and boosts. The other kind ($\chi$) transforms as a (0, 1/2) representation via

$$\chi' = (1 + \mathrm{i}\boldsymbol{\epsilon} \cdot \boldsymbol{\sigma}/2 + \boldsymbol{\eta} \cdot \boldsymbol{\sigma}/2)\chi. \tag{P.3}$$

The Dirac equation itself is then

$$(\mathrm{i}\partial_t + \mathrm{i}\boldsymbol{\sigma} \cdot \boldsymbol{\nabla})\phi = m\chi \tag{P.4}$$

$$(\mathrm{i}\partial_t - \mathrm{i}\boldsymbol{\sigma} \cdot \boldsymbol{\nabla})\chi = m\phi. \tag{P.5}$$

In (P.4) and (P.5), the '$\mathrm{i}\partial_t \pm \mathrm{i}\boldsymbol{\sigma} \cdot \boldsymbol{\nabla}$' factors on the left-hand sides are just what is required to change the transformation character of a $\phi$ into that of a $\chi$ or *vice versa*, since both sides must transform consistently under Lorentz transformations. (In terms of the 'R–L' classification of section 20.4, $\phi$ is $\psi_{\mathrm{R}}$ and $\chi$ is $\psi_{\mathrm{L}}$).

We may ask the question: is it possible to find a spinor $\phi_{\mathrm{c}}$, constructed from the components of $\phi$ (and, hence, without additional degrees of freedom), which transforms like a '$\chi$' rather than a '$\phi$'? Then we could, for example, replace $\chi$ on the right-hand side of (P.4) by $\phi_{\mathrm{c}}$, and the two sides would transform consistently but involve only the degees of fredom in $\phi$.

It is indeed possible to find such a $\phi_{\mathrm{c}}$: namely, consider[1]

$$\phi_{\mathrm{c}} \equiv \mathrm{i}\sigma_2\phi^*. \tag{P.6}$$

Then

$$\begin{aligned}\phi_{\mathrm{c}}{}' = (\mathrm{i}\sigma_2\phi^*)' &= \mathrm{i}\sigma_2(1 - \mathrm{i}\boldsymbol{\epsilon} \cdot \boldsymbol{\sigma}/2 - \boldsymbol{\eta} \cdot \boldsymbol{\sigma}/2)^*\phi^* \\ &= \mathrm{i}\sigma_2(1 + \mathrm{i}\boldsymbol{\epsilon} \cdot \boldsymbol{\sigma}^*/2 - \boldsymbol{\eta} \cdot \boldsymbol{\sigma}^*/2)\phi^* \\ &= (1 - \mathrm{i}\boldsymbol{\epsilon} \cdot \boldsymbol{\sigma}/2 + \boldsymbol{\eta} \cdot \boldsymbol{\sigma}/2)(\mathrm{i}\sigma_2\phi^*)\end{aligned} \tag{P.7}$$

using $\sigma_1^* = \sigma_1$, $\sigma_2^* = -\sigma_2$, $\sigma_3^* = \sigma_3$ and $\sigma_i\sigma_j + \sigma_j\sigma_i = 2\delta_{ij}$. Equation (P.7) shows that '$\mathrm{i}\sigma_2\phi^*$' does transform like a '$\chi$' (the 'i' is inserted for convenience, so as to make $\mathrm{i}\sigma_2$ real). In 'R–L' language, we may say that $\mathrm{i}\sigma_2\psi_{\mathrm{R}}^*$ transforms like $\psi_{\mathrm{L}}$, and similarly with $\mathrm{R} \leftrightarrow \mathrm{L}$.

Consider, then, the two-component wave equation

$$(\mathrm{i}\partial_t + \mathrm{i}\boldsymbol{\sigma} \cdot \boldsymbol{\nabla})\phi = m(\mathrm{i}\sigma_2\phi^*). \tag{P.8}$$

Take the complex conjugate of this equation, multiply from the left by $\mathrm{i}\sigma_2$, and commute the $\sigma_2$ through the bracket on the left-hand side, as in (P.7): we find that

$$(-\mathrm{i}\partial_t + \mathrm{i}\boldsymbol{\sigma} \cdot \boldsymbol{\nabla})(\mathrm{i}\sigma_2\phi^*) = -m\phi. \tag{P.9}$$

It follows from (P.8) and (P.9) that

$$\begin{aligned}(-\mathrm{i}\partial_t + \mathrm{i}\boldsymbol{\sigma} \cdot \boldsymbol{\nabla})[(\mathrm{i}\partial_t + \mathrm{i}\boldsymbol{\sigma} \cdot \boldsymbol{\nabla})\phi] &= m(-\mathrm{i}\partial_t + \mathrm{i}\boldsymbol{\sigma} \cdot \boldsymbol{\nabla})(\mathrm{i}\sigma_2\phi^*) \\ &= -m^2\phi,\end{aligned} \tag{P.10}$$

[1] The reader may usefully bear in mind, at this point, the discussion of the SU(2) representations **2** and **2**$^*$ in the paragraph containing equations (12.53)–(12.57) in chapter 12.

and, hence,

$$(\partial_t^2 - \nabla^2 + m^2)\phi = 0 \tag{P.11}$$

showing that $\phi$ satisfies the KG equation, precisely the condition we imposed in (4.27) on the Dirac wavefunction $\psi$. Hence, even though it has only two components, $\phi$ is a valid relativistic wavefunction for a massive spin-$\frac{1}{2}$ particle.

Plane-wave solutions can be found, having the form

$$\phi = \mathrm{e}^{-\mathrm{i}p\cdot x}\begin{pmatrix} a \\ b \end{pmatrix} + \mathrm{e}^{\mathrm{i}p\cdot x}\begin{pmatrix} c \\ d \end{pmatrix} \tag{P.12}$$

where $\begin{pmatrix} c \\ d \end{pmatrix}$ will be related to $\begin{pmatrix} a \\ b \end{pmatrix}$ via the wave equation (P.8). We find that

$$\begin{pmatrix} c \\ d \end{pmatrix} = \left(\frac{E + \boldsymbol{\sigma}\cdot\boldsymbol{p}}{m}\right) \cdot - \mathrm{i}\sigma_2 \begin{pmatrix} a^* \\ b^* \end{pmatrix}. \tag{P.13}$$

We can also check that $\rho = \phi^\dagger\phi$ and $\boldsymbol{j} = \phi^\dagger\boldsymbol{\sigma}\phi$ satisfy the continuity equation (cf (4.55)):

$$\partial_t \rho + \nabla \cdot \boldsymbol{j} = 0. \tag{P.14}$$

These results were first given by Case (1957).

Comparison of (P.9) and (P.5) reveals that it is consistent to use a four-component wavefunction of the type (P.1) in this case too, identifying $\chi$ with $\mathrm{i}\sigma_2\phi^*$, and writing

$$\psi_{\mathrm{M}} = \begin{pmatrix} \phi \\ \mathrm{i}\sigma_2\phi^* \end{pmatrix}. \tag{P.15}$$

Consider now the charge conjugation operation $C_0$ of section 20.5, assuming that it is the 'right' operation in the present case as well. We find that

$$\begin{aligned} \psi_{\mathrm{M}\,\mathbf{C}} = \mathrm{i}\gamma_2\psi_{\mathrm{M}}^* &= \begin{pmatrix} 0 & -\mathrm{i}\sigma_2 \\ \mathrm{i}\sigma_2 & 0 \end{pmatrix}\begin{pmatrix} \phi^* \\ \mathrm{i}\sigma_2\phi \end{pmatrix} \\ &= \begin{pmatrix} \phi \\ \mathrm{i}\sigma_2\phi^* \end{pmatrix} = \psi_{\mathrm{M}}. \end{aligned} \tag{P.16}$$

Hence $\psi_{\mathrm{M}}$ satisfies the *Majorana condition*

$$\psi_{\mathrm{M}\,\mathbf{C}} = \psi_{\mathrm{M}} \tag{P.17}$$

and, in this sense, may be said to describe a self-conjugate particle.[2]

However, in terms of the 'R–L' classification

$$\psi_{\mathrm{M,R}} = \begin{pmatrix} \phi \\ 0 \end{pmatrix} \qquad \psi_{\mathrm{M,L}} = \begin{pmatrix} 0 \\ \mathrm{i}\sigma_2\phi^* \end{pmatrix} \tag{P.18}$$

[2] The true 'anti-particle' is defined via the operator $\hat{\mathbf{C}}\hat{\mathbf{P}}\hat{\mathbf{T}}$, since this is believed to be conserved by all interactions.

we obtain

$$\psi_{\mathrm{M,R}\,\mathrm{C}} = \psi_{\mathrm{M,L}} \qquad \psi_{\mathrm{M,L}\,\mathrm{C}} = \psi_{\mathrm{M,R}}. \tag{P.19}$$

Inserting (P.12) and (P.13) into (P.15), we find that the four-component wavefunction has the form

$$\psi_{\mathrm{M}} = \mathrm{e}^{-\mathrm{i}p\cdot x}\begin{pmatrix} a \\ b \\ \left(\frac{E-\boldsymbol{\sigma}\cdot\boldsymbol{p}}{m}\right)\begin{pmatrix} a \\ b \end{pmatrix}\end{pmatrix} + \mathrm{e}^{\mathrm{i}p\cdot x}\begin{pmatrix} \left(\frac{E+\boldsymbol{\sigma}\cdot\boldsymbol{p}}{m}\right)\cdot -\mathrm{i}\sigma_2\begin{pmatrix} a^* \\ b^* \end{pmatrix} \\ \mathrm{i}\sigma_2\begin{pmatrix} a^* \\ b^* \end{pmatrix}\end{pmatrix}. \tag{P.20}$$

Let us take $\begin{pmatrix} a \\ b \end{pmatrix} = \phi_+$, such that $\boldsymbol{\sigma}\cdot\boldsymbol{p}\phi_+ = |\boldsymbol{p}|\phi_+$ as in (20.46). Then $\psi_{\mathrm{M}}$ of (P.20) can be written as

$$\psi_{\mathrm{M}}(\lambda = +1) = \mathrm{e}^{-\mathrm{i}p\cdot x}u(p, \lambda = +1) + \mathrm{e}^{\mathrm{i}p\cdot x}(\mathrm{i}\gamma_2 u^*(p, \lambda = +1)) \tag{P.21}$$

using (20.47) and the normalization (20.48). In the form (P.21), it is evident that the Majorana condition (P.16) holds.

A problem arises, however, when we try to describe this theory in terms of a Lagrangian. Mimicking what we learned in the Dirac case, we would expect to derive (P.8) and (P.9) from the Lagrangian

$$\mathcal{L}_{\mathrm{M}} = \bar{\psi}_{\mathrm{M}}(\mathrm{i}\not{\partial} - m)\psi_{\mathrm{M}}. \tag{P.22}$$

But consider the mass term:

$$\begin{aligned} m\bar{\psi}_{\mathrm{M}}\psi_{\mathrm{M}} &= m(\phi^\dagger\ \phi^{\mathrm{T}}(-\mathrm{i}\sigma_2))\begin{pmatrix} 0 & 1 \\ 1 & 0 \end{pmatrix}\begin{pmatrix} \phi \\ \mathrm{i}\sigma_2\phi^* \end{pmatrix} \\ &= \mathrm{i}m\{\phi^\dagger\sigma_2\phi^* - \phi^{\mathrm{T}}\sigma_2\phi\}, \end{aligned} \tag{P.23}$$

which can also be written, as usual, as

$$m\{\bar{\psi}_{\mathrm{M,R}}\psi_{\mathrm{M,L}} + \bar{\psi}_{\mathrm{M,L}}\psi_{\mathrm{M,R}}\}. \tag{P.24}$$

If $\phi$ is an ordinary c-number spinor of the form

$$\phi = \begin{pmatrix} a \\ b \end{pmatrix} \tag{P.25}$$

then each term in (P.23) vanishes identically, since $\sigma_2$ is anti-symmetric, and we seem unable to form the required Lagrangian.

The same point was noticed with respect to (22.119) but resolved there by the fact that $\phi$ was replaced by a two-component fermionic (anti-commuting) field. Let us therefore now consider the quantum field case.

## P.2 Majorana quantum fields

The field-theoretic charge-conjugation operator $\hat{\mathbf{C}}$ was introduced in section 20.5, with the property (20.71):

$$\hat{\psi}_{\mathbf{C}} \equiv \hat{\mathbf{C}}\hat{\psi}\hat{\mathbf{C}}^{-1} = \mathrm{i}\gamma_2\hat{\psi}^{\dagger\mathrm{T}} = \mathrm{i}\gamma_2\gamma_0\left(\bar{\hat{\psi}}\right)^{\mathrm{T}}. \tag{P.26}$$

The standard normal mode expansion of a Dirac field is

$$\hat{\psi}_{\mathrm{D}}(x) = \int \frac{\mathrm{d}^3\boldsymbol{k}}{(2\pi)^3}\frac{1}{\sqrt{2E}}\sum_{\lambda}\left[\hat{c}_{\lambda}(k)u(k,\lambda)\mathrm{e}^{-\mathrm{i}k\cdot x} + \hat{d}^{\dagger}_{\lambda}(k)v(k,\lambda)\mathrm{e}^{\mathrm{i}k\cdot x}\right] \tag{P.27}$$

clearly showing the *four* degres of freedom—namely via the operators $\hat{c}_{\lambda}$ and $\hat{d}^{\dagger}_{\lambda}$, which respectively destroy particles with $\lambda = \pm 1$ and create anti-particles with $\lambda = \pm 1$. The charge-conjugate field is

$$\hat{\psi}_{\mathrm{D\,C}}(x) = \int \frac{\mathrm{d}^3\boldsymbol{k}}{(2\pi)^3}\frac{1}{\sqrt{2E}}\sum_{\lambda}\left[\hat{c}^{\dagger}_{\lambda}(\mathrm{i}\gamma_2 u^*)\mathrm{e}^{\mathrm{i}k\cdot x} + \hat{d}_{\lambda}(\mathrm{i}\gamma_2 v^*)\mathrm{e}^{-\mathrm{i}k\cdot x}\right]. \tag{P.28}$$

Let us write (with a slight abuse of notation)

$$u = \begin{pmatrix} \phi \\ \chi \end{pmatrix} \tag{P.29}$$

as usual. Then

$$\mathrm{i}\gamma_2 u^* = \begin{pmatrix} -\mathrm{i}\sigma_2\chi^* \\ \mathrm{i}\sigma_2\phi^* \end{pmatrix} \sim \begin{pmatrix} \phi \\ \chi \end{pmatrix} \sim u \tag{P.30}$$

where '$\sim$' means 'transforms in the same way as'—a result which follows from the work of section P.1. Similarly,

$$\mathrm{i}\gamma_2 v^* \sim u. \tag{P.31}$$

These results show that, as claimed in section 22.6, the charge-conjugate field transforms in the same way as the original field. In fact, (20.70) shows more: that, with our conventions, $\mathrm{i}\gamma_2 v^*(p, \lambda = +1)$ is actually equal to $u(p, \lambda = +1)$ and, similarly, $\mathrm{i}\gamma_2 u^*(p, \lambda = +1) = v(p, \lambda = +1)$. Thus, in $\hat{\psi}_{\mathrm{D\,C}}$, the operators $\hat{c}$ and $\hat{d}$ are interchanged, as required.

Apart from the spinor factors, (P.27) is analogous to the expansion (7.16) of a complex scalar field, which also has two distinct kinds of mode operator, $\hat{a}$ and $\hat{b}^{\dagger}$. In contrast, the expansion of the real scalar field in (5.116) has only one type of operator, $\hat{a}$ and its Hermitian conjugate $\hat{a}^{\dagger}$. Consider, therefore, the field obtained by replacing $\hat{d}^{\dagger}_{\lambda}$ in (P.27) by $\hat{c}^{\dagger}_{\lambda}$ (compare (P.21)):

$$\hat{\psi}_{\mathrm{M}}(x) = \int \frac{\mathrm{d}^3\boldsymbol{k}}{(2\pi)^3}\frac{1}{\sqrt{2E}}\sum_{\lambda}\left[\hat{c}_{\lambda}(k)u(k,\lambda)\mathrm{e}^{-\mathrm{i}k\cdot x} + \hat{c}^{\dagger}_{\lambda}(k)(\mathrm{i}\gamma_2 u^*(k,\lambda))\mathrm{e}^{\mathrm{i}k\cdot x}\right] \tag{P.32}$$

where we have used $v = \mathrm{i}\gamma_2 u^*$. Then it is easy to see that

$$\hat{\mathbf{C}}\hat{\psi}_{\mathrm{M}}\hat{\mathbf{C}}^{-1} = \hat{\psi}_{\mathrm{M}} \tag{P.33}$$

and the Majorana condition is satisfied for the field $\hat{\psi}_{\mathrm{M}}$. The operators $\hat{c_\lambda}$ and $\hat{c}_\lambda^\dagger$ in (P.32) destroy and create just one kind of particle, which has two possible spin states.

The mass term considered at the end of the previous section is now clearly possible in this quantum field formalism, taking the fermionic operators to anti-commute as usual. But so are other types of mass term. Reverting to the L–R notation, we may have a 'Dirac' mass of the form

$$-m_{\mathrm{D}}\left(\bar{\hat{\psi}}_{\mathrm{M,R}}\hat{\psi}_{\mathrm{M,L}} + \bar{\hat{\psi}}_{\mathrm{M,L}}\hat{\psi}_{\mathrm{M,R}}\right) \tag{P.34}$$

and also two 'Majorana' mass terms, one of the form

$$-\tfrac{1}{2}m_{\mathrm{L}}\left(\overline{\hat{\psi}_{\mathrm{M,L\,C}}}\,\hat{\psi}_{\mathrm{M,L}} + \overline{\hat{\psi}_{\mathrm{M,L}}}\,\hat{\psi}_{\mathrm{M,L\,C}}\right) \tag{P.35}$$

as in (22.116), and also

$$-\tfrac{1}{2}m_{\mathrm{R}}\left(\overline{\hat{\psi}_{\mathrm{M,R\,C}}}\,\hat{\psi}_{\mathrm{M,R}} + \overline{\hat{\psi}_{\mathrm{M,R}}}\,\hat{\psi}_{\mathrm{M,R\,C}}\right). \tag{P.36}$$

These may all be combined into the most general Lorentz-invariant mass term (for one neutrino flavour)

$$-\tfrac{1}{2}\left(\overline{\hat{N}_{\mathrm{L\,C}}}\,M\hat{N}_{\mathrm{L}} + \bar{\hat{N}}_{\mathrm{L}}\,M\hat{N}_{\mathrm{L\,C}}\right) \tag{P.37}$$

where

$$\hat{N}_{\mathrm{L}} = \begin{pmatrix} \hat{\psi}_{\mathrm{M,L}} \\ \hat{\psi}_{\mathrm{M,R\,C}} \end{pmatrix} \qquad \hat{N}_{\mathrm{L\,C}} = \begin{pmatrix} \hat{\psi}_{\mathrm{M,L\,C}} \\ \hat{\psi}_{\mathrm{M,R}} \end{pmatrix} \tag{P.38}$$

and the matrix $M$ is given by

$$M = \begin{pmatrix} m_{\mathrm{L}} & m_{\mathrm{D}} \\ m_{\mathrm{D}} & m_{\mathrm{R}} \end{pmatrix}. \tag{P.39}$$

We shall assume for simplicity that the parameters $m_{\mathrm{L}}$, $m_{\mathrm{R}}$ and $m_{\mathrm{D}}$ are all real, which coresponds to a **CP**-conserving theory in this sector.

The theory of (one-flavour) neutrino mixing may now be developed, by finding the matrix which diagonalizes $M$, and hence obtaining the fields in the 'mass' basis as linear combinations of those in the 'weak interaction (L–R)' basis. A clear discussion is given in Bilenky (2000), for example. One case that has attracted considerable attention is that in which $m_{\mathrm{L}} = 0$ and $m_{\mathrm{D}} \ll m_{\mathrm{R}}$. Then the eigenvalues of (P.39) are approximately given by

$$m_1 \approx m_{\mathrm{R}} \qquad m_2 \approx -m_{\mathrm{D}}^2/m_{\mathrm{R}}; \tag{P.40}$$

where the apparently troubling minus sign in $m_2$ can be absorbed into the mixing parameters. Thus, one eigenvalue is (by assumption) very large compared to $m_\mathrm{D}$ and one is very much smaller. The condition $m_\mathrm{L} = 0$ ensures that the lepton-number-violating term (P.36) is characterized by a large mass scale $m_\mathrm{R}$. It may be natural to assume that $m_\mathrm{D}$ is a 'typical' quark or lepton mass term, which would then imply that $m_2$ of (P.40) is very much lighter than that—as, of course, appears to be true for the neutrinos. This is the well-known 'see-saw' mechanism of Gell-Mann *et al* (1979), Yanagida (1979) and Mohapatra and Senjanovic (1980, 1981). If, in fact, $m_\mathrm{R} \sim 10^{16}$ Gev, we recover an estimate for $m_2$ which is similar to that in (22.122). The reader may consult Bilenky (2000), for example, or Kayser *et al* (1989), for the devlopment of the theory of neutrino masses, mixing, and oscillations. The experimental status is reviewed by Kayser in Hagiwara *et al* (2002).

# APPENDIX Q

# FEYNMAN RULES FOR TREE GRAPHS IN QCD AND THE ELECTROWEAK THEORY

## Q.1 QCD

### Q.1.1 External particles

*Quarks*

The SU(3) colour degree of freedom is not written explicitly: the spinors have 3 (colour) $\times$ 4 (Dirac) components. For each fermion or anti-fermion line entering the graph include the spinor

$$u(p, s) \qquad \text{or} \qquad v(p, s) \tag{Q.1}$$

and for spin-$\frac{1}{2}$ particles leaving the graph, the spinor

$$\bar{u}(p', s') \qquad \text{or} \qquad \bar{v}(p', s') \tag{Q.2}$$

as for QED.

*Gluons*

Besides the spin-1 polarization vector, external gluons also have a 'colour polarization' vector $a^c (c = 1, 2, \ldots, 8)$ specifying the particular colour state involved. For each gluon line entering the graph include the factor

$$\epsilon_\mu(k, \lambda) a^c \tag{Q.3}$$

and for gluons leaving the graph, the factor

$$\epsilon^*_\mu(k', \lambda') a^{c*}. \tag{Q.4}$$

### Q.1.2 Propagators

*Quark*

$$= \frac{\mathrm{i}}{\not{p} - m} = \mathrm{i}\frac{\not{p} + m}{p^2 - m^2}. \tag{Q.5}$$

*Gluon*

 $$= \frac{\mathrm{i}}{k^2}\left(-g^{\mu\nu} + (1-\xi)\frac{k^\mu k^\nu}{k^2}\right)\delta^{ab} \qquad \text{(Q.6)}$$

for a general $\xi$ gauge. Calculations are usually performed in Lorentz or Feynman gauge with $\xi = 1$ and gluon propagator equal to

 $$= \mathrm{i}\frac{(-g^{\mu\nu})\delta^{ab}}{k^2}. \qquad \text{(Q.7)}$$

Here $a$ and $b$ run over the eight colour indices 1, 2, ..., 8.

### Q.1.3 Vertices

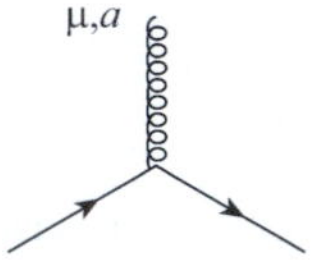


$$-\mathrm{i}g_s\frac{\lambda^a}{2}\gamma^\mu$$

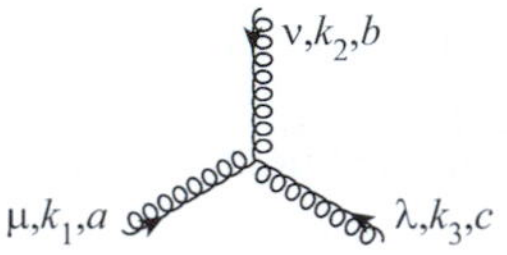


$$-g_s f_{abc}[g_{\mu\nu}(k_1-k_2)_\lambda + g_{\nu\lambda}(k_2-k_3)_\mu + g_{\lambda\mu}(k_3-k_1)_\nu]$$

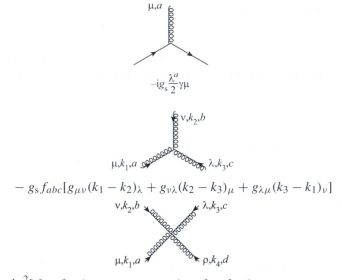

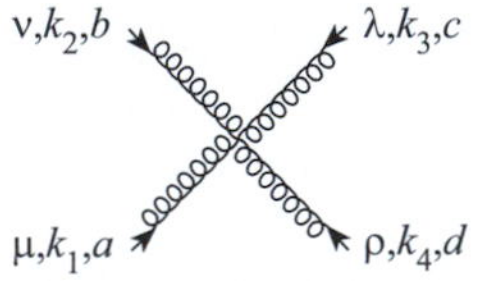


$$-\mathrm{i}g_s^2[f_{abe}f_{cde}(g_{\mu\lambda}g_{\nu\rho} - g_{\mu\rho}g_{\nu\lambda}) + f_{ade}f_{bce}(g_{\mu\nu}g_{\lambda\rho} - g_{\mu\lambda}g_{\nu\rho}) + f_{ace}f_{dbe}(g_{\mu\rho}g_{\nu\lambda} - g_{\mu\nu}g_{\lambda\rho})]$$

It is important to remember that the rules given here are only adequate for tree-diagram calculations in QCD (see section 13.5.3).

## Q.2 The electroweak theory

For tree-graph calculations, it is convenient to use the U-gauge Feynman rules (sections 19.5 and 19.6) in which no unphysical particles appear. These U-gauge rules are given here for the leptons $l = (\mathrm{e}, \mu, \tau)$, $\nu_l = (\nu_\mathrm{e}, \nu_\mu, \nu_\tau)$; for the $t_3 = +\frac{1}{2}$ quarks denoted by f, where f ≡ u, c, t; and for the $t_3 = -1/2$ CKM-mixed quarks denoted by f′ where f′ ≡ d, s, b. The mixing matrix $V_{\mathrm{ff'}}$ is discussed in section 22.7.1.

*Note that for simplicity we do not include neutrino flavour mixing; see section 22.7.2.*

### Q.2.1 External particles

*Leptons and quarks*

For each fermion or anti-fermion line entering the graph include the spinor

$$u(p, s) \qquad \text{or} \qquad v(p, s) \tag{Q.8}$$

and for spin-$\frac{1}{2}$ particles leaving the graph, the spinor

$$\bar{u}(p', s') \qquad \text{or} \qquad \bar{v}(p', s'). \tag{Q.9}$$

*Vector bosons*

For each vector boson line entering the graph include the factor

$$\epsilon_\mu(k, \lambda) \tag{Q.10}$$

and for vector bosons leaving the graph, the factor

$$\epsilon^*_\mu(k', \lambda'). \tag{Q.11}$$

### Q.2.2 Propagators

*Leptons and quarks*

$$= \frac{\mathrm{i}}{\not{p} - m} = \mathrm{i}\frac{\not{p} + m}{p^2 - m^2}. \tag{Q.12}$$

Vector bosons (U gauge)

$$= \frac{\mathrm{i}}{k^2 - M_\mathrm{V}^2}(-g_{\mu\nu} + k_\mu k_\nu / m_\mathrm{V}^2) \tag{Q.13}$$

where 'V' stands for either 'W' (the W-boson) or 'Z' (the $Z^0$).

*Higgs particle*

$$= \frac{\mathrm{i}}{p^2 - m_\mathrm{H}^2} \tag{Q.14}$$

### Q.2.3 Vertices

*Charged-current weak interactions*

Leptons

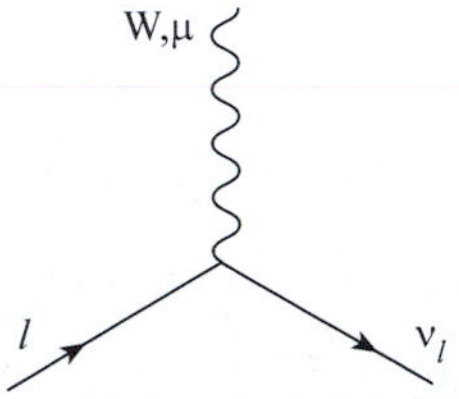


$$-\mathrm{i}\frac{g}{\sqrt{2}}\gamma_\mu\frac{1-\gamma_5}{2}$$

Quarks

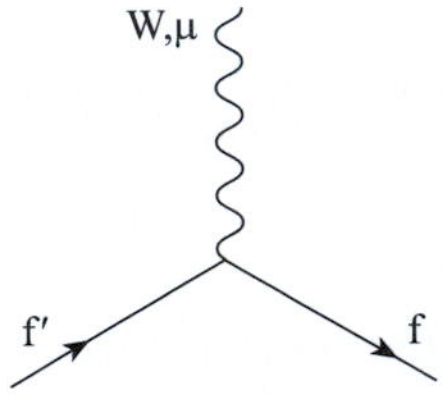


$$-\mathrm{i}\frac{g}{\sqrt{2}}\gamma_\mu\frac{1-\gamma_5}{2}V_{\mathrm{f\,f'}}$$

*Neutral-current weak interactions (no neutrino mixing)*

Fermions

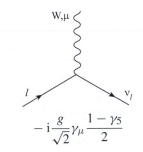

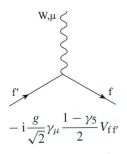

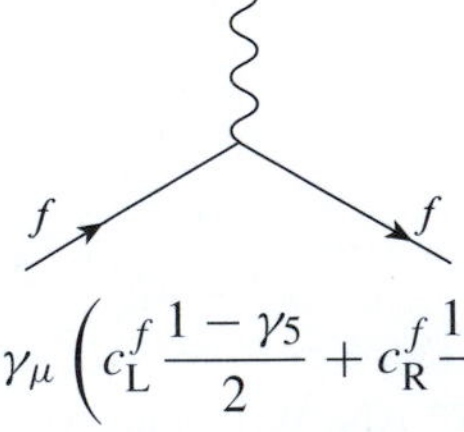


$$\frac{-\mathrm{i}g}{\cos\theta_{\mathrm{W}}}\gamma_\mu\left(c_{\mathrm{L}}^f\frac{1-\gamma_5}{2}+c_{\mathrm{R}}^f\frac{1+\gamma_5}{2}\right)$$

where

$$c_{\mathrm{L}}^f = t_3^f - \sin^2\theta_{\mathrm{W}}Q_f \tag{Q.15}$$

$$c_{\mathrm{R}}^f = -\sin^2\theta_{\mathrm{W}}Q_f \tag{Q.16}$$

and $f$ stands for any fermion.

*Vector boson couplings*

(a) Trilinear couplings:
$\gamma W^+W^-$ vertex

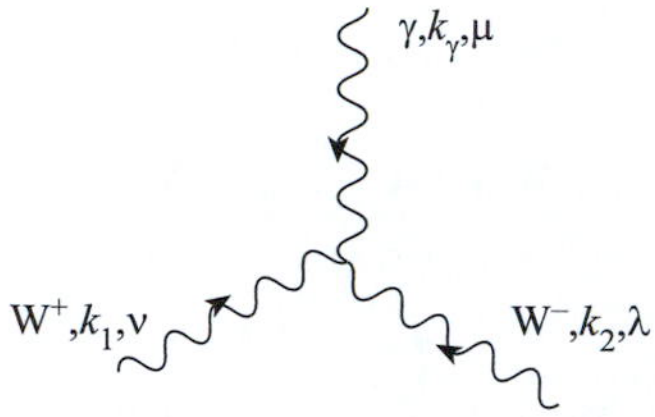


$$ie[g_{\nu\lambda}(k_1 - k_2)_\mu + g_{\lambda\mu}(k_2 - k_\gamma)_\nu + g_{\mu\nu}(k_\gamma - k_1)_\lambda]$$

$Z^0W^+W^-$ vertex

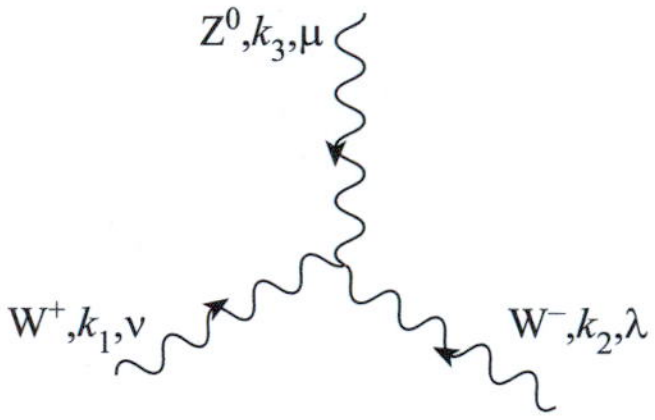


$$ig\cos\theta_W[g_{\nu\lambda}(k_1 - k_2)_\mu + g_{\lambda\mu}(k_2 - k_3)_\nu + g_{\mu\nu}(k_3 - k_1)_\lambda]$$

(b) Quadrilinear couplings:

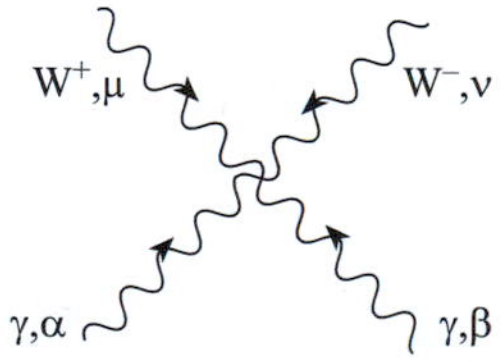


$$-ie^2(2g_{\alpha\beta}g_{\mu\nu} - g_{\alpha\mu}g_{\beta\nu} - g_{\alpha\nu}g_{\beta\mu})$$

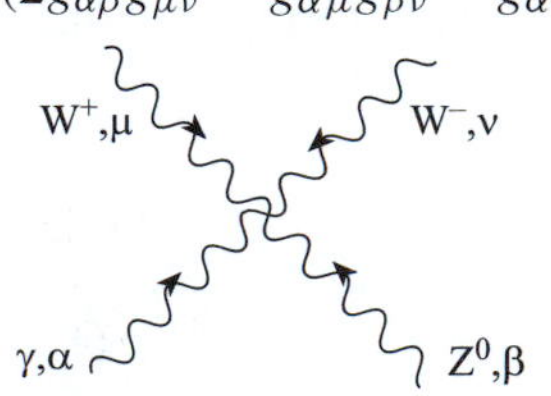


$$-ieg\cos\theta_W(2g_{\alpha\beta}g_{\mu\nu} - g_{\alpha\mu}g_{\beta\nu} - g_{\alpha\nu}g_{\beta\mu})$$

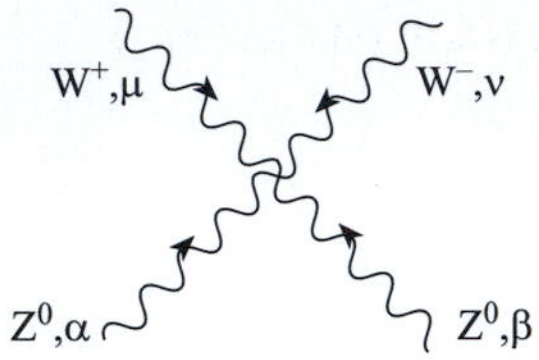


$$-\,\mathrm{i}g^2\cos^2\theta_{\mathrm{W}}(2g_{\alpha\beta}g_{\mu\nu} - g_{\alpha\mu}g_{\beta\nu} - g_{\alpha\nu}g_{\beta\mu})$$

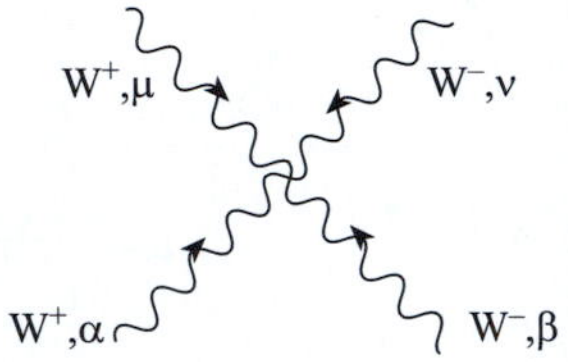


$$\mathrm{i}g^2(2g_{\mu\alpha}g_{\nu\beta} - g_{\mu\beta}g_{\alpha\nu} - g_{\mu\nu}g_{\alpha\beta})$$

*Higgs couplings*

(a) Trilinear couplings:
$\mathrm{HW^+W^-}$ vertex

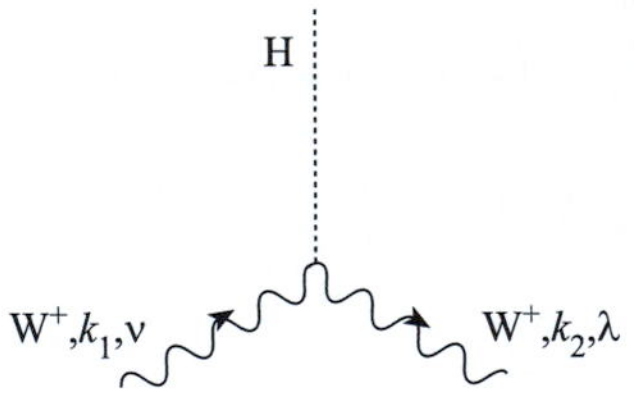


$$\mathrm{i}gM_{\mathrm{W}}g_{\nu\lambda}$$

$\mathrm{HZ^0Z^0}$ vertex

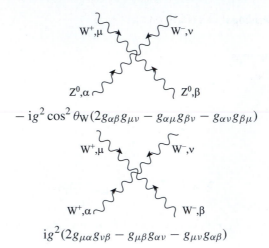

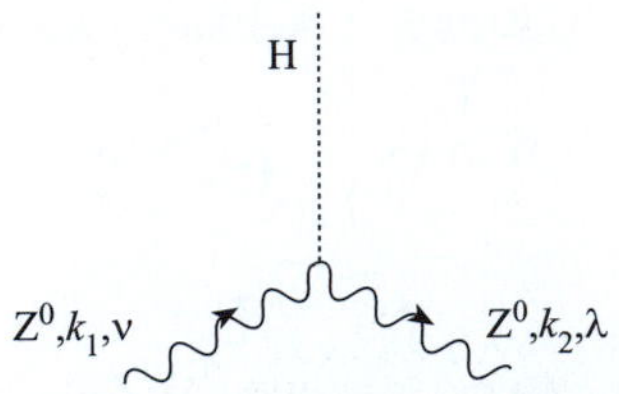


$$\frac{\mathrm{i}g}{\cos\theta_{\mathrm{W}}}M_{\mathrm{Z}}g_{\nu\lambda}$$

Fermion–Yukawa couplings (fermion mass $m_f$)

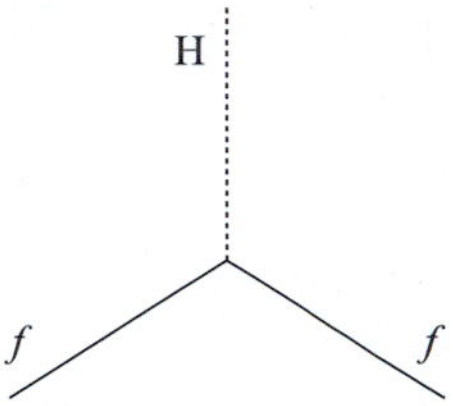


$$-\,\mathrm{i}\frac{g}{2}\frac{m_f}{M_\mathrm{W}}$$

Trilinear self-coupling

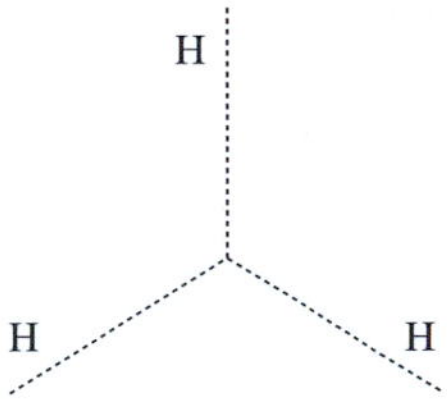


$$-\,\mathrm{i}\frac{3m_\mathrm{H}^2 g}{2M_\mathrm{W}}$$

(b) Quadrilinear couplings:
$\mathrm{HHW^+W^-}$ vertex

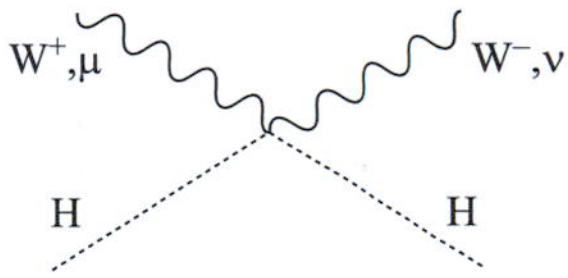


$$\frac{\mathrm{i}g^2}{2}g_{\mu\nu}$$

HHZZ vertex

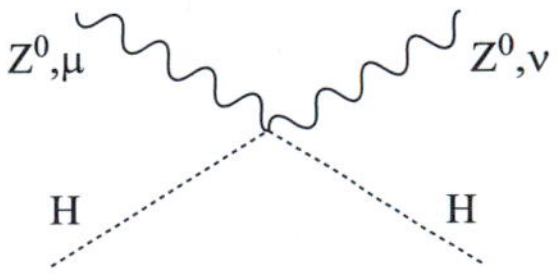


$$\frac{\mathrm{i}g^2}{2\cos^2\theta_\mathrm{W}}g_{\mu\nu}$$

Quadrilinear self-coupling

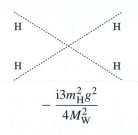

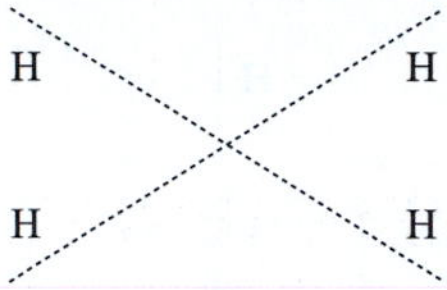


$$-\frac{i3m_H^2 g^2}{4M_W^2}$$

# REFERENCES


Abachi S *et al* 1995a D0 collaboration *Phys. Rev. Lett.* **74** 2422
——1995b *Phys. Rev. Lett.* **74** 2632
Abbiendi G *et al* 2000 *Eur. Phys. J.* C **13** 553
Abbott B *et al* 2000 *Phys. Rev. Lett.* **84** 222
Abe F *et al* 1994a CDF collaboration *Phys. Rev.* D **50** 2966
——1994b *Phys. Rev. Lett.* **73** 225
——1995a *Phys. Rev.* D **52** 4784
——1995b *Phys. Rev. Lett.* **74** 2626
Abe K 1991 *Proc. 25th Int. Conf. on High Energy Physics* ed K K Phua and Y Yamaguchi (Singapore: World Scientific) p 33
Abe K *et al* 2000 *Phys. Rev. Lett.* **84** 5945
——2002 Belle collaboration *Phys. Rev.* D **66** 032007
Abramovicz H *et al* 1982a *Z. Phys.* C **12** 289
——1982b *Z. Phys.* C **13** 199
——1983 *Z. Phys.* C **17** 283
Abreu P *et al* 1990 *Phys. Lett.* B **242** 536
Acciari M *et al* 2000 *Phys. Lett.* B **476** 40
Adler S L 1963 *Phys. Rev.* **143** 1144
——1965 *Phys. Rev. Lett.* **14** 1051
——1969 *Phys. Rev.* **177** 2426
——1970 *Lectures on Elementary Particles and Quantum Field Theory (Proceedings of the Brandeis Summer Institute)* vol 1, ed S Deser *et al* (Boston, MA: MIT)
Adler S L and Bardeen W A 1969 *Phys. Rev.* **182** 1517
Affolder T *et al* 2001 *Phys. Rev.* D **64** 052001
Aitchison I J R *et al* 1995 *Phys. Rev.* B **51** 6531
Akhundov A A *et al* 1986 *Nucl. Phys.* B **276** 1
Akrawy M Z *et al* 1990 OPAL collaboration *Phys. Lett.* B **235** 389
Alitti S *et al* 1992 *Phys. Lett.* B **276** 354
Allaby J *et al* 1988 *J. Phys. C: Solid State Phys.* **38** 403
Allasia D *et al* 1984 *Phys. Lett.* B **135** 231
——1985 *Z. Phys.* C **28** 321
Allton C R *et al* 1995 *Nucl. Phys.* B **437** 641
——2002 UKQCD collaboration *Phys. Rev.* D **65** 054502
Alper B *et al* 1973 *Phys. Lett.* B **44** 521
Altarelli G 1982 *Phys. Rep.* **81** 1
Altarelli G and Parisi G 1977 *Nucl. Phys.* B **126** 298
Altarelli G *et al* 1978a *Nucl. Phys.* B **143** 521
——1978b *Nucl. Phys.* B **146** 544(E)
——1979 *Nucl. Phys.* B **157** 461

——1989 *Z. Phys. at LEP-1* CERN 89-08 (Geneva)
Amaudruz P *et al* 1992 NMC collaboration *Phys. Lett.* B **295** 159
Anderson P W 1963 *Phys. Rev.* **130** 439
Antoniadis I 2002 *2001 European School of High Energy Physics* ed N Ellis and J March-Russell CERN 2002-002 (Geneva) pp 301ff
Aoki S *et al* 2000 CP-PACS collaboration *Phys. Rev. Lett.* **84** 238
Appel J A *et al* 1986 *Z. Phys.* C **30** 341
Appelquist T and Chanowitz M S 1987 *Phys. Rev. Lett.* **59** 2405
Arnison G *et al* 1983a *Phys. Lett.* B **122** 103
——1983b *Phys. Lett.* B **126** 398
——1984 *Phys. Lett.* B **136** 294
——1985 *Phys. Lett.* B **158** 494
——1986 *Phys. Lett.* B **166** 484
Aubert B *et al* 2002 BaBar collaboration *Phys. Rev. Lett.* **89** 201802
Bagnaia P *et al* 1983 *Phys. Lett.* B **129** 130
——1984 *Phys. Lett.* B **144** 283
Bahcall J 1989 *Neutrino Astrophysics* (Cambridge: Cambridge University Press)
Bailin D 1982 *Weak Interactions* (Bristol: Adam Hilger)
Banks T *et al* 1976 *Phys. Rev.* D **13** 1043
Banner M *et al* 1982 *Phys. Lett.* B **118** 203
——1983 *Phys. Lett.* B **122** 476
Barber D P *et al* 1979 *Phys. Rev. Lett.* **43** 830
Bardeen J 1957 *Nuovo Cimento* **5** 1766
Bardeen J, Cooper L N and Schrieffer J R 1957 *Phys. Rev.* **108** 1175
Bardeen W A *et al* 1978 *Phys. Rev.* D **18** 3998
Bardin D Yu *et al* 1989 *Z. Phys.* C **44** 493
Barger V *et al* 1983 *Z. Phys.* C **21** 99
Beenakker W and Hollik W 1988 *Z. Phys.* C **40** 569
Bergsma F *et al* 1983 *Phys. Lett.* B **122** 465
Berends F A *et al* 1981 *Phys. Lett.* B **103** 124
Belavin A A *et al* 1975 *Phys. Lett.* B **59** 85
Bell J S and Jackiw R 1969 *Nuovo Cimento* A **60** 47
Benvenuti A C *et al* 1989 BCDMS collaboration *Phys. Lett.* B **223** 485
Berezin F A 1966 *The Method of Second Quantisation* (New York: Academic)
Berman S M, Bjorken J D and Kogut J B 1971 *Phys. Rev.* D **4** 3388
Bernard C W and Golterman M F 1992 *Phys. Rev.* D **46** 853
Bernstein J 1974 *Rev. Mod. Phys.* **46** 7
Bigi I I and Sanda A I 2000 *CP Violation* (Cambridge: Cambridge University Press)
Bilenky S M 2000 in *1999 European School of High Energy Physics* ed A Olchevski CERN 2000-007 (Geneva)
Binney J J *et al* 1992 *The Modern Theory of Critical Phenomena* (Oxford: Clarendon)
Bjorken J D 1973 *Phys. Rev.* D **8** 4098
Bjorken J D and Drell S D 1965 *Relativistic Quantum Fields* (New York: McGraw-Hill)
Bjorken J D and Glashow S L 1964 *Phys. Lett.* **11** 255
Blatt J M 1964 *Theory of Superconductivity* (New York: Academic)
Bloch F and Nordsieck A 1937 *Phys. Rev.* **52** 54
Bludman S A 1958 *Nuovo Cimento* **9** 443
Boas M L 1983 *Mathematical Methods in the Physical Sciences* (New York: Wiley)

Boehm F and Vogel P 1987 *Physics of Massive Neutrinos* (Cambridge: Cambridge University Press)
Bogoliubov N N 1947 *J. Phys. USSR* **11** 23
——1958 *Nuovo Cimento* **7** 794
Bogoliubov N N *et al* 1959 *A New Method in the Theory of Superconductivity* (New York: Consultants Bureau, Inc.)
Bosetti P C *et al* 1978 *Nucl. Phys.* B **142** 1
Bouchiat C C *et al* 1972 *Phys. Lett.* B **38** 519
Branco G C *et al CP Violation* (Oxford: Oxford University Press)
Brandelik R *et al* 1979 *Phys. Lett.* B **86** 243
Budny R 1975 *Phys. Lett.* B **55** 227
Büsser F W *et al* 1972 *Proc. XVI Int. Conf. on High Energy Physics (Chicago, IL)* vol 3 (Batavia: FNAL)
——1973 *Phys. Lett.* B **46** 471
Cabibbo N 1963 *Phys. Rev. Lett.* **10** 531
Cabibbo N *et al* 1979 *Nucl. Phys.* B **158** 295
Callan C G 1970 *Phys. Rev.* D **2** 1541
Carruthers P A 1966 *Introduction to Unitary Symmetry* (New York: Wiley)
Case K M 1957 *Phys. Rev.* **107** 307
Caswell W E 1974 *Phys. Rev. Lett.* **33** 244
Celmaster W and Gonsalves R J 1980 *Phys. Rev. Lett.* **44** 560
Chadwick J 1932 *Proc. R. Soc.* A **136** 692
Chanowitz M *et al* 1978 *Phys. Lett.* B **78** 285
——1979 *Nucl. Phys.* B **153** 402
Chen M-S and Zerwas P 1975 *Phys. Rev.* D **12** 187
Cheng T-P and Li L-F 1984 *Gauge Theory of Elementary Particle Physics* (Oxford: Clarendon)
Chetyrkin K G *et al* 1979 *Phys. Lett.* B **85** 277
Chetyrkin K G and Kuhn J H 1993 *Phys. Lett.* B **308** 127
Christenson J H *et al* 1964 *Phys. Rev. Lett.* **13** 138
Coleman S 1985 *Aspects of Symmetry* (Cambridge: Cambridge University Press)
——1966 *J. Math. Phys.* **7** 787
Coleman S and Gross D J 1973 *Phys. Rev. Lett.* **31** 851
Collins J C and Soper D E 1987 *Annu. Rev. Nucl. Part. Sci.* **37** 383
Collins P D B and Martin A D 1984 *Hadron Interactions* (Bristol: Adam Hilger)
Combridge B L *et al* 1977 *Phys. Lett.* B **70** 234
Combridge B L and Maxwell C J 1984 *Nucl. Phys.* B **239** 429
Commins E D and Bucksbaum P H 1983 *Weak Interactions of Quarks and Leptons* (Cambridge: Cambridge University Press)
Consoli M *et al* 1989 *Z. Phys. at LEP-I* ed G Altarelli *et al* CERN 89-08 (Geneva)
Cooper L N 1956 *Phys. Rev.* **104** 1189
Cornwall J M *et al* 1974 *Phys. Rev.* D **10** 1145
Cowan C L *et al* 1956 *Science* **124** 103
Dalitz R H 1953 *Phil. Mag.* **44** 1068
——1965 *High Energy Physics* ed C de Witt and M Jacob (New York: Gordon and Breach)
Danby G *et al* 1962 *Phys. Rev. Lett.* **9** 36
Davis R 1955 *Phys. Rev.* **97** 766
Dawson S *et al* 1990 *The Higgs Hunter's Guide* (Reading, MA: Addison-Wesley)

de Groot J G H *et al* 1979 *Z. Phys.* C **1** 143
DiLella L 1985 *Annu. Rev. Nucl. Sci.* **35** 107
——1986 *Proc. Int. Europhysics Conf. on High Energy Physics, Bari, Italy, July 1985* ed L Nitti and G Preparata (Bari: Laterza) pp 761ff
Dine M and Sapirstein J 1979 *Phys. Rev. Lett.* **43** 668
Dirac P A M 1931 *Proc. R. Soc.* A **133** 60
Dokshitzer Yu L 1977 *Sov. Phys.–JETP* **46** 641
Donoghue J F, Golowich E and Holstein B R 1992 *Dynamics of the Standard Model* (Cambridge: Cambridge University Press)
Draper T *et al* 2002 *Talk Presented at 20th Int. Symp. on Lattice Field Theory (LATTICE 2002) Boston, MA, June 2002* hep-lat/0208045
Duke D W and Owens J F 1984 *Phys. Rev.* D **30** 49
Eden R J, Landshoff P V, Olive D I and Polkinghorne J C 1966 *The Analytic S-Matrix* (Cambridge: Cambridge University Press)
Eichten E *et al* 1980 *Phys. Rev.* D **21** 203
Einhorn M B and Wudka J 1989 *Phys. Rev.* D **39** 2758
Ellis J *et al* 1976 *Nucl. Phys.* B **111** 253
——1994 *Phys. Lett.* B **333** 118
Ellis R K, Stirling W J and Webber B R 1996 *QCD and Collider Physics* (Cambridge: Cambridge University Press)
Ellis S D, Kunszt Z and Soper D E 1992 *Phys. Rev. Lett.* **69** 3615
Englert F and Brout R 1964 *Phys. Rev. Lett.* **13** 321
Enz C P 1992 *A Course on Many-Body Theory Applied to Solid-State Physics (World Scientific Lecture Notes in Physics 11)* (Singapore: World Scientific)
——2002 *No Time to be Brief* (Oxford: Oxford University Press)
Fabri E and Picasso L E 1966 *Phys. Rev. Lett.* **16** 408
Faddeev L D and Popov V N 1967 *Phys. Lett.* B **25** 29
Feinberg G *et al* 1959 *Phys. Rev. Lett.* **3** 527, especially footnote 9
Fermi E 1934a *Nuovo Cimento* **11** 1
——1934b *Z. Phys.* **88** 161
Feynman R P 1963 *Acta Phys. Polon.* **26** 697
——1977 in *Weak and Electromagnetic Interactions at High Energies* ed R Balian and C H Llewellyn Smith (Amsterdam: North-Holland) p 121
Feynman R P and Gell-Mann M 1958 *Phys. Rev.* **109** 193
Feynman R P and Hibbs A R 1965 *Quantum Mechanics and Path Intergrals* (New York: McGraw-Hill)
Gaillard M K and Lee B W 1974 *Phys. Rev.* D **10** 897
Gamow G and Teller E 1936 *Phys. Rev.* **49** 895
Gasser J and Leutwyler H 1982 *Phys. Rep.* **87** 77
Gastmans R 1975 *Weak and Electromagnetic Interactions at High Energies, Cargese, 1975* ed M Levy *et al* (New York: Plenum) pp 109ff
Geer S 1986 *High Energy Physics 1985, Proc. Yale Theoretical Advanced Study Institute* (Singapore: World Scientific)
Gell-Mann M 1961 *California Institute of Technology Report* CTSL-20 (reprinted in Gell-Mann and Ne'eman 1964)
Gell-Mann M *et al* 1979 *Supergravity* ed D Freedman and P van Nieuwenhuizen (Amsterdam: North-Holland) p 315
Gell-Mann M and Low F E 1954 *Phys. Rev.* **95** 1300

Gell-Mann M and Ne'eman 1964 *The Eightfold Way* (New York: Benjamin)
Georgi H 1984 *weak Interactions and Modern Particle Theory* (Menlo Park, CA: Benjamin/Cummings)
Georgi H *et al* 1978 *Phys. Rev. Lett.* **40** 692
Georgi H and Politzer H D 1974 *Phys. Rev.* D **9** 416
Ginsparg P and Wilson K G 1982 *Phys. Rev.* D **25** 25
Ginzburg V I and Landau L D 1950 *Zh. Eksp. Teor. Fiz.* **20** 1064
Giusti L 2002 plenary talk at *20th Int. Symp. on Lattice Field Theory (LATTICE 2002) Boston, MA* hep-lat/0211009
Glashow S L 1961 *Nucl. Phys.* **22** 579
Glashow S L *et al* 1978 *Phys. Rev.* D **18** 1724
Glashow S L, Iliopoulos J and Maiani L 1970 *Phys. Rev.* D **2** 1285
Goldberger M L and Treiman S B 1958 *Phys. Rev.* **95** 1300
Goldhaber M *et al* 1958 *Phys. Rev.* **109** 1015
Goldstone J 1961 *Nuovo Cimento* **19** 154
Goldstone J, Salam A and Weinberg S 1962 *Phys. Rev.* **127** 965
Gorishny S G *et al* 1991 *Phys. Lett.* B **259** 144
Gorkov L P 1959 *Zh. Eksp. Teor. Fiz.* **36** 1918
Gottschalk T and Sivers D 1980 *Phys. Rev.* D **21** 102
Greenberg O W 1964 *Phys. Rev. Lett.* **13** 598
Gribov V N and Lipatov L N 1972 *Sov. J. Nucl. Phys.* **15** 438
Gross D J and Llewellyn Smith C H 1969 *Nucl. Phys.* B **14** 337
Gross D J and Wilczek F 1973 *Phys. Rev. Lett.* **30** 1343
——1974 *Phys. Rev.* D **9** 980
Guralnik G S *et al* 1964 *Phys. Rev. Lett.* **13** 585
——1968 *Advances in Particle Physics* vol 2, ed R Cool and R E Marshak (New York: Interscience) pp 567ff
Hagiwara K *et al* 2002 *Phys. Rev.* D **66** 010001
Halzen F and Martin A D 1984 *Quarks and Leptons* (New York: Wiley)
Hamberg R *et al* 1991 *Nucl. Phys.* B **359** 343
Hambye T and Reisselmann K 1997 *Phys. Rev.* D **55** 7255
Hammermesh M 1962 *Group Theory and its Applications to Physical Problems* (Reading, MA: Addison-Wesley)
Han M Y and Nambu Y 1965 *Phys. Rev.* B **139** 1066
Hasenfratz P *et al* 1998 *Phys. Lett.* B **427** 125
Hasenfratz P and Niedermayer F 1994 *Nucl. Phys.* B **414** 785
Hasert F J *et al* 1973 *Phys. Lett.* B **46** 138
Heisenberg W 1932 *Z. Phys.* **77** 1
Higgs P W 1964 *Phys. Rev. Lett.* **13** 508
——1966 *Phys. Rev.* **145** 1156
Hollik W 1990 *Fortsch. Phys.* **38** 165
——1991 *1989 CERN-JINR School of Physics* CERN 91-07 (Geneva) p 50ff
Hughes R J 1980 *Phys. Lett.* B **97** 246
——1981 *Nucl. Phys.* B **186** 376
Isgur N and Wise M B 1989 *Phys. Lett.* B **232** 113
——1990 *Phys. Lett.* B **237** 527
Isidori G *et al* 2001 *Nucl. Phys.* B **609** 387
Itzykson C and Zuber J-B 1980 *Quantum Field Theory* (New York: McGraw-Hill)

Jackiw R 1972 *Lectures in Current Algebra and its Applications* ed S B Treiman, R Jackiw and D J Gross (Princeton, NJ: Princeton University Press) pp 97–254
Jacob M and Landshoff P V 1978 *Phys. Rep.* C **48** 285
Jarlskog C 1985a *Phys. Rev. Lett.* **55** 1039
——1985b *Z. Phys.* C **29** 491
Jones D R T 1974 *Nucl. Phys.* B **75** 531
Jones H F 1990 *Groups, Representations and Physics* (Bristol: IOP Publishing)
Kadanoff L P 1977 *Rev. Mod. Phys.* **49** 267
Kaplan D B *Phys. Lett.* B **288** 342
Kayser B *et al* 1989 *The Physics of Massive Neutrinos* (Singapore: World Scientific)
Kennedy D C *et al* 1989 *Nucl. Phys.* B **321** 83
Kennedy D C and Lynn B W 1989 *Nucl. Phys.* B **322** 1
Kibble T W B 1967 *Phys. Rev.* **155** 1554
Kim K J and Schilcher K 1978 *Phys. Rev.* D **17** 2800
Kinoshita T 1962 *J. Math. Phys.* **3** 650
Kittel C 1987 *Quantum Theory of Solids* second revised printing (New York: Wiley)
Klein O 1948 *Nature* **161** 897
Kobayashi M and Maskawa K 1973 *Prog. Theor. Phys.* **49** 652
Kugo T and Ojima I 1979 *Prog. Theor. Phys. Suppl.* **66** 1
Kunszt Z and Pietarinen E 1980 *Nucl. Phys.* B **164** 45
Landau L D 1957 *Nucl. Phys.* **3** 127
Landau L D and Lifshitz E M 1980 *Statistical Mechanics* part 1, 3rd edn (Oxford: Pergamon)
Langacker P (ed) 1995 *Precision Tests of the Standard Electroweak Model* (Singapore: World Scientific)
Larin S A and Vermaseren J A M 1991 *Phys. Lett.* B **259** 345
——1993 *Phys. Lett.* B **303** 334
Lautrup B 1967 *Kon. Dan. Vid. Selsk. Mat.-Fys. Med.* **35** 1
Leader E and Predazzi E 1996 *An Introduction to Gauge Theories and Modern Particle Physics* vol 2 (Cambridge: Cambridge University Press)
Lee B W *et al* 1977a *Phys. Rev. Lett.* **38** 883
——1977b *Phys. Rev.* D **16** 1519
Lee T D and Nauenberg M 1964 *Phys. Rev.* B **133** 1549
Lee T D, Rosenbluth R and Yang C N 1949 *Phys. Rev.* **75** 9905
Lee T D and Yang C N 1956 *Phys. Rev.* **104** 254
——1957 *Phys. Rev.* **105** 1671
——1962 *Phys. Rev.* **128** 885
LEP 2003 (The LEP working Group for Higgs Searches, ALEPH, DELPHI, L3 and OPAL collaborations) *Phys. Lett.* B **565** 61
LEP Higgs 2001 (LEP Higgs Working Group) CERN-EP/2001-055 (Geneva)
Levine I *et al* 1997 *Phys. Rev. Lett.* **78** 424
Lichtenberg D B 1970 *Unitary Symmetry and Elementary Particles* (New York: Academic)
Llewellyn Smith C H 1973 *Phys. Lett.* B **46** 233
London F 1950 *Superfluids Vol I, Macroscopic theory of Superconductivity* (New York: Wiley)
Lüscher M 1981 *Nucl. Phys.* B **180** 317
——1998 *Phys. Lett.* B **428** 342
Lüscher M *et al* 1980 *Nucl. Phys.* B **173** 365

Maki Z *et al* 1962 *Prog. Theor. Phys.* **28** 247
Mandelstam S 1976 *Phys. Rep.* C **23** 245
Mandl F 1992 *Quantum Mechanics* (New York: Wiley)
Marciano W J and Sirlin A 1988 *Phys. Rev. Lett.* **61** 1815
Marshak R E *et al* 1969 *Theory of Weak Interactions in Particle Physics* (New York: Wiley)
Martin A D *et al* 1994 *Phys. Rev.* D **50** 6734
——2002 *Eur. Phys. J.* C **23** 73
Merzbacher E 1998 *Quantum Mechanics* 3rd edn (New York: Wiley)
Mohapatra R N *et al* 1968 *Phys. Rev. Lett.* **20** 1081
Mohapatra R N and Pal P B 1991 *Massive Neutrinos in Physics and Astrophysics* (Singapore: World Scientific)
Mohapatra R N and Senjanovic G 1980 *Phys. Rev. Lett.* **44** 912
——1981 *Phys. Rev.* D **23** 165
Mohr P J and Taylor B N 2000 *Rev. Mod. Phys.* **72** 351
Montanet L *et al* 1994 *Phys. Rev.* D **50** 1173
Montvay I and Münster G 1994 *Quantum Fields on a Lattice* (Cambridge: Cambridge University Press)
Nambu Y 1960 *Phys. Rev. Lett.* **4** 380
——1974 *Phys. Rev.* D **10** 4262
Nambu Y and Jona-Lasinio G 1961a *Phys. Rev.* **122** 345
——1961b *Phys. Rev.* **124** 246
Nambu Y and Lurie D 1962 *Phys. Rev.* **125** 1429
Nambu Y and Schrauner E 1962 *Phys. Rev.* **128** 862
Narayanan R and Neuberger H 1993a *Phys. Lett.* B **302** 62
——1993b *Phys. Rev. Lett.* **71** 3251
——1994 *Nucl. Phys.* B **412** 574
——1995 *Nucl. Phys.* B **443** 305
Ne'eman Y 1961 *Nucl. Phys.* **26** 222
Nielsen N K 1981 *Am. J. Phys.* **49** 1171
Nielsen H B and Ninomaya M 1981a *Nucl. Phys.* B **185** 20
——1981b *Nucl. Phys.* B **193** 173
——1981c *Nucl. Phys.* B **195** 541
Noether E 1918 *Nachr. Ges. Wiss. Göttingen* 171
Okada S *et al* 1998 *Phys. Rev. Lett.* **81** 2428
Pais A 2000 *The Genius of Science* (Oxford: Oxford University Press)
Parry W E 1973 *The Many Body Problem* (Oxford: Clarendon)
Pauli W 1934 *Rapp. Septième Conseil Phys. Solvay, Brussels 1933* (Paris: Gautier-Villars), reprinted in Winter (2000) pp 7, 8
Pennington M R 1983 *Rep. Prog. Phys.* **46** 393
Perkins D H in *Proc. Int. Symp. on Lepton and Photon Interactions at High Energies, Stanford, CA* p 571
——2000 *Introduction to High Energy Physics* 4th edn (Cambridge: Cambridge University Press)
Peskin M E 1997 in *1996 European School of High-Energy Physics* ed N Ellis and M Neubert CERN 97-03 (Geneva) pp 49-142
Peskin M E and Schroeder D V 1995 *an Introduction to Quantum Field Theory* (Reading, MA: Addison-Wesley)
Politzer H D 1973 *Phys. Rev. Lett.* **30** 1346

Pontecorvo B 1947 *Phys. Rev.* **72** 246
——1948 *Chalk River Laboratory Report* PD-205
——1967 *Zh. Eksp. Theor. Phys.* **53** 1717 (Engl. transl. *Sov. Phys.–JETP* **26** 989)
Prescott C Y *et al* 1978 *Phys. Lett.* B **77** 347
Puppi G 1948 *Nuovo Cimento* **5** 505
Quigg C 1977 *Rev. Mod. Phys.* **49** 297
Reines F and Cowan C 1956 *Nature* **178** 446
Reines F, Gurr H and Sobel H 1976 *Phys. Rev. Lett.* **37** 315
Renton P 1990 *Electroweak Interactions* (Cambridge: Cambridge University Press)
Richardson J L 1979 *Phys. Lett.* B **82** 272
Ross D A and Veltman M 1975 *Nucl. Phys.* B **95** 135
Rubbia C *et al* 1977 *Proc. Int. Neutrino Conf., Aachen, 1976* (Braunschweig: Vieweg) p 683
Ryder L H 1996 *Quantum Field Theory* 2nd edn (Cambridge: Cambridge University Press)
Sakurai J J 1958 *Nuovo Cimento* **7** 649
——1960 *Ann. Phys., NY* **11** 1
Salam A 1957 *Nuovo Cimento* **5** 299
——1968 *Elementary Particle Physics* ed N Svartholm (Stockholm: Almqvist and Wiksells)
Salam A and Ward J C 1964 *Phys. Lett.* **13** 168
Samuel M A and Surguladze L R 1991 *Phys. Rev. Lett.* **66** 560
Schiff L I 1968 *Quantum Mechanics* 3rd edn (New York: McGraw-Hill)
Schrieffer J R 1964 *Theory of Superconductivity* (New York: Benjamin)
Schutz B F 1988 *A First Course in General Relativity* (Cambridge: Cambridge University Press)
Schwinger J 1951 *Phys. Rev.* **82** 664
——1957 *Ann. Phys., NY* **2** 407
——1962 *Phys. Rev.* **125** 397
Shaevitz *et al* 1995 CCFR collaboration *Nucl. Phys.* B *Proc. Suppl.* **38** 188
Sharpe S R 1992 *Phys. Rev.* D **46** 3146
Shaw R 1995 The problem of particle types and other contributions to the theory of elementary particles *PhD Thesis* University of Cambridge
Slavnov A A 1972 *Teor. Mat. Fiz.* **10** 153 (Engl. transl. *Theor. and Math. Phys.* **10** 99)
Sikivie P *et al* 1980 *Nucl. Phys.* B **173** 189
Sirlin A 1980 *Phys. Rev.* D **22** 971
——1984 *Phys. Rev.* D **29** 89
Sommer R 1994 *Nucl. Phys.* B **411** 839
Staff of the CERN $\bar{p}p$ project 1991 *Phys. Lett.* B **107** 306
Stange A *et al* 1994a *Phys. Rev.* D **49** 1354
——1994b *Phys. Rev.* D **50** 4491
Steinberger J 1949 *Phys. Rev.* **76** 1180
Sterman G and Weinberg S 1977 *Phys. Rev. Lett.* **39** 1436
Stueckelberg E C G and Peterman A 1953 *Helv. Phys. Acta* **26** 499
Sudarshan E C G and Marshak R E 1958 *Phys. Rev.* **109** 1860
Susskind L 1977 *Phys. Rev.* D **16** 3031
——1979 *Phys. Rev.* D **19** 2619
Sutherland D G 1967 *Nucl. Phys.* B **2** 433
Symanzik K 1970 *Commun. Math. Phys.* **18** 227

Tarasov O V *et al* 1980 *Phys. Lett.* B **93** 429
Tavkhelidze A 1965 *Seminar on High Energy Physics and Elementary Particles* (Vienna: IAEA) p 763
Taylor J C 1971 *Nucl. Phys.* B **33** 436
——1976 *Gauge Theories of Weak Interactions* (Cambridge: Cambridge University Press)
't Hooft G 1971a *Nucl. Phys.* B **33** 173
——1971b *Nucl. Phys.* B **35** 167
——1971c *Phys. Lett.* B **37** 195
——1976a *Phys. Rev.* D **14** 3432
——1976b *High Energy Physics, Proc. European Physical Society Int. Conf.* ed A Zichichi (Bologna: Editrice Composition) p 1225
——1980 *Recent Developments in Gauge Theories, Cargese Summer Institute 1979* ed G 't Hooft *et al* (New York: Plenum)
——1986 *Phys. Rep.* **142** 357
't Hooft G and Veltman M 1972 *Nucl. Phys.* B **44** 189
Tiomno J and Wheeler J A 1949 *Rev. Mod. Phys.* **21** 153
Tournefier E 2001 LEP collaborations *Proc. 36th Rencontre de Moriond: Electroweak Interactions and Unified Theories* (Les Arcs)
Valatin J G 1958 *Nuovo Cimento* **7** 843
van der Bij J J 1984 *Nucl. Phys.* B **248** 141
van der Bij J J and Veltman M 1984 *Nucl. Phys.* B **231** 205
van der Neerven W L and Zijlstra E B 1992 *Nucl. Phys.* B **382** 11
van Ritbergen T and Stuart R G 1999 *Phys. Rev. Lett.* **82** 488
Veltman M 1967 *Proc. R. Soc.* A **301** 107
——1968 *Nucl. Phys.* B **7** 637
——1970 *Nucl. Phys.* B **21** 288
——1977 *Acta Phys. Polon.* B **8** 475
von Weiszäcker C F 1934 *Z. Phys.* **88** 612
Weerts H 1994 D0 collaboration *Proc. 31st Rencontre de Moriond: QCD and High-energy Hadronic Interactions* (Les Arcs) FermiLab-Conf-96-132-E
Wegner F 1972 *Phys. Rev.* B **5** 4529
Weinberg S 1966 *Phys. Rev. Lett.* **17** 616
——1967 *Phys. Rev. Lett.* **19** 1264
——1973 *Phys. Rev.* D **8** 605, especially footnote 8
——1975 *Phys. Rev.* D **11** 3583
——1979 *Phys. Rev.* D **19** 1277
——1996 *The Quantum Theory of Fields Vol II Modern Applications* (Cambridge: Cambridge University Press)
Weisberger W 1965 *Phys. Rev. Lett.* **14** 1047
Williams E J 1934 *Phys. Rev.* **45** 729
Wilson K G 1969 *Phys. Rev.* **179** 1499
——1971a *Phys. Rev.* B **4** 3174
——1971b *Phys. Rev.* B **4** 3184
——1974 *Phys. Rev.* D **10** 2445
——1975 *New Phenomena in Subnuclear Physics, Proc. 1975 Int. School on Subnuclear Physics 'Ettore Majorana'* ed A Zichichi (New York: Plenum)
Wilson K G and Kogut J 1974 *Phys. Rep* **12C** 75
Winter K 2000 *Neutrino Physics* 2nd edn (Cambridge: Cambridge University Press)

Wolfenstein L 1983 *Phys. Rev. Lett.* **51** 1945
Wu C S *et al* 1957 *Phys. Rev.* **105** 1413
Yanagida T 1979 *Proc. Workshop on Unified Theory and Baryon Number in the Universe* ed O Sawada and A Sugamoto (Tsukuba: KEK)
Yang C N and Mills R L 1954 *Phys. Rev.* **96** 191
Yosida K 1958 *Phys. Rev.* **111** 1255

# INDEX